DATE			

*BIOPHYSICAL
SCIENCE*

BIOPHYSICAL SCIENCE

second edition

EUGENE ACKERMAN
LYNDA B. M. ELLIS
LAWRENCE E. WILLIAMS

Medical School
University of Minnesota

PRENTICE-HALL, INC.
Englewood Cliffs, New Jersey 07632

Library of Congress Cataloging in Publication Data

ACKERMAN, EUGENE (date)
    Biophysical science.

    Bibliography: p.
    Includes index.
    1. Biological physics. I. Ellis, Lynda B. M.,
joint author. II. Williams, Lawrence E., joint
author. III. Title.
QH505.A25 1979   574.1'91   78-18682
ISBN   0-13-076901-0

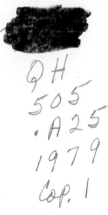

© 1979, 1962 by PRENTICE-HALL, INC.
Englewood Cliffs, New Jersey 07632

Printed in the United States of America

10 9 8 7 6 5 4 3 2 1

PRENTICE-HALL INTERNATIONAL, INC., *London*
PRENTICE-HALL OF AUSTRALIA PTY. LIMITED, *Sydney*
PRENTICE-HALL OF CANADA, LTD., *Toronto*
PRENTICE-HALL OF INDIA PRIVATE LIMITED, *New Delhi*
PRENTICE-HALL OF JAPAN, INC., *Tokyo*
PRENTICE-HALL OF SOUTHEAST ASIA PTE. LTD., *Singapore*
WHITEHALL BOOKS LIMITED, *Wellington, New Zealand*

This new edition is dedicated to our spouses:

Dorothy Hopkirk Ackerman
John Emmett Ellis
Sonia Bell Williams

Contents

Preface   xv

PART A
Special Sensory Systems

1 *Sound and the Ear*   2

1.1 Hearing, 2/   1.2 Acoustics, 4/   1.3 Hearing Tests, 9/
1.4 Anatomy and Action of the Ear, 16/   References, 22/

2 *Light and the Eye*   24

2.1 Vision, 24/   2.2 Optics, 26/   2.3 Anatomy of the Eye, 30/
2.4 Thresholds and Acuity, 40/   References, 46/

3 *Insect Communication*   48

3.1 Introduction, 48/   3.2 Visual Communication, 50/
3.3 Communication by Sound, 53/
3.4 Chemical Communication, 54/
3.5 Interspecies Communication, 56/   3.6 Summary, 57/
References, 58/

4 *Alternative Methods of Orientation*   **59**

    4.1 INTRODUCTION, 59/   *4.2* THE BATS, 59/
    4.3 THE DOLPHINS AND WHALES, 65/
    4.4 INFRARED SENSITIVITY OF SNAKES, 67/
    4.5 ELECTRIC FISH, 68/   *4.6* BIRD NAVIGATION AND HOMING, 71/
    4.7 SUMMARY, 73/   REFERENCES, 74/

PART B
Nerve and Muscle

5 *Conduction of Impulses by Nerves*   **78**

    5.1 THE ROLE OF THE NERVOUS SYSTEM, 78/
    5.2 A BRIEF GLANCE AT ELECTRICITY, 80/
    5.3 ANATOMY AND HISTOLOGY OF NEURONS, 83/
    5.4 THE SPIKE POTENTIAL, 87/
    5.5 CLAMPED NERVE EXPERIMENTS, 91/
    5.6 SYNAPTIC CONDUCTION, 97/   *5.7* SUMMARY, 100/
    REFERENCES, 101/

6 *Electrical Potentials of the Brain*   **102**

    6.1 ELECTROENCEPHALOGRAPHY, 102/
    6.2 THE CENTRAL NERVOUS SYSTEM, 103/
    6.3 ORIGINS OF THE ELECTROENCEPHALOGRAPHIC POTENTIALS, 107/
    6.4 THE ELECTROENCEPHALOGRAPHIC PATTERNS, 110/
    6.5 SLEEP STAGES, 113/   *6.6* EVOKED RESPONSE AVERAGING, 115/
    6.7 SUMMARY, 117/   REFERENCES, 117/

7 *Neural Mechanisms of Hearing*   **119**

    7.1 THE TYPES OF THEORIES, 119/   *7.2* SPATIAL THEORIES, 120/
    7.3 TEMPORAL THEORIES, 129/
    7.4 AUDITORY SYSTEM PATHWAYS, 130/
    7.5 ACOUSTIC ENCODING, 132/   *7.6* SUMMARY, 135/
    REFERENCES, 135/

8 *Neural Aspects of Vision*   **137**

    8.1 INTRODUCTION, 137/   *8.2* COLOR DISCRIMINATION, 138/
    8.3 ISOLATED NEURAL RESPONSES, 142/
    8.4 COORDINATED NEURAL RESPONSES, 144/
    8.5 CORTICAL REPRESENTATION, 146/
    8.6 SUMMARY OF VISION, 148/   REFERENCES, 149/

9 *Muscles* 151

9.1 INTRODUCTION, 151/   9.2 MUSCLE STRUCTURE, 152/
9.3 PHYSICAL CHANGES DURING MUSCULAR CONTRACTION, 157/
9.4 MUSCLE CHEMISTRY, 162/   9.5 SUMMARY, 167/
REFERENCES, 168/

10 *Mechanical and Electrical Character of the Heartbeat* 169

10.1 ROLE OF THE VERTEBRATE CIRCULATORY SYSTEM, 169/
10.2 BLOOD PRESSURES AND VELOCITIES, 170/
10.3 THE VERTEBRATE HEART, 172/
10.4 THE HEART SEQUENCE, 174/
10.5 VENTRICULAR CONTRACTILITY AND ENERGY, 177/
10.6 ELECTROCARDIOGRAPHY, 180/
10.7 PHYSICS OF DIPOLES, 183/
10.8 VECTOR ELECTROCARDIOGRAPHY, 187/
10.9 SUMMARY, 189/   REFERENCES, 190/

PART C
Physical Biology

11 *Ionizing Radiation in Tissue* 196

11.1 INTRODUCTION, 196/   11.2 QUANTITATION OF DOSE, 197/
11.3 CONTEMPORARY DOSE LEVELS, 198/
11.4 ANNUAL DOSE LIMITS, 200/
11.5 CELLULAR CONSIDERATIONS, 201/
11.6 MACRO BIOLOGICAL EFFECTS, 209/   11.7 SUMMARY, 213/
REFERENCES, 213/

12 *Biological Effects of Electromagnetism* 215

12.1 INTRODUCTION, 215/   12.2 ELECTRICAL IMPEDANCE, 216/
12.3 BIOLOGICAL IMPEDANCE MEASUREMENTS, 217/
12.4 THE ATTENUATION OF ELECTROMAGNETIC RADIATION, 222/
12.5 LOW-FREQUENCY EFFECTS, 223/
12.6 MICROWAVE RADIATION, 224/
12.7 LASER IRRADIATION, 227/
12.8 ULTRAVIOLET RADIATION, 229/   12.9 SUMMARY, 230/
REFERENCES, 231/

13 *Sonic Irradiation*   **233**

13.1 INTRODUCTION, **233**/   13.2 SOUND AND ULTRASOUND, **234**/
13.3 SONIC CHARACTERISTICS OF BIOLOGICAL MATERIALS, **236**/
13.4 CAVITATION, **242**/
13.5 CAVITATION DAMAGE TO BIOLOGICAL CELLS, **245**/
13.6 HIGH-INTENSITY IRRADIATION, **251**/   13.7 SUMMARY, **254**/
REFERENCES, **254**/

14 *X-Ray Analysis*   **256**

14.1 X-RAY DIFFRACTION, **256**/   14.2 PROTEIN STRUCTURE, **263**/
14.3 NUCLEIC ACIDS, **269**/   14.4 SUMMARY, **273**/
REFERENCES, **275**/

15 *Viruses*   **276**

15.1 INTRODUCTION, **276**/
15.2 PHYSICAL METHODS USED IN VIRUS STUDIES, **278**/
15.3 PHYSICAL BIOCHEMISTRY OF VIRUSES, **283**/
15.4 PHAGE GENETICS, **286**/   15.5 SUMMARY, **289**/
REFERENCES, **290**/

PART D
Molecular Biology

16 *Proteins and Nucleic Acids*   **294**

16.1 INTRODUCTION, **294**/   16.2 ULTRACENTRIFUGATION, **294**/
16.3 DNA REPLICATION, **297**/
16.4 TRANSCRIPTION OF RNA, **298**/
16.5 TRANSLATION OF RNA INTO PROTEIN, **299**/
16.6 THE GENETIC CODE, **302**/
16.7 CONTROL OF PROTEIN SYNTHESIS, **303**/
16.8 SUMMARY, **304**/   REFERENCES, **305**/

17 *Molecular Effects of Ionizing Radiation*   **306**

17.1 INTRODUCTION, **306**/   17.2 ABSORPTION OF RADIATION, **307**/
17.3 DIRECT EFFECTS OF RADIATION ON WATER, **311**/
17.4 RADIATION EFFECTS ON SYNTHETIC POLYMERS, **312**/
17.5 TARGET THEORY, **317**/
17.6 INACTIVATION OF DRIED PROTEIN FILMS, **319**/
17.7 INDIRECT EFFECTS, **322**/   17.8 SUMMARY, **323**/
REFERENCES, **324**/

18 *Enzyme Kinetics* **325**

18.1 INTRODUCTION, 325/   18.2 ENZYMES, 327/
18.3 MICHAELIS–MENTEN KINETICS, 329/
18.4 ACTION OF INHIBITORS, 337/
18.5 CATALASE AND PEROXIDASE, 341/
18.6 BIOLOGICAL OXIDIZING AGENTS, 347/
18.7 GLYCOLYSIS AND THE CITRIC ACID CYCLE, 348/
18.8 OXIDATIVE PHOSPHORYLATION, 350/
18.9 SUMMARY OF ENZYME KINETICS, 354/   REFERENCES, 355/

19 *Molecular Basis of Vision* **357**

19.1 VISION AND PHOTOPIGMENTS, 357/   19.2 RHODOPSIN, 358/
19.3 OTHER PHOTOPIGMENTS, 363/
19.4 COLOR VISION DEFECTS, 364/
19.5 ORIGIN OF THE EARLY RECEPTOR POTENTIAL, 366/
19.6 SUMMARY, 367/   REFERENCES, 367/

20 *Photosynthesis* **369**

20.1 INTRODUCTION, 369/
20.2 A LITTLE PLANT HISTOLOGY, 371/
20.3 BASIC PROCESSES OF PHOTOSYNTHESIS, 373/
20.4 THE PHOTOSYNTHETIC PIGMENTS, 375/
20.5 THE LIGHT REACTIONS, 381/
20.6 THE PATH OF CARBON IN PHOTOSYNTHESIS, 383/
20.7 SUMMARY, 387/   REFERENCES, 388/

PART E
Thermodynamics and Transport Systems

21 *Thermodynamics and Biology* **393**

21.1 THE ROLE OF THERMODYNAMICS IN BIOLOGY, 393/
21.2 THE LAWS OF THERMODYNAMICS, 394/
21.3 OTHER THERMODYNAMIC FUNCTIONS, 398/
21.4 EQUILIBRIUM, 400/   21.5 COLLISION THEORY, 405/
21.6 ABSOLUTE RATE THEORY, 409/   REFERENCES, 412/

22 *Irreversible Thermodynamics* **413**

22.1 NONEQUILIBRIUM AND IRREVERSIBLE SYSTEMS, 413/
22.2 DISSIPATION FUNCTIONS, 417/
22.3 PHENOMENOLOGICAL EQUATIONS, 421/
22.4 APPLICATIONS OF IRREVERSIBLE THERMODYNAMICS, 424/
22.5 SUMMARY OF THERMODYNAMICS, 430/   REFERENCES, 431/

23 *Diffusion, Permeability, and Active Transport* **432**

23.1 INTRODUCTION, 432/ 23.2 DIFFUSION EQUATIONS, 434/
23.3 PERMEABILITY AND THE RED BLOOD CELL, 439/
23.4 ACTIVE TRANSPORT, 444/
23.5 IRREVERSIBLE THERMODYNAMIC INTERPRETATION OF
PERMEABILITY, 447/
23.6 SUMMARY, 451/ REFERENCES, 452/

24 *Biological Membranes* **453**

24.1 INTRODUCTION, 453/
24.2 DONNAN MEMBRANE POTENTIALS, 454/
24.3 MEMBRANE STRUCTURE, 456/
24.4 MEMBRANE FUNCTION, 460/ 24.5 SUMMARY, 463/
REFERENCES, 464/

25 *Information Theory and Biology* **465**

25.1 LANGUAGES, 465/
25.2 INFORMATION THEORY AND COMPUTING, 466/
25.3 INFORMATION THEORY AND SENSORY PERCEPTION, 470/
25.4 INFORMATION THEOREMS AND BIOMEDICAL COMPUTING, 474/
25.5 INFORMATION THEORY AND PROTEIN STRUCTURE, 475/
25.6 THE CODING OF GENETIC INFORMATION, 477/
25.7 SUMMARY, 479/ REFERENCES, 479/

PART F
Specialized Instrumentation and Techniques

26 *Absorption Spectroscopy* **485**

26.1 INTRODUCTION, 485/ 26.2 QUANTUM THEORY, 486/
26.3 QUANTUM THEORY AND SPECTROSCOPY, 490/
26.4 LIGHT ABSORPTION, 496/ 26.5 OPTICAL ACTIVITY, 503/
26.6 MAGNETIC RESONANCE, 506/ 26.7 SUMMARY, 513/
REFERENCES, 513/

27 *Emission Spectroscopy* **515**

27.1 INTRODUCTION, 515/
27.2 FLUORESCENCE AND PHOSPHORESCENCE, 515/
27.3 X-RAY FLUORESCENCE, 520/ 27.4 STIMULATED EMISSION, 523/
27.5 MÖSSBAUER SPECTROSCOPY, 526/ 27.6 SUMMARY, 529/
REFERENCES, 531/

28 *Medical Ultrasonography* **532**

28.1 INTRODUCTION, **532**/ 28.2 METHODS, **533**/
28.3 WAVE REFLECTION TECHNIQUES, **537**/
28.4 DOPPLER TECHNIQUES, **542**/
28.5 ACOUSTIC HOLOGRAPHY, **543**/ 28.6 SUMMARY, **545**/
REFERENCES, **545**/

29 *Tracer Methods* **547**

29.1 INTRODUCTION, **547**/ 29.2 RADIOACTIVITY, **548**/
29.3 OTHER TRACERS, **551**/ 29.4 RADIOACTIVE IMAGING, **553**/
29.5 APPLICATIONS, **558**/ 29.6 SUMMARY, **562**/
REFERENCES, **562**/

30 *Biomedical Computation* **563**

30.1 BACKGROUND, **563**/ 30.2 DATA PROCESSING, **567**/
30.3 SIMULATION, **570**/ 30.4 ON-LINE APPLICATIONS, **572**/
30.5 INFORMATION SYSTEMS, **575**/ 30.6 SUMMARY, **578**/
REFERENCES, **578**/

Appendices

A *Acoustics* **583**

REFERENCES, **588**/

B *Electrical Terminology* **589**

REFERENCES, **594**/

C *The Fast Fourier Transform* **595**

C.1 THEORY, **595**/ C.2 APPLICATIONS, **598**/ REFERENCES, **599**/

D *Biochemistry* **600**

D.1 AMINO ACIDS AND PROTEINS, **600**/
D.2 PURINES, PYRIMIDINES, AND NUCLEIC ACIDS, **605**/
D.3 LIPIDS AND CARBOHYDRATES, **608**/
D.4 BIOCHEMICAL TAXONOMY, **613**/ REFERENCES, **613**/

E *Cellular Physiology* **614**

E.1 THE CELL, **614**/   *E.2* CELLULAR REPRODUCTION, **616**/
REFERENCES, **619**/

Index **621**

Preface

Since the appearance of the first edition of *Biophysical Science* in 1962, major progress has been made in many areas within biophysics. Accordingly, the earlier edition has been completely revised. An attempt has been made, however, to retain the structure and philosophic outlook of the earlier edition.

This book represents an introduction to many of the topics which are considered part of biophysics. Biophysics deals with biological problems; thus, the various chapters have been grouped by the type of problem described rather than by the methodology employed. The mathematical level required has been limited, in most cases, to elementary calculus. However, sections of a few chapters assume a higher level of mathematics. An attempt has been made to write those sections in such a fashion that the conclusions reached can be understood even if the mathematical developments must be omitted by some readers.

Different types of readers will have varied training and knowledge. Accordingly, background material from physics and biology is introduced at appropriate points. Such material will give the reader an appreciation of the importance of both physics and biology. Biophysics is as unsuited to people who know no biology as to those who know no physics.

In terms of the nature of the material covered, biophysics is closer to conventional biology than to traditional physics. Nonetheless, most physics and engineering majors can equip themselves, by extra reading, with sufficient biological knowledge to understand all the topics of biophysics. However, certain students majoring in the biological sciences must accept on faith the conclusions of many mathematical proofs. In terms of methodology, as opposed to content, biophysics is closer to physics than to biology.

As a separate discipline, biophysics is a recent addition to the subdivisions of natural science. Until the mid-nineteenth century, it was quite common for investigators to be natural scientists contributing to many diverse fields. A well-known scientist who exemplified this wide range of interest was Hermann von Helmholtz, who was trained as a medical doctor and practiced medicine. He not only conducted histological studies of the eye and ear but also worked on theories of vision and hearing. In addition to being an excellent biologist, Helmholtz was an outstanding physicist. He developed acoustic instrumentation which he used for frequency analyses of speech and music. His contributions to thermodynamics are emphasized by the term *Helmholtz free energy*. His name is associated with a law in geometrical optics as well as with a differential equation for sinusoidal waves.

With the growth of factual knowledge, it became more difficult for a person to do significant work in both the physical and biological sciences. Within each division of natural science, large numbers of subdivisions appeared; each small field had its own textbooks, its own theories, and its own part of human knowledge.

However, there is a group of problems for which extreme specialization is not desirable. Many of the problems of biophysics fall into this category and require a knowledge of several specialized fields. For example, a complete background for the study of vision must include geometrical optics, spectroscopy, quantum biochemistry, physiology, psychology, neurophysiology, and electronics.

A certain group of research topics, all of which involve both biology and advanced physics, has come to be called biophysics. However, there is no general agreement concerning the topics properly belonging to this field. For example, some scientists consider the material in Chapters 11 and 17 on radiobiology to be synonymous with biophysics, whereas others feel that the effects of ionizing radiation should not even be considered as part of biophysics.

To develop a branch of natural science as a logical structure, it is desirable to describe the behavior of highly organized systems in terms of the properties of simpler systems. This is not always feasible and in some cases cannot be followed at all. For example, in physics one discusses electric currents before attempting to present electronic conduction bands in metals. In this text we have tried to start from more general topics with which all varieties of readers will be intuitively familiar, and proceed to more abstract ideas. Thus, Part A on special sensory systems includes a chapter on vision and the eye; the neural aspects of vision are presented in a chapter following discussion of nerve activity; the molecular actions which convert light into nerve impulses form the basis of a chapter in Part D which deals with molecular biology; finally, the eye as a coding mechanism is discussed in a chapter on information theory located in Part E. Specialized physical instruments, necessary for these and other studies, are discussed in the last part of the text.

The areas of the text taken together demonstrate the broad range of biophysics. Within each part a careful selection has been made from a variety of topics, all of which are part of biophysics. The topics included in this text were chosen not only for their relative importance, but also for their suitability for presentation in a one-year course for students with a variety of backgrounds. Other possible topics are included in the discussion questions at the end of each section of the text.

It is the authors' hope that the readers of this book will gain an insight into the nature of the topics included in biophysics, recognizing the attempt to quantify and develop biological problems in terms of physical models wherever this is practical. The readers should become acquainted both with the biological basis of the various areas of biophysics and also with the essential role of mathematical analysis in most biophysical problems. The projected values of this text for several types of readers are detailed in the following paragraphs.

Biophysics has become a field that is important for all physicists to study. For the prospective college physics teacher, it presents a variety of examples which can make general physics more interesting and of greater direct personal appeal. Accordingly, students from biological and premedical curricula may learn more physics in a general course taught by an instructor well versed in biophysics. Similarly, the industrial physicists may find that a knowledge of biophysical science will broaden their appreciation of the applications of physics.

Biophysics is likewise valuable for seniors or graduate students majoring in biological sciences or medicine. Representing a different approach to topics they may study in other courses, biophysics can make an important contribution to a well-rounded training in biology. The premedical student may find that a study of biophysical science helps to understand normal and abnormal physiology.

This text on biophysical science may also prove useful for students in a variety of engineering disciplines, especially biomedical engineering. While an overlap exists between the topics covered in biomedical engineering courses and biophysical science, the latter represents a different emphasis and orientation. This text should also supplement the conventional training in other engineering specialties.

The authors wish to thank their many students and friends, all of whom have had a profound influence on the material in this text. It is not possible to name all who have helped. In the first edition, special mention was made of some of those who contributed an extra amount of their time and ideas. In addition, Dorothy Hopkirk Ackerman prepared the original sketches for several of the figures used in that edition, some of which have been carried over to the current edition. For the revised edition, Dr. Laël Gatewood, at the University of Minnesota, provided much needed constructive criticism. We also wish to express our thanks to the editors of Prentice-Hall for encouraging the preparation of this new edition. The permission of numerous publishers and authors to reprint their figures is also gratefully acknowledged. Without secretarial help, this revised text would never have been completed; special thanks are due Margie Henry, Lynn Ruggiero, and Diane Henning. We also appreciate Kathleen Seidl's assistance with the galley proofs, page proofs, and index.

Minneapolis, Minnesota   EUGENE ACKERMAN
LYNDA B. M. ELLIS
LAWRENCE E. WILLIAMS

BIOPHYSICAL
SCIENCE

PART **A**

SPECIAL SENSORY SYSTEMS

Introduction to Part A

The first two chapters were chosen as biophysical topics that are intuitively familiar to a wide group of potential readers. These two chapters, on hearing and vision, emphasize basic concepts, such as the physical nature of the stimuli and the anatomical character of the receptors. The ideas of Chapters 1 and 2 are extended in Chapter 3 to insect communication and in Chapter 4 to alternative methods of orientation in bats, snakes, and other animals. The uses of sensory systems in these two chapters include ones that human beings can duplicate only with electronic equipment.

Sensory systems form links between the central nervous system and the external world. Biophysical scientists study other sensory systems, such as taste and balance. In addition, they are concerned with discussing hearing and vision at the levels of the nervous system and of information processing. These and related topics are presented in Parts B, D, and E.

CHAPTER 1

Sound and the Ear

1.1 Hearing

The study of hearing is one of the oldest fields in biophysics. The reception and analysis of sound by the human ear has interested scientists who studied either physics or biology and has appealed especially to persons who have a background in both the physical and biological sciences. The hearing mechanisms form one of the major sensory systems through which animals are stimulated by their environment. Vertebrates, in particular, have complicated sensory receptor systems which analyze incident sound waves for tone, quality, and loudness.

Human beings rely heavily on visual information when they want accuracy, such as is required in recording scientific data. However, in communicating daily with the people around them, humans rely principally on hearing. As a result of this major role of hearing in social intercourse, persons with a hearing deficiency suffer more social disapproval than do those with visual deficiencies. Hearing is important not only for communicating with other persons, but also for avoiding dangers, such as being struck by an automobile. In addition, we learn to recognize certain living creatures and many types of events by their noises, for example, the cat's meow and the telephone's ring. Human emotions, too, are influenced by the sense of hearing. Many of our forms of entertainment—concerts, the theatre, movies, radio, and even television—depend upon our sense of hearing.

Hearing can be studied from different points of view. Physicists have learned how sound waves are generated and how they are transmitted. Anatomists have probed into the structure of the ear on a gross level and also on a microscopic level. They have traced the pathways by which auditory nerve impulses travel from the ear to the brain. Psychologists, physiologists, and physicists all have studied the thresholds of sensitivity of the hearing system and the way in which we understand speech. Most of these groups, especially biophysicists, have been interested in the manner in which the hearing organ operates, how sounds are analyzed, and how they are converted into nervous impulses and then separated according to pitch, quality, and loudness. In this chapter and in Chapter 7 an attempt has been made to synthesize these different avenues of approach, while emphasizing those parts of each that have the greatest interest to the biophysicist.

The first careful study of the ear and attempt to relate its structure to hearing was carried out by Helmholtz. Before that period, various theories of hearing existed, but few have had more significance than the one which has survived in our colloquial speech. This was the idea that the ears were connected to a common hollow region within the head where the sound was somehow stored. If we were not careful, so this theory went, the sound would go in one ear (through the storage chamber) and out the other.

Since the middle of the last century, hearing has been the subject of many scientific investigations. The nature of these studies was radically altered around 1930 by the introduction of electronic techniques. These techniques have completely changed the study of hearing; they have dramatically influenced the interpretation of all phases of hearing from pure acoustics to the final analyses of sounds within the brain. So complete is the dependence on electronic techniques today that it is hard to realize that Helmholtz and Lord Rayleigh could do acoustic experiments successfully without electronic instrumentation. Experimental studies of hearing have made use of improved electronic equipment as it became available. Since about 1965 the introduction of digital computational techniques has provided added facilities which have enabled the careful pursuit of studies delineating hearing phenomena. While these refinements have contributed much detail, they have not appreciably altered the biophysical overview of sound and the ear presented in this chapter.

Hearing is the response to mechanical, vibratory stimuli. Not all such stimuli evoke the sensations of hearing. The sound must be loud enough to be heard and also be of a suitable pitch. The latter condition is physically equivalent to saying that the vibration must be within the audible frequency range. Vibrations outside this frequency range may be detected by human sensory systems other than hearing. At frequencies too low to be heard, vibrations are perceived through the sense of touch; much greater amplitudes are needed for touch than are needed for hearing. Frequencies higher than the audible range are not sensed until the energy becomes so great as to cause local heat and pain. Between these two extremes lie the frequencies to which the ear is sensitive. The exact frequency range depends on the person; it usually changes as the individual ages and is influenced by the environment.

All vertebrates have a hearing apparatus homologous to the human ear. In fishes and amphibians the acousticolateralis system responds not only to sound but also to chemical stimuli, fluid motion, and, in some cases, electric fields (see Chapter 4). Reptilian and avian hearing systems are closer to the mammalian ear. The latter differ among one another in details such as size and frequency range. Many other animals, such as insects, are sensitive to vibratory energy over a wide range of frequencies, but their receptors are different, and the mechanisms involved in their response may be different. (Insect communication is discussed further in Chapter 3.) Even the single-celled animal paramecium can respond to vibratory energy in some fashion. Thus, there are many different types of sensory systems excited by vibratory mechanical energy. One of these, the human hearing apparatus, has been chosen for presentation in this chapter and in its sequel, Chapter 7.

1.2   Acoustics

The physical aspects of sound transmission and the vibration of the ear are a subdivision of acoustics. The latter, in turn, is a branch of physical mechanics. In order to understand journal articles dealing with the ear and hearing, it is very helpful to be familiar with the terminology of acoustics and with the electroacoustic analogs often used. The various acoustic terms useful in describing studies of hearing are defined and discussed briefly in Appendix A. In contrast, this section of this chapter contains only a few of the acoustic terms used most frequently in studies of the ear and hearing.

Perhaps most familiar is the term *frequency*, which describes how many times a second the sound pattern is repeated. The simplest possible case is one in which the *sound pressure*, p, can be described by an equation such as

$$p = p_0 \sin 2\pi v t \qquad (1\text{-}1)$$

where p_0 is the acoustic pressure amplitude, t is the time, and v is the frequency. This is referred to as a *pure tone*. The latter term is applicable since tone (or pitch) is the sensation associated with frequency. Most sounds consist of a mixture of frequencies, which gives the sound its characteristic quality and timbre. A tuning fork comes close to producing a pure tone. One can come even closer by using an electronic oscillator and loudspeaker.

Any complicated tone can be represented as a sum of simpler pure tones. This is known as a *Fourier representation*. In many cases, only a finite or a discrete set of frequencies is necessary; then one refers to the representation as a Fourier series. Speech and the character of musical instruments are determined by the frequencies present and their relative amplitudes. In the most general case, the sound is represented by an amplitude distribution, which is a continuous function of frequency. This amplitude function is called a *Fourier transform*. (See Appendices A and C.) The amplitude distribution for a sound "ee" is shown in Fig. 1.1.

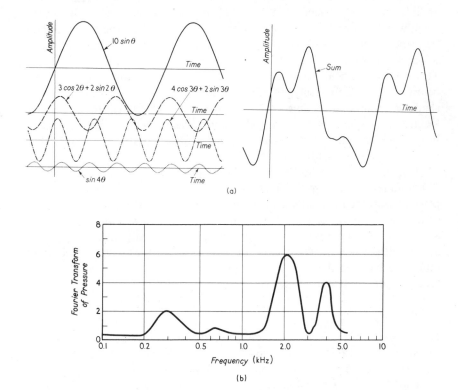

Figure 1.1 (a) Fourier series. The complicated wave form labeled "sum" can be formed by adding relative amounts of the four pure tones shown. (b) Fourier transform (or spectrum). The spectrum of the sound "ee" has the general form shown. The Fourier transform is a complex number; only its absolute value is shown.

A term closely related to frequency is *wavelength*, λ. This is the distance between the two nearest wave fronts with the same displacement and particle velocity in a plane sound field. If one knows the frequency and the velocity of sound propagation, c, the wavelength may be determined by the relationship

$$\lambda = c/\nu \qquad (1\text{-}2)$$

The wavelength is important in discussing diffraction, a phenomenon common to all wave motion. Diffraction patterns are significant when the wavelength is comparable to the object which the sound wave encounters. At shorter wavelengths, specular reflection and shadows are produced, whereas at longer wavelengths, the wave is transmitted as though the object were not there.

In air, a low tone of frequency 35 Hz has a wavelength of about 10 m, which is comparable in size to a house. At the other end of the human audible range, a frequency of 9×10^3 Hz (9 kHz) has a wavelength of about 3 cm, which is small compared to a person's head. Thus, the lowest audible frequencies will be diffracted around a house; in other words, the sound waves at the lowest frequencies will appear to bend around most obstacles. This makes it difficult to localize the source of the very low frequency tones below 100 Hz. Conversely, the highest audible frequencies will form sharp shadows around small objects; the source of a 5 to 10 kHz tone is easy to locate. At frequencies around 1 kHz, the wavelength is comparable to the head. The diffraction pattern has the effect of increasing the amplitudes at the ear above those in the incident wave.

This increased amplitude makes the sounds near 1 kHz seem extra loud.[1] The loudness is not simply determined by the particle velocity v or the displacement in the incident wave. Rather, the loudness is most readily related to another physical characteristic, the *sound pressure amplitude*. The latter and not the particle velocity or displacement is actually measured in most acoustic experiments. The sound, or acoustic, pressure, p, is defined as the difference between the average (or equilibrium) pressure, P_0, and the instantaneous total pressure, P; that is,

$$p = P - P_0 \qquad (1\text{-}3)$$

Diagrammatically, one may represent this as shown in Fig. 1.2. The acoustic pressure p is a scalar that will vary with both position and time.

Two waves of the same amplitude and wavelength traveling in opposite directions give rise to what is known as a *standing wave pattern*. Under some conditions, the wavelengths correspond to the characteristic dimensions of a physical system, and the phenomenon of *resonance* arises. This is illustrated in Fig. 1.3 for organ pipes. Note that in each case a series of *characteristic (eigen) frequencies* exists. Vibrations at these frequencies are particularly easy to excite. The lowest possible frequency is called the *fundamental frequency* or *first harmonic*. The next highest frequency is called the *first*

---

[1] Other effects discussed later in the chapter also contribute to the increased loudness of sounds at 1 to 3 kHz.

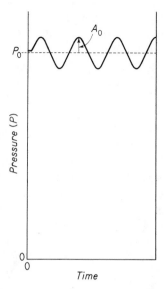

Figure 1.2   The dotted line shows the average pressure P_0 and the solid line indicates the absolute pressure P. The difference between P_0 and P is the acoustic pressure p. The maximum of p is A_0, the acoustic pressure amplitude. The figure is drawn for a pure tone showing simple harmonic dependence of p on time. The form of p is usually more complicated than is shown here.

			λ_n	ν_n	ψ_n
	Fundamental	*Fundamental*	$4l$	$\frac{c}{4l}$	$\sin\frac{\pi x}{2l}$
	1st Overtone	*3rd Harmonic*	$\frac{4}{3}l$	$\frac{3c}{4l}$	$\sin\frac{3\pi x}{2l}$
	2nd Overtone	*5th Harmonic*	$\frac{4}{5}l$	$\frac{5c}{4l}$	$\sin\frac{5\pi x}{2l}$

Figure 1.3   Particle velocity for various overtones of a closed-end organ pipe. The characteristic wavelengths and frequencies are indicated by λ and ν, respectively. The velocity of sound is represented by c. When the particle velocity has a node, the acoustic pressure has an antinode, and conversely. The external ear canal resembles a closed-end organ pipe with a fundamental around 3 kHz.

overtone. If it is an integral multiple of the fundamental, it is called a *harmonic.* For example, an overtone five times the fundamental is the fifth harmonic. The standing wave pattern in the outer ear is discussed further in Sec. 1.4.

It was noted above that the loudness of a given pure tone is determined primarily by the sound pressure amplitude. Often, another physical term, intensity, is associated with loudness. *Intensity* is the energy per unit time (power) transmitted across a unit area. In practice, intensity is difficult to measure and not too useful as a concept for studies of hearing. For a plane wave, the intensity, I, is related to the pressure by

$$I = \frac{\bar{p}^2}{\rho c} \qquad (1\text{-}4)$$

where $\bar{p}$ is the root-mean-square (rms) acoustic pressure, ρ is the density of the air, and c is the wave velocity. For other wave shapes, the expression is more complicated (although the term ρc always appears). The intensity for a given value of $\bar{p}$ varies with the temperature, since ρc is temperature dependent. Loudness depends only on $\bar{p}$, not on the temperature.

Instead of presenting data in terms of the rms sound pressure magnitude, it is customary to use the sound pressure level, L, measured in decibels (dB). This is defined by

$$L = 20 \log \left(\frac{\bar{p}}{p_0} \right) \qquad \text{dB} \qquad (1\text{-}5)$$

In air it is customary to use for p_0 the arbitrary value $2 \times 10^{-5} \, \text{N/m}^2$. The 20 in the definition of decibels (dB) arose for historical reasons; from the properties of logarithms, one might equally well write this as

$$L = 10 \log \left(\frac{\bar{p}^2}{p_0^2} \right) \qquad (1\text{-}6)$$

or, for a plane wave,

$$L = 10 \log \left(\frac{I}{I_0} \right) \qquad (1\text{-}7)$$

The latter is a sound intensity level rather than a sound pressure level; Eq. (1-7) was the historic definition of the decibel. However, I depends on temperature, the medium, and the wave shape, so the sound pressure level defined by Eq. (1-5) is really a more convenient quantity.

The use of decibels implies measuring sound pressure levels on a logarithmic scale. This has a number of advantages. For example, it is helpful in plotting graphs. It gives an indication of relative loudness since, at a fixed frequency, L as defined by Eq. (1-5) is to some extent proportional to loudness. Moreover, the use of the logarithmic scale for sound pressures allows a comparison of two pressures without knowing the absolute levels of either one. It also makes it appear as if many acoustic measurements were more precise than they actually are. Table 1-1 gives the sound pressure level of several common sounds, as well as the sound pressure amplitude.

TABLE 1-1
VARIOUS SOUND PRESSURE LEVELS

Sound pressure amplitude (N/m^2)	Sound pressure level (dB)	Example
2×10^3	160	Mechanical damage to human eardrum
2×10^2	140	Pain threshold
2×10^1	120	Discomfort threshold
		Human hearing loss (prolonged exposure)
2	100	Average factory; automobile
2×10^{-1}	80	Class lecture; loud radio
2×10^{-2}	60	Typical office; conversational speech
2×10^{-3}	40	Average living room
2×10^{-4}	20	Very quiet room
2×10^{-5}	0	Threshold of hearing

In this section, the physical quantities important in hearing have been introduced, and their application to a study of hearing has been indicated. The measurement of the typical values of these quantities, significant in human hearing, has given rise to a variety of types of tests. Some of these are discussed in the following section.

1.3   Hearing Tests

There are various ways of studying hearing. Tests on humans which do not involve any surgical techniques are discussed in this section. Clinically, the most widely employed tests measure the threshold of hearing. The observed thresholds are then compared with the normal threshold. The simplest of these tests uses pure tones. However, the exact sound pressure levels of the normal thresholds seem to be rather difficult to determine. The graph in Fig. 1.4 shows the results of several investigations. These emphasize that the threshold depends to some extent on who measures it. Notice that the ordinate is in decibels; thus a difference of 20 dB means a factor of 10 in the sound pressure. All the curves show the same general shape with a minimum threshold, that is, maximum sensitivity, in the frequency range 1–4 kHz. When the tests are conducted under controlled laboratory conditions, with carefully screened young people, the thresholds are lower than those found in mass surveys.

There exist various types of limits of hearing, none of which is very precise. These limits include a minimum pressure threshold and an upper pressure limit at each frequency, as well as a highest and a lowest frequency

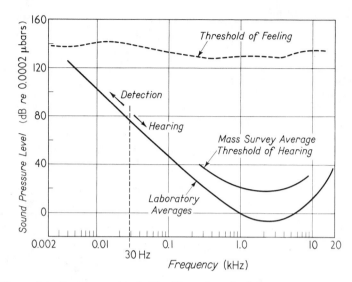

Figure 1.4 Pure tone thresholds. Note that the laboratory averages with trained, selected personnel are consistently lower than the mass survey averages. Studies at The Pennsylvania State University by J. Corso and his associates gave mass survey values between the two curves shown. Notice that the threshold of feeling is not near the threshold of hearing at either 30 Hz or 20 kHz. The latter are limits of hearing, in the sense that people can no longer distinguish tones outside these limits. After J. C. R. Licklider, in *Handbook of Experimental Psychology*, S. S. Stevens, ed. (New York: John Wiley & Sons, Inc., 1951).

limit at which one can hear. Of these, the threshold sound pressure level is most precise, but even it is a statistical limit. If a pure tone stimulus close to the subject's acoustic pressure threshold is presented repetitively, the tone will be heard some of the times and not at other times. It is customary in tests of this nature to choose the halfway point where the subject hears the tone 50 percent of the time as the limit of hearing.

The upper limit of hearing is an even less clear concept. As the sound pressure level is raised toward 110 dB, one becomes aware of feeling the sound in the external ear. At a still higher sound pressure level, perhaps 130 dB, one begins to experience pain. If the sound pressure level is raised to 145 dB, the pain becomes very severe. It has been shown in accidents due to carelessness that at sound pressure levels of about 155–160 dB the human eardrum is ruptured. (The eardrum will eventually heal.)

It is instructive to convert these sound pressure levels for eardrum rupture to actual sound pressures. Rewriting Eq. (1-5),

$$\bar{p} = p_0 \times 10^{L/20}$$

If $p_0 = 2 \times 10^{-5}$ N/m^2 and $L = 160$ dB,

$$\bar{p} = 2 \times 10^3 \text{ N/m}^2$$

This is the root-mean-square acoustic pressure. The acoustic pressure amplitude will be $\sqrt{2}$ times greater for a pure tone. This gives an acoustic pressure amplitude, A_0, of

$$A_0 \simeq 3 \times 10^3 \text{ N/m}^2$$

The average atmospheric pressure is about 10^5 N/m^2, so that 160 dB may also be written

$$A_0 \simeq 0.03 \text{ atm}$$

The sound pressure level at which the eardrum is ruptured puts an upper limit on the loudness that one can hear. The low-frequency limit to hearing is due to a different type of phenomenon. It used to be stated that the upper and the lower frequency limits of hearing were at the frequencies where the thresholds of pain and hearing crossed. At the low-frequency end of the human hearing range, this does not seem to be the case. Rather, the limit at about 30 Hz is due to the inability to identify tones or direction of frequency change.

In the audible range, a person recognizes the direction in which a frequency change occurs, provided that it is sufficiently great. For example, if the frequency is lowered from, say, 1,000 Hz to 500 Hz, the listener hears a decrease in frequency of 1 octave. (A frequency ratio of 2 is called an octave in music.) Below about 30 Hz, the listener cannot really distinguish tones or tell whether the frequency is being raised, lowered, or held constant. If the frequency is lowered to, say, 1 Hz, the tone identified is not the applied sound frequency but rather something in the neighborhood of 1,000 Hz.

Likewise, at high frequencies a point is reached above which a person can no longer distinguish tones. In addition, the threshold sound pressure rises very sharply. This latter effect limits experiments at the high-frequency end of the spectrum. The exact frequency range in which this sharp rise occurs varies widely from individual to individual. For one graduate student who was tested extensively, this sharp rise occurred around 25 kHz. He was consistently able to tell that 23 kHz was higher in pitch than 22 kHz. Other young adults fail to respond to reasonable sound pressure levels above 17 kHz. Older individuals may have normal hearing for their age but fail to hear any tones above 6 kHz.

The highest frequency heard varies from one normal, young listener to another by a factor greater than 3.[2] This range may seem large, but it is small compared to the individual variations of pure tone thresholds.

_____

[2] Eight kHz to higher than 25 kHz.

Variations from one individual to another may be as high as 40 dB within the normal range of hearing. These numbers, when translated into actual acoustic pressures, represent a pressure ratio of a hundredfold, truly an enormous variation.

In an ordinary room, the lowest sound pressure levels one can hear are limited by the ambient noise. In a very quiet room, where all the ambient noise is below the hearing threshold, the physiological noise level is approximately at the threshold of hearing. This physiological noise is due to a variety of causes: the pulse in the ear, the muscles contracting, breathing, and any motion of the joints. Physiological noise is effective only at those frequencies where the ear is most sensitive, that is, the range 1–4 kHz.

Most sounds come to the ear from the air. Some, such as a few of the physiological noises, are transmitted by bone conduction. The entire structure of the middle and inner ear discriminates strongly in favor of airborne vibration as opposed to bone conduction. However, a sufficiently strong signal can be conducted by the bone. The bone conduction threshold can be observed by blocking the ears[3] or by applying a vibrator directly to the head. The sound pressure levels necessary for hearing by bone conduction are about 40 dB higher than by air conduction, and the threshold curve is much flatter.

The pure tone hearing threshold tests described above depend on the accuracy of the apparatus and the technique of the operator, as well as on the hearing of the person being tested. By suitable calibrating techniques, the equipment can be standardized so that the sound pressure levels are accurately known to within 1 dB (that is, about ± 10 percent in the actual sound pressure). The effect of the operator is harder to remove. The latter must present successively lower and then higher sound pressure levels to the subject. If the operator starts far above the threshold, the subject becomes familiar with the tone and will distinguish it at lower sound pressure levels than if the operator started below the threshold. The operator must cross and recross the threshold until it is the operator's judgment that a stable value has been found.

One very ingenious attempt to remove the effect of the operator was introduced by Békésy. His *audiometer* includes the person being tested as part of a feedback loop in an automatic control device designed to keep the sound pressure level at the ear close to the threshold. The system is illustrated in block diagram form in Fig. 1.5. The output of an oscillator is fed through a variable attenuator to the earphones. The subject is given a switch and told to depress it when a tone is heard and release it when a tone cannot be distinguished. The switch is connected to a reversible motor which drives the variable attenuator in such a fashion that the sound pressure level increases with the switch released and decreases with the switch depressed. The entire

[3] This may raise the threshold.

however, the principles inherent in the Békésy audiometer have served to limit the effects of observer bias.

The information obtained from a speech audiometer is different from that found by using a pure tone audiometer. In a speech audiometer various test words are presented at a constant sound pressure level. Some persons who have appreciable pure tone hearing losses at certain frequencies do not show any hearing loss for speech. Conversely, other people, with normal pure tone thresholds, have marked speech-hearing deficiencies. The problem of recognition of speech is much more complex than hearing a pure tone. Understanding speech involves the function of several parts of the brain. Indeed, speech can still be understood if any two continuous octaves of the audible spectrum are presented and the rest of the energy filtered out. Even up to 50 percent of every syllable or word can be removed. The remainder when compressed to eliminate the blank times is still understandable.

The speech threshold measures a person's ability to participate in a conversation or listen to a lecture. It depends as much on the functional condition of the brain as it does on the action of the ear. In contrast, the pure tone threshold indicates to a greater extent the action of the ear itself.

As people grow older, the pure tone thresholds are raised, particularly at higher frequencies. For people of all ages, these thresholds are raised by exposure to loud noises. The latter effect is reversible if only occasional exposures occur but is quite irreversible after years of continuous exposure. It is not worthwhile here to go into the details of current estimates on criteria for levels at which, say, 5 percent of the persons will be appreciably deafened after years of exposure. The currently accepted levels are lower than those which exist in many factories.

The relationship of pleasure to audible frequency range is very complicated. In these days of high fidelity and extended frequency ranges, one might guess that the wider the frequency range, the more pleasing. This does not seem to be the case. Older people find that hearing aids which correct their high-frequency losses make music sound harsh and unpleasant but that flat-response amplifiers increase their satisfaction in listening to music. In other words, what the listener is used to hearing is enjoyable.

Other types of information can be gleaned from experiments similar to those used to obtain the pure tone threshold curves. One test is to ask the subject to match, in loudness, tones of different frequencies. On the basis of these results, equal loudness curves can be drawn. They are illustrated in Fig. 1.6. The lowest is the threshold curve itself. As the sound pressure level is raised, the equal loudness curves tend to flatten out, approaching straight lines by the time the sound pressure level at 1 kHz has reached 100 dB.

Another test is to ask the subject to choose just noticeable differences in loudness. A change of this nature is sometimes referred to as a *difference limen* (DL). When the sound pressure level is 60 dB or more above the

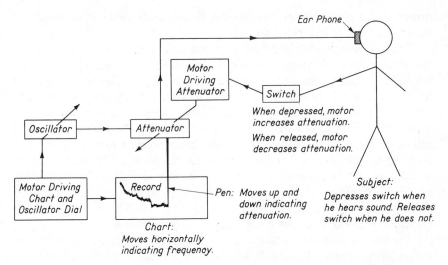

Figure 1.5   Block diagram of the Békésy audiometer, which records the threshold of hearing without the influence of any operator other than the subject.

setup then hunts for the threshold, continuously crossing and recrossing it. A recording pen is attached to the variable attenuator. The pen writes on a calibrated chart, recording the instantaneous setting of the attenuator. Another motor drives both the chart and the oscillator so that a record is obtained of threshold level versus frequency. This level is recorded without any influence of the examiner. Numerous modifications exist, such as ones that operate at discrete frequencies rather than a continuum.

The Békésy continuous audiometer is very successful in limiting the role of any operator other than the subject. It also gives an uninterrupted graphical record of threshold versus frequency instead of values only at discrete points. However, it has several disadvantages. It is slower than an audiometer operated at discrete frequencies by an experienced technician. Also, one cannot distinguish between losses in a certain frequency range and apparent losses due to extraneous physiological noises such as swallowing. Accordingly, discrete-frequency Békésy audiometers are preferred for screening purposes. Finally, the results obtained with any Békésy audiometer reflect not only the characteristics of the subject's auditory system but also the mechanical skill of the subject and the subject's understanding of the instructions given. Both of these latter vary from person to person, introducing variables into the apparent threshold.

The Békésy audiometer accordingly must be regarded as a research tool. As such, it has found application in a number of hearing studies. It has suffered from a number of confounding influences on the apparent thresholds and has remained impractical for clinical applications. In a research setting,

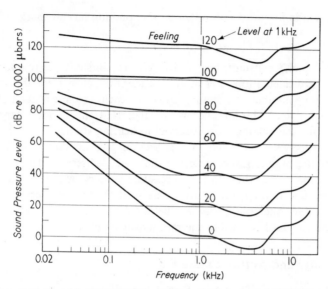

Figure 1.6 Equal loudness contours, after the American Standards Association (1936). There is no general agreement on the exact shape of these curves, but the general flattening at higher sound pressure levels is always observed. After J. C. R. Licklider, in *Handbook of Experimental Psychology*, S. S. Stevens, ed. (New York: John Wiley & Sons, Inc., 1951).

threshold of hearing, the DL is of the order of 0.5 dB[4] throughout most of the auditory range. At lower sound pressure levels, the DL's are greater. At 30 dB above threshold they are about 1 dB; they are as large as 6 dB near threshold.

Similarly, difference limens, or just noticeable differences, exist as the frequency is varied. At very low frequencies, a 0.5-Hz change is detectable. In the middle-frequency range (around 1 kHz), the normal person can notice a 3-Hz change. At the very high frequency end of the audible range, changes greater than 25 Hz are necessary before a change of pitch is noticed. These difference limens for frequency change are not independent of the sound pressure level. As the latter is lowered, the size of the difference limen for frequency changes increases.

The presence of these finite steps, dignified by the term "difference limens," resembles phenomena well known in many phases of chemistry and physics, usually grouped under the classification quantum mechanics. Similarly, pure tone thresholds measured on individuals at very low frequencies suggest some type of quantum effect. Quantum effects do occur in acoustics, but the physical quanta of sound energy, known as *phonons*, are far too small to

---

[4] That is, it is between 0.25 and 1.0 dB.

associate them in any way with hearing. The phonon has an energy, E, such that

$$E = h\nu$$

where h is Planck's constant, 6.67×10^{-34} J·sec, and ν is the frequency. A straightforward calculation will show the reader that, even at the threshold of hearing, a huge number of phonons must be reaching the ear each second, or even each cycle. This number is so large that the phonon cannot be responsible for the quantization observed in hearing studies.

The hearing tests described in this section give no direct clues to the location of the organs responsible for the effects observed. These hearing tests are simple in that they do not necessitate surgery or putting electrodes into people. By contrast, the studies described in the next section and in Chapter 7 allow one to determine whether the effects are mechanical or nervous and to gain insight into the mechanism of hearing.

1.4 Anatomy and Action of the Ear

The ear is the organ of hearing. Sound waves impinge on the ear which couples them to the endings of the sensory nerve associated with hearing. It is customary to divide the mammalian ear into three major divisions: the outer ear, the middle ear, and the inner ear. The outer and middle ear are filled with air; their primary purpose seems to be to conduct sound to the inner ear. The inner ear consists of several parts, some of which are concerned with balance, and one of which is part of the hearing apparatus.

The incident sound waves in the air surrounding the head enter the outer ear first. This consists of three parts, an *external auricle* (or pinna), a narrow tube called the *external auditory meatus* (or ear canal), and the *tympanic membrane* (or eardrum). These are illustrated in Fig. 1.7. The auricles are

Figure 1.7   The outer ear. After A. J. Carlson and V. Johnson, *The Machinery of the Body* (Chicago: The University of Chicago Press, 1941).

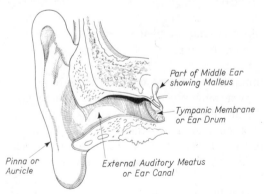

almost vestigial in humans and play a very minor role in the phenomenon of hearing. In most mammals, they are large and can be directed so as to help locate the origin of a given sound (see Chapter 4). In rodents and some other mammals, the auricle is at times laid down across the opening to the meatus to give some protection against very loud sounds.

In humans, the external auditory meatus is somewhat circular in cross section and more or less a straight tube. In an average adult, it is about 1.04 mℓ in volume and about 2.7 cm long. As in many other biological measurements, variations of ± 10 percent from the mean are quite usual but variations as great as ± 20 percent are rare. The meatus is terminated by a thick fibrous membrane called the *tympanum* or *tympanic membrane*. Along the edges of the membrane are glands which secrete a waxlike substance called *cerumen*. This forms a protective coating. In cases of irritation, an excess of this wax is secreted, often causing a temporary loss of hearing.

The external auditory meatus may be thought of somewhat as a closed-end organ pipe. The tympanum at the end of the meatus is relatively stiff. Here, the particle velocity should be a minimum and the acoustic pressure a maximum. The opening to the air should be just the opposite, a pressure node and particle velocity antinode. Figure 1.4 shows that the external auditory meatus at resonance is a quarter-wavelength long. At this frequency, about 3 kHz, there will be a maximum acoustic pressure delivered to the inner ear for a given incident pressure. This resonance corresponds to the minimum in the pure tone threshold curve. Studies with probe tubes attached to microphones show that the maximum pressure amplification in the ear canal is about 10 dB. This is not sufficient to account for the threshold minimum from 1 to 4 kHz but definitely contributes to it.

In humans the tympanic membrane is oval in shape, about 66 mm² in area, and about 0.1 mm thick. It couples the vibration of the air molecules in the outer ear to the small bones of the middle ear. At extreme intensities the tympanic membrane is a nonlinear device; that is, it produces harmonics and subharmonics of the frequencies exciting it. These nonlinear effects, however, are only important at very high sound pressure levels. In some mammalian species, the tympanum vibrates as an elastic membrane. In other species, including the human, the motion of the tympanum is more like that of a piston.

Various techniques have been used to observe the motion of the tympanum. The simplest is to glue a long light stick to the tympanum and observe the motion of the end of the stick. Both laser and Mössbauer techniques (see Chapter 27) have been used to attain better resolution. Nonetheless, most of the observations have been possible only at low frequencies and high sound pressure levels; the results can be extended only by extrapolation. Tests of this type show that the particle velocity of the tympanum is of the same order of magnitude as that in a plane wave in air. Applying this result to 0 dB, the approximate threshold at 1 kHz, one finds for the particle velocity, v, using the plane wave relationship (see Appendix A), that

$$v = \frac{p}{\rho c}$$

$$v \simeq 5 \times 10^{-6} \text{ cm/sec}$$

For the displacement, ξ, assuming a sinusoidal vibration,

$$\xi = \frac{V}{2\pi v} \simeq 10^{-9} \text{ cm} = 0.01 \text{ nm}$$

This displacement is smaller than an atomic radius!

Since this displacement was obtained by extrapolation, it could conceivably be too small, owing to nonlinearities. Even if this resulted in a factor-of-10 error, so that the displacement was really 0.1 nm, it still would be almost unbelievably small. It is necessary to note that this is not the displacement of an individual atom or electron but rather the average displacement of a large number of atoms. Moreover, there would be no response to a slow average displacement of this magnitude.

The tympanic membrane forms the outer boundary of the middle ear. The latter is an air-filled space in the temporal bone; this space is referred to as the *tympanic cavity*. It has a volume of about 1 mℓ and an irregular shape. Within this cavity are three small bones or *ossicles*, which are named according to their shapes. These are the *malleus* (hammer), the *incus* (anvil), and the *stapes* (stirrup). They are illustrated in Fig. 1.8. The ossicles act as a mechanical transformer and increase the fraction of the incident energy available to excite the mechanisms of the inner ear.

The bones of the middle ear are so pivoted that they are particularly insensitive to vibrations of the head and to bone-conducted sound waves. One action of the ossicles is to amplify the acoustic pressure of vibrations transmitted from the air via the tympanum, while discriminating against vibrations reaching them via the skull. This insensitivity of the ossicles to bone conduction, as well as the symmetry of the vocal cords, restricts most of the hearing of one's own voice to sound transmitted in the air from the mouth around to the ears. (This can be demonstrated by covering one's ears while talking and noting the changes in loudness and quality.)

The ossicles are believed to have an additional function. This is to decrease the amount of energy fed into the inner ear at high sound levels. Part of this is thought to be accomplished by changes in the tension of the tensor tympani and stapedius muscles which hold the ossicles in place. The action may be compared to automatic volume control in a radio. In both cases, when a large signal enters the system and is detected, the amplification of an earlier portion

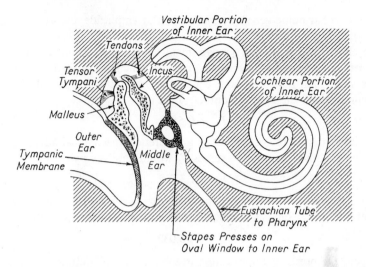

Figure 1.8 The middle ear, which is filled with air, is connected by two membranes, called windows, to the fluid-filled canals of the inner ear. The eustachian tube connecting it with the pharynx is even smaller in diameter than is indicated here. Adapted from G. G. Simpson, C. S. P. Hendrigh, and L. H. Tiffany, *Life: An Introduction to Biology* (New York: Harcourt Brace Jovanovich, Inc., 1957).

of the system is decreased. These are specific examples of "feedback systems" or "automatic control," as this type of phenomenon is called by physicists and engineers. In the case of the middle ear, one may describe this action in teleological terms as trying to maintain a constant sound level incident to the inner ear. Although this response is too slow to protect the ear from damage due to sudden noises, it is of the proper nature to explain the flattening of the equal loudness contours at high intensities.

In physical form the outermost ossicle, the malleus, is pressed against the tympanic membrane. The innermost one, the stapes, pushes against a membrane called the *oval window*, which separates the air-filled middle ear from the liquid-filled channels of the inner ear. The oval window forms one end of one of these channels, the *scala vestibuli*. Another channel, the *scala tympani*, also ends in a membrane separating it from the middle ear. This second membrane is called the *round window*.

High signal transmissions are limited by a shift in the mode of vibration of the stapes. In one of its two possible modes of vibration, the stapes pushes uniformly on the oval window. In the other, it rocks in such a fashion that it causes a negligible net displacement of the oval window. The latter type of motion is believed more important at higher intensities. Both the variable coupling and the two possible modes of vibration are nonlinear effects. Both contribute to harmonic generation as well as to amplitude distortion and to beat frequency generation.

The effective area of the tympanum in a human is about 0.66 cm², of which perhaps 0.55 cm² is in contact with the malleus. The force, F_m, on the malleus, due to the acoustic wave, equals the product of the pressure, p_t, on the typanum times the area of contact. That is

$$F_m = 0.55p_t$$

Models indicate that the ossicles have a theoretical mechanical advantage of 1.3. Therefore, the force on the stapes F_S would be given by

$$F_S = 1.3F_m$$

if friction were absent. Likewise, the pressure, p_w, exerted by the stapes on the oval window, which it contacts for 0.032 cm², can be computed from

$$p_w = F_s/0.032$$

Solving for the pressure amplification in the absence of friction,

$$A = \frac{p_w}{p_t} = \frac{1.3 \times 0.55p_t/0.032}{p_t} \simeq 22$$

Actual measurements have shown that the physiologically correct value is 17. This latter number is a 25-dB gain in acoustic pressure. This value is believed valid throughout most of the auditory range, although it is based on extrapolation from low frequencies and high sound pressures.

Since the middle ear is filled with air, any difference in pressure on the two sides of the tympanic membrane will tend to displace the membrane. Small differences in pressure at frequencies to which the ear responds cause the vibrations of the tympanic membrane during normal hearing. In contrast, large slow changes in pressure, due to atmospheric variations or altitude changes, could distort the shape and position of the tympanic membrane. To avoid this distortion, a connection is necessary between the middle ear and the ambient air; but to allow low-frequency sounds to be heard, this connection must be unable to transmit changes that take place in less than a tenth of a second. A small narrow tube will do exactly this. Such a tube does connect the middle ear with the pharynx; it is called the *eustachian tube.*

The soft walls of the eustachian tube are easily collapsed by an excess pressure outside the tube. This leads to a very unpleasant feeling, sometimes experienced when descending in an airplane. Swallowing, chewing gum, or attempting to blow with the mouth and nose held shut will open the eustachian tube, permitting equalization of the pressure outside and within the middle ear.

The outer and middle ear together produce a maximum pressure amplification of about 35 dB. They tend to reduce the hearing of sounds that are conducted through the bones, to make one insensitive to one's own voice except inasmuch as it is heard through air conduction outside the head, and also to act as an automatic control unit. None of these are essential for hearing, although all are desired effects. It is possible to hear without a tympanic membrane and without ossicles. There is a hearing loss under these conditions, but this loss is comparable to the variations in the normal range of hearing thresholds. However, the two windows to the inner ear, one of which is driven much more than the other by the incident wave, are necessary for hearing.

The mammalian inner ear consists of several portions all having two common fluids and all served by the eighth cranial nerve. Only the cochlear portion of the inner ear is associated with hearing. Grossly, the *cochlea* is a spiral; in the human there are two and a half complete turns. Around this spiral run three parallel, fluid-filled tubes; the tympanic, vestibular, and cochlear ducts (Figs. 1.9 and 1.10). The fluid in the tympanic and vestibular ducts is called the *perilymph*. These two ducts (or *scalae*) are connected at the apex of the spiral through a small duct called the *helicotrema*. The *cochlear duct* is sandwiched between these two ducts and is filled with a fluid, similar to perilymph, called the *endolymph*. The endolymph and perilymph are anatomically and electrically separated from each other. Between the cochlear duct and the vestibular duct is a very thin fibrous membrane known as *Reisner's membrane*. Between the cochlear duct and the tympanic duct is a thicker membrane called the *basilar membrane*. The basilar membrane gets progressively broader and thicker as one proceeds towards the apex of the spiral.

The basilar membrane is the seat of the *organ of Corti*, shown in detail in

Figure 1.9   (a) The cochlea or inner ear removed from the bone. (b) Cross section through one turn of the cochlea. The tympanic and vestibular ducts are filled with perilymph and the cochlear duct with endolymph. After A. J. Carlson and V. Johnson, *The Machinery of the Body* (Chicago: The University of Chicago Press, 1941).

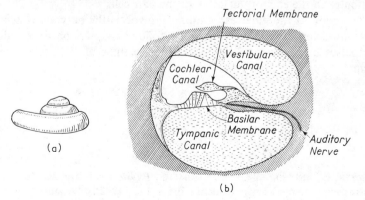

(a)

(b)

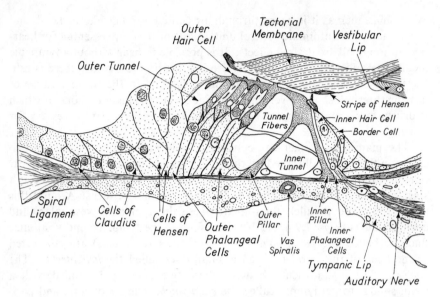

Figure 1.10 Histology of the organ of Corti. After A. A. Maximow and W. Bloom, *Textbook of Histology* (Philadelphia: W. B. Saunders Company, 1957).

Fig. 1.10. This organ contains the nerve endings. Thus one may think of the organ of Corti as a neuromechanical *transducer*. (A transducer is a device that converts one form of energy to another form.) Histologists have studied the organ of Corti in great detail. It seems as if almost every cell has its own name, many of which are shown in Fig. 1.10. The organ of Corti includes Claudine cells, Hensen cells, inner and outer hair cells, and the tectorial membrane. It is believed that the bending of the hair cells in some way excites the nerve endings, which are also located in this organ. The tectorial membrane rests on the hair cells; its motion relative to the basilar membrane contributes to the excitory bending of the hair cells.

The action of the inner ear intimately involves the nervous system. The details are deferred to Chapter 7, which follows chapters on the conduction of impulses by nerves and the electrical potentials of the central nervous system.

REFERENCES

1. MEYER, E., AND ERNST-GEORG NEUMANN, *Physical and Applied Acoustics: An Introduction* (New York: Academic Press, Inc., 1972), 412 pages.

2. DAVIDOVITS, P. *Physics in Biology and Medicine* (Englewood Cliffs, N.J.: Prentice-Hall, Inc., 1975), 298 pages.

3. WEVER, E. G., AND MERLE LAWRENCE, *Physiological Acoustics* (Princeton, N.J.: Princeton University Press, 1954), 454 pages.

4. STEVENS, S. S., ed., *Handbook of Experimental Psychology* (New York: John Wiley & Sons, Inc., 1951).
    (a) VON BÉKÉSY, GEORG, AND W. A. ROSENBLITH, "The Mechanical Properties of the Ear," pp. 1075–1115.
    (b) LICKLIDER, J. C. R., "Basic Correlates of the Auditory Stimulus," pp. 985–1039.

5. TOBIAS, J. V., ed. *Foundations of Modern Auditory Theory*, Vols. I and II (New York: Academic Press, Inc., 1970 and 1972), 466 and 508 pages.

6. MOLLER, A. R., AND P. BOSTON, eds., *Basic Mechanisms in Hearing* (New York: Academic Press, Inc., 1973), 941 pages.

7. POLYAK, S. L., GLADYS McHUGH, AND D. K. JUDD, *The Human Ear in Anatomical Transparencies* (published under auspices of Sonotone Corporation, Elmsford, N.Y., 1946; distributed by T. H. McKenna, Inc., New York).

8. CORSO, J. F., *The Experimental Psychology of Sensory Behavior* (New York: Holt, Rinehart and Winston, 1967), 628 pages.

9. HOAR, W. S., *General and Comparative Physiology* (Englewood Cliffs, N.J.: Prentice-Hall, Inc., 1966), 815 pages.

CHAPTER 2

Light and the Eye

2.1 Vision

In many aspects of human life, vision is far more important than any other sensation. History, legal agreements, and knowledge of the universe are all recorded in written form. Without vision these records would be of little value. In most measurements in physics, it is customary to base sensitive, precise observations on visual information. In mechanics, attributes such as weight, length, and pressure are measured visually by examining pointers on scales, meters or dials. In acoustics, precise data are often based on visual readings of electrical meters. Indeed, in almost all of natural science, the reading of electrical meters is an important means of gathering data. However, even before the advent of electronics, the data of the biologist and the chemist, as well as those of the physicist, were based primarily on what could be measured by visual means.

Vision plays other roles in life besides data gathering. Many of our aesthetic pleasures come from objects that are viewed. Furthermore, vision acts to protect us from many dangers, such as those which beset us in crossing a street, driving a car, or climbing the stairs. For other types of activity, vision is not necessary but nonetheless plays an important role in normal human beings; most outstanding of these is the sense of balance. Human beings use visual cues more frequently than any other type of sensory information.

Vision depends on light. During most of the evolutionary development of animals, light came primarily from the sun. It is only in recent times that artificial lighting has been used. Since all animals were exposed to similar physical light stimuli in their development, it is not surprising that all animals have similar visual ranges. This uniformity contrasts sharply with the spread of the frequency ranges of hearing, which vary by more than an order of magnitude from one species to another.

It is necessary to understand something about the physical character of visible light to have an appreciation of the phenomena of vision. Light may be discussed, depending on the problem under consideration, from three different avenues of approach. The first of these, and historically the oldest, is called *geometrical optics*. It applies to many problems in optics which can be solved by treating light as if it were propagated as bundles of rays, each normal to the wave front. Most of geometrical optics dealing with lenses can be discussed from this point of view. The optical properties of the eye as a focusing lens system are most simply described by geometrical optics.

The second approach to the study of light places its emphasis on *wave aspects*. Light waves are electromagnetic in character; the properties of the waves are used to describe the transmission of light through a medium. In particular, the wave theories are useful in discussing such phenomena as diffraction, interference, polarization, and resolving power. The wave theories are also useful in discussions of visual acuity and color vision.

From the point of view of physics, the most basic approach to a study of light is that of *quantum mechanics*. It is used in problems dealing with the emission or absorption of light. In the quantum theory, light is considered to be made of packets (or quanta) of energy called *photons*. The probability of finding a photon at a given place can be described by a mathematical form called a *wave function*. This quantum view of light is necessary for studies of visual thresholds described in this chapter and for the discussions in Chapter 19 of the absorption of light on a molecular scale.

The next section presents several of the physical phenomena of light which apply directly to vision. These include the three avenues of approach outlined above: geometrical optics, electromagnetic waves, and the quantum theory of light. This is followed by Sec. 2.3, on the anatomy of the eye. The optical properties of the eye considered as a thick lens, as well as visual defects, are included in that section. Biophysicists have also been interested in visual thresholds and in measurements of visual acuity; these are discussed in the final section.

Many aspects of vision will be deferred to later chapters. For example, color vision and the neural mechanisms making vision possible are described in Chapter 8, which follows chapters on the operation of the nervous system. The properties of the retinal pigments which absorb light are easier to understand following a study of enzymes. The visual pigments are therefore discussed in Chapter 19, Part D. Finally, Chapter 25, on information theory, contains a section that includes visual information.

2.2 Optics

A. GEOMETRICAL OPTICS

Many properties of lens and mirror systems can be treated by regarding light as bundles of rays each of which moves at right angles to the wave front. This approach is utilized in this section in the discussion of the properties of thick lenses. These properties are applied to the eye in Sec. 2.3.B.

From the point of view of geometrical optics, the most important property of a medium is the velocity at which light is propagated. In free space, the velocity of light is usually designated by the symbol c, and in SI units, it has the value

$$c = 3 \times 10^8 \text{ m/sec}$$

It is customary to specify the velocity, v, in any other medium by the index of refraction, n. This is a dimensionless number defined by the ratio

$$n = \frac{c}{v} \qquad (2\text{-}1)$$

Strictly speaking, n is always the index of refraction referred to the velocity of light in free space. However, one may also use the relative index of refraction, n_{12}, between any two media, where n_{12} is defined by

$$n_{12} = \frac{v_2}{v_1} \qquad (2\text{-}2)$$

In the eye, the luminous energy passes through a series of curved surfaces of refraction. All these surfaces may be approximated by sections of spheres whose centers lie on a common line. This general case has been shown to be mathematically equivalent to a single thick lens, which separates two media of different indices of refraction. For a thin lens at such an interface, all images can be constructed knowing only three parameters, the optic center and the two focal points on either side of the interface. However, six cardinal points are necessary to completely specify thick lens action: two focal points, two principal points, and two nodal points. This general case is illustrated in

Fig. 2.1. These cardinal points will be used in the next section to describe the eye. (The more detailed use of geometrical optics to describe the properties of thick lenses will not be pursued here; the interested reader is referred to the reference by Meyer-Arendt.)

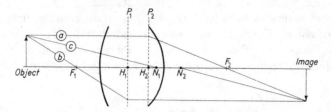

Figure 2.1   A thick lens immersed in different media on its two sides. F_1 and F_2 are focal points. Note that F_1 does not equal F_2. The principal points are H_1 and H_2, and the nodal points are N_1 and N_2.

The strength of a lens (or its power), L, is defined as the reciprocal of the focal length, f, measured from the corresponding principal plane; that is,

$$L = \frac{1}{f} \tag{2-3}$$

When f is measured in meters, L will be expressed in diopters. A lens with a shorter focal length can produce a real image for closer objects than a lens with a longer focal length. Thus, the former lens produces a greater algebraic change in curvature of an incident light front. In this sense, a lens of shorter focal length is indeed stronger. In any case, increasing the radius of curvature of a converging surface will increase the focal length and decrease the lens strength. In a system of a series of spherical surfaces, such as is found in the eye, the forward and backward focal lengths will be different.

B. LIGHT AS AN ELECTROMAGNETIC WAVE

Although many actions of lens systems may be adequately described by geometrical optics, others cannot be. In Chapter 1, reference was made to the phenomena of diffraction and interference. *Diffraction* refers to the fact that a wave will not behave as a bundle of rays, especially in the neighborhood of objects comparable in linear dimension to the wavelength of the light. (See Chapter 1 for a definition of wavelength.) In discussing sound, it was noted that the wavelengths of many audible sounds were comparable to the sizes of rooms and buildings. Thus, speech sound waves are diffracted by (or bent around) the furniture and other objects. The wavelength of visible light is much smaller than most common objects; hence, diffraction effects are not a usual part of everyday experience. However, experiments with slits, fine wires, small spheres, and so forth show that diffraction effects do occur. For similar reasons, interference effects in the form of standing waves are familiar in sound experiments but demand special equipment in order to be demonstrated for light. These and many other experiments make it impossible to avoid the conclusion that light is a wave motion represented to a sufficient approximation by rays only in limited circumstances.

The wave nature of light has two very important consequences for the sensation of vision. The first is that there is a theoretical limit to the resolution of any lens system, including the eye; that is, there is a minimum separation of two points whose images are resolvable. Figure 2.2 shows the diffraction

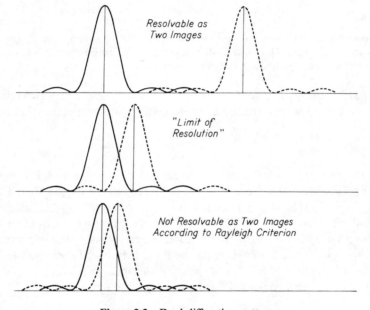

Resolvable as Two Images

"Limit of Resolution"

Not Resolvable as Two Images According to Rayleigh Criterion

Figure 2.2   Dual diffraction patterns.

patterns of the light originating from two point light sources. If one computes the dimensions of the diffraction patterns of the light originating at the two points and asks that the central maximum of one coincide with the first minimum of the second, one finds that the angular separation θ in radians of the lines from the lens center to the two points is given by

$$\theta = \frac{1.22\lambda}{a} \tag{2-4}$$

where λ is the wavelength of the light and a is the radius of the aperture of the lens. It is often assumed that this is about the minimum separation at which two points can be distinguished. The reciprocal of θ in minutes of arc is called the *resolving power*. This result was first developed by Lord Rayleigh and is often called the *Rayleigh criterion*; it is used in Sec. 2.4.C in the discussion of human visual acuity.

In addition to its use in predicting resolving power, the wave nature of light is necessary to discuss color vision. If light of a narrow wavelength band is present, it is said to be monochromatic; that is, it gives the sensation of a single color. Only about 1 octave (i.e., a factor of 2 in the frequency) is visible to humans. In wavelength terms, the visible spectrum runs from about 760 nm (red) down to about 380 nm (violet), although the exact limits quoted by different experimenters vary. One octave seems a narrow band when compared with the sense of hearing where musical tones are audible in at least 9 octaves. The resolution of different wavelengths by the eye is much poorer than the sharp tone discrimination of the ear. Combinations of different wavelengths of light produce complex color sensations, because the eye does not analyze frequencies in a fashion analogous to that of the ear.

A number of different types of experiments have left no doubt that light is a form of electromagnetic radiation. Two of these experiments will be mentioned here. First, one can compute, using Maxwell's equations, that an electromagnetic wave should be transverse and have a velocity that can be determined by electrostatic and magnetic measurements. Polarization experiments confirm that light waves are transverse. Optically determined values of the velocity of light, c, agree with those predicted for electromagnetic waves to better than one part in a million. Second, light is continuous with radiation produced by other methods. Using techniques that overlap at their wavelength limits, one may produce radio waves, microwaves (radar), heat waves (infrared), light waves, ultraviolet rays, X rays, and gamma rays. Thus, all of these are examples of the same basic phenomenon: electromagnetic radiation. No explicit use will be made of the electromagnetic properties of light waves in the chapters on the eye or on vision in this text.

C. LIGHT AS PHOTONS

The electromagnetic wave theory correctly describes the transmission of light, but a number of other effects are impossible to understand without the quantum theory. These include the characteristic spectra of atoms, the absorption spectra of atoms and molecules, the photoelectric effect, blackbody radiation, and the failure of the equipartition of energy for electrons in a metal and for the vibrations of diatomic gases at room temperatures. These and many other phenomena have been explained only in the terms of quantum mechanics, wherein energy is postulated to occur in packets or *quanta*. In particular, for electromagnetic radiation, these quanta are called *photons*. Each has energy, E, such that

$$E = \frac{hc}{\lambda} \tag{2-5}$$

where h is Planck's constant, 6.67×10^{-34} J·sec, c is the velocity of light, and λ is the wavelength. The relative probability of finding a photon at a given place computed by quantum mechanics is essentially identical to the intensity computed on the basis of the electromagnetic wave theory. (Strictly speaking, in quantum mechanics, a measurable quantity is specified by a probability.)

The photon nature of light is important in describing the threshold of vision. It is likewise necessary in Chapter 19, where vision is discussed on the molecular level. In the latter case one may ask: How many photons react with a molecule; how do the photons change the sensitive molecules; and how are the resulting small bursts of energy transduced to neural impulses? Unfortunately, it will appear that one cannot give a complete answer. Nonetheless, the language of photons and of quantum mechanics is the only one in which these topics can be discussed. The reader with a background in biology, or even an undergraduate physics major, may feel that this topic of quantum mechanics has been introduced too lightly, but only the concepts of quantum mechanics which are needed for a discussion of vision have been included above.

Quantum mechanics is necessary for an understanding of characteristic spectra. Accordingly, quantum theory is discussed more thoroughly in Chapters 26 and 27. Even there, the authors must make several statements which are foreign to everyday experience and certainly are not proved in this text. It is hoped that, in spite of this, readers will at least gain a feeling of what quantum mechanics is and how it is used, even though they may be unable to use it themselves.

2.3 Anatomy of the Eye

A. GROSS ANATOMY

Eyes are found in many arthropods, a few worms, some mollusks, and most vertebrates. The compound arthropod eye, shown in Fig. 2.3, is found in insects, crabs, and spiders. It contains lenses, light-sensitive *retinal cells*, and nerves to the brain. The eye is composed of many tube-shaped units called *ommatidia*, each with its own lens, retinal cells, and nerve fibers, arranged on a spherical surface. Each individual unit can focus only a small point of light; together all form a fairly good composite image.

The eye in all other species is similar to the human eye discussed below. This is called a *camera eye*, since there is an opening in a diaphragm which admits light through a lens into a dark "developing chamber." The human eyeball is roughly a sphere approximately 2.4 cm in diameter. It is supported in a special socket in the cranium. The orientation of the eyeball is controlled by six sets of muscles. These rotate the eyeball quite freely because the socket

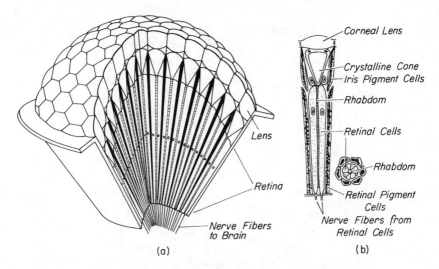

Figure 2.3   Image-forming eye of the compound type. (a) The compound eye of an insect cut away to show the hundreds of individual ommatidia. (b) An individual ommatidium. The corneal lens and crystalline cone of each ommatidium focus incoming light rays onto the rhabdom, a clear, rod-shaped structure. Light passes from the rhabdom into eight retinal cells that surround it. Each retinal cell contributes a nerve fiber to the optic nerve. The whole ommatidium is surrounded by pigment cells, which prevent leakage of light from one ommatidium to another. After R. Buchsbaum, *Animals Without Backbones,* Copyright 1938 by The University of Chicago, Chicago, Ill. All rights reserved.

is well lubricated. The muscles are controlled by three pairs of nerves. Many binocular judgments of distance, size, and orientation could be "computed" by the central nervous system from data on the relative tensions of these muscles. However, Bindley has clearly demonstrated that if an eye is blinded, so that no visual image is transmitted, it could be passively rotated through arcs of 30 degrees or more with no sensation to the subject. That is, humans have no sense organs at or near the eye muscles which measure these muscle tensions.

The external covering of the eyeball is made up of three spherical layers, as shown in Fig. 2.4. The outermost is the *sclera.* It is a white fibrous coat commonly called the white of the eye. At the very front portion of the eye, the sclera leads into the cornea, a clear transparent structure that admits light into the eye. The human cornea is about 12 mm in diameter and has a radius of curvature of about 8 mm. A major part of the refractive power occurs at the cornea. Inside the sclera is another thin layer, called the *choroid layer.* It contains the blood vessels and a pigmented substance. The choroid layer does not continue all the way around to the cornea, as shown in Fig. 2.4. The third and innermost layer of the eyeball is the light-sensitive

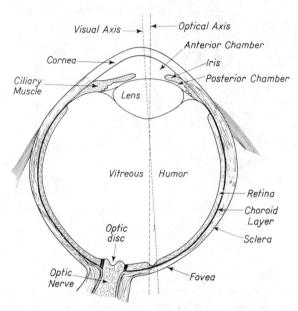

Figure 2.4 The human eye. After A. A. Maximow and W. Bloom, *Textbook of Histology* (Philadelphia: W. B. Saunders Company, 1957).

retina. The active photoreceptors, called *rods* and *cones*, are located in the retina. It is convenient to divide the retina into 10 layers. Light must pass through eight of these before reaching the rods and cones located in the ninth layer. Slightly displaced from the intersection of the optic axis of the eye with the retina is a yellow spot (*macula lutea*) known as the *central fovea*. It is a slight depression on the surface of the retina. The active elements in the fovea are all cones; they are very closely packed. For maximum acuity, the eye is directed so that the image falls on the fovea.

Somewhat on the nasal side of the fovea is the optic disk. Here the optic nerve pierces through the sclera, the choroid layer, and the retina; in the center of the optic nerve are a vein and an artery. From this disk, nerve fibers and blood vessels branch out over the surface of the retina. Objects focused on this disk cannot be seen since there are no rods or cones in it. Thus, this disk is referred to as the *blind spot*. One may put two marks on a piece of paper as indicated in Fig. 2.5, cover one eye, and fixate one of the marks. If one then alternately moves the head toward and away from the paper, the other mark will disappear when its image falls on the blind spot. In Fig. 2.5, the *x*, the ., and the : disappear at different distances.

Within the eye there are additional, optically important structures. One of these is the *iris*, which acts as a light diaphragm. In bright light, the iris has a minimum opening. This is desirable for several reasons. A smaller opening

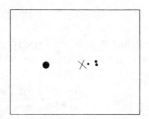

Figure 2.5 Pattern to observe the blind spot in the eye. Fixate the right eye on the large dot and bring the face very close to the figure. Now slowly move the face away while keeping the right eye fixated on the large dot. The other symbols will disappear and then reappear as their images cross the blind spot on the retina.

means fewer light photons entering the eye, thereby decreasing the overloading of the retinal system. In addition, since just a small section of the spherical lens is used, a small iris opening limits such distortions as spherical aberration, field curvature, and coma associated with finite sections of spheres. Finally, a small iris opening increases the depth of focus. The reason for this can be seen from a simple ray diagram, such as is shown in Fig. 2.6. At night,

Figure 2.6 Effect of aperture on depth of focus. A point focused at q will appear as a circle of diameter δ on the retina. As shown in (a), if the aperture of the iris diaphragm is wide, the diameter of δ will be large; hence, one image will blur into the next unless q is very close to the retina. Thus, increasing the aperture decreases the depth of focus. As shown in (b), a narrower aperture increases the depth of focus but decreases the luminous energy reaching the retina.

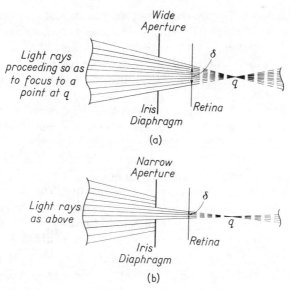

maximum acuity and depth of focus are less important than maximum sensitivity. At this time the iris is opened to its widest.

Another optically significant structure within the eye is the *crystalline lens*. In spite of its name, this is actually a cellular structure. The rear face is curved more sharply than the front. The eye accommodates to objects at different distances by changing the curvature of the front face of this lens. When the object is farther away, a weaker lens is needed to focus the image on the retina than when the object is closer. Hence, for more distant objects, the lens must be flatter, whereas for closer objects it must become more curved.

The shape of the crystalline lens is controlled by a ring of muscles surrounding the lens. These are called the *ciliary muscles*. Most physiologists believe that the lens is normally held in a strained position by the ciliary fibers. These fibers hold the lens in a flattened condition suitable for viewing distant objects. When the ciliary muscle contracts, it moves the base of the fibers forward, permitting the lens to relax into a more curved shape. When the muscle relaxes, the lens is again placed under tension.

The space between the lens and the retina is filled with the *vitreous humor*, a jellylike mass of material traversed by fibrils. Staining techniques indicate that the vitreous humor does contain some sort of structure. Optically, the vitreous humor is indistinguishable from the *aqueous humor*, which fills the space in the eyeball between the cornea and the crystalline lens. The aqueous humor, as its name implies, is a waterlike solution containing the normal solutes of a body fluid.

B. GEOMETRICAL OPTICS OF THE EYE

Light enters the eye through the transparent cornea. It then passes through the aqueous humor, through the crystalline lens, and into the vitreous humor. It is received on the photosensitive retina, where there must be an image in focus if the object is to be seen clearly. The dimensions, radii of curvatures, distances apart, and positions of the six cardinal points are shown in Fig. 2.7 for a schematic eye.

The greatest part of the refractive power of the eye occurs at the cornea. Individuals lacking a lens can still see, but their vision is much less sharp than that of a normal person because the image on the retina is out of focus. By changing the exact shape of the lens, the eye can accommodate for objects at different distances. The young person with normal vision can accommodate for objects nearer than 250 mm. An object distance of 250 mm corresponds to about 16 focal lengths. Accordingly, to compensate for the change in image distance as the object is moved from about 16 focal lengths to infinity, the effective posterior focal length of the eye must change about 6 percent. In terms of the radius of curvature of the crystalline lens, this corresponds to a change of around 20 percent. The posterior focal length of the average human

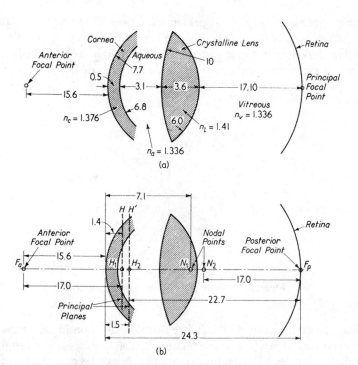

Figure 2.7 Optical properties of the eye. All distances shown are in milli-
meters. The values are averages and will vary from individual to individual.
These drawings, not to scale, show Ogle's modification of Gullstrand's
schematic eye. Notice that although the lens of the eye appears to be strong in
air, it is much weaker *in situ* since the difference in index of refraction between
the lens and the surrounding media is much smaller. After K. N. Ogle, *Optics,
An Introduction for Ophthalmologists* (Springfield, Ill.: Charles C Thomas,
Publisher, 1961).

eye from the second principal point, H_2, to the posterior focal point, $+F$, is
2.2 cm. Thus, the eye has a strength of about 48 diopters.

   If the eye is stronger than this, images of distant objects will be focused in
front of the retina. Such an eye is called *near-sighted* or *myopic* because near
objects will be focused on the retina. This ocular defect can be corrected by
placing a negative (diverging) lens in front of the eye. If the refractive power
of the eye is too weak, the image of near objects will be formed behind the
retina, and positive lenses are needed for correction. Such eyes are called
far-sighted or *hyperopic*. By and large, it is not possible to design a corrective
positive lens for objects at all distances, and so bifocals or trifocals are
necessary.

   Another frequent defect which can be corrected by glasses is called
astigmatism. This defect consists of having different focal lengths for lines in

different directions. A normal person would see all the lines of a fan chart (Fig. 2.8) as equally black, whereas one with astigmatism will see lines in one meridian darker than those in the meridian at right angles. Astigmatism is due to the fact that some of the refractive surfaces of the eye, especially the cornea, are not spherical but have different curvatures in two meridians.

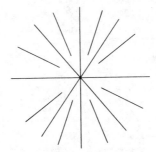

Figure 2.8   Pattern for observing astigmatism.

To recapitulate, the eye lends itself to a description in the terms of geometrical optics. The eye is a system of spherical surfaces separated by media of different indices of refraction. Optically, it can be described in terms of six cardinal points. The common defects easily corrected by glasses can also be described in the language of geometrical optics.

C. HISTOLOGY OF THE EYE

Each gross structure of the eye can be described on a microscopic scale. This is the role of histology. The evidence from histology, in turn, forms part of the basis of the biophysics of vision. Without a knowledge of the histology of the retina, there can be no neural interpretation of vision, such as is discussed in Chapter 8.

Light enters the eye through the cornea, whose microscopic structure is shown in Fig. 2.9. First, the light passes through an outer layer of epithelial cells. These cells are separated by a thin membrane from an inner fibrous layer which in thickness comprises most of the cornea. These fibers are very similar to the fibers in the sclera. Those in the cornea are unique in that they are arranged in an orderly fashion. It appears that it is this orderliness of the fibers of the cornea that is responsible for its transparency, as contrasted with the opacity of the sclera. Inside the fibrous layer of the cornea is another very thin limiting membrane and finally a lining of cells called *endothelial cells*.

As noted previously, the shape of the cornea is responsible for the major

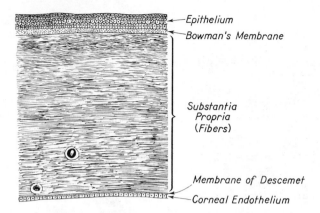

Figure 2.9   Histology of the cornea. After Schaffer, in A. A. Maximow and W. Bloom, *Textbook of Histology* (Philadelphia: W. B. Saunders Company, 1957).

refraction of the eye. Any large irregularities or abrasions would reduce the acuity of vision. The usefulness of the eye depends on keeping the cornea clear and transparent. If a large object approaches the cornea, the eyelids are closed by a reflex action. Smaller particles are removed by blinking and through tear formation. The outer epithelial layer of the cornea is very highly innervated; the nerves terminate in bare nerve endings. Any slight disturbance stimulates these endings, resulting eventually in the blinking reflex. All persons normally blink quite frequently; this cleans and moistens the outer surface of the cornea, which otherwise would become dehydrated and lose its transparency.

It always appears surprising when one first encounters the idea that light can pass through several layers of cells and fibers and still retain its original form. If these layers are arranged in a sufficiently orderly fashion, there is relatively little scattering or absorption of light as it passes through the tissue.

The *crystalline lens* is also a cellular structure. The cells are long hexagonal columns. Most of the cell nuclei are grouped in a restricted region of the lens which is not active in vision. A typical cross section of a lens is shown schematically in Fig. 2.10.

The last cellular structure of the eye through which incoming light must pass is the retina. Here are located the active photoreceptors. There are two types of receptors, called *rods* and *cones*. It is technically correct to refer to these rods and cones as *transducers*, devices which convert one form of energy into another. These transducers convert light energy into electrical impulses which travel along the nerve fibers. As noted earlier, the retina may be divided

(a)   (b)

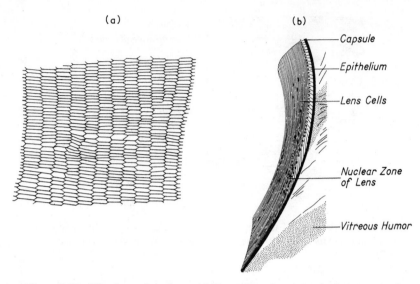

Figure 2.10   Histology of the lens. (a) Frontal section through the equator of
the lens, showing the regular arrangement of the cells. (b) Transverse section
through the lens. After Schaffer, in A. A. Maximow and W. Bloom, *Textbook
of Histology* (Philadelphia: W. B. Saunders Company, 1957).

into 10 layers. These are diagrammed in Fig. 2.11. Starting from the innermost layer, closest to the light, one can list the layers shown in Table 2-1.

TABLE 2-1

LAYERS OF THE RETINA

Layer	Function or structure
1. Inner limiting membrane	
2. Optic nerve fibers	Also some blood vessels, connective tissue, and so forth
3. Layer of ganglion cells	Neuron cell bodies
4. Inner plexiform layer	Synapses between processes from cells of layers 5 and 3
5. Inner nuclear layer	Neuron cell bodies
6. Outer plexiform layer	Synapses between processes from rods and cones and cells of layer 5
7. Outer nuclear layer	Cell bodies of rods and cones
8. Outer limiting membrane	
9. Rods and cones	The photoreceptors
10. Pigmented epithelium	Absorbs light, limits reflection

(*Path of light* — indicated along the left side, pointing downward)

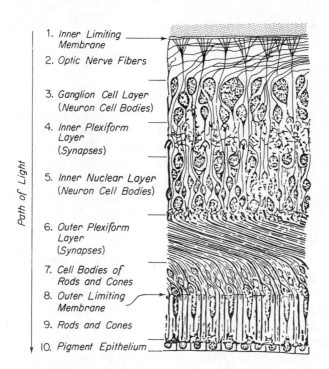

1. *Inner Limiting Membrane*
2. *Optic Nerve Fibers*
3. *Ganglion Cell Layer (Neuron Cell Bodies)*
4. *Inner Plexiform Layer (Synapses)*
5. *Inner Nuclear Layer (Neuron Cell Bodies)*
6. *Outer Plexiform Layer (Synapses)*
7. *Cell Bodies of Rods and Cones*
8. *Outer Limiting Membrane*
9. *Rods and Cones*
10. *Pigment Epithelium*

Path of Light

Figure 2.11 Histology of the retina. After A. A. Maximow and W. Bloom, *Textbook of Histology* (Philadelphia: W. B. Saunders Company, 1957).

The neurons in the retina are similar to those in other parts of the nervous system. Their detailed form and action are discussed in Chapter 5. For the purposes of this chapter, one should note that the neurons are the functional units of the nervous system. Each consists essentially of a cell body, a long process called an *axon*, and shorter processes called *dendrites*. The rod and cone cell bodies are similar to neuron cell bodies except that they are attached to photoreceptors in lieu of axons.

It should be emphasized that light goes through layers 1–8 before being useful for vision in layer 9. The arrangement of two layers of neuron cell bodies, with their connections to the rod and cone cell bodies, as well as almost innumerable connections between neuron cell bodies, is indeed complex. To those who have looked behind the front panels of a digital computer, the retinal structure suggests strongly that the output of the rods and cones is analyzed in a computerlike fashion by these layers of nerve cell bodies. And indeed, it will be shown in Chapter 8 that electrophysiological evidence supports this suggestion.

Within the layers of nerve cell bodies, a number of different types of cells have been discovered. Whenever the mechanism of color vision is being discussed, it is important to bear in mind that these different types must be taken into consideration.

2.4   Thresholds and Acuity

In this section, three different types of measures of the sensitivity of the eye are discussed. The first is the *quantum threshold*, that is, the minimum number of photons necessary to elicit a sensory response. The second is the *relative sensitivity* of the eye to light of varying wavelengths. The last measure, the *acuity*, represents the keenness of vision and is measured by the minimum angular separation of two objects that can just be discriminated as two and not one.

A. QUANTUM THRESHOLDS

Vision occurs when light is absorbed by the photosensitive rods and cones. At the threshold of vision, only a minimum of light is necessary. The absorption of light is best described in terms of quantum theory. A natural question then is: How many photons must be absorbed by a visual receptor (rod or cone) for the subject to see a flash of light? This problem was first investigated in detail by the biophysicist S. Hecht.

His first approach was to use light of wavelengths to which the eye was most sensitive and to expose the eye to short flashes. The eyes were dark-adapted to make their sensitivity a maximum. The number of photons striking the cornea for a just noticeable flash was measured. The number was reduced by the fraction (about four fifths) which he found to be absorbed in the eye. The final number, then, should be the minimum or number of photons necessary for threshold vision. At least it would be if this number were much larger than unity, in which case all pulses could be considered as having equal numbers of photons. Otherwise, the entire data would have to receive a probability-type interpretation.

Early estimates based on this method indicated that almost 150 photons were necessary at the cornea, and about 30 of these reached the retina for a just visible flash. As this number was redetermined during the 1920's and 1930's, it decreased steadily from 30 down to 1 or 2. The small number violates the original basis of the determinations, because the number of photons in a light pulse, the number absorbed along the way, and even the fraction absorbed in the retina of those which get there are subject to probability considerations. In general, one cannot measure these probabilities separately. However, the average number of photons, b, absorbed by a single receptor of the retina will be proportional to the intensity, I, provided that the eye does not move; that is,

$$b = kI \qquad\qquad\qquad (2\text{-}6)$$

The proportionality constant, k, will vary with many factors, including the size of the test patch, the pupil opening, the wavelength, and the length of the flash. It is clearly desirable to carry out an experiment to measure the threshold number of photons independently of k. The following mathematical manipulations indicate how to design an experiment that satisfies this criterion.

The number of photons absorbed by a photoreceptor in the retina during a given flash is an integer. It may have any positive value, or it may be zero. However, the average number of photons need not be an integer but will have a definite value, b. The probability, P, that exactly m photons will be absorbed during a flash by the photoreceptor will be given by the *Poisson probability distribution:*

$$P(m) = \frac{e^{-b}b^m}{m!} \tag{2-7}$$

Vision will occur if some given integral number n or more photons are absorbed during the exposure. The probability, P_n, that n or more photons will be absorbed in a flash is given by

$$P_n = \sum_{m=n}^{\infty} P(m) = 1 - \sum_{0}^{n-1} P(m) \tag{2-8}$$

Now, one may plot computed values for P_n against $\log b$, giving curves such as those shown in Fig. 2.12. Notice that each of these has a different slope. Although the value of b is not known, the value of the intensity, I, can be measured. Therefore, a plot of the fraction of number of correct responses

Figure 2.12 P_n versus $\log b$ for quantum threshold calculation. In this graph, P_n is the Poisson distribution probability for n or more events occurring, and b is the average number of events occurring.

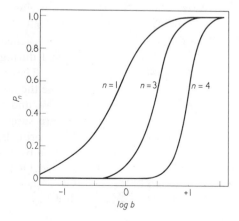

when the light was perceived by the subject against log I should have the same shape as one of the curves shown in Fig. 2.12. By adding an arbitrary constant to log I, it should be possible to show that the experimental points correspond best to one value of n.

This experiment satisfies the criterion of not needing to measure the constant k in Eq. (2-6) and gives unique data for the determination for any individual value of the integer n in Eq. (2-8). The value for this constant for some human subjects indicates that n is as high as 8. For other subjects, consistent values as low as 1 or 2 have been found for the number of photons necessary to elicit a visual response. In spite of these individual variations, the human data support the idea that the quantum threshold n is a very low number. Most of these measurements are for rod vision, but there is nothing to indicate that the threshold number of photons absorbed is different for cones.

For the human eye, it is impossible to determine whether the response measured is that of a single receptor, but it is possible in experiments using invertebrates eyes, such as those of the horsehoe crab, *Limulus*. These eyes have only rodlike receptors called ommatidia, described in Sec. 2.3.A. There is one receptor per nerve fiber. For threshold experiments, the eye, with the optic nerve attached, is removed from the animal. The nerve is then dissected until only one nerve fiber remains intact. It then becomes possible to measure electrically the response of only one receptor.

Such experiments indicate that one or two photons are necessary to initiate an electrical response in the nerve fiber. Most investigators today use the number 1: that is, one photon *absorbed* in the receptor, one response. Note that this is very different from the statement: one photon *reaching* the receptor, one response. Also, though one quantum can *excite* a rod, the excitation of more than one rod may be necessary to evoke a *sensation* of light.

This quantum threshold seems surprisingly small, since one photon has so little energy. It is instructive to compute the size of a photon of visible light. Applying Eq. (2-5) to the energy, E, of a photon of green light, wavelength about 500 nm, one finds that

$$E \simeq 4 \times 10^{-19}\ J$$

In terms of 1 mole of photons (called an *einstein*), this becomes

$$E \simeq 40\ \text{kcal/mole}$$

Readers familiar with chemical thermodynamics (see Chapter 21) will recognize that these numbers imply that a photon of green light can break only a small number of molecular bonds when it is absorbed. It is indeed impressive that such a small change can alter the electrical state of the photo-receptor in such a fashion as to initiate a nervous pulse which results in the sensation of vision.

B. LUMINOSITY THRESHOLDS

The sensitivity discussed in the preceding section is based on the number of photons absorbed. This absorption is the result of the action of certain photosensitive pigments found in the rods and cones of the retina. The relative fraction of light that reaches the rods and cones and is absorbed varies markedly with wavelength. It is convenient to separate the effect of wave-length from the numerous other factors altering threshold intensities. To do this, a set of threshold data is taken, varying only the wavelength. The entire set is multiplied by a normalizing constant chosen to reduce the minimum threshold to an arbitrary value. The reciprocal of the normalized threshold is known as the *relative luminosity*. Relative luminosity curves have been measured both for dark-adapted eyes and for light-adapted eyes. Vision under conditions of dark adaptation is called *scotopic*; vision with light-adapted eyes is called *photopic*.

In either case, one may interpret the thresholds as the intensity at which a response is obtained 50 percent of the time. For short flashes, less than 10 msec, the product of the intensity and exposure length determines the observed threshold, whereas for long exposures, say more than 50 msec, only the intensity is important. The exact size of the test patch used becomes very important if it subtends less than 0.5 mrad. With very small test patches, the exact location of the test patch markedly affects the shape of the relative luminosity versus wavelength curve. For larger test patches, threshold curves are obtained which do not depend specifically on the particular area of the retina which is illuminated.

The general shape of the relative luminosity curves for photopic and scotopic vision is shown in Fig. 2.13. Owing to the definition of relative luminosity, the absolute height of the curve does not have any significance. Much greater intensities are needed for photopic vision than for scotopic vision. Luminosity threshold measurements are not easy to perform. More than half an hour is necessary for dark adaptation. Care must be taken to illuminate the same area of the retina, and many other precautions must be observed as well. However, the relative luminosity curves do lead to repro-ducible results.

The separation of the maximum points of the scotopic and the photopic curves can be interpreted as an indication that luminosity depends on at least two types of receptors. The simplest interpretation might be to assign scotopic vision to the rods and photopic vision to the cones. This choice would be indicated by the fact that scotopic sensitivity is greater in the periphery, where

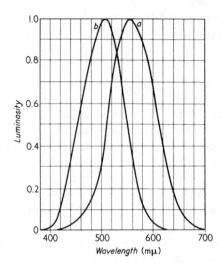

Figure 2.13   Relative luminosity curves. The curve for the dark-adapted eye is labeled *b* and for the light-adapted eye, *a*. After Committee on Colorimetry, The Optical Society of America, *The Science of Color* (New York: Thomas Y. Crowell Company, 1953), p. 225.

there are more rods; whereas photopic response is greatest in the fovea, where there are no rods. However, this separation of function is definitely over-simplified; the rods appear to be active in both dark-adapted and light-adapted eyes, whereas the cones are active only in light-adapted eyes.

C. ACUITY

Studies of the acuity of vision also indicate that the rods are the active elements in the dark-adapted eye. The acuity of the eye adapted to scotopic vision is a minimum at the fovea, where there are no rods. Thus, the rods seem to be the active elements in scotopic vision. The acuity of scotopic vision shows a maximum for light at the retinal region where the rod density is highest—about $\frac{1}{2}$ rad from the fovea. Acuity under scotopic conditions is lower than under photopic conditions in any region of the retina. The neural basis for this is discussed in Chapter 8. However, in photopic vision there is a sharp maximum in the ability to resolve two spots of light when the images fall on the fovea. The acuity in the foveal region is much greater than in the remainder of the retina.

The acuity of vision may be expressed in terms of the minimum angular separation of two equidistant points of light which can just be resolved. The angular separation θ between two points, when expressed in radians, is approximately equal to the distance between the points divided by the distance from the eye, provided that θ is less than 0.1 rad. The angle θ will also be equal to the separation of the two images on the retina divided by the distance from the second nodal point. From Eq. (2-4), one can calculate a minimum value of θ, according to the Rayleigh criterion, for green light ($\lambda = 500$ nm) and an iris diameter of about 0.5 cm. Rounding off to one significant figure, the limit, according to this criterion, would be 0.1 mrad. This is a theoretical lower limit for the resolution of two points of light.

Experiments have shown that most people cannot resolve two points of light if their separation is as small as 0.5 mrad. Persons with the most acute vision can resolve an angular separation of about 0.2 mrad under optimum conditions. Because this is higher than the Rayleigh criterion, it seems that visual resolution must be limited by other factors, such as scattering, spherical aberration, and the separation of the receptors in the retina.

In the center of the fovea, where the resolution is greatest, the cones are separated by about 2 μm from center to center. In order to resolve two points of light as separate images, it must be necessary to excite at least two cones while leaving one in between unexcited. Thus, the images on the retina would have to be separated at least 4 μm from center to center. If it were necessary to have two cones unexcited between the images of the two spots, this number would be increased to 6 μm. The maximum resolution observed of 0.2 mrad corresponds to a separation between the image centers on the retina of 5 μm. In other words, the discrete structure of the retinal receptors could be responsible for the lower limit of resolution for persons with the most acute vision.

The psychophysical processes of recognizing shapes are very complex. However, a minimum requirement for small objects is that the angular separation of their different parts be larger than the limit of resolution. At 25 cm from the eye, an angle of 0.5 mrad would correspond to about 100 μm. This is about the length of the large protozoan *Paramecium caudatum*, which can be recognized as having a rod shape at that distance. In contrast, a smaller species, *Paramecium aurelia*, has to be brought closer to the unaided eye before its shape can be recognized.

In a camera, resolution in white light is often limited by chromatic aberration, that is, the different wavelengths focus at different planes. The resolution can be improved to some extent by using a system of positive and negative lenses made of different types of glass.[1] The index of refraction of each will vary in a different fashion with wavelength. By a proper choice, a combination can be made which has a positive focal length that is almost independent of wavelength throughout the visible region.

Chromatic aberration in the eye is minimized by limiting the wavelengths of light to which the eye will respond. A bare retina from which the vitreous humor has been removed will respond far into the ultraviolet. However, in

---

[1] These are called *achromatic* lenses.

the intact eye, the cornea absorbs most energy at wavelengths shorter than 300 nm. Accordingly, energy at these wavelengths does not contribute to vision, although it can produce corneal damage.

The crystalline lens has a very sharp cutoff at about 380 nm. Persons without this lens cannot accommodate to different object distances and so lack acuity; they can see objects using ultraviolet radiations only. They have a sensation of violet when viewing ultraviolet. Persons with a lens do not receive any appreciable energy at the retina at wavelengths shorter than about 380 nm. Thus, the lens (and cornea) limit the photons reaching the retina to wavelengths greater than 380 nm.

On the long-wavelength side, the water molecules in the cornea and aqueous humor eventually absorb most of the energy at wavelengths longer than 1,200 nm. However, the eye pigments become very insensitive to light above 700 nm. Technically, to find the long-wavelength limit, one should go to such high intensities that the eye is heated but not badly burned; this experiment is rarely performed.

Thus, the filter action of the lens and cornea plus the response character-istic of the optically active pigments in the photoreceptors tend to restrict the wavelength band, thereby reducing chromatic aberration. In addition, the greatest acuity occurs in photopic vision at the fovea. In this region, there are only cones, which probably do not respond to blue light. Also in this region is a yellow pigment believed by many to further eliminate the blue end of the spectrum. Accordingly, the acuity at the fovea is greatest not only for objects viewed with monochromatic green light, but also for those seen in white light.

REFERENCES

1. DAVIDOVITS, P., *Physics in Biology and Medicine* (Englewood Cliffs, N.J.: Prentice-Hall, Inc., 1975), 298 pages.
2. STEVENS, S. S., ed., *Handbook of Experimental Psychology* (New York: John Wiley & Sons, Inc., 1951).
   (a) JUDD, D. B., "Basic Correlates of the Visual Stimulus," pp. 811–867.
   (b) GRAHAM, C. H., "Visual Perception," pp. 868–920.
3. GLASSER, OTTO, ed., *Medical Physics*, Vol. 1 (Chicago: Year Book Publishers, Inc., 1944).
   (a) LUCKIESH, MATTHEW, AND F. K. MOSS, "Light, Vision, and Seeing," pp. 672–684.
   (b) SHEARD, CHARLES, "Optics: Ophthalmic, with Applications to Physiological Optics," pp. 830–869.

For a more thorough discussion of optics at an intermediate physics level, see:

4. HORRIDGE, G. A., "The Compound Eye of Insects," *Sci. Am.* **237** (July 1977): 108–120.
5. MEYER-ARENDT, J. R., *Introduction to Classical and Modern Optics* (Englewood Cliffs, N.J.: Prentice-Hall, Inc., 1972), 558 pages.
6. MAXIMOW, A. A., WILLIAM BLOOM, AND D. W. FAWCETT, *A Textbook of Histology*, 9th ed. (Philadelphia: W. B. Saunders Company, 1968), 858 pages.

For a presentation from the point of view of medical physiology, see:

7. BEST, C. H., AND N. B. TAYLOR, *Physiological Basis of Medical Practice*, 9th ed. (Baltimore, Md.: The Williams & Wilkins Company, 1973).
8. OGLE, K. N., *Optics: An Introduction for Ophthalmologists* (Springfield, Ill.: Charles C Thomas, Publisher, 1961), 265 pages.

CHAPTER 3

Insect Communication

3.1 Introduction

Human beings use several sensory systems for communication, such as hearing, vision, touch, and the chemical senses of smell and taste. There is no reason to suppose that other living organisms are restricted to these means of communication, but they appear to be the important ones there as well. The differences that exist for the most part involve quantitative aspects, such as the frequency range of hearing, the wavelength band of vision, and the particular chemicals to which an organism responds. (Some alternative senses are discussed in Chapter 4.)

Communication can be defined as the act of imparting information (see also Chapter 25); thus, all methods of communication involve signal (information) generation, transmission, and reception. Once such an information exchange is postulated, it is relatively straightforward to examine its generation and transmission, but first a recognizable effect of the information transfer must be observed. This usually means a noticeable change in behavior or physiology of the recipient. The effects of signals can be grossly separated into two overlapping groups: *immediate* (occurs with little or no lag time after reception) or *latent* (there is a lag measurable in minutes, hours, days, or longer before a response is noted). Latent effects are more difficult to measure; thus, most researchers have limited their studies to signals that produce immediate effects.

While many parts of this book emphasize the human, mammalian, or vertebrate aspects of biophysics, here the emphasis is on the invertebrate, arthropod world and the methods by which insects communicate. Since there are more species of insects (several hundred thousand) than all other species of organisms combined, this chapter of necessity can touch on only a small fraction of a percent of known insect communication. Selected examples of visual (Sec. 3.2), acoustic (Sec. 3.3), and chemical communication (Sec. 3.4) between insects of the same species are included; signal generation, transmission, and reception of all of these is discussed. Section 3.5 briefly examines communications between different species.

Before continuing this discussion, a short review of insect physiology is in order. An insect's body is divided into head, thorax, and abdomen, as shown in Fig. 3.1. The head of a typical insect has one pair of antennae, simple and

Figure 3.1   Diagram of a typical insect. Most insects have fewer abdominal segments, owing to loss or fusion at the posterior end. After R. Buchsbaum, *Animals Without Backbones.* Copyright 1938 by the University of Chicago Press, Chicago, Ill. All rights reserved.

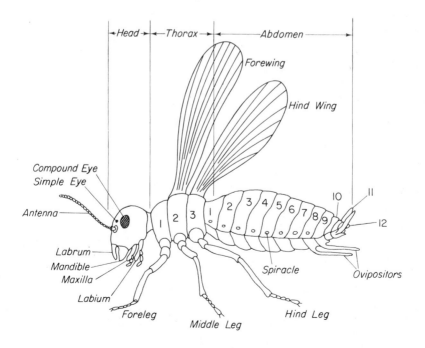

compound eyes, and a mouth composed of several interlocking parts. The thorax has three pairs of legs and two pairs of wings. The abdomen is segmented and contains small holes, or spiracles, which allow air to enter the body. Ovipositors in the female, as the name implies, aid in egg placement.

3.2 Visual Communication

A visual signal can impart information based on its size, shape, intensity, hue, movement, and variation with time. Any insect, by existing, transmits a visual signal to others close enough to see it, and many go to great lengths to disguise this signal through camouflage, mimicking rocks, plants, or even other insects. In this section two specialized visual signals used for intraspecies communication will be examined in more detail: *bioluminescence* and the *waggle dance* of honeybees.

A. BIOLUMINESCENCE

Several insect families are known which "glow" or otherwise transmit a light signal to others of their species. The biochemistry of this reaction has been studied extensively by McElroy and co-workers, who suggest that it is an offshoot of a process used by anaerobic organisms to remove toxic oxygen. The most well-characterized bioluminescent reaction is found in fireflies, where the enzyme leuciferinase catalyzes the oxidation of leuciferin to oxyleuciferin as shown. (ATP is hydrolyzed in this reaction; see Appendix D.)

The glands that produce the light are usually abdominal and vary in position, size, and shape, which, in turn, varies the signal produced.

Information can be encoded in a luminescent signal using these and other factors, and species are known that can distinguish signals differing only in size, shape, intensity, hue, movement, or timing. The latter is especially important, since most bioluminescent signals are not continuous but rather pulsate, with a variable number of pulses per unit time, duration of each pulse, and so on.

Transmission of most bioluminescent signals is inhibited by daylight or even a bright full moon; thus, most are produced at night or in special habitats such as caves or ravines. The signal can be directed at sites where potential recipients may lurk; for example, flying males hoping to attract sedentary females may preferentially flash down rather than all around their flight path. Similarly, the females may climb to higher perches and rotate their abdomens to increase visibility of their own signals.

The reception of bioluminescence, and indeed all visual signals, depends on the insect eye. The anatomy of this organ, and of eyes in general, is discussed in Chapter 2 and is not repeated here. Luminescent insects, especially the motile males mentioned above, have eye modifications, such as increased sensitivity, which facilitate visual communication. After the signal is received, it may trigger several responses. Since most signals are sent between potential mating pairs, reception of a signal may cause mutual attraction or, more commonly, homing in on a sedentary female by a mobile male. A second possible reaction involves change in a generated signal in response to a received signal; for example, a sedentary female may change to signals that aid in homing or landing by the male. Finally, since these are precopulatory signals, they may have a latent effect on the insect's reproductive physiology, but this area has received little study to date.

B. HONEYBEES

If food is placed near a beehive, it may remain undiscovered for hours or even days. However, once a single bee discovers the food and returns to the hive, many new bees soon appear. This occurs even if the initial forager is captured before it can lead others to the food. The forager must somehow communicate food position to the other bees.

Some of this communication depends on odor (see Sec. 3.4), but the German biologist Karl von Frisch showed that there was a significant visual component in the transmission of the information. The forager bee performs a complicated "waggle dance" on the side of the hive (see Fig. 3.2). Many aspects of the dance transmit information. The amplitude of the dance pattern communicates the time of flight, the predominant angle with the vertical reveals the direction, the rhythm indicates the distance, and the vigor of the dance indicates the quality of the source.

The information content of the dance has been interpreted in various ways (see also Chapter 25). From 4 to 9 bits of direction information and 3 to 6 bits of distance information are transferred. If "quality" can be coded on a scale of 1 to 4, this adds 2 additional bits. Thus, 9 to 17 bits of information can be communicated. When allowance is made for the information transmitted through odors, some workers feel that bees can communicate with 25-bit "sentences." The corresponding minimum estimate for a short sentence using a 10,000-word human vocabulary is 74 bits.

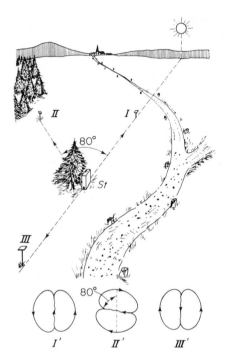

Figure 3.2   Three examples of the indication of direction on a vertical comb surface: St, beehive; I, II, III, feeding stations in three different directions; I′, II′, III′, the corresponding tail-wagging dances on the vertical comb. After Karl von Frisch, *The Dance Language and Orientation of Bees*, Belknap Press of Harvard University Press, Cambridge, Mass., 1967.

The eye and higher-level vision apparatus of the bee are well adapted to interpreting these dances. For example, the direction indication given in the dance, as shown in Fig. 3.2, is the angle between the flight direction and the sun. Bees can orient on the sun even when in shade themselves, for their compound eyes can perceive the differences in polarization of the clear sky, which is directly related to the sun's position. They have a biological clock that can correct for the hourly change in the sun's position as the day proceeds and they also have a "sun-compass" reaction, which allows them to correct daily for the sun's seasonal movement. The ocelli, or simple eyes, are important only in low-light situations, such as those early in the morning or at twilight.

The response to the visual dance signal is usually immediate foraging or food collection. However, when time for migration or swarming occurs, scout bees can transmit information about new nesting sites in a similar manner for many hours prior to the actual swarm. The direction indicated by this preswarming dance inside the hive changes as the sun's position changes, and continues to change even after sunset.

3.3 Communication by Sound

As anyone who has encountered a "singing" cicada realizes, even a relatively small insect can easily produce quite loud sounds. As mentioned in Chapter 1, a sound signal is a pressure wave set in motion by some type of vibration. If an insect sound generator is simplified to a rigid disk vibrating in a flat baffle (see the Michelson and Nocke article), then for a given vibration velocity the frequency, v, for maximum signal power is related to r, the radius of the disk, and to c, the speed of sound in the medium, as follows:

$$v \geqq \frac{c}{\pi r} \tag{3-1}$$

If $c = 340$ m/sec (air) and $r = 3$ mm (the size of a cricket's sound radiator), v must be greater than or equal to 36 kHz. Thus, insects can only signal effectively in the kHz range, and small insects must use even higher frequencies.

An insect's muscles, however, cannot contract faster than 1 kHz; thus, the frequency of muscular contractions must be multiplied to produce the signal. For example, the cicada mentioned above has a curved platelike tymbal which vibrates after being deformed or bent by the tymbal muscle, and after rebounding from that deformation. Thus, the tymbal multiplies the frequency of the contraction and also emits the sound. A second example of such multiplication is *stridulation*. This is vibration generated by the movement of a scraper over the teeth of a file. In most cases, sound is not produced by the frictional process itself. Instead, the vibrations are transmitted to a sound radiator. This may be part of the wing (cricket) or even the whole body cuticle (some ants).

The final element of a sound generator is often some type of *sound guide*, an anatomical (or environmental) structure to channel and amplify the sound. Such a guide may approximate a closed box (cicada), baffle (cricket), or horn (mole cricket).

The range of transmission of an audible signal depends on the temperature, humidity, and frequency of the signal. If temperature and humidity are held constant, then the higher the signal frequency, the greater will be its absorption in air. This puts an upper limit on frequencies, which are bounded at the lower end by Eq. (3-1); for example, absorption will limit transmission of a 50-kHz signal to 10–20 m. Ground absorption is also important for those insects which generate sounds at or near ground level but has not been well studied. Similarly, although temperature and temperature gradients are known to affect sound transmission, little is understood about their effect on insect sounds.

Finally, signal reception requires insect ears. The plural is intended since, unlike the insect eye or mammalian ear, there are several different kinds of insect ear. Readers are referred to the Michelsen and Nocke article (see the references at the end of the chapter) for further information; here we shall mention two of the most common types, both related to structures in the human ear (Chapter 1). *Hair receptors* are flexibly attached to move with every sound vibration; they primarily measure movement. *Tympanal organs* alternatively use membranes to record pressures and pressure gradients (see Fig. 4.1 for the frequency range of hearing in several insect groups). When sounds are received and interpreted, they can produce a variety of responses, including sexual attraction, territorial defense, alarm behavior, and even change in flight path to maintain swarm cohesion.

3.4 Chemical Communication

The human senses of smell and taste lose their distinction in lower organisms. Environmental molecules react with appropriate receptors, information is transferred, and the process is most simply called *chemical communication*. While such communication is known in almost all species, insects have received the most extensive study due in part to the economic importance of control of some insect species.

Insect sex attractants are very obvious chemical signals. In 1959, Butenandt and co-workers determined the chemical structure of bombykol, *trans*-10-*cis*-12-hexadecadienol, the sex-stimulating molecule in silkworm moths (*Bombyx mori*):

$$CH_3-(CH_2)_2 \quad \overset{H}{\underset{}{C}}=\overset{H}{\underset{}{C}} \quad \overset{H}{\underset{}{C}}=\overset{H}{\underset{(CH_2)_8-CH_2OH}{C}}$$

This compound is a remarkably powerful attractant. The sensory hairs on the male moth's antennae need only react with one molecule of bombykol to activate a receptor nerve cell. When about 200 nerve cells on each antenna are activated, the moth will begin to fly toward the target.

The moth receptor cell will respond to virtually no other molecule. This extreme selectivity may play a role in the origin of new species; two closely related species of moths are known whose sex attractants differ only at the configuration of one carbon atom, yet males of one species will respond only to their own attractant. In general, molecules such as these which trigger certain behavior in restricted species are termed *pheromones*, indicating that they are to the collective social organism what hormones are to the individual organism.

The pheromones for reproduction are perhaps most spectacular, yet chemical communication plays a role in feeding, assembly, and defense as well. Most of the chemically characterized pheromones such as bombykol are related structurally to natural fatty acids. Thus, signal generation begins with some modifications of the biochemical pathways involved in fatty acid synthesis. The pheromone signal is released through an externally ducted gland, often located in the abdomen (sex pheromones) or mandibles and sting apparatus (alarm and defense). The information being transmitted influences the physical characteristics of the pheromone as discussed below, but it may influence its chemical characteristics as well. For example, mandibular alarm or defense pheromones may be chemical irritants, designed to aggravate wounds inflicted by the mandibles. Similarly, an alarm pheromone of certain ants, formic acid, is a potent defense as well.

The signal is released and transmitted to other individuals of the same species. Here the physical characteristics of the pheromone are most important, especially its volatility as measured by its vapor pressure and diffusion coefficient. Wilson (1970) has defined K, the behavioral threshold concentration of pheromone in molecules/cm^3, as

$$K = \frac{Q}{2Dr} \, \mathrm{erfc} \left(\frac{r}{\sqrt{4Dt}} \right)$$

where Q is the emission rate in molecules/second, t is the time to the onset of the response, r is the distance between generator and receptor, D is the diffusion coefficient, and erfc (x) is 1 minus the error function. If the ratio O/K is calculated for a number of pheromones, they can be grouped in three classes. Those with Q/K ratios between 0.1 and 10, such as the fire ant's trail odor pheromone, are effective only at the millimeter distance and fade out after a few seconds. Alarm substances, with Q/K from 10^3 to 10^5, have longer ranges and are more persistent. Certain sex pheromones, such as bombykol, have ratios as high as 10^{12}. These are theoretically capable of transmitting information over several kilometers which could (ignoring weather and adsorption) endure for years.

Since diffusion coefficients of most compounds are much smaller in water than in air, the Q/K ratio must be increased by about 10^6 for a similar effective radius and persistence time in an aqueous environment.

The pheromone must be intercepted before it can produce a response. Chemoreceptors in insects are usually hairs, pegs, or plates found on legs or antennae. All have a pore-tubule system which connects the outside world to an interior fluid-filled lumen containing a nerve cell dendrite (see Chapter 5).

Proteins may also interact with the pheromone; the actual dendrite activator might be a tightly bound pheromone-receptor protein complex. Alternatively, the binding protein might be used to destroy or inactivate excess pheromone.

Behavioral responses to this nerve message are varied. Sex attractants cause movement and, in close proximity to the source, mating behavior. At the other extreme, trail or territorial pheromones serve as markers for less intense behavior. Alarm pheromones cause flight or aggression. Reception of alarm pheromones, unlike some other recruiting or aggregation pheromones, does not trigger the receiver to propagate the signal.

As a summary of pheromones, the following briefly describes a few instances of their use in bees. Forager bees are known to secrete assembly pheromones to attract other bees when they return to the hive. They also transport the odor of the food they discovered. Thus, the bees they recruit to return to the food site go to objects scented with the food the forager found and ignore nearby objects that have other scents. There is also evidence that the odors of the location around the food, the "olfactory landscapes," are also transmitted back to the hive. While the dance language of bees is important, they can in addition use odors to communicate quite complicated information. The queen honeybee substance, 9-oxo-2-decenoic acid, is an immediate sex pheromone which also has a delayed effect in suppressing the development of worker ovaries. Finally, for defense, honeybees can secrete an alarm pheromone. This low-molecular-weight compound has not been completely characterized, but it is known to attract and excite close neighbors to defend the alarmed bee. Bumblebees also use pheromones defensively. They regularly perfume small areas in their routine flight paths and use these as territorial markers.

3.5 Interspecies Communication

Insects use visual communication across species. For example, certain flowers attract insect pollinators with ultraviolet markers visible only to those insects. Birds also avoid ill-tasting butterflies (and their mimics) because of the visual information (patterns, markings) on their wings.

Sound communication between insects and other species is not common but has been noted in several instances. For example, human researchers have reported dropping captured cicadas when startled by their loud squawks. (It is not certain that this protects against the more usual predators.) Also, it has been observed that certain flying moths avoid the approach of bats by dropping or diving to lower altitudes (see Chapter 4). The insect may then attempt concealment among the various ground-cover objects. The reverberation of returning echoes apparently can confuse the pursuing bat, which becomes discouraged and leaves. Two types of cells involved in one species have shown vivid response to cruising bat signals. At large distances the nearer of the two bilateral tympanic organs (ears) has the larger output, so that, to a limited degree, the moth can establish the direction of the feeding bat. If one of the organs is surgically destroyed, the subsequently stimulated insect will deviate in the direction of the mutilation. This avoidance response is thought to be keyed to the signal level and, in particular, to the repetition rate of the ultrasonic source. If 10 to 15 pulses per second are utilized, the moth will begin evasive maneuvers while a few pulses per second will generally be ineffective. Some moth families, it should be noted, actually advertise their presence by means of their own ultrasonic radiation. Bats tend to avoid these species when they are broadcasting. Certain observers have considered these signals to be either direct or mimicked acoustic statements as to the insect's unpleasant taste. Other observers, however, feel that they are actually a form of jamming signal meant to confuse the attacker.

Chemical communication across species was mentioned briefly above. Obviously, it is the sweet odor of flowers that attracts the forager bee in the first place. Certain other plants smell of decayed meat to attract flies, and at least one orchid family mimics the sex attractants of bees. The initial copulatory activity of the bee on the orchid causes pollination. Insect repellents for skin and clothing are chemical communicators that work by disorienting insects when they approach for the final landing, but there is also a measure of tactile communication; the insect does not like the "feel" of the repellent-treated skin.

Interspecies communication is also known between insect species. While sexual attractants are not used as such between different species, certain insect females are known which mimic signals of other species and then eat the attracted males. Other insects can follow the pheromone trail ants lay down to food. Finally, species are known which, primarily through chemical communication, "fool" ants into rearing their young and taking care of the adults as well.

3.6 Summary

Visual, auditory and chemical communication are quite well developed in the insect world. Tactile communication, while possible for hive organisms, is effective only at short distances.

Communication can be between members of one species or across species lines. Visual communication between members of one species is best developed in the dance language of the honeybee. Sounds are generated, transmitted, and received by a large number of insect species. Chemical communication in a single species is by means of pheromones, special behavior-inducing molecules. Pheromones are also used by man in an attempt to selectively attract and dispose of certain noxious insects while minimally interfering with other organisms.

Interspecies communication was mentioned briefly. Again, man has used such attractants (yellow "insect lights") or repellents (citronella candles) to help control potential pests.

REFERENCES

General

1. WILSON, E. O., "Animal Communication," *Sci. Am.* **227** (September 1972): 52–60.

Visual

2. LLOYD, J. L., "Bioluminescent Communication in Insects," *Ann. Rev. Entomol.* **16**:97–122 (1971).
3. FRISCH, KARL VON, *Dancing Bees: An Account of the Life and Senses of the Honey Bee*, Dora Ilse, trans. (London: Methuen & Co. Ltd., 1954), 183 pages.
4. GOULD, J. L., "Honey Bee Recruitment: The Dance-Language Controversy," *Science* **189**:685–693 (1975).
5. LINDAUER, M., "Recent Advances in Bee Communication and Orientation," *Ann. Rev. Entomol.* **12**:439–470 (1967).

Sound

6. MICHELSEN, A., AND H. NOCKE, "Biophysical Aspects of Sound Communication in Insects," *Advan. Insect Physiol.* **10**:247–296 (1974).

Chemical

7. KULLENBERG, B., AND G. BERGSTROM, "Chemical Communication Between Living Organisms," *Endeavour* **34**:59–66 (1975).
8. LAW, J. H., AND REGNIER, "Pheromones," *Ann. Rev. Biochem.* **40**:533–548 (1971).
9. LEONARD, J. E., L. EHRMAN, AND A. PRUZAN, "Pheromones as a Means of Genetic Control of Behavior," *Ann. Rev. Genet.* **8**:179–193 (1974).
10. WILSON, E. O., "Chemical Communication within Animal Species," in *Chemical Ecology*, E. Sondheimer and J. B. Simeone, eds. (New York: Academic Press, Inc., 1970), pp. 133–155.

CHAPTER 4

Alternative Methods of Orientation

4.1   Introduction

This chapter discusses some of the biological variations in vertebrate methods of orientation. Although the visual sense is dominant in many of the vertebrates, lack of adequate light may have contributed to the development of alternative organs and the more specialized development of hearing, touch, smell, and/or taste. Since darkness prevails for one half of the average day, an ecological niche exists for those animals which place less reliance on visual cues. It is possible to discuss only a subset of such variations in this text. These include the specialized acoustic sense in bats and toothed whales and alternative sense organs in electric fish and infrared-sensitive snakes. The chapter concludes with an examination of bird migration and homing.

4.2   The Bats

Compared to most other animals, man has an abbreviated range of acoustic sensitivity and a relatively low upper-frequency limit at approximately 20 kHz. This is illustrated in Fig. 4.1, where the logarithmic scale should be noted. We shall use the term "ultrasonic" to describe radiation at frequencies greater than 20 kHz. The drawing reveals that bats, in particular, have their greatest sensitivity in the ultrasonic range, as well as a strikingly elevated high-frequency cutoff. Actual uses of such signals by these extraordinary mammals have only been documented with the use of electronic apparatus.

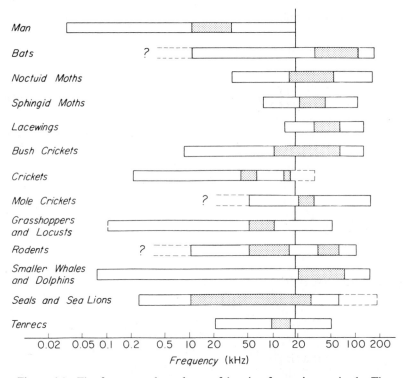

Figure 4.1 The frequency dependence of hearing for various animals. The dark bands refer to the intervals of greatest sensitivity. The vertical line near 20 kHz identifies the high-frequency limit in man. Note that the scale is logarithmic. After G. Sales and D. Pye, *Ultrasonic Communication by Animals* (New York: John Wiley & Sons, Inc., 1974).

There are two basic suborders of bats, the Megachiroptera[1] and the Microchiroptera. These Greek words distinguish the larger, fruit-eating types from the smaller, insectivorous species. Among the larger bats, only one genus is found to use echolocation as an orientation mechanism. The nocturnal feeding of the fruit-eaters is aided by their exceptional eyes. These have

[1] Chiro signifies hand; so, large-handed.

many rods per unit visual (solid) angle, due to a folding of the retinal surface. The Microchiroptera, on the other hand, are primarily insect eaters which almost always use ultrasonic echolocation to navigate as well as to hunt. Their eyesight is notoriously poor and has given rise to the expression "blind as a bat." Some species have such degenerate vision that an experimenter literally has to place the bat in a tray containing food for it to commence feeding. Rather than assisting the animal in locating food, bright lights inhibit the activities of insect-hunting bats, and they retreat into the darkest region of a cave or cage if disturbed in this fashion.

A. TYPES OF SIGNALS

The use of sounds for bat navigation was first suspected by Spallanzani in the late eighteenth century. He actually blinded several small bats and followed their subsequent behavior. These animals continued to thrive as well as their unmutilated fellows, implying an almost complete lack of dependence on the visual sense. D. R. Griffin and his co-workers at Harvard in the early 1940's demonstrated that bats can avoid obstacles in a darkened room and that they emit ultrasonic frequencies during this process. Three basic types of frequencies vs. time histograms (spectographs) are shown in Fig. 4.2. The insect hunters primarily use either fast, 1-octave sweeps from high to low frequency (Fig. 4.2(a)) or, alternatively, *Doppler-shift* frequency compensation (Fig. 4.2(b)). Doppler shift refers to the apparent change in detected frequency caused by relative velocity between source and detector. Our interpretation of the animals' true orientation method is, of course, only an estimate based on spectral data. Some species may use both methods. Those large bats that do use sound for localization exhibit an audible click-type pulse encompassing a range of frequencies in a very short time interval (Fig. 4.2(c)).

Most insect catchers employ frequency sweeping whereby the finite length of the output pulse (τ) does not obscure two spatially adjacent reflectors. To show this, assume a single frequency such that two returning echoes could be confused into one if they return to the transmitter with time delays differing by τ. Thus, the *least* detectable range differential for a single frequency pulse is, using c for the speed of sound in air,

$$\Delta r = \frac{c\tau}{2} \tag{4-1}$$

If, on the other hand, the sweep method is employed, all echoes are simply a replay of the source spectrogram (see Fig. 4.2(a)). When the bat is listening to the first monotonically decreasing echo, the second reflection is clearly heard as a sharp and unexpected rise in tone as that returning signal begins to be detected. With this method, a relatively long pulse may be used so as to improve the signal-to-noise characteristics of the system. The term "chirp" sonar is sometimes used to describe this technique.

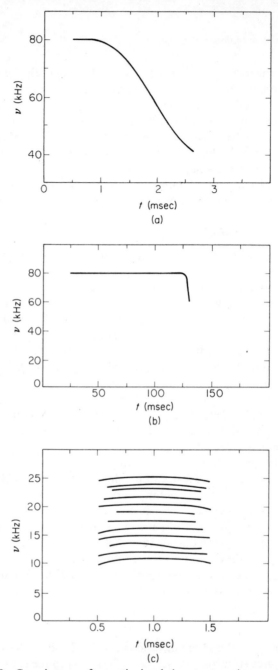

Figure 4.2 Generic types of acoustic signals (spectrograms) generated by bats. (a) The relatively rapid, 1-octave frequency sweep is often termed a chirp signal. (b) Essentially constant frequency signal; this type of vocalization characterizes Doppler bat species. The short terminal downsweep may be used as a chirp (ranging) signal by these species. (c) Click signal, typically produced by those Megachiroptera that echolocate (e.g., the Egyptian tomb bat *Rousettus aegyptiens*).

Doppler techniques are rarer in insectivorous bats. It has been shown that the greater horseshoe bat shifts its essentially monochromatic signal to compensate appropriately for the motion of a swinging pendulum. A similar result holds when the flying animal alights on a fixed platform. Because the sound is produced by a moving source and detected by (the same) moving receiver, the observed Doppler frequency (ν') is given by

$$\nu' = \nu \frac{1 + v/c}{1 - v/c} \tag{4-2}$$

where c is the speed of sound in air and v is the bat's speed toward the target. In many presumably Doppler species, the shifted frequency is probably made to correspond to a sharp resonance in the auditory response curve. An example of this for the greater horseshoe bat is given in Fig. 4.3. The magnitude

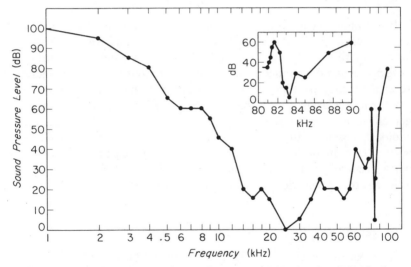

Figure 4.3 Auditory response curve of the greater horseshoe bat (*Rhinolophus ferrumequinum*). The sharp minimum near 83 kHz corresponds very closely to the constant vocal frequency of this species. (See Fig. 4.2(b).) Data were obtained by electrodes placed in the inferior colliculus. After G. Neuweiler et al., *Z. Vergl. Physiol.* **74**:57–63 (1971).

of the frequency shift may be calculated from Eq. (4-2). If we assume that $\nu = 80$ kHz, $v = 5$ m/sec, and $c = 332$ m/sec, the detected frequency would be increased by around 3.0 percent. Pitch discriminations down to a few tenths of 1 percent have been measured. Such discrimination is quite striking in an animal and is reminiscent of the "perfect pitch" of a musician. The use of

Doppler sonar is probably associated with predicting motion of the prey prior to interception. These animals also appear to rely on independent ear motion as well as signal intensity to assist in their pursuit strategies.

Click signals for localization are associated with specific fruit-eating bats as well as certain birds that nest in dark caves. The latter include the fruit-eating Venezuelan oil bird (*Steatornis caripensis*) and various cave swiftlets. These species rely on echolocation only when ambient light conditions are below a certain level, as all have quite adequate vision. Here we may be observing an example of convergent evolution in that several radically dissimilar creatures rely on the same means for navigation. Other birds have been suspected of using a click-locating method—one of the more interesting candidates being the nighthawk. It must be emphasized that birds cannot hear in the ultrasonic range, so that audible clicks are required.

B. SOUND PRODUCTION AND DETECTION

Insect-hunting bats emit ultrasound via nose and/or mouth. The nasal and ear configurations of these animals are fantastic in their variation and have caused primitive people to regard some species with awe and terror. Presumably the focusing and directional properties of the various antenna configurations are advantageous to the animals possessing them. Sound production is believed to occur in the larynx for all the small bats, and the signal repetition rate is generally some harmonic of the respiratory cycle. The fruit eaters using echolocation employ their lips, tongue, and oral cavity to generate the output signal. The brain of the bat is especially adept at pulse pair resolution, so two signals occurring within 1 msec can be distinguished. This ability is essential to the use of echolocation and is quite rare in other mammals.

C. RECEPTION RATES AND SPATIAL RESOLUTION

As a bat approaches its objective, the repetition rate of the signal will increase with a corresponding decrease in the temporal length of each individual pulse. Griffin has referred to this phenomenon as the interception "buzz." In the small brown bat (*Myotis lucifugus*), a frequency-sweep species, the bat will emit approximately 10 pulses per second (pps) while cruising; an attacking animal can achieve several hundred pps. Signal spectrographs change somewhat in this process, generally becoming narrower in bandwidth.

Similar increases appear to hold in the Doppler species, whereby the repetition rate may jump by one or two orders of magnitude upon landing or attack. In the latter case, no frequency change is observed as the pulse rate rises, a result in keeping with their Doppler detection mechanism. Those species using click localization have also been shown to produce changes in repetition rate upon landing. Since these fruit-eating bats only use their sonar to locate roosting sites inside darkened caves, consideration of interception is not pertinent.

Spatial resolution has generally been measured by means of wire obstacles of variable diameter. A small brown bat can detect wires down to 0.12-mm diameter using sweep sonar, and Doppler-type species can avoid obstacles that are as small as 0.05–0.10 mm. The range at which those bats detected the wire was on the order of 1 m or less in either case. An Egyptian fruit bat, using clicking, can detect a 0.5-mm wire. Ranging sensitivity is usually limited by the pulse-length considerations given in the preceding discussion. Specifically, for a 1-msec pulse at a single frequency, Eq. (4-1) indicates that the uncertainty, Δr, in the distance from the bat to the insect may approach 0.8 m. The sweep nature of the chirp pulse decreases Δr by about 2 orders of magnitude (i.e., down to approximately 1 cm). The bat's wingspread is more than adequate to catch the flying insect within this ambiguity. By strobo-scopic photography, the bat is found to catch the insects in its tail and wing membranes prior to ingestion.

4.3   The Dolphins and Whales

Many of the echolocation features cited for bats hold for both the small and the large toothed whales. The baleen whales have not been investigated extensively enough to define their use, if any, of such a facility. On the other hand, Fig. 4.4 compares the spectrograph of a small bat and the bottlenose dolphin (*Tursiops truncatus*). Among a variety of signals, this dolphin uses a click type of output pulse which incorporates a modulated downswing in frequency so as to aid in the differential distance measurement. As in the insectivorous bats, repetition rates rise as the bottlenose dolphin moves closer to an object of interest. A cruising animal will emit on the order of a few pulses per second but may achieve 500 pps while investigating a detail in its tank. Pulse-pair resolution is comparable to the maximum repetition rate. Dolphins have one advantage over bats in that the ambient noise levels at ultrasonic frequencies are much lower in a marine environment. The lack of rapidly moving small objects and the high-frequency attenuation of water both contribute to this effect. The equivalence of the external and internal acoustical impedance also allows the whale to dispense with the elaborate external ear forms of the bats. While most whales do possess adequate eyesight, one freshwater species has degenerate eyes and hunts almost completely by sonar in the silt-filled streams of the Indian subcontinent. This animal exhibits the

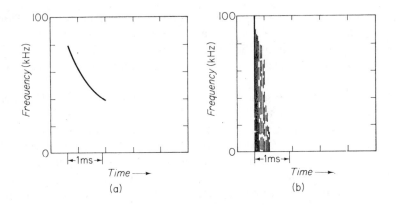

Figure 4.4 Comparison of vocal spectrograms of two echolocating species. (a) The small brown bat. (b) The bottlenose dolphin. After R. G. Busnel, ed., *Animal Sonar Systems, Biology and Bionics* (Jouy-en-Josas, France: Laboratoire de Physiologie Acoustique, 1967).

same evolutionary trend in its visual sense as is characteristic of the insectivorous bats. Eyesight is essentially redundant in its environmental niche and has correspondingly become nonselective.

The bottlenose dolphin, the common porpoise, and the killer whale all appear to use the click signal for echo soundings. Of these three examples, the first two have been blindfolded and yet are still able to negotiate wire channels set up in their tanks. Spatial resolution can be as good as parts of a millimeter if the animal is required to resolve a metallic wire; poorer results hold for nylon. The bottlenose dolphin may also be able to utilize the phase of the returning signal to differentiate air-filled targets from solid ones. If the acoustic impedance of the target is less than that of the water, the reflected pressure wave will be inverted, the converse being true if the target impedance is greater. An alternative explanation is that the bottlenose dolphins may recognize intrinsic reverberations in the target. This is supported by experiments showing that they can distinguish between disks of the same diameter made of different metals.

Besides the click pulses, the bottlenose dolphin also produces various complex signals as well as tones of relatively pure frequency. The distress call, for example, will effectively bring any other bottlenose dolphin within hearing toward the victim. Notice the heavy reliance on low frequencies in this signal (Fig. 4.5), as might be expected if long-distance transmission is to be important. The pure tones are thought to be more specific intragroup messages, although their meaning is as yet unclear. Some work has even suggested that the click signal itself may be effective in communication between cooperating dolphins that must associate on some task.

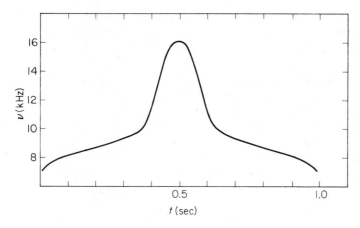

Figure 4.5   Distress call of the bottlenose dolphin. The uniqueness of such important communication signals makes them relatively easy to detect amid background noise. After Lilly, J. C., "Distress call of the bottlenose dolphin: stimuli and evoked behavioral responses," *Science*, Vol. 139, pp. 116–118, Fig. 1. Copyright 1963 by the American Association for the Advancement of Science.

4.4   Infrared Sensitivity of Snakes

Cells sensitive to infrared (IR) electromagnetic radiation have been evolved by two snake families: the Crotalidae (pit vipers) and the Boidae (constrictors). In the vipers, the two bilaterally symmetric pits are overtly placed on either side of the head between eye and mouth. The constrictors use a specialized arrangement of cells extending along either jaw. The detector is essentially energy sensitive (i.e., integrates the infrared (or other) radiation over a period of time). Bullock and his co-workers at San Diego have shown that the central nervous system response to IR is proportional to temperature increases in the pit organ. Mitochondrial changes can be observed microscopically in the organ itself upon irradiation.

Gamow and Harris have investigated the neural reaction of constrictors to 8-msec pulsed infrared radiation from a CO_2 laser (see Chapter 27). In this experiment, the entire animal is irradiated by a monochromatic heat source. The snake's integration time appears to be around 35 msec, as determined by the delay between stimulus and electroencephalographic response (see Chapter 6 for a discussion of eeg's). No brain wave response is seen with other snake families. The ambient radiation level critical to trigger a response in constrictors is around 5×10^{-6} J/cm^2, although this number may be uncertain by a factor of 2 or so.

The pit vipers present an example of the usefulness of such a sensing system. If the ambient light level decreases or is zero, the heat radiation from small mammals and birds can be used to locate prey in a relatively cool desert evening. Some authors have suggested that the pits are a stereoscopic imaging system which helps to further localize the warm-blooded animal. The boas can use similar clues in a darkened tree environment. Thermal comfort may be a second use of heat sensors. If the snake finds its ambient temperature unsatisfactory, it can find more congenial surroundings by simply "looking" at various nearby possibilities, much as an infrared sensitive earth satellite detects cold ocean upwellings. The survival value of such an attribute is great since tactile interrogation of the surroundings is no longer required. Temperature maintenance has been a long-standing problem of the reptiles; the demise of the dinosaurs has been ascribed to thermal problems.

4.5   Electric Fish

Since prehistoric times, aquatic species have been recognized that are capable of generating very strong electric fields. Medical treatments using the large electric ray *Torpedo nobiliana* were reported by the Romans. This marine animal is now known to be able to provide 50 A at 60 V. The existence of a group of strongly electric fish was of some concern to Darwin since lower-voltage species were unknown during his time. The efficacy of an electrical defense and attack mechanism was clear to most of the nineteenth-century biologists who ventured to touch a strongly electric fish.

Since 1945, a second category called weakly electric fish has been described. In the early work, the classification assignment of a given species would follow the investigator's putting his hand into the water near the animal. Those species that were not particularly distressing (i.e., having output voltages between 0.1 and a few volts) would be placed under the weakly electric heading. The utility of such small fields remained in question until it was suggested that these species used electric fields to detect inhomogeneities in the ambient electrical conductivity so as to orient and find prey in rather turbid waters. Such a hypothesis presumes a detector mechanism; several organs sensitive to electric potential gradients have been subsequently discovered. These appear to be developments of the lateral line receptor system (part of the ears of the fish), which detect electric current instead of pressure.

It is now apparent that there exist families of both cartilaginous and bony fish which have developed an electric broadcasting and receiving apparatus. Table 4-1 summarizes these as well as the relatively rare strongly electric fish.

Lissmann has intensively investigated one weakly electric species, the *Gymnarchus niloticus*, for many years. This bony fish shown in Fig. 4.6 is quite unusual in several ways. For one thing, it is such a rare African specimen that no common name has ever been associated with it. *G. niloticus* swims by undulation of a dorsal fin while keeping its backbone perfectly

TABLE 4-1
PARTIAL LIST OF ELECTRIC FISH[a]

Name	Discharge type	Habitat
Electric eel	Strong (600 V); also weak signals	Freshwater, South America
Electric rays	Strong (60 V; 1 kW)	Ocean
Electric catfish	Strong (about 300 V)	Freshwater, Africa
Gymnarchus	Weak	Freshwater, Africa
Mormyrids	Weak	Freshwater, Africa
Gymnotid eels	Weak	Freshwater, South America
Knifefish	Weak	Freshwater, South America
Skates	Weak	Ocean

[a] See the Electric Fish references for a more complete tabulation.

straight. This property is particularly impressive if one observes it negotiating a corner or going backward. Like many insectivorous bats, the fish has extremely degenerate eyes and is thought to be active only at night. *G. niloticus* produces a relatively fixed frequency 260-Hz signal that has a longitudinal dipolar field distribution, the head of the fish being positive. The rigid spinal alignment is thought to be necessary so that the animal is not electrically confused by its own swimming motions. Charge detectors are located on the head of the fish and may be sensitive to as few as 10^3 positive ions. Minimal detectable field strengths are on the order of 10 μV/m. Although these values for *G. niloticus* seem quite impressive, they may be representative of all sentient electrical systems.

Figure 4.6   The African freshwater electric fish *Gymnarchus niloticus*. The fish swims by undulating its dorsal fins while maintaining a rigid posture. Discharge rates are approximately 300 Hz with the tail being negative relative to the head of the animal.

A. TYPES OF SIGNALS

Colloquially, it is possible to divide the species of Table 4-1 into "bummers, hummers, and buzzers." The bummer produces a voltage pulse of one phase (sometimes termed a *dc signal*) and of considerable power. The electric rays and South American electric eel (*Electrophorus electricus*) are examples of this class, which includes only the strongly electrical fish.

A hummer has, by definition, a weak voltage output of constant frequency but with no net current flow. This is sometimes termed an *ac signal*. *G. niloticus* is a good example.

Buzzers also produce no net current flow. They are weakly electric fish capable of shifting their firing rate so as to suit local conditions. A frequency increase may follow the disturbance of an individual by some aspect of its environment and as such is reminiscent of the bat's interception buzz (see Sec. 4.2.C.). As in the Doppler bats, some buzzer species can frequency-shift if a second signal of their original repetition rate is broadcast into the tank. Thus, confusion may be resolved and many sentinent individuals can coexist socially.

Hummers and buzzers both operate in the range between 10 Hz and 1,500 Hz, with a particular frequency range being characteristic of a given species. Each individual has its own signature rate, and can be recognized in a group of its fellows. The social uses of these weakly electrical signals are as yet unclear, however.

B. ELECTROCYTES

The cells responsible for the electric field production are termed *electrocytes* (or *electroplaques*); they appear to be evolutionary variants on ordinary muscle and nerve cells. The basic rule of thumb appears to be that lower-frequency fish (e.g., *G. niloticus*) have cells derived from muscle while the higher-rate animals possess electrocytes attributable to nervous origins. An individual electrocyte will have an emf of about 0.1 V. The bummers produce their powerful discharge with an extensive array of electric cells in series and parallel. In the electric eel, for example, tens of thousands of electrocytes may be arrayed in a series arrangement to yield voltages approaching 500 V. The discharge mechanism of these cells has been described as follows. Originally, the membranes are such that two opposing emf's are present. Upon the nervous signal, the single innervated membrane becomes reversed in polarity, the two emf's add, and the output voltage is produced. The many electrocytes must fire in synchrony. If not, the fish will not be effective in generating the requisite voltage. Variable lengths of nervous tissue (delay lines) may be utilized to accomplish the temporal coordination in a long electric organ. A pacing center in the brain is thought to be responsible for the overall master frequency of the weakly electric fish. This pacemaker has been shown to be temperature sensitive and, in the case of one gymnotid, *Eigenmannia virescens*, modulated by a frequency-difference neuron capable of driving the frequency of the fish away from the ambient frequencies of its neighbors.

4.6   Bird Navigation and Homing

A. BACKGROUND

One of the classical biophysical problems is an explanation of the navigational abilities of various animals, particularly birds. A very common rationalization involves visual memories as being the most significant clue to the bird making a migratory journey. The route is assumed to be learned by eye while going along with a flock of older animals. This concept might explain very short migratory excursions, such as up and down a mountain slope. However, the phenomenon of migration appears more complex for flight paths exceeding a few hundred kilometers. The primary difficulty of ascribing such abilities to visual memory is the enormous number of data in the visual field. Climatic conditions can blow the birds off course into unknown territory or cover certain strategic landmarks with snow. Thus, the landscape would be incoherent relative to the stored topographic images. Instances are known in which a flock has been diverted many kilometers from their basic route and yet can still attain the objective. In addition, birds flying totally, or in part, over open water would have little unique visual data. One route would look very much like any other, so visual dependence would be completely nonspecific.

Until miniaturized transmitters were produced, the only data available on a returning flight were its total time and the initial direction from the point of release. Workers at several universities are now routinely and discreetly following radio-equipped birds with light aircraft in controlled experiments. The pigeons so studied, for example, do not appear to navigate from landmark to landmark; instead, a course is set and followed until within visual range of the home roost. Frosted contact lenses fitted to the birds' eyes do not change these results, except that the animals cannot see well enough to land (!) and simply drop to earth near their home. They are then hand-carried into the loft.

It has been well documented that experimentally displaced birds can return over open-water distances comparable to continental size. For example, Laysan albatrosses have been released at various sites around the Pacific Ocean away from their Midway Island home. Some 14 of 18 released birds returned, one coming back 5,100 km in 10 days. Others successfully relocated their individual nests from distances approaching 6,400 km. In a second controlled relocation, a Manx shearwater (*Procellaria puffinus*) was taken from Wales to Boston, Massachusetts, a distance of 5,100 km over the north Atlantic. The bird returned home in 12.5 days. These experiments imply a rather precise navigational system accurate to a small part of a mile when the animal has had *no* previous experience over the chosen route. No visual clues could be considered likely until the bird was within a few yards of its nesting site.

B. THEORY AND EXPERIMENT

In order to explain bird navigation, two theoretical viewpoints were developed in the early 1950's. Both hypotheses depended on the animals' use of the sun as well as an internal clock mechanism. We shall discuss these in turn. Mathews proposed his solar-arc theory in 1951. Each bird is assumed to have an internal clock which maintains intrinsic time to perhaps 15-min accuracy. In addition, the solar arc is detected and extrapolated to its maximum (local noon) elevation above the horizon. The combination of intrinsic time and maximum solar height gave the bird longitude and latitude, respectively. If a bird were displaced due east, for example, it would notice that the sun was much advanced in its angular motion compared to the intrinsic clock, but that the extrapolated arc would not be any higher in the sky than at home. Thus, a westerly flight was called for until the intrinsic and local times coincided. The technique is tantamount to the mariner's technique of navigation and allows the animal to return home without reference to any data stored in a memory system.

On the other hand, Kramer allowed his avian model to possess an intrinsic clock, but employed the solar position only as a directional or compass aid. In this system, if the internal clock said noon, the sun's disk in the northern hemisphere would define a due-south compass bearing. The knowledge of direction is, however, not sufficient to direct the animal home, so that Kramer introduced the concept of "map" carried in the memory of the animal. From its map, the displaced bird might ascertain that it was north of home and that it should therefore fly into the supposedly noontime sun in order to return. Neither the physical nor the physiological nature of the map was described.

Since 1950, various experiments have been devised to discriminate between the two models. These experiments involve changing the animals' intrinsic clock by artificial lighting, and then releasing them at some distance from the nest. The birds' initial flight path is used as an indicator of the correct model. Details of such trials are found in the reference by Keeton. Most are consistent with the map-compass theory rather than the solar-arc theory.

As might be anticipated, any theory that relies heavily upon the sun would predict that birds would not attempt flight under heavy cloud conditions. This is indeed the case for many species, but pigeons are known to home even in rainy weather. The possibility of a secondary guidance system has been investigated by other researchers. Birds released when the sun was not visible would fly in the correct direction unless bar magnets were attached to their heads. With the magnets in place, the birds left at random points of the compass. Similar effects were seen using Helmholtz coils placed on the animal. Apparently, a magnetic field sensor is used in these conditions when the sun cannot be seen. A map concept is, of course, still critical, as the bird only gets directional data from the earth's field, which is approximately 0.5×10^{-4} T.

Several other species have been shown to have magnetic field sensitivities as low as 10^{-7} T, but the nature of the pertinent organs is unknown. Just as in animals that use echolocation, if the species can use its visual sense, it will. Otherwise, a secondary system must be called into play.

Several problems still exist with the map-compass theory. Most notably, the nature of the map is a complete mystery. Coriolis effects on the birds' inner ear cannot account for the map, because of the magnitude of Brownian motion. Inertial reactions also seem unlikely to contribute to such a map, since pigeons have no trouble returning to their roosts even if they are carried to the release point in rotating cages. In spite of the preceding, the map-compass theory still remains the best model for bird navigation.

4.7   Summary

Absence of light has occasioned the evolutionary development of alternative techniques of localization and orientation. Bats and toothed whales have been found to detect their own ultrasonic radiation. Various acoustic spectra are utilized, although a given species probably uses only a single type. Weakly electric fish also appear to possess broadcasting and receiving apparatus for use in murky waters. Many of these fishes are nocturnal as well. Their detection mechanism is a development of a lateral line organ—part of the primitive "ear" of a fish.

Passive detection of ambient signals occurs in the infrared-sensitive snakes. Environmental temperature data can then be used to select prey or more congenial surroundings. In view of the general reptilian problems with temperature control, such an adaptation could be of enormous advantage.

Bird navigation is also associated with passive electromagnetic radiation detection. The use of stellar and solar motions is of paramount importance, but many species, including pigeons, can discern the earth's magnetic field. The mechanism for the latter process remains unclear. Similarly, the nature of the intrinsic geographic "map" carried by the bird is not understood. Biophysical science faces a very complicated challenge in the comprehension of the bird's orientation system.

REFERENCES

Echolocation

1. SALES, G., AND D. PYE, *Ultrasonic Communication by Animals* (New York: John Wiley & Sons, Inc., 1974), 281 pages.
2. GRIFFIN, D. R., *Listening in the Dark: The Acoustic Orientation of Bats and Men* (New Haven, Conn.: Yale University Press, 1958), 413 pages.

Electric Fish

3. LISSMANN, H. W., "Electric Location by Fishes," *Sci. Am.* **208** (March 1963): 50–59.
4. STEINBACH, A. B., "Behavior of Weakly Electric Fishes," *Biol. Bull.* **138**:200–210 (1970).
5. BENNETT, M. V. L., "Comparative Physiology: Electric Organs," *Ann. Rev. Physiol.* **32**:471–528 (1970).

Infrared-Sensitive Snakes

6. BULLOCK, T. H., AND R. BARRETT, "Radiant Heat Reception in Snakes," *Commun. Behav. Biol.*, *Part A*, **1**:19–29 (1968).
7. GAMOW, R. I., AND J. F. HARRIS, "The Infrared Receptors of Snakes," *Sci. Am.* **228** (May 1973):94–100.

Bird Orientation and Homing

8. MATTHEWS, G. V. T., *Bird Navigation*, 2nd ed. (London: Cambridge University Press, 1968), 197 pages.
9. KEETON, W. T., "Orientation by Pigeons: Is the Sun Necessary?" *Science* **165**:922–928 (1969). Also see "Mystery of Pigeon Homing," *Sci. Am.* **231** (December 1974):96–107.
10. WALCOTT, C., AND R. P. GREEN, "Orientation of Homing Pigeons Altered by a Change in the Direction of an Applied Magnetic Field," *Science* **184**:180–182 (1974).
11. SCHMIDT-KOENIG, K., *Migration and Homing in Animals* (Berlin: Springer-Verlag, 1975), 99 pages.

DISCUSSION QUESTIONS—PART A

1. Chemical senses are known as taste and olfaction (smell) in higher animals. What is the present state of biophysical knowledge of these senses?

2. The term "music" is used for combinations of sounds that obey certain restrictions. What are the physical relations between tones in a Western musical major scale? Minor scale? Twelve-tone scale? Scales used in Oriental music?

3. The human vocal tract is as specially designed to produce speech and song as the ear is to receive them. Describe the physics of the vocal tract as it affects pitch, overtones, and loudness of the sounds produced.

4. Certain whales produce sounds that can travel many hundreds of miles through the ocean. This "whale music" is different from the echolocation noises discussed in Chapter 4. Describe this music and the physics of its production and reception by another whale.

5. The echolocation used by bats may help humans suffering from blindness. Biophysicists have experimented with helmetlike arrangements that would transmit high-frequency sounds, receive their echos, and relay them, at lower audible frequencies, to the person wearing the helmet. Describe the present state of such research.

6. Discuss the evidence concerning any possible role of polarized light in the vision of man and other vertebrates and compare this to its use by insects.

7. Fiber optics is a method of transmitting light without the usual lenses. Describe how light is transmitted in light fibers. Does this method have any biological analogies?

8. The theory of lenses discussed in Chapter 2 used the small-angle approximation

$$\sin \theta \sim \theta$$

Derive third-order equivalent formulas and discuss in terms of these formulas: spherical aberration, coma, field curvature, astigmatism, and image distortion. What is the importance of these effects for the eye?

9. What is known about the effect of temperature, temperature gradients, and ground absorption on the transmission and reception of insect sounds?

10. The octopus and related species have developed a third type of eye, distinct from the insect and vertebrate eyes. Describe the structure and function of this cephalopod eye.

11. Describe the equipment necessary to study the sensory physiology of insects, including their vibration and sound thresholds.

12. Chemical communication occurs in liquid as well as gaseous environments. It has been used to explain homing behavior in certain fish. Discuss the limits on such communication, including pheromone transmission and reception in water.

13. Relate the pulses emitted by two or more species of bats to the physical structure of the ears, nose, and mouth of each species and its feeding habits.

14. Certain hawks and eagles may use vision in the infrared region to detect small warm-blooded prey at great distances. What is the evidence that they have this extended range of vision?

PART B

NERVE AND MUSCLE

Introduction to Part B

The following six chapters are devoted to biophysical studies of nerves and muscles and to the interpretation of other phenomena in terms of the properties of these two tissues. The first chapter of this part (Chapter 5) contains a discussion of the conduction of information by nerve fibers in the form of electrical impulses. Several concepts of basic electrical theory, needed in various chapters throughout the text, are summarized in Appendix B; it is hoped that readers unfamiliar with these terms will read that appendix.

In Chapter 6, electroencephalographic waves are described. Their interpretation and relationship to nerve impulse conduction are also discussed. Chapters 7 and 8 discuss the neural mechanisms associated with hearing and vision, respectively. The ideas presented in Chapters 1 through 5 are used in these two chapters about the neural aspects of hearing and vision.

The physical and chemical nature of muscular contraction forms the basis for Chapter 9. Some biochemical concepts, presented more fully in later chapters, are introduced in order to restrict the discussion of muscles to Chapter 9. Finally, the last chapter in Part B, "Mechanical and Electrical Character of the Heartbeat," applies to the mammalian heart many of the ideas presented in Chapters 5 and 9.

CHAPTER 5

Conduction of Impulses by Nerves

5.1 The Role of the Nervous System

The nervous system is composed of units called neurons which transmit information in the form of electrical pulses from one place within the organism to another. This action is essential for the rapid responses of animals to external stimuli. Animals respond more rapidly than plants to conditions outside themselves. For instance, certain plants have flowers that are open only in bright sunlight, and deciduous trees shed their leaves during the fall season. However, these are comparatively slow responses involving time intervals from minutes to days. In contrast, the responses of a motorist to a red light or of a fly to an approaching swatter are both very much quicker. These rapid responses of animals timed in milliseconds or, at most, seconds are mediated by the nervous system.

The rapid coordinations and responses of animals strongly suggest that the nervous system must transmit information in an electrical or magnetic form. One might reach this conclusion without detailed knowledge of the structure and properties of the neurons. Studies of nerves have shown that they consist of bundles of long processes called *axons* or nerve fibers. The axons are each a part of an individual neuron. Along the nerve fiber, the information is coded and transmitted in the form of an "all-or-none" or "on–off" electrical pulse called an *action potential* or *spike potential.* (The latter name arose from the appearance of these impulses on the screen of a cathode ray oscilloscope.)

On a teleological basis, the problems of the nervous system are similar to those of transmitting telephone messages over long distances. Either there must be many parallel low-frequency channels, or fewer high-frequency channels, each modulated by many separate signals. Anatomic and physiologic studies show that animals have varying numbers of parallel low-frequency electrical channels. The number of channels tends to increase with the complexity of the animal. Along each of these channels (nerve fibers), information is transmitted by electrical pulses. The individual channel, with its energy supply and its connections, is called a *neuron.*[1] Its distinguishing features involve the biological generation and transmission of electrical potentials.

The earliest experiments that could be called bioelectrical occurred toward the end of the eighteenth century. Galvani put two dissimilar metals into a frog's leg muscle and observed a twitch. He correctly associated the response with electricity but assumed that the electricity was generated within the muscle by a vital process. Volta proved that Galvani's electricity was not of biological origin; the existence of true biological electrical generators was not discovered for almost another century. Today, it is known that all nerve fibers, in fact, probably all cell membranes, are charged electrically. The membrane charges, as well as the spike potentials, are so small that they could not be observed with the instrumentation of Galvani and Volta. The field of bioelectricity is a fertile one for the application of physical instrumentation; it has attracted many persons with a background in physics who welcomed a challenging biological problem to which they could apply their previous training. A major application of bioelectricity is the study of the conduction of impulses by nerves.

Animals possess other mechanisms, besides the bioelectrical actions of the nervous system, for transmitting information from one part to another. These include immune responses, mechanical and hydrodynamic changes, and metabolic signals. The *endocrine* system involves transmission of specialized chemical signals called *hormones.* The latter alter metabolic rates, dilation of blood vessels, and secretory rates at specific target organs. Similarly, when information is transmitted from one neuron to another, there is often a chemical intermediate. The process differs from the endocrine system only in the length of time involved. Hormones act, in general, over a period of hours or days, whereas the transmission from one neuron to the next takes only

---

[1] Some giant invertebrate fibers are fusion products of several embryonic neurons.

milliseconds. Some hormones act faster, so that there is no sharp dividing
line between the hormones and the neurochemical transmitters. Plants also
possess chemical transmitters. The distinguishing feature of higher animals is
their nervous system, which transmits information far more rapidly than do the
endocrine systems.

Biophysicists have studied the nervous system as well as the endocrine
systems. Both lend themselves to the application of physical techniques and
analyses. This is particularly true of the interactions between groups of
neurons, of interactions between groups of endocrine glands, and also of
neuron–endocrine interactions. In all of these, *feedback* loops exist in which
the effect produced alters the behavior of the neurons or endocrine glands
that produce the effects. Physicists and electrical engineers refer to these
types of control mechanisms as *negative feedback*; physiologists have called
many of them *homeostatic* mechanisms, because they tend to keep the state
of the organism constant.

It is the aim of this chapter to present, insofar as possible, a picture of the
physical properties of nervous tissues and a description of how nerve fibers
conduct spike potentials. Because each reader will have a different background, an attempt has been made first to present the fundamentals of
electricity. A more detailed discussion of electrical terminology can be found
in Appendix B. The electricity section of this chapter is followed by a brief
description of certain salient features of the vertebrate neuron. Details of the
physical characteristics of the action potential are then presented. Next are
the bioelectric characteristics of the membrane. The final section of this
chapter deals with conduction from one neuron to the next, called *synaptic
transmission*. Readers with various backgrounds may skip Secs. 5.2, 5.3, or
5.5 with minimal loss of continuity.

Many aspects of the nervous system are discussed in other chapters.
Chapter 6 describes the electrical potentials of the brain and contains a
discussion of feedback mechanisms. Chapters 7 and 8 deal with the neural
aspects of vision and hearing. Chapter 9 includes the stimulation of muscles
by nerves, and Chapter 10 the neural control of the heart rate. The special
properties of membranes are discussed in Chapter 24.

5.2 A Brief Glance at Electricity

Physicists consider all matter to be made up of neutral atoms, which, in turn,
are made up of positively and negatively charged particles. Although large
chunks of matter are electrically neutral, on a subatomic scale many particles
have a net charge. In a liquid or a crystal, there are often ions or groups of

atoms that are likewise charged. For instance, NaCl splits into Na^+ and Cl^- ions in a water solution. In a NaCl crystal, the sites are occupied by Na^+ and Cl^- ions. Even water has measurable H^+ and OH^- concentrations. Thus, on an atomic or molecular scale, charges frequently do not balance, even though a volume containing many molecules is approximately electrically neutral.

Likewise, when a metal is placed in a liquid, or when two dissimilar metals are placed in contact, the surfaces at the discontinuity become oppositely charged. The "dry" cell and storage battery are examples of two unlike electrodes in a liquid. Unlike charges are separated at the metal–liquid interfaces. If two dissimilar metals are used, the charge separation will be unequal; charges will flow when these two metals are connected by an external conductor. The thermocouple is an example of a practical use of the charge separation at the junction of two metals.

If charge is not allowed to flow after equilibrium has been established, the actual charge separation is very small in each of the cases above; the net charge separation is negligible compared to the SI unit of charge, which is called a coulomb. This leads one to suspect that although matter is approximately neutral, the charges do not balance to the last electron. Biological cells and parts of cells are not exceptions. The net charge on any cell measured in coulombs is infinitesimal, but measured in the units of the charge on an electron e $(1.6 \times 10^{-19} \text{ C})$, it is appreciable. The neurons are distinguished from most other cells in that they are specialized to transmit changes in their surface potential rapidly. (Muscle fibers are similar to neurons in this respect.)

The flow of electrical *charge* is known as *electrical current*. Currents are measured in units of charge per unit time; a current of 1 C/sec is called an ampere. Early investigators of bioelectrical phenomena regarded the current as the fundamental event in the conduction of impulses by neurons. Hence, they referred to these as action currents. Considerable experimental evidence, however, indicates that the *electrical potential* changes are the most unique property of the neuron membrane. In any case, the potential is the parameter actually measured in most experiments. The *electrical potential difference* between two points is defined in elementary physics as the energy received by a unit charge when it is carried between these two points. Thus, it is a potential energy per unit charge. Electrical potential, V, is usually measured in volts.

Qualitatively, one may think of the potential as being similar to an electrical pressure or force that drives positive charges to regions of lower potential (and negative ones in the opposite direction). The ratio of the potential difference to the current flowing through a conductor is called the *resistance*, R. For many substances, R is a constant independent of the current and potential difference; substances of this type are said to obey Ohm's law. In these cases, one can easily analyze direct-current circuits, such as those shown in Fig. 5.1.

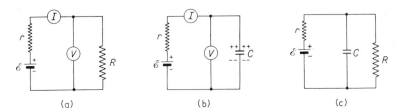

Figure 5.1 (a) Direct-current circuit. (b) Direct-current circuit with capacitor. (c) Simplified circuit representing a resting axon. In each circuit, r = the internal resistance of the battery, $\mathscr{E}$ = the emf of the battery, R = the load resistance, V = the potential difference across R, and I = the current through R. C is the capacitor; current flows only while the capacitor is being charged.

Most bioelectrical phenomena involve changes that occur quite rapidly in time. As stated in Chapter 1, events with complex time dependence can be analyzed in terms of simple harmonic changes (alternations) at one frequency. In electricity, alternating current (ac) circuits are more complex than direct current (dc), inasmuch as elements other than resistances can impede the flow of an alternating current. In an ac circuit of fixed frequency, the ratio of the potential to the current is called the *impedance*. The ratio of the component of the potential in phase with the current, to the current, is called the *resistance*, whereas the ratio of the out-of-phase component of the potential to the current is the *reactance*. Reactances arise due to *capacitors*, C, which do not pass a direct current, and *inductors*, L, which do not impede a direct current. An ac circuit is illustrated in Fig. 5.2. Inductors are not as frequently encountered in biological systems as are capacitors. Most biological membranes act as leaky capacitors in an equivalent electrical circuit. As such, they may be charged, maintaining a fixed potential difference between their two sides, or

Figure 5.2 An ac series circuit. Note in the symbolism used that $\mathscr{E}$, $\mathscr{E}_0$, I, I_0, V_1, V_2, and V_3 are all complex numbers and e is base of natural logarithms. The use of the complex notation for ac circuits is discussed in Appendix B.

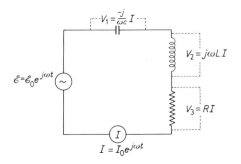

they may conduct a rapid change in potential. These ideas are applied directly to neurons in Sec. 5.4.

In addition, most membranes can generate a potential difference between their two sides, thereby expending chemical energy. Such a generator is called an *electromotive force*, or *emf*. These generator properties of neuron and muscle membranes are discussed in this chapter and in Chapter 9, as well as in Chapters 23 and 24. Particularly large emf's are produced by the electrical organs of certain fish. The cells of these organs are evolutionary modifications of muscle and nerve fibers (see Chapter 4).

5.3   Anatomy and Histology of Neurons

The functional unit of the nervous system is called the *neuron*. It consists of a nerve *cell body*, small processes called *dendrites*, and one large process called an *axon*.[2] Outside the central nervous system, many of the larger axons are surrounded by a thick, fatty *myelin sheath*. The sheath is interrupted some-what periodically at the *nodes of Ranvier*. Along the side of the sheath are satellite cells called *Schwann cells*. Some axons are more than 1 m long. In spite of these long extensions, vertebrate axon diameters are in the range 1–20 μm. A diagram of such a neuron is presented in Fig. 5.3. Smaller axons, although not as thickly myelinated, are always surrounded by a lipid layer as well as by extremely small satellite cells. The neuron[3] is a single cell. The dimensions of the cell body, of its nucleus, and of the diameters of the axons and dendrites are all typical of other cells, ranging for vertebrates from 1 to 100 μm. The length of the axon is the outstanding exception, being far longer than typical cellular dimensions. For example, the neurons that control the muscles in the human finger have their cell bodies in the spinal cord, whereas their axons run the entire length of the arm. In the elephant and giraffe, the lengths of the axons are as great as 3 or 4 m.

The axons are grouped together in bundles called *nerves*. Each nerve consists of a large number of axons of various sizes carrying impulses in their respective physiologically important direction. Just as the axons are grouped in nerves, the nerve cell bodies usually occur in compact clusters known as *ganglia*. Within the central nervous systems of vertebrates, clusters of nerve cell bodies are referred to as *nuclei* (which is a confusing term) and axon bundles are called *fiber tracts*.

---

[2] Some authors use axon and dendrite to indicate direction of transmission. In this text, axon and nerve fibers are used as synonyms. Some neurons have the cell body in the middle of the nerve fiber rather than at one end.

[3] The satellite cells along the myelinated fibers are actually separate cells but are, nonetheless, considered part of the neuron.

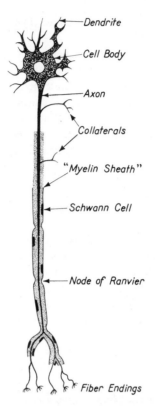

Figure 5.3 Large myelinated axon. Dark-staining material in cell body is called *Nissl substance.* In some neurons, the axon is off to one side of the cell body; in others, the cell body is in the middle of the axon. In both of these types the dendrites are attached directly to the axon; that is, both extreme ends of the axon appear similar. After A. A. Maximow and W. Bloom, *Textbook of Histology* (Philadelphia: W. B. Saunders Company, 1957).

Connections between two neurons are called *synapses.* These occur between the branched ends of axons, dendrites, and collaterals of different neurons. Other axons end at receptors, such as the hair cells of the organ of Corti in the ear. Still others terminate at controlled targets, as the motor end plates of a muscle fiber. The nervous system of vertebrates is organized into a central nervous system enclosed within the skull and spinal column and a peripheral nervous system which includes both nerves and ganglia. Throughout the entire system the basic element is the neuron. There is nothing inherent in the axon to control the direction in which it conducts. This is a property of the synapse only. In general, a peripheral nerve will contain both sensory (or afferent) axons, which conduct toward the central nervous system, and motor (or efferent) axons, which conduct away from it.

The sensory axons conduct impulses from the receptors to the nerve cell bodies. In the vertebrate nervous system, there are multineuron afferent pathways which extend from sensory receptors to higher centers of the central nervous system; along such pathways, the axons conduct toward the cell bodies. In these, there is always at least one cell body in a ganglion outside the central nervous system. For example, the sensory pathways entering the spinal cord have their first cell body in ganglia at the dorsal roots of the spinal cord, just outside the spinal column. Synapses occur in these ganglia; the axons entering the spinal cord are those of the second neuron.

In efferent pathways leading to muscles, glands, and other target organs, the axon conducts impulses from the nerve cell body. Vertebrate efferent pathways outside the central nervous system are customarily divided into the voluntary motor pathways and the autonomic nervous system. The former have all their cell bodies within the central nervous system. For example, the cell bodies for the axons that control the human toe muscles are within the spinal cord.

In contrast to the voluntary pathways, the autonomic pathways almost always have a synapse and cell body outside the central nervous system. The autonomic system is divided into two parts on a functional and anatomical basis. Those pathways leaving the spinal cord in the nerves between the thoracic and lumbar vertebrae comprise the *thoracolumbar* division, or sympathetic system. The other part of the autonomic system, with pathways that leave the central nervous system via the sacral region of the spinal column or the cranium itself, is called the *craniosacral* division, or parasympathetic system. Most organs are supplied by both the craniosacral division and the thoracolumbar division.

The thoracolumbar division, in most of its activities, prepares an animal for "fight or flight." The effects of stimulation of the thoracolumbar system include accelerating heart and respiratory rates, suppressing digestion, increasing blood flow to striated muscles, increasing blood pressure, and decreasing blood flow to the skin and smooth muscle. In general, stimulation of the craniosacral system produces effects which are opposite to those produced by the thoracolumbar division.

Although the neurons within the central nervous system look similar to those without, they are different in several respects. If a nerve fiber is injured within the central nervous system, the entire neuron degenerates. On the contrary, if a peripheral nerve is severed, the fibers will regrow out of the old nerve trunk from the central ends. Another difference is the sensitivity to oxygen. If a neuron within the central nervous system (of an adult) is deprived of oxygen for a short period of time, it will be irreversibly destroyed. In contrast, axons outside the central nervous system will continue to conduct impulses for more than an hour in the absence of oxygen. This is in part due to the difference between the sensitivity of the nerve cell bodies and the axons.

These differences hold for the larger axons, which are heavily myelinated outside the central nervous system, as well as for the smaller, less myelinated axons.

The neuron cell bodies are, in several ways, similar to the secretory cells of endocrine glands. (A general discussion of cellular structure and function can be found in Appendix E.) The dark-staining material within the nerve cell body, known as *Nissl substance*, has been shown to be chemically and morphologically identical (in electron micrographs) to that found in cellular organelles known as ribosomes and Golgi bodies (see Appendix E). The ribosomes are associated with protein synthesis, and the Golgi bodies with secretion. The nerve cell body appears to "secrete" the axon and dendrites. The metabolic rate and the protein synthesis in the nerve cell body are both greatest when the axon is being formed or replaced after injury and are at a minimum in normal adults. By contrast, the axon appears to lack ribosomes and the ability to synthesize proteins. One may consider the nerve cell body to be an intracellular secreting cell, whereas endocrine gland cells produce extracellular secretions.

Axons do not possess the organelles necessary for protein synthesis. They do contain mitochondria, which are intracellular organelles associated with metabolism. No qualitative differences are known between the metabolism of axons and that of other animal cells.

The axons are surrounded by myelin sheaths and satellite cells. It is not known if these are important in providing the proper chemical media for axon metabolism. However, it is quite well established that for all axons—vertebrate and invertebrate, "myelinated" and "unmyelinated," large and small—the satellite cells extend out and coil around the axon forming a double lipid layer known as *myelin*. In the larger vertebrate axons, outside the central nervous system, such as those diagrammed in Fig. 5.3, the satellite cells, called *Schwann cells*, have processes which are wrapped many times around the axons. In these, the myelin sheath is very easily visible.

Where the processes from two neighboring Schwann cells meet, the myelin layer is much thinner and is pierced by canals about 30 nm in diameter. This region is known as a node of Ranvier. The nonmyelinated fibers appear to have similar "canals" through their myelin sheaths all along the axon. Thus, their axons are in more-or-less continuous chemical contact with the intercellular fluids. In contrast, the large myelinated fibers appear to communicate chemically with the intercellular fluids only at the nodes. The possible electrical significance of the myelin sheaths is discussed in Sec. 5.4.

At the end of the neural fibers, the surface layer appears in the electron microscope to contain small vesicles. These are filled with the chemical carriers active in synaptic transmission. This idea is referred to again in Sec. 5.6. These tiny vesicles are so small that they cannot be conveniently observed except in the electron microscope.

As a vertebrate, man's greatest interest has been in vertebrate nerves. However, certain invertebrate axons are easier to work with because of their greater diameter. The maximum diameter of the vertebrate axon is about 20 μm. By contrast, squids have axons larger than 200 μm in diameter; these are visible to the unaided eye. Insects, too, sometimes have large axons; the cockroach has some as large as 50 μm. The extreme size of the squid axon has made it of special interest and importance in all studies of the conduction of impulses by nervous tissues. The synapses of the large invertebrate neurons have also proved convenient to study. The material in this chapter is based in part on studies of invertebrate axons.

5.4 The Spike Potential

Current technology permits placing the tip of a microelectrode within an axon and measuring the electrical potential relative to the medium surrounding the axon. This potential, in normal axons, ranges from -50 to -100 mV. It is comparable in size to the contact potentials and polarization potentials that can occur at electrodes. However, when these artifacts are reduced to the microvolt range, the true resting potential of the nerve axon remains. Diagrammatically, the axon may be represented by an insulator shaped as a cylindrical shell. The inner and outer faces of the shell are charged. The hollow shell is filled with one conducting medium (cytoplasm) and immersed in another (intercellular fluid). This picture applies to all axons except the thickly myelinated ones, which are discussed further later in this section.

The existence of these potentials across the extremely thin axon membrane indicates the ability of these membranes to withstand very high electrical field strengths (potential gradients) of about 10^8 V/m. Dry air breaks down at 3×10^6 V/m. Many insulators (including corrosion on spark plugs) raise the field strength necessary for breakdown of air as high as 10^8 V/m. The potentials at surfaces of cells are generated in such small regions that numbers for air prove misleading. The cell membranes are more nearly analogous to the junctions between two dissimilar metals. At the latter, field strengths as high as 5×10^9 V/m are known without sparking or breakdown of any sort. Thus, it is not too surprising that the neuron membrane can withstand field strengths at which dry air breaks down.

When the axon is stimulated, its surface potential changes in a characteristic fashion to an action potential or a spike potential. Axons may be stimulated by any of a wide variety of means. Electrical pulses of various shapes, heat, cold, chemical changes, and mechanical pressures all lead to the same phenomena. The local membrane polarization disappears, reverses in polarity very quickly, then returns to normal over a series of "bumps." The spike potential formed in this fashion travels down the axon in both directions from the point of stimulation. (Owing to the nature of the synapses, only one of these directions is usually effective when an intact nerve is stimulated.) The variations of the transmembrane potential at a fixed position along an axon are diagrammed in Fig. 5.4.

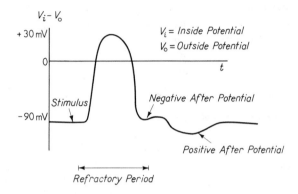

Figure 5.4 Diagrammatic representation of the time course of the spike potential at a fixed point along the axon. During the refractory period, another impulse cannot be started. The threshold for stimulation is lowered during the negative after potential and raised during the positive after potential. The magnitude and duration of these effects are characteristic of the particular nerve fiber.

In laboratory experiments, spike potentials are usually excited by electrical stimuli because they are easier to control in time, space, and strength than are any other type of stimuli. For very weak stimuli, a local response occurs which is similar to, but smaller than, the spike potential. As the stimulus is increased, a certain threshold is reached where a transmitted spike potential is generated. The spike potential then travels along the axon at a characteristic velocity.

The spike potential is an all-or-none response. Either there is a transmitted spike or there is not. If the spike is present, its height and shape are independent of the stimulus strength. The neuron acts in a similar manner to a bistable electronic circuit such as that used in counters and in digital computers. That is, the neuron is either in the conducting or nonconducting state; nothing is transmitted in between. This analogy seems so strong that it is hard to avoid describing the computer in anthropomorphic terms and the nervous system in terms of a digital computer.

If an axon is cut and the two pieces insulated electrically, no impulse travels from one part to the other. However, if the two are connected by a metallic conductor, or a salt bridge, the spike potential crosses readily from one part of the axon to the other. This emphasizes the essentially electrical nature of the action potential. These spike potentials occur in tissues, which are fluidlike media. Currents in fluids are carried by ions; therefore, it is appropriate to consider the resting potential as well as the spike potential as being due to ionic distributions.

While the spike potential is present at the axon, another one cannot be started. By contrast, several subthreshold stimuli may be summed to give a response if they come close enough together in time. During the positive after-potential, the threshold is increased. The lengths of time for these potentials and the rate of conduction of the spike potentials led to the classification of vertebrate axons presented in Table 5-1.

Particular attention should be called to the giant squid axon. From a comparative point of view, it is a huge axon. It is possible to shove all sorts of electrodes and shafts inside this axon. Experiments with squid axons confirm that the resting potential and the spike potential depend only on the membrane, not on the bulk of the axoplasm. Similar experiments have shown that this pattern is valid also for all other axons, for muscle fibers, and for many long algal cells (*Nitella*).

The local response shown in Fig. 5.5(a) has the same form as the spike potential shown in Fig. 5.4. A small depolarization applied externally results

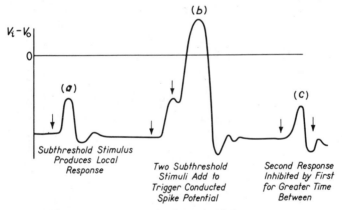

Figure 5.5   Temporal summation of subthreshold responses.

in a flow of ions, so that a greater depolarization of the membrane occurs. Figure 5.5(b) shows a second subthreshold stimulus following closely after a first one. These add and give rise to a conducted spike potential. Figure 5.5(c) illustrates two stimuli slightly farther apart in time. In this case, the first local response inhibits the second one.

The velocity at which a spike potential travels along a fiber is limited by the diameter of the fiber. Larger diameters correspond to larger velocities. In

TABLE 5-1
PROPERTIES OF AXON AND FIBERLIKE CELLS

Property	Units	Vertebrate Axon Type			Fiber or filament type		
		Myelinated			Skeletal	Squid	
		A	B	C	Muscle	Axon	Nitella
Diameter	μm	10–20	3	4	10	200	200
Conduction velocity	m/sec	50–160	3–15	0.7–2.3	3	20	0.01
Absolute refractory period	msec	0.4–1.0	0.12	2.0	5	1	1,000
− After potential							
Size	percent of spike	3–5	None	3–5	None[a]	None	?
Duration	msec	12–20	None	50–80	None[a]	None	?
+ After potential							
Size	percent of spike	0.2	1.5–4.0	1.5	10	5	?
Duration	msec	40–60	100–300	300–1,000	500	5–15	?
Resting potential	mV	90	90	90	90	70	100
Spike height	mV	140	140	140	140	90	100

[a] Strictly speaking, this is wrong; see Chapter 9.

the large myelinated vertebrate fibers, the spike travels at a rate in excess of that predicted from the axon diameter. An explanation for this is discussed at the end of Sec. 5.5.

The myelin sheaths of the nonmyelinated axons may act primarily as electrical insulation between fibers. This limits the probability that a spike potential along one axon will stimulate its neighbor. Muscle fibers have similar spike potentials but lack myelin insulation. In the muscle, unlike the nerve, it may be desirable for one fiber to stimulate other parallel ones, although this has never been demonstrated to occur.

5.5   Clamped Nerve Experiments

Detailed knowledge of the ionic changes that accompany nerve excitation and conduction resulted in a large measure from experiments originated by Hodgkin, Huxley, and Katz using the giant axons of squid and cuttlefish. In these studies a special electrode arrangement, illustrated in Fig. 5.6, makes it possible to measure separately the voltage across and current through a sizable segment of nerve membrane. An external current generator supplies current between electrodes *a* and *e*. The potential difference from *b* to *c* gives the membrane potential, whereas the potential difference from *c* to *d* is proportional to the membrane current. In this fashion one can measure the membrane potential and current. Either of these can be used to operate an electronic feedback system which controls the current generator in such a fashion as to clamp the current or voltage at a value preset by the operator. Thus, this experimental arrangement is called a *clamped nerve.*

The experimental arrangement can also be used to follow current and potential changes if a current pulse is supplied. The results of a series of experiments showed that when a negative current pulse was applied (i.e., a current pulse in the direction that would tend to increase the magnitude of the resting potential), the membrane potential rose rapidly and then fell slowly to its resting value. If the membrane behaved as a passive circuit element with voltage- and current-independent values of its impedance, the potential should have reached the resting value in about 8 μsec. Instead it took over 1,000 μsec.

Two quite different types of results, also unexplainable by membrane impedances independent of the membrane voltage, were obtained when positive current pulses were used. For small pulses, the curve shape was similar to that obtained with negative current pulses. For pulses above a sharp threshold around 18 pC/cm^2, a dramatically different type of response occurred which resembled a spike potential (except that a longer segment of membrane was involved). Thus, the membrane potential changed to a new equilibrium value determined by the membrane itself but at a rate related to the magnitude of the applied current pulse.

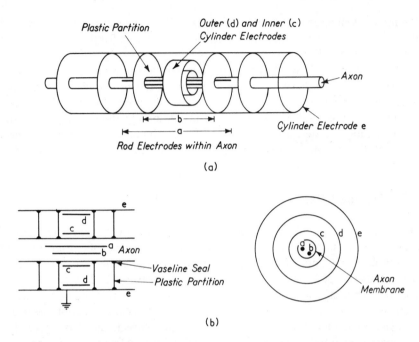

Figure 5.6 (a) Pictorial sketch of electrodes used in voltage clamp experiments. The hollow cylinder electrode *e* is filled with saline or other solution. Electrodes *a* and *b* are actually wires wrapped around a glass cylinder; the wires are insulated except in the region shown as electrodes in the sketch. (b) Schematic arrangement of electrodes used in voltage clamp experiments. The electrodes *a, b, c, d,* and *e* are all metallic. The axon is sealed to the plastic insulators with Vaseline. Current flows from electrode *a* to electrode *e*; potential and current measurements are made only between the two central insulators to eliminate end effects. The electrode and insulator sizes are not to scale. After A. L. Hodgkin, A. F. Huxley, and B. Katz, "Current–Voltage Relation in Nerve," *J. Physiol.* **116**:424 (1952).

To study the membrane behavior in more detail, the voltage across the membrane was changed rapidly and clamped to one of a series of new voltages while the necessary currents were monitored. Typical curves showing resultant currents when the membrane voltage was increased (A) and decreased (B) by 65 mV are shown in Fig. 5.7. The curves indicate that the steady-state currents necessary to maintain a decreased membrane potential (A) are much smaller than those for an increased membrane potential (B). Even more remarkable is that for a range of changes, the initial current change is in the opposite direction from that expected from any passive system.

By removing the Na^+ from the outside medium and replacing it with an organic ion unable to penetrate the membrane of the nerve axon, it was shown that the initial current is due to the flow of sodium ions. The concentration

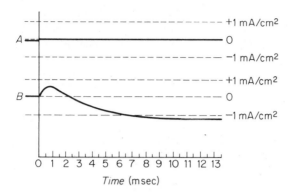

Figure 5.7 Records of membrane current under a voltage clamp. At zero time, the membrane potential was increased by 65 mV (record A) or decreased by 65 mV (record B); this level was then maintained constant throughout the record. The inward current is shown as an upward deflection. Temperature, 3.8°C. In a passive system, the current in record B should always have been outward (downward deflection). After A. L. Hodgkin, A. F. Huxley, and B. Katz, "Current–Voltage Relation in Nerve," *J. Physiol.* **116**:424 (1952).

of the latter is greater outside the membrane, leading to a transmembrane sodium electromotive force. The latter is an example of a more general phenomenon, the Donnan membrane potential, discussed in Chapter 24. Because the axon is relatively impermeable to Na^+, the sodium potential is quite different from the membrane resting potential; in fact, it has the opposite sign. At the start of the adjustments to a clamped decrease in the membrane potential, the sodium permeability increases rapidly, thereby admitting the current indicated by the initial hump in Fig. 5.7.

This may be expressed symbolically as follows, using V for the change from the resting transmembrane potential and V_{Na} for the sodium potential difference from the transmembrane potential:

$$J_{Na} = g_{Na}(V - V_{Na})$$

where J_{Na} is the current density due to sodium and g_{Na} is the areal conductance for sodium. The experiments in the preceding paragraph can then be summarized as indicating a time and voltage dependency for g_{Na}. Similar equations for potassium (subscript K) and all other ions (subscript L for leakage) can be also written as

$$J_K = g_K(V - V_K)$$
$$J_L = g_L(V - V_L)$$

The leakage conductance is assumed to be constant, but g_K varies with V and time. These equations may be summarized by the electrical model in Fig. 5.8, where E is used for the absolute potential difference across the membrane. Note that instead of g, the figure uses its reciprocal R, the areal resistance.

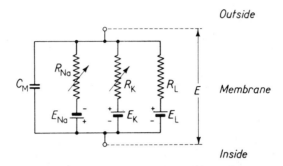

Figure 5.8 Equivalent membrane circuit. The symbol E is used for the absolute value of the potentials; V is used for the differences from the resting potential. The symbol R is an areal resistance in ohm·cm² and C_m is an areal capacitance in F/cm². After A. L. Hodgkin and A. F. Huxley, "A Quantitative Description of Membrane Current and Its Application to Conduction and Excitation in Nerve," *J. Physiol.* **117**:500 (1952).

The time dependance of g_K and g_{Na} can be represented by the following five equations:

$$g_K = \bar{g}_K n^4 \qquad\qquad 0 < n < 1 \qquad\qquad (5\text{-}1)$$

$$\frac{dn}{dt} = \alpha_n - (\alpha_n + \beta_n)n \qquad\qquad\qquad (5\text{-}2)$$

$$g_{Na} = \bar{g}_{Na} m^3 h \qquad\qquad 0 < m < 1 \qquad\qquad (5\text{-}3)$$

$$\frac{dm}{dt} = \alpha_m - (\alpha_m + \beta_m)m \qquad\qquad\qquad (5\text{-}4)$$

$$\frac{dh}{dt} = \alpha_h - (\alpha_h + \beta_h)h \qquad\qquad\qquad (5\text{-}5)$$

The functions n, m, and h are defined by these differential equations. These equations, with the data in Table 5-2, summarize books of data. Qualitatively, these equations predict three types of events which follow a decrease in membrane polarization. First, the membrane permeability to Na⁺ increases markedly, although after a finite time delay. Second, the permeability to K⁺ increases. Finally, the membrane permeability to Na⁺ decreases, but not to the original value.

TABLE 5-2

A. Functional form of constants in Hodgkin and Huxley's differential equations

$$\alpha_n = \frac{(0.01)(V + 10)}{e^{(V+10)/10} - 1}$$

$$\beta_n = 0.125e^{V/80}$$

$$\alpha_m = \frac{(0.1)(V + 25)}{e^{(V+25)/10} - 1} \qquad \alpha_h = 0.07e^{V/20}$$

$$\beta_m = 4e^{V/18} \qquad \beta_h = (e^{(V+30)/10} + 1)^{-1}$$

Units of V are millivolts.

B. These values are all at 6°C. The constants all increase about threefold for a 10° rise in temperature.
C. It is doubtful if the functional forms of α and β have any theoretical significance.
D. Alternative forms: If the membrane potential is changed from V_0 to V at $t = 0$, the equations (5-2), (5-4), and (5-5) have as solutions

$$n = n_\infty - (n_\infty - n_0)e^{-t/\tau_n}$$

$$m = m_\infty - (m_\infty - m_0)e^{-t/\tau_m}$$

$$h = h_\infty - (h_\infty - h_0)e^{-t/\tau_h}$$

In these, the constants are given by

$$n_\infty = \frac{\alpha_n}{\alpha_n + \beta_n} \qquad m_\infty = \frac{\alpha_m}{\alpha_m + \beta_m} \qquad h_\infty = \frac{\alpha_h}{\alpha_h + \beta_h}$$

$$\tau_n = (\alpha_n + \beta_n)^{-1} \qquad \tau_m = (\alpha_m + \beta_m)^{-1} \qquad \tau_h = (\alpha_h + \beta_h)^{-1}$$

Even if Eqs. (5-1) through (5-5) had no further utility, their success in data reduction would warrant their inclusion in this text. However, an extension of the reasoning above can be used to predict the behavior of the conducting axon. To demonstrate this, consider the currents and potentials illustrated in Fig. 5.9, where subscript 1 refers to outside the membrane and 2 inside. Then Ohm's law implies, using r for resistance per unit length, that

$$r_1 I_1 = -\frac{\partial V_1}{\partial x} \quad \text{and} \quad r_2 I_2 = -\frac{\partial V_2}{\partial x}$$

The internal and external currents can only be altered by the transmembrane current I per unit length; thus,

$$I = \frac{\partial I_1}{\partial x} = -\frac{\partial I_2}{\partial x}$$

Similarly, the transmembrane potential V is given by

$$V = V_2 - V_1$$

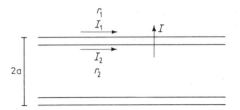

Figure 5.9 Current flow along an axon. Some of the symbols used in the text to develop Eq. (5-6) are illustrated.

Combining the five preceding relationships leads to

$$I = \frac{1}{r_1 + r_2} \frac{\partial^2 V}{\partial x^2}$$

Converting to current J per unit area of membrane, resistivity R', and noting that

$$r_1 \ll r_2$$

leads, using a for the axon radius, to

$$J = \frac{a}{2R_2'} \frac{\partial^2 V}{\partial x^2}$$

This, in turn, may be combined with earlier definitions and Eqs. (5-1) and (5-3) to give

$$\frac{a}{2R_2'} \frac{\partial^2 V}{\partial x^2} = C_m \frac{\partial V}{\partial t} + \bar{g}_K n^4 (V - V_K) + \bar{g}_{Na} m^3 h(V - V_{Na}) + g_L(V - V_L)$$

$$(5\text{-}6)$$

Now Eqs. (5-2), (5-4), (5-5), and (5-6) form a set of simultaneous equations to be solved for $V(x, t)$. However, an additional relation is needed. For this it is assumed that V obeys a wave equation, with wave velocity θ:

$$\frac{\partial^2 V}{\partial t^2} = \theta^2 \frac{\partial^2 V}{\partial x^2} \qquad (5\text{-}7)$$

Only for θ within a very narrow range will the entire set of equations yield finite solutions. Agreement between experimentally determined θ values and those estimated from the preceding equations tends to reinforce belief that even though the form of Eqs. (5-1) through (5-6) was arbitrary, they do represent the phenomena occurring at the squid axon membrane. This view is further augmented by good agreement between the shapes of computer-simulated transmitted action potentials and those observed directly.

The ionic changes accompanying the conduction of a spike potential are schematically shown in Fig. 5.10. This indicates that ionic movement is coupled with metabolism. The changes during the passage of a spike are divided for convenience into a regenerative phase and a recovery phase.

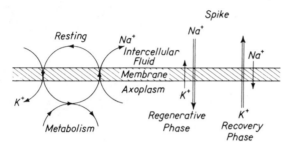

Figure 5.10 Ion movements across axon surface. After A. L. Hodgkin and R. D. Keynes, "Active Transport in Nerve," *J. Physiol.* **128**:28 (1955).

For myelinated nerves, the rate of conductance is faster than that suggested by the mechanism described above. This is explained by conduction, as in a passive cable between nodes of Ranvier. Such cable conduction is far faster but would result in attenuations of a factor of 2 in peak amplitude. Accordingly, the action potential must be regenerated at each successive node.

This general scheme of ionic and electrical changes in neuron conduction appears to be well verified. It leaves many questions unanswered, however, such as the underlying significance of Eqs. (5-2), (5-4), and (5-5). Also, explicit inclusion of Ca^{2+} and other ions seems needed for a more complete picture. Nonetheless, the clamped nerve experiments have revealed many details of the ionic changes accompanying nerve conduction of information.

5.6   Synaptic Conduction

Along the axon, the information is transmitted as an electrical spike potential. This transmitted spike is maintained at a constant height by renewal and amplification, either continuously or at certain nodes. There are thus two symbols for coding transmitted information: either a spike or its absence. In other words, all neural information is coded as binary digits (see Chapter 25).

The axon transmits equally well in either direction, but in the intact animal it is used in one direction only. This limitation is imposed by the synapses between neurons and by their junctions with the sensory receptors. Similar limitations exist for muscle fibers which conduct spike potentials in either direction along the fiber but are only stimulated in life at the junction between the nerve terminals and the muscle. At this point, the muscle fiber has a special structure called an *end plate*. The neuromuscular junction is homologous to the synapses between neurons; much of our knowledge of neural synaptic conduction is based on studies of the neuromuscular junction. Accordingly, synaptic conduction is interpreted in this section to include the transmission of spike potentials from nerve to muscle.

Two different modes of synaptic conduction occur: electrical and chemical. These are illustrated in Fig. 5.11. The electrical conduction may be very rare;

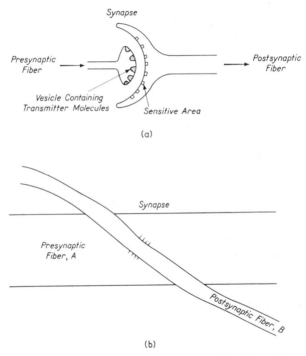

(a)

(b)

Figure 5.11   (a) Chemical transmission. The incoming spike releases packets of molecules, which diffuse across synapse to produce local excitation at sensitive areas. Diagrammatic representation. (b) Electrical transmission. With suitable geometry, a large field strength can be created in synapse by a spike potential on fiber A, thereby exciting B. The geometry and synaptic rectification prohibits conduction in the other direction. Diagrammatic representation.

it has been demonstrated positively for the giant synapse in the crayfish. At this synapse, impulses travel electrically from one axon to another with negligible time delay. Conduction can occur in one direction only. Similar giant synapses in the squid, however, exhibit appreciable time delay and no electrical transfer of charge. Conduction across the squid giant synapses, just as across all vertebrate synapses studied, is mediated by a specific chemical.

There is no reason to believe that the same chemical is involved at all the synapses lacking direct electrical transfer. On the contrary, there is considerable evidence to indicate that different substances act at different synapses. Most nerve fiber terminals are so small that it is impossible to make direct observations and hence to determine the transmitter substance. As a result of the small size, only very small amounts of the transmitting chemicals are necessary. At the neuromuscular junction in vertebrates about 10^{-18} mole of the transmitter acetylcholine (ACh) produces a spike potential. Within the central nervous system and within autonomic ganglia, considerable evidence indicates that gamma-amino butyric acid, as well as selected other catecholamines such as norepinephrine, act as synaptic transmitters.

In chemical transmission, the transmitter, such as ACh, is released when a spike potential reaches the appropriate nerve fiber terminal. It then diffuses across the synapse. This distance is of the order of 1–2 μm, and diffusion can occur in 1–2 msec. The diffusing substance is then absorbed at the receiving terminal or motor end plate, where it changes the ionic permeability of the membrane. Finally, the absorbed molecules are enzymatically destroyed.

Experiments with vertebrate motor end plates have shown that the area sensitive to ACh is extremely small; it is confined specifically to the outer surface of the motor end plate nearest to the nerve endings. This outer surface may be regarded as a chemoreceptor. Furthermore, these studies showed that ACh does not produce a depolarization of the membrane but increases its permeability to all small cations, such as Li^+, Na^+, and K^+. Finally, the ACh is destroyed by a specific protein catalyst, acetylcholinesterase, located in the end plate.

The response across the synapse is a local response. The spike potential may originate near there, as in the case of vertebrate muscle fibers and giant synapses in squid. In contrast, in a sensory neuron (whose axon runs toward the cell body) the spike is formed at the distal end of the axon, where several fiber terminals join together. In motor neurons, where the local response occurs in dendrites, the transmitted spike potential is formed at or past the nerve cell body.

One spike potential on a presynaptic fiber may excite one spike potential on the postsynaptic fiber. In some cases, a spike on any one of several different presynaptic fibers may excite a spike on a given postsynaptic fiber. At other synapses, one spike on a presynaptic fiber will produce only a local subthreshold response. In this case, there exists the possibility of adding subthreshold responses from several synapses to produce a spike potential. Thus, the neuron can act as an adder. Likewise, two, three, or more local responses in a short time at one synapse may be necessary to produce a transmitted impulse. Then the synapse is acting as a "divider." If several terminals from one neuron cell synapse with differing time delays at the same second neuron, the original spike could be "multiplied."

Similarly, neurons can subtract. This is possible because not all post-synaptic membranes are similar. For instance, ACh produces a spike potential at motor end plates but inhibits heart muscles. (This inhibitory effect of the ACh secreted by the vagus nerve endings in the heart led to its original discovery.) At inhibitory junctions, the transmitter substance increases the permeability to K^+ and larger cations but does not alter the Na^+ or Li^+ permeability. The net result is a change in the transmembrane potential and an increase in the local response necessary at other synapses to start a transmitted spike. This produces, effectively, subtraction of the impulses from two different incoming neurons.

At the synapses, then, the arithmetic processes of addition, subtraction, multiplication, and division can occur. Because the local responses exhibit a complex time pattern, the calculus operations of integration and differentiation can also be produced. However, the neurons are not as simple as electronic circuits, and the various numerical processes are also much more complicated. This situation may be described in mathematical terms by saying that the system is nonlinear. For example, a dividing synapse, if presented with three impulses, may transmit one; but 7 may be necessary for two transmitted spikes, and 14 or more may be needed for three transmitted spikes.

In addition, the synaptic conduction is altered by slow potential fluctuations which are small compared to the membrane potentials, and by changes in the ionic content of the intracellular fluid. Aside from the direct effects of K^+ and Na^+, the Ca^{2+} and Mg^{2+}, and particularly their ratio, alter the synaptic conduction. At the neuromuscular junction, it has been shown that ACh is released in packets of the order of 1,000 molecules from small vesicles in the nerve endings. The probability of a given packet entering the inter-cellular fluid is a function of the Ca^{2+}/Mg^{2+} ratio.

To summarize this section, transsynaptic conduction usually occurs in one direction. It may be mediated by electrical charge conduction or by special chemical transmitters. The latter alter the permeability of the surface of the second neuron (or the muscle fiber) at specific receptor spots. Depending on the receptor, and perhaps on the chemical nature of the transmitter, this may result in stimulation or inhibition. All manner of arithmetic and calculus operations can occur at synapses between neurons. The behavior is similar to that in digital computers but is far more complicated.

5.7 Summary

This chapter starts with an introduction to the nervous system and its fundamental structural unit, the neuron. Other introductory sections review electrical terminology and the anatomy and histology of the nervous system.

Emphasis is placed on the electrical potential across the membrane of the axon, which is an elongated process of the neuron. Nerves transmit information in the form of transient changes in the transmembrane electrical potentials of the axons. Physical techniques have been used to observe these characteristic changes, to measure their speed of transmission, and to relate them to changes in membrane permeability for Na^+ and K^+ ions. Conduction from one neuron to the next is usually accomplished by chemical transmitter substances that travel across the small intersynaptic gaps. Other biophysical aspects of neural action are discussed in the next three chapters.

REFERENCES

1. MAXIMOW, A. A., WILLIAM BLOOM, AND D. W. FAWCETT, *A Textbook of Histology*, 9th ed. (Philadelphia: W. B. Saunders Company, 1968), 858 pages.
2. RAY, C. D., ed., *Medical Engineering* (Chicago: Year Book Medical Publishers, 1974), 1256 pages.
3. BEST, C. H., AND N. B. TAYLOR, *The Physiological Basis of Medical Practice*, 9th ed. (Baltimore, Md.: The Williams & Wilkins Company, 1973).
4. GRUNDFEST, H., "Mechanism and Properties of Bioelectric Potentials," in *Modern Trends in Physiology and Biochemistry*, E. S. G. Barron, ed. (New York: Academic Press, Inc., 1952), pp. 193–228.
5. STEVENS, S. S., ed., *Handbook of Experimental Psychology* (New York: John Wiley & Sons, Inc., 1951).
   (a) BRINK, FRANK, JR., "Excitation and Conduction in the Neuron," pp. 50–93.
   (b) BRINK, FRANK, JR., "Synaptic Mechanisms," pp. 94–120.

CHAPTER 6

Electrical Potentials of the Brain

6.1 Electroencephalography

Electroencephalography is a study (or graphing) of the electrical potentials on the surface of the head. In terms of its derivation, electroencephalography (electro + encephalon + ography) could refer to any electrical potentials of the head. Actually, it is restricted to those potentials, other than neuron spikes, that are associated with the brain's action. At the outer surface of the scalp, these *electroencephalographic* (eeg)[1] *potentials* are small compared to the potentials due to the heartbeat and are comparable to the potentials associated with the motion of the muscles controlling the eye, jaw, neck, and so on. The small eeg potentials can be observed only with electronic amplifiers which discriminate both against other potentials of physiological origin and against electrical noise.

[1] Throughout this chapter, the abbreviation "eeg" will be used as an adjective or noun, as appropriate, to refer either to these potentials, to the recording apparatus, or to the graphic record of these potentials as a function of time. The eeg potentials arise from the action of nervous tissue. The student will find it profitable to have read Chapter 5 thoroughly before starting this one. A knowledge of that material is presupposed in this chapter.

The characteristic form of the eeg pattern has been used clinically and experimentally. Various types of epilepsy have typical eeg patterns which are useful for diagnosis and occasionally in treatment. Brain tumors may also be located from an eeg if the tumor is sufficiently close to the brain's surface. Many brain injuries can be diagnosed from alterations in the patterns of the potentials near the injury. Behavioral experiments use eeg patterns to indicate alarm reactions, sensory responses, and so forth. As eeg technology has advanced, new applications to research and to health care have appeared. These have included sleep monitoring, pharmacological testing, and the diagnosis of metabolic disorders.

From the viewpoint of this text, the more significant application of these "brain waves" is that they may indicate the operation of the central nervous system. Many theories have been proposed, based on the form of these brain potentials. To date, none of these theories has been altogether successful. Further studies of the eeg potentials may eventually lead to a better understanding of the function of the brain. Some progress has been made in this direction; thus, it is possible to associate particular eeg patterns with specific regions of the brain. In addition, evoked response averaging discussed in Section 6.5 has contributed to an understanding of sensory pathways through the nervous system. Evoked response averages also provide added data for the diagnostician.

The potentials on the scalp of the adult human may reach 200- or 300-μV peak-to-peak amplitude during deep sleep. Normal potentials observed in the awake eeg are an order of magnitude smaller but can still be observed and quantitated with suitable electronic equipment. In small laboratory animals, the eeg potentials on the surface of the scalp tend to be still smaller and more difficult to observe. In both humans and laboratory animals, electrodes on the surface of the brain or even within the brain reveal potentials severalfold larger than those on the scalp, but still possessing similar time patterns. Such potentials on or within the brain sometimes are also referred to as eeg potentials. The instrument used to record the potentials is called an *electroencephalograph* and the record is called an *electroencephalogram*. Where it is desired to distinguish the potentials measured on the surface of the brain, their study is called *cortiography*.

6.2   The Central Nervous System

The eeg potentials result from the action of the central nervous system (CNS). To aid in discussing these brain potentials, an outline of the anatomy of the CNS is given in this section. In Sect. 6.3, some of the actions of the CNS are interpreted by analogy with electronic feedback networks.

The CNS, as is the case with all other nervous tissue, is made up of neurons. Some carry information into the central nervous system; these are *sensory* or *afferent* neurons. Others carry spike potentials out of the central nervous system and are called *motor* or *efferent neurons.* The great majority of the units within the CNS start and end there; these are called *interneurons.* Thus, many neurons form links between other neurons. As was pointed out in Chapter 5, one neuron may receive impulses from several neurons, and it may excite or inhibit more than one other neuron. Each neuron follows an all-or-none law; that is, it either is or is not conducting a spike potential. This assemblage of neurons connecting with other neurons is very similar in form to a digital computer whose units are in one of two possible states.

In addition to the spike potentials, there are also more diffuse changes in electrical potential in various areas of the brain. These may also play an important role in the CNS function, for example, by altering the synaptic transmission from one neuron to the next. These diffuse, slower potential changes are similar to what one might expect to find in an analog computer. They indicate the difficulty of trying to use any electronic model for the central nervous system.

The vertebrate CNS is easily divided into two major parts: the brain and the spinal cord. Both are surrounded by three membranes, or *meninges,* which serve to protect the CNS from injury. Between the various meninges are layers of cerebrospinal fluid that cushion the CNS from shock. There are also fluid-filled chambers within the CNS itself: four ventricles in the brain and the central canal in the spinal cord. All four ventricles and the spinal canal are interconnected.

Various nerves leave (or enter) the CNS. Along the spinal cord, a pair of nerves passes between each pair of vertebrate. These supply sensory, motor, and autonomic fibers to all parts of the body other than the head. In addition, 12 pairs of nerves originate in the brain itself.

The spinal cord and brain consist of white matter and gray matter. The white color is due to the myelin around the large nerve fibers; the white matter is made up of fiber tracts. The gray matter contains most of the cell bodies. Some of these are arranged in compact volumes referred to as *nuclei.* Many nuclei can be associated with specific functions or actions, such as control of respiration, or conducting impulses from muscular proprioceptors, and so forth. However, the overall action of the nervous system, particularly with respect to subjective phenomena as thinking or memory, is still in the realm of speculation.

Figure 6.1 shows the structure of a medial section through the human brain. The portion of the brain joining the spinal column is called the *brain stem.* In lower vertebrates such as fishes, there are two small bumps called the *cerebral hemispheres* near the olfactory area. In mammals, and to the greatest extent in human beings, these cerebral hemispheres are a major part of the brain. The cerebral cortex which covers the hemispheres is so folded around

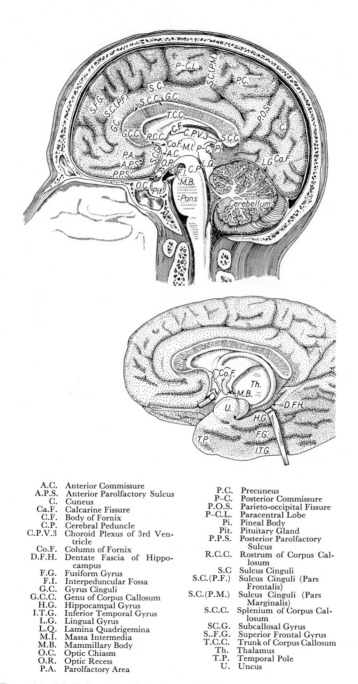

A.C.	Anterior Commissure
A.P.S.	Anterior Parolfactory Sulcus
C.	Cuneus
Ca.F.	Calcarine Fissure
C.F.	Body of Fornix
C.P.	Cerebral Peduncle
C.P.V.3	Choroid Plexus of 3rd Ventricle
Co.F.	Column of Fornix
D.F.H.	Dentate Fascia of Hippocampus
F.G.	Fusiform Gyrus
F.I.	Interpeduncular Fossa
G.C.	Gyrus Cinguli
G.C.C.	Genu of Corpus Callosum
H.G.	Hippocampal Gyrus
I.T.G.	Inferior Temporal Gyrus
L.G.	Lingual Gyrus
L.Q.	Lamina Quadrigemina
M.I.	Massa Intermedia
M.B.	Mammillary Body
O.C.	Optic Chiasm
O.R.	Optic Recess
P.A.	Parolfactory Area

P.C.	Precuneus
P–C.	Posterior Commissure
P.O.S.	Parieto-occipital Fissure
P–C.L.	Paracentral Lobe
Pi.	Pineal Body
Pit.	Pituitary Gland
P.P.S.	Posterior Parolfactory Sulcus
R.C.C.	Rostrum of Corpus Callosum
S.C	Sulcus Cinguli
S.C.(P.F.)	Sulcus Cinguli (Pars Frontalis)
S.C.(P.M.)	Sulcus Cinguli (Pars Marginalis)
S.C.C.	Splenium of Corpus Callosum
SC.G.	Subcallosal Gyrus
S..F.G.	Superior Frontal Gyrus
T.C.C.	Trunk of Corpus Callosum
Th.	Thalamus
T.P.	Temporal Pole
U.	Uncus

Figure 6.1 Medial aspects of the human brain. Copyright *The CIBA Collection of Medical Illustrations*, Vol. 1, *The Nervous System*, by Frank H. Netter, M.D. (Summit, N.J.: CIBA Pharmaceutical Products, Inc. 1953).

and over the brain stem in mammals that the eeg potentials on the skull are related to the cerebral cortex only, and probably only to the outermost layers of the cerebral cortex. (By placing electrodes within the brain, eeg potentials can be measured as a function of the part of the brain nearest the electrodes, rather than of the outer layers of the cortex.)

The portion of the brain stem connected directly to the cerebral cortex is called the *thalamus*. The sensory pathways all have synapses in the thalamus. Certain thalamic regions are believed associated with emotional responses. For example, if an electrode is placed in the appropriate spot in a rat's thalamus, it will pull a lever to shock itself in preference to eating food. Other areas in the thalamus produce other effects when stimulated. It appears proper to consider all mammals, and possibly all vertebrates, as having emotions homologous to ours and represented by thalamic centers.

Thought, memory, conscious sensations, and conscious motor activity are all associated with the cerebral cortex. In the relative size and complexity of the cerebral cortex, humans are unique among the animals. As illustrated in Fig. 6.2, certain areas can be associated with specific functions. However, the role of many areas of the cerebral cortex is not known, nor is it known in detail

Figure 6.2 Functions of the human cerebral cortex. Copyright *The CIBA Collection of Medical Illustrations*, Vol. 1, *The Nervous System*, by Frank H. Netter, M.D. (Summit, N.J.: CIBA Pharmaceutical Products, Inc. 1953).

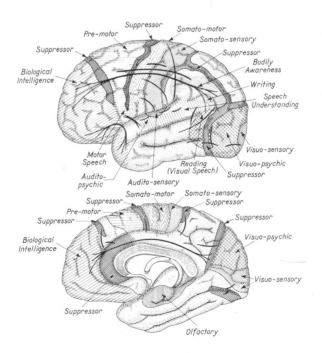

how we analyze, or think, or remember. Because it reflects, in some sense, the activity of this part of the brain, the eeg has attracted the interest of many investigators.

However, if the cerebral cortex is removed, somewhat similar eeg patterns remain. Even fishes, whose cerebral cortices are negligible, possess typical eeg patterns that grossly resemble ours. The eeg is a vertebrate phenomenon; insect ganglia do not exhibit comparable potentials. The eeg must, in some way, be related to the structure and function of the vertebrate central nervous system. It is necessary that electrodes used for observing the eeg potentials be large enough to respond to an average of many neurons, particularly if the electrodes are used on the cortex or within the central nervous system. The individual neurons produce local and spike potentials whose nature is discussed in Chapter 5. The eeg potentials are slower, more complicated, and associated with larger regions of the brain than a single neuron.

6.3 Origins of the Electroencephalographic Potentials

In the earlier studies of eeg potentials, time, effort, and talent were expended in trying to find simple mechanisms or isolated regions of the central nervous system which served as generators or scanners or oscillators giving rise to the observed patterns. Although these studies have become less popular, owing in large measure to a realization of the complexities of the origins of the eeg, nonetheless the general belief remains that some sort of feedback loops are involved. While their exact nature remains unclear, it is well demonstrated that neuronal networks do play an important role in the overall action of the nervous system. To clarify this point, brief discussions of feedback and control systems are included in the following paragraphs.

The basic elements of a feedback loop are diagrammed in general and as two specific examples in Fig. 6.3. The loop may be specified in terms of the quantity being controlled (e.g., room temperature), a sensing device (e.g., thermostat), and a device whose rate can be regulated (e.g., furnace). If, as in the room-temperature example, the control opposes changes, it is said to be negative feedback. Such control systems find numerous engineering applications, one of which is shown in Fig. 6.3(b). More complicated control circuits may have many interacting loops.

Positive feedback causes the controlling circuit to augment any small changes. This results frequently in oscillations, although both positive and negative feedback may be combined for greater stability. By and large, the amplitude of the oscillations will be controlled by nonlinear properties of the control system while the frequency may be characteristic of the feedback circuit elements. Such oscillatory behavior can be demonstrated in chemical

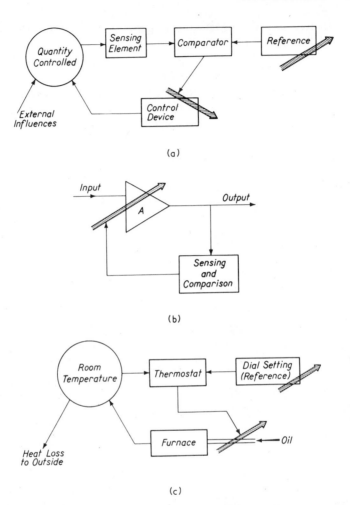

Figure 6.3 (a) General scheme of negative feedback control loop (servo-mechanism). Large crosshatched arrows indicate variable elements. (b) Block diagram of negative feedback loop to keep amplifier output constant. Such amplifiers can be used to present subjects with speech of constant (average) sound pressure level. (c) Negative feedback control loop used to regulate room temperature. In practice, an additional loop is used to keep the furnace below some set temperature.

systems and suggests that positive feedback could give rise to eeg patterns. To account for the complexity of the eeg, it is necessary to assume that numerous loops are involved. It is also highly suggestive that several well-identified neurophysiological control loops involve many neurons and interlocking loops (or networks).

For example, the feedback control of the iris diaphragm of the eye (see Chapter 2) is sketched in Fig. 6.4. Light enters the eye through the iris diaphragm and is encoded as neural impulses in the retina. In some fashion the retina or the brain (or some combination of these two) computes both the total light flux reaching the retina and the maximum light flux per receptor area. These are then compared to certain built-in values (sometimes called *set points*). Following this comparison, either the craniosacral or thoracolumbar autonomic nervous system is activated in such a fashion as to decrease the differences from the set points.

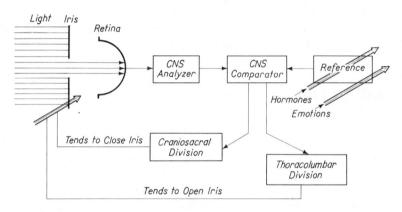

Figure 6.4   Feedback control loop regulating iris opening.

The latter reference points can be varied by a variety of internal factors such as the emotional state of the subject or the hormonal composition of the blood. The control system of the iris has been subject to a variety of physical studies which show its essential nonlinear characteristics, sometimes called *rein control*. This term is used because the quantitative behavior of the two sets of nerves controlling iris opening and closing differ in a manner analogous to using reins to control a horse's speed. The iris will also oscillate if the system is driven at a frequency higher than those usually encountered. In this regard it is similar to many physical control systems, the negative feedback becoming positive away from the normal frequency range. Such an effect might also contribute to the generation within the central nervous system of the eeg potentials.

Other feedback control systems which have received much attention include conscious muscular control, balance, and reflex reactions. In all these cases, it is possible to study the overall control system and demonstrate that feedback does indeed occur. It is also possible to trace, at least to a limited degree, the neurons involved in the individual pathways. Such studies, in addition to confirming the existence of feedback loops and neural networks, also demonstrate that the action potentials in the individual neurons do not have a form that is simply related to the eeg, but rather consist of spike potentials of the nature discussed in Chapter 5. Moreover, grosser electrodes placed on an entire nerve or fiber tract show a summation of the individual neuronal activities but still do not look like eeg potentials. Thus, eeg potentials represent the electrical activity of still larger regions of brain tissue.

Specific regions of the human brain apparently involved with various eeg patterns are described in the following section. Suffice it to note here that the detailed mechanisms of the origin of the eeg patterns remain unknown in spite of dedicated work spread out over decades. Yet, in many ways, this was and is the most important question that a biophysicist could raise about the brain-related potentials measured on the human scalp. Analogies between brain and digital computer have proved to be of little value, and most studies of the eeg have turned to more immediate questions which can be answered with current technology and concepts and which can be applied directly to the delivery of health care.

6.4   The Electroencephalographic Patterns

The eeg patterns are always present on the scalp of living individuals. Their absence is used as a legal indicator of death. The electrocortiographic potentials tend to spread over the surface of the cortex. Thus, there will be eeg patterns changing in time on any given electrode and also spatial patterns over the scalp at any given time. Various attempts have been made at summarizing this information in a format that is easy to use. The most frequent approach is to use a large number of electrodes on the surface of the skull and record 8 or 16 or even 24 simultaneous records. Many of these are measured relative to an indifferent electrode such as the ear. To detect some abnormalities, it is convenient to record differences between pairs of electrodes on the scalp. Figure 6.5 shows one arrangement of electrodes that could be used.

Not all the electrodes need be recorded at once. Examples of resulting records are shown in Figs. 6.6 and 6.7. The potentials in sleep may rise as high as 200 or 300 mV, although the normal awake values are 20 mV or less. (Values on the surface of the cortex are three- or fourfold greater.) Frequency analyses show little energy above 100 Hz; most of the useful information is within a frequency band of 0–35 Hz. On visually scanning a record such as that shown in Fig. 6.6, it is apparent that there are pronounced frequencies present. However, frequency analyses hide some of the information in a record of this type, since one cannot distinguish from the spectrum between a continuous wave of unit amplitude and one of twice the amplitude which is present only half the time.

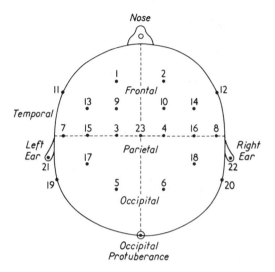

Figure 6.5 Electrode arrangement on the scalp is indicated by dots. The differences between pairs of these electrodes are recorded to obtain graphic records such as those shown in Figs. 6.6 and 6.7.

Figure 6.6 Normal eeg patterns illustrating abolition of the α rhythm with eyes open. If eyes remained open for longer period of time, the α rhythm would build up again. Original figure of R. G. Bickford, M.D.

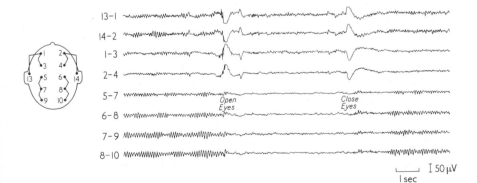

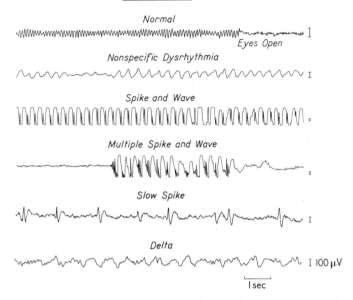

E.E.G. Patterns

Normal

Eyes Open

Nonspecific Dysrhythmia

Spike and Wave

Multiple Spike and Wave

Slow Spike

Delta

I 100 μV

I sec

Figure 6.7 The various abnormal eeg patterns each have specific clinical implications. For example, the "spike and wave" is characteristic of petit mal seizures. Original figure of R. G. Bickford, M.D.

Nonetheless, frequency analyses are used to characterize several of the types of patterns. These spectra change in time, so a presentation is desired which shows, in a closed format, the spectrum for a short interval repeated at appropriate longer intervals. One presentation that meets this goal is the *compressed spectral analysis* developed by Bickford and his colleagues. Alternative presentations divide the eeg spectrum into four frequency regions and use the dominant frequency and the amplitude in each of these.

For historic reasons, the different eeg frequency ranges are assigned the lowercase Greek letters delta, δ (0.5–3 Hz); theta, θ (4–7 Hz); alpha, α (8–13 Hz); and beta, β (above 14 Hz). Some eeg-ers restrict β to 14–30 Hz and designate 30–50 Hz by gamma, γ. Usually, the dominant frequency is called a wave (such as α wave or δ wave). The exact frequency for the cutoff of each of these ranges is arbitrary and varies from one laboratory to the next.

The α rhythm is the most pronounced one in the awake individual with eyes closed. It is believed that the α rhythm is associated with some underlying pacemaker in the thalamus (see Fig. 6.1). Because of this, apprehension and unexpected stimuli can block α rhythms. The usual α wave is quite sinusoidal. However, under a variety of conditions in selected individuals other shapes, called lambda, λ, and mu, μ, are seen. These are associated with some types of stimulation and with visual activity, although their absence in many individuals is hard to rationalize. The α rhythm, as other eeg patterns, varies with age, state of awakeness, and various types of abnormalities.

Both β and γ waves are usually present in adults. There is less energy in these frequency regions, and so these are harder to study and there is less information concerning them. By contrast, δ waves may exceed 200 mV in deep sleep. They are also indicative of cortical abnormalities such as unconsciousness, anesthesia, blows to the head, and tumors.

Alpha rhythms tend to be most prominent in occipital regions. By way of contrast, θ waves are more pronounced in the frontal and temporal regions of the skull. They are thought to be somehow associated with the hippocampal and limbic systems (see Fig. 6.7). Theta waves make up the sleep spindles (see Sec. 6.5) and are also associated with certain abnormalities.

The utility of the eeg in clinical diagnosis has expanded gradually as more and more experience has been amassed. It is not intended to pursue this type of application in this text. Nonetheless, the most dramatic abnormality, the high-voltage spike, should be mentioned. Spiky records are associated over 90 percent of the time with epilepsy; over 90 percent of persons suffering from epilepsy show records with characteristic spikes (even though they may be receiving chemotherapy which completely suppresses all their seizures). Various spike patterns are shown in Fig. 6.7.

To summarize briefly, various eeg patterns have been classified most often in terms of their frequency spectra. This has contributed to delivery of health care. However, these classifications are all ad hoc and there is only minimal understanding of the origin or biophysical basis for the observed eeg patterns. Automated computational techniques have been used to partially process and display the data. These have helped human users but none have met with general widespread acceptance.

6.5 Sleep Stages

Electroencephalographic studies have contributed markedly to the classification of the various stages of sleep. The nature and characteristics of sleep and its relationship to the functions of animals in general and humans in particular have been a topic for debate and study for many years. This lack of understanding still remains the case today, although numerous investigators have measured metabolic changes, heart and respiratory rates, and endocrine levels during sleep. The most fruitful of these studies have used classification schemes based on the eeg and the electrooculogram (eog). Sleep-associated eeg patterns, as well as others, suggest that the electrical measurements do indicate the fundamental, underlying processes.

In general, for sleep staging it is necessary to record both the eeg and the eog. It is also customary in many laboratories to measure the electromyogram (emg; see Chapter 9) from some head area such as under the chin (sublingual). Using these, it is possible to set up rules to classify the stage of sleep. Although there is not complete agreement on the details, all investigators recognize five stages in which dreaming is not believed to occur (0–4), and one in which dreaming always occurs, called the *rapid eye movement* (rem) *stage*.

Stage 0 represents an awake state in which the subject is becoming drowsy. During this stage the α waves build in amplitude, gradually increasing in dominant frequency. Then the α waves tend to disappear, leaving a relatively flat appearance called stage 1. As in all sleep staging, the investigator looks at epochs of the order of 1 min and classifies the entire epoch, trying to avoid or omit transient changes in the sleep stage assigned. During stage 1, the individual is not soundly asleep and can be easily awakened.

During stage 1, the rem periods also occur. In these the eeg is depressed, almost completely flat, and the eog becomes exaggerated as a result of the rapid eye movements. Major cardiovascular and respiratory changes, as well as emg activity, also occur. An individual awakened during a rem period can always remember a dream, whereas during any other sleep stage this is not possible. The result is interpreted as rem being associated with dreaming, although a more accurate statement would be that being awakened during a rem episode is associated with dreaming.

As individuals pass into deeper sleep they go into stage 2, which is characterized by the appearance of spindles of θ waves. During initial sleep the subject then passes rapidly into stages 3 and 4. These later stages are recognized by the appearance of large δ waves, 75 μV or more in amplitude. Stages 3 and 4 are differentiated by the percentage of the time that large delta waves are present. During a night's sleep the stages change from time to time, stage 4 usually occurring only during the first hour or two.

As persons age, the amount of stage 3 and stage 4 sleep decreases. Also, the rem periods decrease somewhat, being greatest in infancy. If an individual is awakened every time that a rem pattern appears, the time spent in rem episodes can be decreased. Such subjects report various discomforts and when allowed to return to undisturbed sleep exhibit increased rem periods. Thus, it appears that rem sleep (dreaming?) is necessary for humans (and probably for other mammals). The physiological, psychological, or teleogical significance of rem sleep is unknown.

The percentages of time spent in each stage of sleep, the frequency of changing stages, and the general pattern as illustrated in Fig. 6.8 are all of clinical significance. Their variations give objective measures of insomnia and

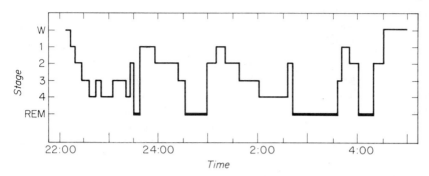

Figure 6.8   Sleep stages for normal individual. (Original material courtesy of Dr. G. E. Chatrian, University of Washington Hospital, Seattle.)

narcolepsy, and of the action of pharmaceuticals. This essentially biophysical classification of sleep leads to speculation as to whether added biophysical techniques could determine more about the nature of sleep and its relationship to human life.

6.6   Evoked Response Averaging

Another use of eeg potentials has been to monitor transient responses to sensory stimuli. By and large, these transients are so small that they can be barely distinguished in the single eeg tracing. However, the transient signal should be time-locked to the stimulus, while the background eeg is uncorrelated with the stimulus provided that the latter is not repeated at completely regular intervals. Thus, if one adds repetitively the eeg potentials, starting at the stimulus, the transient should be emphasized and the background deemphasized.

In particular, if one assumes that the background eeg potential, V_B, at any instant after the stimulus will appear as a random Gaussian signal, it can be shown that the sum is expected to increase as the square root of the number of stimuli. Expressed analytically, using $E(V_B)$ to symbolize the expected value of V_B, this states that

$$E(V_B) = k \sqrt{N}$$

where V_B is the background voltage sum, N is the number of stimuli, and k is a constant (the rms voltage in the background signal). Similarly, for the transient one may write, since all the transients will be time-locked,

$$E(V_T) = NT(t)$$

where V_T is the transient voltage sum, and $T(t)$ is the transient value at t time units after the stimulus. Thus, as the number of repetitions N increases, the ratio of signal to noise would rise as $\sqrt{N}$. If one divides the sums by N, one obtains averages rather than sums.

A record of this type, based on visual stimuli, is shown in Fig. 6.9. One might think that increasing N indefinitely would give progressively better results. However, this is not truly the case. For one thing, the subject both accommodates to the stimulus and also tires. For another, each transient need not be the same as the previous one, so the variation in the transients may limit the average. Finally, nonlinearities in the overall system may limit the growth of the signal-to-noise ratio.

Figure 6.9 Visually evoked response average. The subject was exposed to three sets of flashes; each set was 40 sec long and consisted of 10-msec flashes at a rate of 1 Hz. The zero on the central axis indicates the light flash. The vertical lines along the horizontal axis mark 100-msec intervals, which is also shown by the horizontal calibration line. The vertical calibration line indicates 10 μV. The numbered points on the central axis indicate significant peaks.

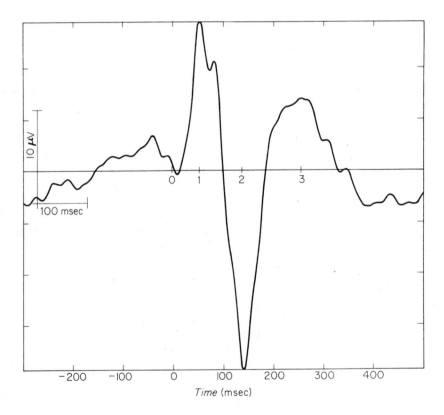

Evoked response averaging has been useful in clinical testing. It also has demonstrated that such diverse stimuli as light, sound, and pain all produce changes within the central nervous system which are reflected in the eeg. The latter observation has had numerous applications in neurology, physiology, and psychology. From the point of view of the biophysical scientist, evoked response averaging has provided convincing evidence that the eeg potentials are closely related to the overall state of activity of the central nervous system.

6.7 Summary

The eeg patterns are tantalizing in that they seem to be intimately associated with the overall action of the brain. They are useful for clinical purposes, just as patient's temperature may be of interest to a physician with no knowledge of temperature-control mechanisms on the cellular level. The fundamental question of interest to the biophysicist is: In what way are the eeg patterns related to the actions of the neurons of the brain? This question has not been answered.

It may be that extending measurements to lower frequencies, small regions of the brain, and so forth, may provide more clues. It seems more probable that what is needed are new ideas concerning the interpretation of the data and the planning of additional experiments.

REFERENCES

The form and action of the central nervous system are described in many texts. The following were used in writing this chapter.

1. BEST, C. H., AND N. B. TAYLOR, *The Physiological Basis of Medical Practice*, 9th ed. (Baltimore, Md.: The Williams & Wilkins Company, 1973).

2. RANSOM, S. W., AND S. L. CLARK, *The Anatomy of the Nervous System: Its Development and Function*, 10th ed. (Philadelphia: W. B. Saunders Company, 1959), 622 pages.

3. NETTER, F. H., *CIBA Collection of Medical Illustrations*, Vol. 1, *Nervous System* (Summit, N. J.: CIBA Pharmaceutical Products, Inc., 1953).

4. THOMPSON, R. F., AND M. M. PATTERSON, eds., *Methods in Physiological Psychology*, Vol. 1, *Bioelectric Recording Techniques*, Part B, "Electroencephalography and Human Brain Potentials" (New York: Academic Press, Inc., 1974), 327 pages.

5. KILOH, L. G., A. J. McCOMAS, AND J. W. OSSELTON, *Clinical Electroencephalography*, 3rd ed. (London: Butterworth & Co. Ltd., 1972), 239 pages.

6. BICKFORD, REGINALD, "Quantitative Techniques in EEG," in *Medical Engineering*, Charles D. Ray, ed. (Chicago: Year Book Medical Publishers, 1974), pp. 385–405.

7. KELLAWAY, PETER, AND INGEMAR PETERSÉN, eds., *Automation of Clinical Electroencephalography: A Conference* (New York: Raven Press, 1973), 318 pages.

8. KLEITMAN, NATHANIEL, *Sleep and Wakefulness* (Chicago: The University of Chicago Press, 1963), 552 pages.

9. RECHTSCHAFFEN, A., AND A. KALES, eds., *A Manual of Standardized Terminology, Techniques and Scoring System for Sleep Stages of Human Subjects* (Washington, D.C.: Government Printing Office, 1968), 57 pages.

10. REGAN, DAVID, *Evoked Potentials in Psychology, Sensory Physiology and Clinical Medicine* (London: Chapman Hall Ltd., 1972), 328 pages.

11. PERRY, N. W., AND D. G. CHILDERS, *The Human Visual Evoked Response: Method and Theory* (Springfield, Ill.: Charles C Thomas, Publisher, 1969), 187 pages.

CHAPTER 7

Neural Mechanisms of Hearing

7.1   The Types of Theories

Hearing may be approached from various viewpoints. Some of these have been so completely studied it is unlikely that our concepts will have changed appreciably 50 years from now. These aspects of hearing were presented in Chapter 1. They included the nature of sound transmission through the atmosphere, the gross anatomy and histology of the ear, and the role of the outer and middle portions of the ear as pressure amplifiers and mechanical transformers.

Other aspects of hearing are far less well understood. Specifically, the conversion of acoustic energy to neural spikes in the inner ear and the analysis of these spikes in the central nervous system continue to be areas of research. They are discussed in the present chapter. It is assumed that the reader is familiar with the material in Chapter 1 on hearing as well as in Chapters 5 and 6 on nerve conduction and on the electrical potentials of the brain.

The physicist regards the inner ear as a *transducer*, that is, as a device that converts one form of energy into another. The inner ear converts mechanical energy into electrical spikes on nerve fibers. It was only in the 1940's that a reasonable understanding of this action was developed. At present, two different types of hearing theories are available; for convenience, these can be labeled *spatial* and *temporal*. A purely spatial or place theory attributes major importance to the correspondence between a particular point on the basilar membrane and one particular audible frequency. Time or, equivalently, frequency theories explain hearing by means of an analysis dependent upon the time separation of the periodic maxima of an acoustic wave. In the older literature, these are often referred to as *telephone theories*. Neither purely spatial nor purely temporal theories will explain the results of all experiments. Thus, it is unlikely that mammalian hearing is exclusively of either type; rather both spatial and temporal information appear important. In the following sections, important examples of each type of theory will be presented, together with some of the pertinent experimental data.

7.2   Spatial Theories

A. HELMHOLTZ RESONANCES

One of the earlier and certainly one of the more complete theories of hearing ever developed was due to Helmholtz (1863). He attributed hearing to a hypothetical resonant behavior of the basilar membrane similar to that exhibited by the strings of a piano. If one considers a wire of linear mass density ρ and length L being subjected to a tensile force T, the fundamental resonant frequency is given by

$$\nu = \frac{1}{2L}\left(\frac{T}{\rho}\right)^{1/2} \tag{7-1}$$

Thus, light, short wires having high tension resonate at a relatively high frequency compared with massive, long wires. The basilar membrane itself gradually becomes broader and more massive with distance in going from the oval window (base) to the helicotrema (apex). In the Helmholtz theory, each transverse resonant fiber was thought to be directly coupled to a specific nerve leading to the brain. Thus, pitch would be detected by the particular fiber most strongly activated, loudness (or sound pressure level) by the amplitude of the fiber motion, and quality by the relative amplitudes of various fibers.

The completeness of the Helmholtz theory was an illusion, even from its inception. For example, a sharp resonance would be both difficult to excite and quite persistent. The problem of persistence of the vibration is associated with the lack of damping; this implies that the pertinent membrane fiber might continue to "ring" for such extended intervals that the hearing system would be ineffective. The model can be changed to avoid this problem by introducing a damping mechanism after a certain prescribed time. This action would be analogous to the tympanist of an orchestra, who manually holds the drumheads to prevent undesired reverberations. In Helmholtz's wirelike membrane model, a change in tension or length would certainly suffice. No such mechanisms have been observed in the extensive physiological investigations of the inner ear. In fact, excision of intact basilar membranes has revealed no explicit transverse fibers of the type envisioned by Helmholtz.

Eventually, Békésy was able to quantitatively evaluate the intrinsic tension across the basilar membrane by inserting thin glass tubes through the abraded wall of a cadaver cochlea. The bending of the glass and the shape of the openings cut into the basilar membrane were indicative of the amount of restoring force supplied by the structure. The results were consistent with there being no transverse tension; the membrane is now considered to be a flaccid object fixed along either edge. Thus, Eq. (7-1) is not applicable to the action of the ear.

The significance of any model, however, is not necessarily equal to its ultimate scientific truth. Rather, the model should lead to direct experimental and theoretical investigations to test its applicability in the physical world. Various investigations of Helmholtz's resonance theory have led to valuable physical and physiological research. In the following, the actual vibrational motions of the basilar membrane and the cochlear fluid are considered in more detail. They have proved to be far more complicated than the Helmholtz picture.

B. BÉKÉSY WAVES

Békésy, like Helmholtz, attached great significance to the role of the basilar membrane and its surrounding cochlear fluid. He showed that, in addition to the compressional waves traditionally treated in acoustics, in a structure such as the cochlea, there could also exist slow hydrodynamic waves. These waves bear certain similarities to surface waves on a large body of water or interfacial tension waves at an oil–water interface. All such hydrodynamic waves occur in dispersive media, that is, ones in which the wave velocity is a function of the frequency. In such media the waves may "pile up" to form maxima in certain regions. This phenomenon can be observed in the buildup and decrease of surface waves on the ocean (see Chapter 26). The various wave analyses of the cochlea, using this type of model, are beyond the scope of this text but should be studied by readers with sufficient mathematical preparation. The reference by Dallos contains a good exposition of the technique.

Békésy demonstrated that these hydrodynamic waves existed not only in the mammalian cochlea but also in physical models. For the simplest model, he used two rigid walls (microscope slides) resting on a solid surface and covered with a dental dam (rubber sheet). A slightly more refined system is shown in Fig. 7.1, where two channels and two windows are included. At low frequencies, the actual motion of the dental dam can be observed. There is a characteristic region of maximum displacement for each frequency; this position is more or less independent of the shape of the channels and varies only slightly for major changes of the thickness or tension of the elastic membrane or of the dimensions of the channel. It is important, however, that one window be driven and the other free. Békésy used large acoustic amplitudes to show that higher-frequency pure tones produced membrane maxima closer to the oval window.

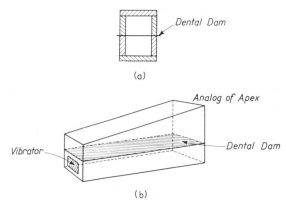

Figure 7.1 Békésy's hydrodynamic analog of the cochlea. (a) Transverse cross section. (b) Perspective view. The two end windows and the partition are of dental dam; this is essentially a rubberlike material in sheet form.

The variation of amplitude and membrane position of several representative sinusoidal signals is contained in Fig. 7.2. For a given sound pressure level, the displacement decreases as the frequency rises. These measurements, when extrapolated to the limit of audibility at 10^3 Hz, implied that maximal displacements of the basilar membrane can be less than 10^{-14} m, a dimension comparable in size to atomic nuclei. (Also see Chapter 1 for a discussion of the displacement of the tympanum.) Because of the necessarily high sound levels used by Békésy in his optical observations, it may not be possible to linearly extrapolate his results to lower intensities. The various theoretical analyses and model experiments agree that all that is essential for the maxima of hydrodynamic waves spatially separated according to frequency are rigid walls, two parallel tubes separated by an elastic membrane, and two windows, one driven and the other "open" to the air of the middle ear. The maxima of these waves give only a crude place localization of different tones. They occut, however, in very similar places in the physical model and in the excised

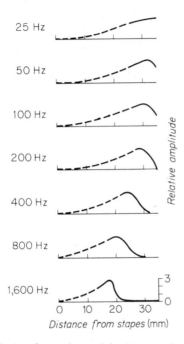

Figure 7.2 Amplitude of the observed basilar membrane displacements. A loudspeaker source is driven at various sinusoidal frequencies for these data. After G. von Békésy, *J. Acoust. Soc. Am.* **21**:245 (1949).

cochlea. Audiometrists have often reported lesions along the cochlea of accidently or experimentally deafened mammals. These could be explained by a persistent hydrodynamic wave of very large amplitude, causing localized destruction of the basilar membrane and its lateral support.

Pitch discrimination remains a problem with the waves as seen by Békésy, since their maxima appear far too broad to allow the spectral resolutions demonstrated by humans, which approach 1 Hz at 10^3 Hz. One could imagine that the large acoustic inputs required for the visual recording of membrane oscillations might cause distortions of the actual motion involved. One way to measure movement at physiological sound levels involves Mössbauer spectroscopy, described in Chapter 27. For these studies gamma-ray emitters are coupled to the basilar membrane; the velocity of these emitters is then measured. The degree of emitter coupling to the membrane is open to question, but Mössbauer spectral analyses have shown traveling waves that are much sharper than those in Fig. 7.2.

The wave motions of these hydrodynamic oscillations are probably best understood if individual particles of the medium can be followed. Using a two-channel cochlear model of the Békésy type, Tonndorf (as reported in the reference by Tobias) placed small aluminum dust particles in the model's vestibular canal and observed them under magnification using stroboscopic illumination. These test bodies were perceived to move in elliptical orbits, completing one circuit in the period of the driving force. The general term for such movement is *trochoidal*, "wheel-shaped," from *trokhos*, the Greek for *wheel*.

The shape of these orbits is strongly dependent upon the distance away from the oval window as well as displacement relative to the cochlear partition. Beyond the point of greatest membrane amplitude, very little fluid motion can be observed. In Fig. 7.3 the recorded motions are represented for a 50-Hz

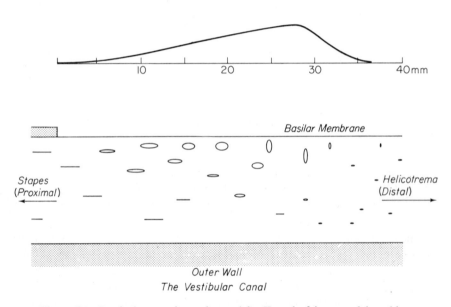

Figure 7.3 Particulate motions observed by Tonndorf in a model cochlea. These orbits become linear near the outer wall and windows, but are generally of an elliptical form. Particles move in clockwise fashion around the orbit. A 50-Hz signal was used to excite the oval window. After J. V. Tobias, ed., *Foundations of Modern Auditory Theory*, Vol. I (New York: Academic Press, Inc., 1972).

sinusoidal tone. Close to the outer wall of the model's cochlea, the wave motion is essentially longitudinal (i.e., along the length of the vestibular canal). This is the direct, compressional wave expected in any fluid. A velocity component orthogonal to the partition arises near the basilar membrane. This part of the motion is relatively pronounced at the maximum of the membrane amplitude curve and is the surface or shear wave due to the existence of a fluid–membrane interface. Basilar membrane waves appear to "break" and fall at a particular distance from the stapes, analogous to surf on a sloping beach. Thus, the orbits in Fig. 7.3 degenerate to points at displacements much beyond 30 cm from the window. As shown in Fig. 7.2, the position of the break moves closer to the stapes as the frequency increases. If the amplitude of the driving signal is sufficiently large, eddy currents are induced in the cochlear fluid, which, in turn, distort the orbits given in Fig. 7.3. The ocean-wave analog of eddy currents is the occurrence of beach undertow. Such currents were originally detected in the 1920's by Wilkinson and Gray, who constructed cochlear models based on dimensional analyses some years prior to Békésy's more famous work in this area. Direct measurements on Tonndorf's model have shown that, at any given point in the eddy pattern, the particle speed is proportional to the square of the exciting stimulus. This distinctly nonlinear response to acoustic vibrations is a relationship that had been predicted earlier by theoretical calculations. Harmonic distortion and peak clipping of the trochoidal motion are two consequences of the appearance of eddy currents.

In contrast to the traveling waves of the Békésy type, the eddy motion has a fixed center of oscillation at a particular position on the cochlear partition. This point closely corresponds to the place where the trochoidal wave is at a maximum. The various distortions produced by these eddies is further evidence for a spatial theory of hearing.

The temporal behavior of the waves that bear his name was also extensively investigated by Békésy using models and mammalian cochlea. In the previous discussions, emphasis has been placed on the envelope of the disturbance and the correlation between the distance from the oval window and the frequency of a sinusoidal audio stimulus. By contrast, Fig. 7.4 shows the time variation in wave form as a function of that distance for a step function input.

Observations of the basilar membrane's motion were originally made by Békésy using a deposition of silver to reflect light from the membrane surface. At a given position, the disturbance requires a finite time, on the order of tens of milliseconds, to arrive and a corresponding interval to decay. The phase of the basilar membrane disturbance was seen to lag quite appreciably behind the stapes stimulus; values approaching -5π rad and lower have been reported. These data stand as further evidence against a Helmholtz picture, since, in the latter theory, the membrane's phase should approach 0 rad as the exciting frequency sweeps through the resonance value. In any event, the Helmholtz membrane would always be within $\pm\pi/2$ rad of the stapidal motion independent of the frequency, in distinct contradiction to the Békésy measurements.

Békésy determined that the traveling waves progress toward the helicotrema regardless of the position of their excitation. Experimenters can drill

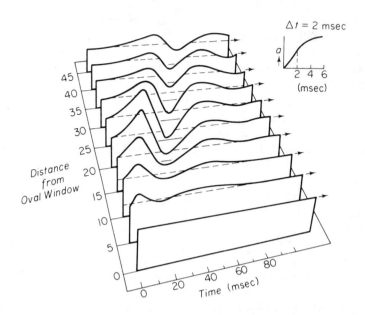

Figure 7.4 A traveling Békésy wave, as seen in a model cochlea. The fast-rising input function is shown as an insert in the upper right corner. As distance from the oval window increases (at a given time after the input) the wave envelope resembles that of Fig. 7.2. At a given position, the temporal sharpness of the oscillation decreases with time. After J. V. Tobias, ed., *Foundations of Modern Auditory Theory*, Vol. I (New York: Academic Press, Inc., 1972).

holes into the cochlea and stimulate the membrane by a pseudo-stapes inserted therein or, even more simply, rely on the bony conduction of the skull and external wall of the cochlea and induce the excitation from the outside. The explanation of this unidirectional property depends on the measured exponential increase in basilar elasticity with distance from the oval window. In electrical terminology, the membrane appears as a progressively lower bandpass filter as distance from the helicotrema decreases.

The basilar membrane may be represented by an electrical analog as shown in Fig. 7.5. Each inductance and resistance shown in the model is identical, but the capacitors become progressively larger as one proceeds from the oval window (base) to the helicotrema (apex). In the analog, in the various hydrodynamic models, and in the intact ear, the group speed of the acoustic disturbance depends not only on its central frequency but also decreases from base to apex. Typical base values are around 45 m/sec, while at the apex, the group speed decreases to around 2 m/sec.

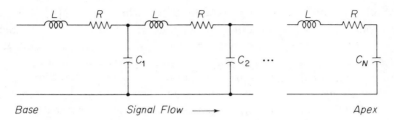

Base   Signal Flow ⟶   Apex

Figure 7.5 Transmission-line analog of the basilar membrane. The circuit is a progressively lower band-pass filter as one approaches the apex. This is accomplished electrically by having C_i *increase* with i (distance) while inductance and resistance remain fixed.

C. NEURAL SHARPENING

As may be seen in Figs. 7.2 and 7.4, the Békésy wave is quite broad. However, small differences in frequency, which produce very similar wave envelopes, are distinguishable. The earlier Helmholtz theory also had trouble explaining the perception of a single frequency when many adjacent nerve fibers were stimulated simultaneously. To explain such frequency discrimination, one may postulate an interpretive or sharpening mechanism in the inner ear or at one or more levels of the auditory nervous system (see Fig. 7.8). Such processing is termed *neural sharpening*.

It is a familiar fact that the nervous system does sharpen many types of stimuli. Thus, when a bright spot is focused on the retina, the sensitivity of the eye to surrounding areas is decreased. The surrounding darker rings are referred to in the example as *Mach bands*. In bright illumination this has the advantage of eliminating the effects of stray light. This is discussed more completely in Chapter 8. Similarly, if two compass points are pressed against the skin of the forearm at distances greater than about 2.5 cm apart, two sensations are received. At around 2.5 cm the two sensations weaken each other, whereas, at still closer distances, the two sensations add to each other. In the latter case, the person feels a single stimulus midway between the two actual compass points. This is illustrated in Fig. 7.6.

Figure 7.6 Neural sharpening and funneling in a spatial context as observed with the human arm as sensory organ. The results are described in the text.

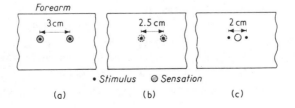

Forearm

3 cm   2.5 cm   2 cm

• *Stimulus*   ◎ *Sensation*

(a)   (b)   (c)

128   *Nerve and Muscle   Part B*

Another example of neural sharpening occurs when locating the position of two clicklike stimuli applied on opposite sides of the finger. For long time delays between the two, separate clicks are felt. If the time delay is decreased, the second click is no longer felt. As the time interval approaches zero, a single sensation is felt which approaches midway between the two stimuli and has a larger apparent area. The results of an experiment of this nature are shown in Fig. 7.7. This same type of phenomenon occurs when one locates a

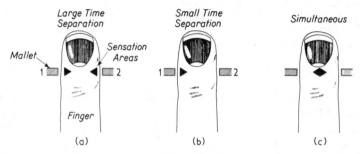

Figure 7.7 Neural sharpening and funneling in a temporal context as observed with the human finger. The results are described in the text.

sound by the difference in the times of its arrival at the two ears. Addition of more than one stimulus into a single, stronger sensation was termed *funneling* by Békésy and his co-workers.

These sharpening and locating effects which occur in the senses of touch and sight as well as hearing are very interesting. They emphasize that the nervous system does act as a computer with a great deal of feedback. They also emphasize that many of the types of neural action essential to a theory based on broadly peaked traveling waves occur in other senses. It was indeed possible for Békésy to make an enlarged physical model of the cochlea using the skin of the forearm as the processing organ.

The model consists of a series of resonant vibrators of varying frequency running along the arm. When these are electrically driven, several neighboring ones respond, the central one most strongly. The person senses the resonant frequency at a much more sharply located spot than is indicated by the behavior of the vibrators. Very sharp frequency and amplitude discrimination are possible with this analog model of the cochlea. Funneling is readily demonstrated as is *masking*, the making of one tone inaudible by means of a second audio signal. The nature of neural processing involved remains unknown, however, and the model is of little help in determining it. Thus, the pitch discrimination within the spatial theories depends in part on neurological actions rather than resonances or maxima of hydrodynamic waves.

The arm model directly supports the idea that pitch discrimination may be an *intrinsic* attribute of the nervous system, either locally or at higher cerebral levels. Many experiments with vertebrates and even invertebrates such as the horsehoe crab (*Limulus*) have shown that the nervous system can carry out complicated data processing (e.g., neural sharpening and funneling). One can conclude that, although it is possible to demonstrate an association between basilar membrane position and audible tone, the actual process of pitch detection in a spatial theory is probably due to neural processing. As will be seen, temporal theories can be simpler, relying instead on processing as a *direct* explanation of frequency perception.

7.3   Temporal Theories

A. THE EVIDENCE FOR FREQUENCY ANALYSIS

Previous discussion has demonstrated that, for sinusoidal inputs, the spatial distribution of the maximum of the Békésy wave is well correlated with the frequency. In the case of more complicated signals, certain rather confounding effects have been well documented. One specific case in point is the general problem of *periodicity pitch*, whereby the listener perceives a definite frequency not present in the Fourier analysis of the stimulus. The phenomenon is well known in the perception of such complicated inputs as violin or vowel sounds.

A signal may be periodically interrupted in several ways. An experimenter can, for example, electronically generate acoustic pulses separated by a time interval (ΔT). The listener then perceives a tone of frequency $1/\Delta T$. If the pulse shaping is carefully done, there will be no appreciable amplitude at that particular frequency. An alternative is to generate white noise in a loudspeaker and control the output by a simple on–off switch. In this case, the appropriate tone is heard if the switch is thrown in a periodic fashion. Amplitude modulation has also been used. In this the amplitude of a sinusoidal imput with frequency ν is varied at a second frequency, ν'. Tones of both these frequencies can then be identified by the listener.

In order to determine the physical and neurological origin of periodicity pitch, various experiments and model evaluations have been performed. It is known, for example, that attempts to *mask* the periodicity signal by appropriate frequency noise are not successful. Masking is generally quite effective in obliterating a pure sinusoid by causing the basilar membrane to vibrate over an extended spatial distance. The implication is that the periodicity pitch is not caused by a Békésy wave. It has been shown using electrodes attached to specific frequency neurons in the cat brain (see Sec. 7.5.B) that the impulses on the higher frequency neurons are modulated at the relatively low periodicity pitch; the lower frequency neurons show no spike potentials. Extrapolating to man would imply that periodicity pitch is generated in the central nervous system without having any physical waves present at that frequency.

Békésy, and later Tonndorf, used cochlear models such as the arm analog described above to investigate the perceived pitch when an amplitude-modulated or interrupted signal was presented to the observer. The two appropriate frequencies were felt and could be observed as Békésy vibrations at the expected spatial positions. It would thus appear that some periodicity pitch experiments are consistent with both types of theories.

One of the first alternatives to Helmholtz's resonance picture was the proposition that all pitch analysis was the function of the brain. This idea, put forward by W. Rutherford in 1886, came to be called the *telephone theory*. At that time, the telephone was a device of great scientific and even social impact, so its association with the theory may be purely circumstantial. There are results, however, discussed in the next section, which can be interpreted as establishing at least the receiver properties of the cochlea.

B. COCHLEAR MICROPHONICS

Early experimental work by Wever and Bray in the 1930's on voltages detected in and around the cochlea was partially confused by the simultaneous presence of nerve spike signals. They did report a small (several hundred microvolts) signal whose time dependence very closely resembled that of the acoustic disturbance. This phenomenon was later labeled a "microphonic," to imply that it might be an artifact of the mechanical motion of the cochlea. Although the term *cochlear microphonic* has persisted, the exact role of the phenomenon in the theory of hearing remains uncertain.

The microphonic can, however, be looked upon as a necessary input for the eventual cerebral processing of the acoustic vibration. It is known to be caused by motion of the hair cells on the organ of Corti and, unlike a nerve spike, to have no threshold level. Thus, an air vibration at the limit of detectability produces a just detectable cochlear microphonic. The amplitude of the microphonic has also been found proportional to the sound level of the stimulus. In addition, the cochlear microphonic is maximal at the point where the Békésy wave has its greatest amplitude.

Thus, it is not possible to exclude the cochlear microphonic as a significant part of the process of hearing. With sufficient central nervous system processing, it is not hard to imagine an eventual recognition of both tone and amplitude by the brain. In this picture, the Békésy wave does little but serve as a presumably necessary intermediate step in the sensory process.

7.4   Auditory System Pathways

In either a spatial or temporal theory of hearing, the ultimate problem is the transmission of the acoustic data into the higher cerebral centers. The organization of the efferent transmission system is contained in this section.

Subsequently, the encoding of frequency and intensity data in terms of spike potentials is considered.

The action potentials eventually generated via the oscillations of the basilar membrane of the cochlea travel along the fibers of the acoustic nerve. As has been stated in Chapter 6, most sensory nerve cell bodies are located in compact groups called *ganglia*. The acoustic nerve, however, has a diffuse set of cell bodies spread out along its path through the spiral bony partition which supports the cochlea. These nerve cell bodies are called the *spiral ganglion*. The pulses in the second set of axons in the acoustic nerve enter the brain. The acoustic nerve is the eighth one (counting from the front end) to enter the brain; it is often called the *eighth cranial nerve*. As shown in Fig.

Figure 7.8   Auditory pathways of the central nervous system. Copyright *The CIBA Collection of Medical Illustrations*, Vol. I, *The Nervous System*, by Frank H. Netter, M.D. (Summit, N.J.: CIBA Pharmaceutical Products, Inc., 1953).

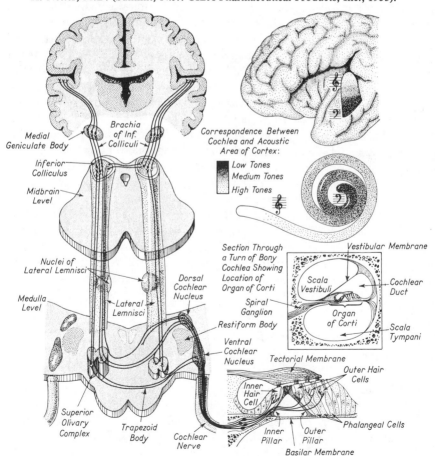

7.8, several additional synapses occur within the brain stem. Some of the pulses cross over to the opposite half of the brain stem so that those starting at either ear are represented in both halves of the brain. Finally, at least in unconscious animals, the pulses are conducted to specific areas on the surface of the temporal lobes of the cerebral hemisphere. This latter projection is believed to be necessary for conscious hearing.

In humans and other primates, this auditory area on the temporal lobe of the cerebral cortex is buried deep in one of the folds of the cortex and is hard to study. In other mammals, the cerebral projection is on or near the exposed portions of the cortex. In the latter group, there are always two, and in some cases, three areas where responses appear (in the unconscious animal) when the ear is stimulated. Each of these areas is connected to both ears. Within each cerebral projection area, specific smaller areas correspond to specific spots along the basilar membrane.

The detailed examination of the acoustic pathway shows that several neurons are involved. The first is located in the spiral ganglion within the inner ear. The nerve fibers leaving this ganglion join those from the vestibular portion of the ear to form the eighth cranial nerve. Within the brain, the vestibular and auditory fibers separate. Those from the cochlea go to one of two nuclei in the lower brain stem known as the *dorsal* and *ventral cochlear nuclei*. Some fibers leaving these have synapses with other neurons associated with reflex actions and balance. Others go to synapses in another nucleus in the lower brain stem, called the *superior olivary complex*. Some fibers synapse in the superior olivary complex on the same side, others on the opposite side of the brain, and still others pass through without interruption, joining fibers from the superior olivary complexes and passing up the brain stem. In the nuclei of the *lateral lemniscus* farther along the brain stem, some of the auditory fibers end, and others pass through uninterrupted.

In the midbrain level, some of the auditory fibers end at synapses in the *inferior colliculus*. From here, some fibers cross over to synapses in the opposite inferior colliculus. All the fibers of the auditory tract have synapses in another nucleus of the midbrain, the *medial geniculate body*. Fibers of these neurons finally reach the auditory areas of the cerebral cortex.

7.5 Acoustic Encoding

A. THE ORGAN OF CORTI

The first step in the transformation of information into neural action potentials probably occurs in the organ of Corti, lying atop the basilar membrane (see Fig. 7.8 insert). Outer and inner hair cells are considered to be the detectors. These specialized cells are thought to be evolutionary descendants of the primitive spatial orientation detector of jellyfish and other marine animals. The action potential trigger is fired by moving the cilia with respect to the body of the cell. It is known that the cochlear microphonic is largely due to the outer hair cells, which have their ciliae directly touching the tectorial membrane. Inner hair cells are probably more indirectly affected by wave disturbances (i.e., are thought sensitive to the first and perhaps higher derivatives of the motions in the fluid). A dependence on the derivatives of the Békésy waves could help account for a mechanical sharpening of the frequency sensitivity in that model. This dependence could result from their stimulation by bending due to the motion of the tectorial membrane (see Fig. 1.10), which is in intimate contact with the hair cells. The tectorial membrane's motion is similar to that of a loaded mechanical beam. In mechanical engineering, it is shown that the latter motion can be described by a fourth-order differential equation. The solutions for such an equation can exhibit the sharp maxima implied by pitch discrimination. However, the exact role of the tectorial membrane and the detailed behavior of the hair cells in the organ of Corti remain unresolved.

B. TUNED NEURONS

The existence of frequency-specific auditory nerve cells distal to the spiral ganglion have been clearly demonstrated in the cat. The tuning is quite sharp, certainly far narrower than the Békésy wave envelope would indicate. Békésy had speculated earlier that each detected Fourier component should be associated with one specific cluster of neurons going to the brain. This can be considered to be the end result of a hydrodynamic traveling wave with lateral sharpening. The encoding of intensity in such a model is given by the actual frequency of spike potentials and/or the number of tuned neurons active at a given time.

From Chapter 5, it may be recalled that the refractory or dead time of a neuron is on the order of 2 msec. This implies a limiting rate of 500 Hz for a single cell. Presumably the central nervous system would be able to establish, through training, the correlation of sound level and firing rate. Although it is tempting to accept the tuned neurons concept as it stands, other approaches are possible and have been considered. One of these is presented in the next section.

C. VOLLEY PRINCIPLE

If one wishes to allow any neuron to carry any frequency, the refractory period given above imposes a rather striking limitation of 500 Hz, since the stimulated nerve cell must fire at least once per amplitude maximum. Wever suggested that several neurons can be considered to function in parallel and thereby to allow more action potentials per unit time. Figure 7.9 shows the method whereby 15 axons could coordinate to fire no more than once per acoustic cycle and yet reproduce the temporal form of the input signal. Where this synchronization would start is not known. Wever has proposed that it occurs in the cochlea, that in some fashion the nerve fibers fire in volleys to reproduce the overall form of the incident pressure wave. However, the synchrony could just as well originate at the first or even second synapse.

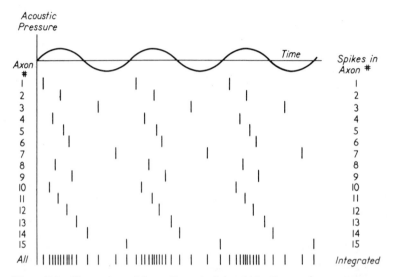

Figure 7.9 Illustration of the volley principle which allows axons to fire once per cycle and still reproduce the shape of the sound wave.

This semisynchronous action is called the *volley theory*. It states, in its simplest form, that below some frequency, say 100 Hz, the number of nerve fibers excited varies with the instantaneous pressure. From 100 to perhaps 3,000 Hz, the volley-type effect reproduces the frequency of the incident sound wave, whereas above 3,000 Hz, submultiples of the stimulus frequency are generated. Somehow the brain is thought to carry out a frequency analysis on the overall electrical signal. In other words, the ear carries out a crude frequency analysis in terms of exciting preferentially certain nerve fibers. The central nervous system then performs a finer analysis. Intensity may be coded via the degree of synchrony or the number of neurons involved in the group.

There appears to be no lack of data to support the volley principle. Synchronous fibers have been detected at frequencies up to 4 kHz. Their ability to give rise to coordinated action potentials is apparently not related to the specific frequency of the neuron as exemplified in the cat experiments. Synchrony can be found at acoustic stimulation frequencies far below the appropriate one for the individual cell. One of the many experimental problems in this area is the spontaneous action potentials that are seen on all auditory neurons. The reason for this seemingly useless activity is quite unclear.

Since data that support both the frequency-specific and volley theories are available, the exact mechanism of neural encoding may be considered a combination of both strategies. Some auditory physiologists feel that several different learning mechanisms intermingle with each other inside the brain. These have evolved toward their present somewhat complementary state.

7.6 Summary

Historically, two different types of theories of hearing were proposed; one assigning frequency discrimination to spatial effects, the other to direct actions of the central nervous system. More recent studies have included both of these types of effects. The actual form and maxima of the hydrodynamic waves, called Békésy waves, traveling along the cochlea have been observed in various ways. While providing some degree of frequency selectivity, the maxima of these waves appear to be too broad to account for pitch discrimination. Closely associated with the hydrodynamic wave is an electrical signal called the cochlear microphonic; its role in hearing, if any, is obscure.

A variety of mechanisms such as neural sharpening have been proposed to account for pitch discrimination. These include a general action of the numerous synapses along the auditory pathway in the central nervous system. In addition, frequency-tuned neurons and volley principles may contribute to pitch discrimination. There may be some degree of sharpening in the actions of the basilar membrane and tectorial membrane in initiating neural impulses. It appears probable that a number of these mechanisms contribute to human hearing, thereby utilizing the mechanical characteristics of the anatomic structures of the inner ear as well as the synaptic patterns of the auditory pathway.

REFERENCES

1. TOBIAS, J. V., ed., *Foundations of Modern Auditory Theory*, 2 vols. (New York: Academic Press, Inc., 1972).
2. LITTLER, T. S., *The Physics of the Ear* (Oxford: Pergamon Press, 1965), 378 pages.

3. BÉKÉSY, GEORG VON, *Experiments in Hearing*, trans., and E. G. Wever, ed. (New York: McGraw-Hill Book Company, 1960), 745 pages. (An almost complete account of the important works by the Nobel Laureate of 1961.)

4. BRIDGMAN, P. W., *Dimensional Analysis*, rev. ed. (New Haven, Conn.: Yale University Press, 1931), 113 pages.

5. DALLOS, PETER, *The Auditory Periphery: Biophysics and Physiology* (New York: Academic Press, Inc., 1973), 548 pages.

6. KIANG, N. Y.-S., T. WATANABE, E. C. THOMAS, AND L. F. CLARK, *Discharge Patterns of Single Fibers in the Cat's Auditory Nerve* (Cambridge, Mass.: The MIT Press, 1965), 154 pages.

7. STEVENS, S. S., AND FRED WARSHOFSKY, *Sound and Hearing* (New York: Time, Inc., 1965), 200 pages. (This is a popular exposition with many fine drawings of the hearing physiology.)

CHAPTER 8

Neural Aspects of Vision

8.1 Introduction

The anatomical and physical features of the eye are described in Chapter 2. That chapter is terminated without a discussion of color discrimination, neural sharpening, and other aspects of vision which depend on the action of nerve cells and the central nervous system. In this chapter, the neural mechanisms necessary for vision are examined in more detail. To review briefly, the retina acts as a "photoneural" transducer, converting incoming electromagnetic energy to spike potentials on nerve fibers. The potentials travel along the optic nerve, enter the central nervous system, and eventually reach specific areas of the cerebral cortex. The information is "analyzed" at a series of synapses, both within the retina and within the central nervous system proper. Out of this analysis there are, in some way, created the sensations of color, acuity, brightness, shape, and so forth. (It is assumed that the reader is familiar with the ideas of Chapters 2, 4, and 5.)

One step in the overall process of vision, the photomolecular reactions in the rod and cone cells of the retina, is of extreme importance to an understanding of vision. At the same time, it is not necessary to understand these reactions before discussing the neural aspects of vision. Accordingly, the molecular reactions are deferred until Chapter 19.

8.2   Color Discrimination

A fundamental test of any theory of the neural aspects of vision is the explanation of color discrimination. The subjective sensations of color are familiar to most humans. However, at the lowest intensities where objects are barely visible to the dark-adapted eye, there is no sensation of color. At light intensities just slightly greater than this, colors begin to be sensed.

A. DEFINITIONS

The sensation of color is a complicated function of the wavelengths of light reaching the eye. Just how complicated has been emphasized by a set of experiments referred to at the end of Sec. 8.4. However, when a large patch of light of the same wavelength is presented to the eye, it is identified as a single color.

Such a light, consisting of a very narrow wavelength band, is called *monochromatic*. The other extreme, equal intensities at all wavelengths, is called *white*. The sensation of white can be evoked by many compositions simpler than uniform intensity throughout the visible spectrum. Certain pairs of colored lights, such as blue and yellow, appear white when mixed in equal proportions. The pairs of colors producing white are called *complementary*. Certain sets of three colors, such as red, green, and blue, as well as other sets of four or more colors, also give a sensation of white. Likewise, varied groups of colored lights can produce any given color sensation.

Psychophysicists distinguish several different qualities of sensations associated with color vision. These include *luminosity* (or luminous intensity), *hue*, and *saturation*. Luminosity is a measure of the amount of radiant energy given off by a source. Its units are the *candela*, the luminous intensity of $(6 \times 10^5)^{-1} \, m^2$ of a blackbody at the temperature of freezing platinum $(2045°K)$. Light flux is measured in *lumens*. A source with an intensity of 1 candela in all directions radiates a light flux of 4π lumens. As an example, a 100-W light bulb emits about 1,700 lumens.

While all lights have luminosity or brightness, a colored light will, in addition, have a certain hue. This is a function of the wavelength of the radiant energy. A given colored light may not be a pure hue, but may be mixed with white light. Saturation is a measure of the purity of the hue. For example, pink represents a mixture of red and white; it is said to be less saturated than a pure red color. Hue and saturation taken together constitute *chromaticity*.

It has been known for many years that sets of three visual stimuli existed, so that by choosing the proper amounts of these, an observer could match the chromaticity of a given light in terms of the sensation it evoked. If the relative amounts of each of the three standards are indicated by X, Y, and Z respectively, one may represent the observed chromaticity of light A by

$$A = X_A + Y_A + Z_A$$

and a light B by

$$B = X_B + Y_B + Z_B$$

If one now adds equal amounts of A and B to form a new light C, which may be represented as

$$C = X_C + Y_C + Z_C$$

it is found that

$$X_A + X_B = X_C \qquad Y_A + Y_B = Y_C \qquad Z_A + Z_B = Z_C$$

In general, any algebraic combination of colored lights is matched by the corresponding algebraic combination of the amounts of the standards matching these lights.

In order to standardize the description of chromaticity, the International Congress on Illumination agreed on three artificial standards as shown in Fig. 8.1. These were chosen so that a monochromatic light of unit intensity

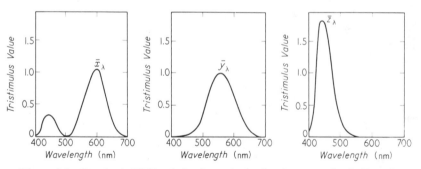

Figure 8.1 Standard C.I.E. tristimulus values of unit energy for indicated wavelengths. After Committee on Colorimetry, The Optical Society of America, *The Science of Color* (New York: Thomas Y. Crowell Company, 1953), pp. 242–243.

at any wavelength λ in the visible spectrum was matched by an average observer with the amounts of the three standards $\bar{x}_\lambda$, $\bar{y}_\lambda$, and $\bar{z}_\lambda$. The curve $\bar{y}_\lambda$ has the same shape as the average photopic luminosity curve. The curves were normalized so that a white light (of equal spectral density at all wavelengths) is matched by equal amounts of the three standards. The area under each curve is the same. Symbolically, this may be stated as[1]

$$\int_{380nm}^{780nm} \bar{x}_\lambda \, d\lambda = \int_{380nm}^{780nm} \bar{y}_\lambda \, d\lambda = \int_{380nm}^{780nm} \bar{z}_\lambda \, d\lambda$$

Any colored light can be analyzed spectrophotometrically to give its spectral intensity I_λ. This is defined so that the total intensity I is given by

$$I = \int_{380nm}^{780nm} I_\lambda \, d\lambda \qquad \left(\text{in other words } I_\lambda = \frac{dI}{d\lambda} \right)$$

Many spectral intensity distributions will give the same sensation. To specify the sensation, the three numbers X, Y, and Z are again needed; in terms of the artificial standards above,

$$X = \int \bar{x}_\lambda I_\lambda \, d\lambda \qquad Y = \int \bar{y}_\lambda I_\lambda \, d\lambda \qquad Z = \int \bar{z}_\lambda I_\lambda \, d\lambda$$

where all three integrals are evaluated from 380 to 780 nm.

These color-matching experiments are based on human response. Because they require subjective information, similar experiments are difficult to perform on laboratory or wild animals. Nonetheless, considerable evidence indicates that many vertebrates, and even insects, have color vision. The primate eyes are all very similar. However, since so much of the available data on color vision comes from the human, most attempts to explain color vision on cellular levels emphasize human vision.

B. THEORIES

Two general types of theories of color vision have been maintained in the past. One of these, *tricolor theories*, were supported historically by Young, Helmholtz, and Maxwell. The other type of theory, *opponent* or *antagonist theories*, have appealed to many psychologists; their variations are often associated with a person's name, such as Hering, Ladd-Franklin or Adams.

Briefly, the tricolor theories assumed that in the retina there were three pigments having maximum absorption in the blue, green, and red regions of the spectrum. These pigments were postulated to exist in separate receptors each of which sent distinguishable impulses to the brain. While the original forms of the tricolor theory had difficulties explaining several types of color blindness, parts of it are used in a molecular understanding of this affliction (see Chapter 19). A more serious objection is that even the best refinements

---

[1] These integrals are usually written as extending over the wavelength range from zero to infinity. However, $\bar{x}_\lambda$, $\bar{y}_\lambda$, and $\bar{z}_\lambda$ are zero at all wavelengths outside the range 380–780 nm.

of this theory failed to use the detailed neural structure of the retina (see Fig. 2.10).

In contrast to the tricolor theories, which minimize retinal action, the antagonist (or opponent) theories regarded the retina as the site of considerably more complex activities. The antagonist theories postulated that there were six retinal responses that occurred in antagonistic pairs. Excitation leading to any single response was supposed to suppress the action of the other member of the pair. These six retinal responses were identified as blue–yellow, red–green, and black–white. Various forms of the antagonist theories had less trouble explaining black–white vision and several forms of color blindness than did the tricolor theories. Most of the antagonist theories did not specify in any detail how the retina actually produced these responses.

Recent models of color vision take parts from both the tricolor and antagonist theories. There are three distinct channels through which visual color information is transmitted from the cones which first receive the light to the ganglion cells which transmit it through the optic nerve to the brain. The simplest explanation for this would require three distinct visual pigments (see Chapter 19) and three distinct ganglial receptors. The transmission of this information from the ganglia to the brain is discussed in Sec. 8.3. These neural experiments could support an antagonist theory. Thus, color vision appears to be at least a two-stage process, consistent with a tricolor theory at the receptor level and an antagonist theory at the optic nerve and beyond.

The tricolor and antagonist theories were originally based almost exclusively on psychophysical evidence. There is considerable other information available, in terms of which any theory of vision must ultimately stand or fall. The evidence from histology, electrophysiology, biochemistry, and communication must all be included before a theory of vision can be considered complete. For example, the extremely small sizes of the rods and cones, combined with their relatively high indices of refraction indicate that they might serve as optical wave guides, selectively transmitting energy only in a narrow wavelength band characteristic of the particular rod or cone. Theoretically, the light energy in a fiberlike guide is transmitted in various characteristic modes; the fraction in each mode can be calculated using a dimensionless parameter V, defined by

$$V = \frac{\pi D}{\lambda} (n_1^2 - n_2^2)^{1/2}$$

where D is the fiber diameter, λ is the wavelength of light, and n_1 and n_2 are the relative indices of refraction in the fiber and its surround, respectively. For values of V between 1 and 2.4, only one mode (and hence one color) will be selectively transmitted.

Vertebrate rods and cones can be shown to have such values for V. Direct microscopic observation of light passing through excised retinas has also demonstrated such color selectivity. Hence, many individual rods and cones theoretically and experimentally transmit characteristic colors. From this one might propose that the transmitted color is also the one at which light energy is transduced to a neural form. Such action could lead to a far more complicated process of color discrimination than described by the tricolor and antagonist theories. However, both physiologically and psychologically, there are no data to indicate whether optical guiding is important for vertebrate vision.

8.3   Isolated Neural Responses

Measurements of neural spike potentials were made by Hartline and his co-workers, who recorded impulses from the optic nerves of the horseshoe crab, *Limulus*. The eye of this crab is particularly simple because it consists of many individual ommatidia, described in Chapter 2. Each of these receptors is connected to an individual nerve fiber. When the nerve is dissected until just one fiber remains intact, a slow natural firing rate is observed in the dark. This is illustrated in Fig. 8.2. If a threshold stimulus is applied to this single

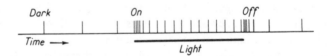

Figure 8.2 Diagrammatic representation of response of a single *Limulus* ommatidium. The vertical lines represent spike potentials. Solid horizontal line represents light on. Note dark rate, on-burst, steady rate in light, off-burst, and return to dark rate. After H. K. Hartline, H. G. Wagner, and F. Ratliff, "Inhibition in the Eye of *Limulus*," *J. Gen. Physiol.* **39**:651 (1956); H. K. Hartline and F. Ratliff, "Inhibitory Interaction of Receptor Units in the Eye of *Limulus*," *J. Gen. Physiol.* **40**:357 (1957).

ommatidium, an extra spike is observed. If light stimuli considerably above threshold are used, the response is somewhat more complicated, as is also shown in Fig. 8.2. Initially, there is a very rapid (transient) burst of spikes as the light is turned on. This is followed by a slower steady-state "firing" rate far faster than the dark rate. The steady-state rate is a function of the intensity of the light stimulus. When the stimulus is removed (i.e., the light is turned off), there is another transient burst of spikes, followed by a gradual return to the dark rate. There is no reason to doubt that individual retinal rods and cones of vertebrates would follow this same pattern.

A second type of experiment carried out by Hartline and his co-workers involved the vertebrate eye. These experiments were more difficult to perform and also much more difficult to interpret in a quantitative fashion. Nonetheless, the results molded the thinking of everyone who has worked in the field of vision since then. In these experiments, the vertebrate eye was removed with the optic nerve intact. The nerve was dissected until just one fiber remained. Through many experiments a variety of types of fibers were found. All showed a spontaneous, rhythmic background firing. On stimulation by a light, some fibers increased their rate of firing or discharge. These were termed "on" fibers. Others decreased their rate of firing ("off" fibers). And still others were almost completely silent during constant stimulation but showed strong discharge bursts upon start and termination of the light ("on and off" fibers).

Another method of obtaining electrophysiological data is to remove the cornea, lens, and vitreous humor of an intact eye. Electrodes are passed over the surface of the retina until the response is that of a single nerve fiber. Granit, in Sweden, was the first to use this method. In snakes, rats, frogs, and guinea pigs, he found that most fibers gave the normal photopic threshold curve. These are called *dominators*. Other fibers, having sharper, more selective spectra, are called *modulators*. In most animals, Granit found at least three and sometimes four such modulators.

The dominators appear to indicate, by their frequency of discharge, the intensity of the light stimulus. The modulators, with their varying sensitivities to different parts of the spectrum, give information about its spectral composition. For example, a red modulator will respond to red light while a blue or green modulator would be much less affected. The existence of the modulators shows very clearly the need for an inhibition mechanism during continuous lumination.

Still a further complexity was introduced when it was discovered that in goldfish, the "on and off" fibers in white light could behave as either "on" *or* "off" fibers, depending on the wavelength of monochromatic light. With some fibers red light was stimulating and blue–green light inhibitory; with others the reverse was true.

In summary, the direct neural measurements indicate that vertebrate nerve fibers of the optic nerve show more response when a light intensity changes than during continuous illumination. In many cases, the rate of spike formation is depressed or abolished during strong illumination. This is in direct contrast to the response of individual receptors whose rate is apparently increased on direction stimulation. Complex neural interaction (i.e., computation) is an important part of retinal function. The retina can sum the responses of several receptors to indicate intensity and wavelength of the incident light. In this respect, the retina acts like a part of the brain. The retina is a subdivision of the brain in terms of its embryological formation. It further resembles the brain in giving rise to electrical potentials, measured in electro-retinograms, which are similar in some ways to the electroencephalographic potentials (see Chapter 6).

8.4 Coordinated Neural Responses

A. NEURAL SHARPENING

Inhibition in the retina can be demonstrated in other ways. One of the more striking is the process called *neural sharpening*. Possibly similar effects in the senses of touch and hearing were discussed in Chapter 7. Sharpening within the retina was demonstrated directly in the experiments of Hartline and co-workers with *Limulus* eyes. The nerve fibers from the ommatidia go through a complex *plexus*, not clearly understood anatomically, in which the various fibers apparently synapse with one another. If two receptors are stimulated instead of one, as described in Sec. 8.3, their responses can be shown to be interrelated. These relationships exist at the ommatidia themselves but are abolished if the nerve fibers are dissected free (i.e., removed from the plexus) from the ommatidia to the points of observation (and cut thereafter). Thus, the interrelationships depend on the neural plexus. As a result, the stimulation of one ommatidium raises the threshold and decreases the steady-state firing rate of the second ommatidium used. These effects are reciprocal and are important only for very close neighbors.

The response of an individual receptor, then, depends on the state of stimulation of its neighbors (or, more correctly, on the firing rate of its neighbors). For example, one may choose three receptors, A, B, and C, such that A and B inhibit each other and B and C inhibit each other, but A and C are too far apart to have an appreciable mutual effect. The results of this experiment are illustrated in Fig. 8.3. If one stimulates A and observes a firing rate, it can be reduced by simultaneously stimulating B. If now C is also stimulated, the firing rate of B will be reduced, thereby permitting the rate of A to rise toward its original value. Thus, the response of any receptor, although affected directly only by its neighbors, depends in a complicated manner on the responses of all the other receptors.

Similar mutual inhibitions have been observed in vertebrate eyes between the receptors exciting one ganglion cell. It is tempting to hypothesize that in *Limulus*, these mutual interactions are the result of direct interfiber synapses, but in the human retina they are mediated by other cells. This mutual inhibition of neighboring receptors serves to increase acuity by decreasing the effects of glare and of scattering within the eye. It also makes the threshold

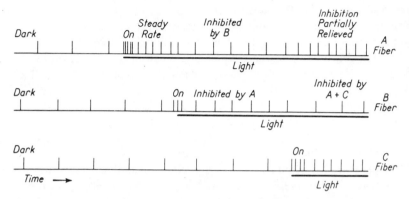

Figure 8.3 Diagrammatic representation of three ommatidia. A and C are so far separated that there is no mutual interaction. However, both A and C interact with B. After H. K. Hartline, H. G. Wagner, and F. Ratliff, "Inhibition in the Eye of *Limulus*," *J. Gen. Physiol.* **39**:651 (1956); H. K. Hartline and F. Ratliff, "Inhibitory Interaction of Receptor Units in the Eye of *Limulus*," *J. Gen. Physiol.* **40**:357 (1957).

much higher near a bright object. Thus, gradation at the edge of a bright light appears much sharper to the eye than to a series of independent photocells.

B. MACH BANDS

If a white paper is uniformly exposed to light, it will appear uniformly bright or luminous. Similarly, if light is completely excluded (e.g., if it is placed in deep shadow), it will appear black or dark grey. If light is partially occluded, such as by holding a smaller card between the light source and the paper 1 inch or so above the paper, there should be three areas of illumination. The first will appear white (light), a second, in shadow, will be dark gray, and the third would be a half-shadow transition zone, between the first two. This last zone would be expected to shade uniformly from the light to the dark luminosity, and indeed physical luminosity measurements will show the expected linear decrease.

Human observers, however, will report a nonlinear transition zone, with two additional bands or lines, one light and one dark, on either side of the half-shadow. These unexpected lines are known as *Mach bands*, after Ernst Mach, who first reported them in 1865.

The Mach bands are a subjective rather than objective effect, yet they can be seen in a photograph. They may even be seen in halftone reproductions, where microscopic measurements can show no difference between the size of the dots that compose the Mach bands and the adjoining areas. (See the reference by Ratliff for illustrations.) Moreover, they can be reproducibly demonstrated wherever nonuniform changes in luminosity occur; that is, they are dependent on the second derivative of the light intensity.

Mach concluded that these bands originated in the retina, through interrelations between neural elements so as to enhance visual contours and borders. Various related edge and contrast effects are now known. All are similarly presumed to result from complicated neural processing in the retina or elsewhere in the visual nervous pathway (see Fig. 8.4).

This processing occurs in neural networks whose activity is the sum of numerous individual spike potentials, some excitatory and some inhibitory. How these are arranged to produce the phenomena mentioned above is still unknown. Several theories, reviewed in the reference by Ratliff, have involved mathematical models to account for the Mach bands.

C. COLOR ANALYSES

A different type of neural analysis has been demonstrated by Land and his associates. They found that, although the description given previously in this chapter for color discrimination was valid for large patches of color or for one or two colors in the visual field, it was very misleading for color vision as it normally occurs. To show this, they used two black and white photographic slides, one exposed in the short-wavelength region of visible light and the other in the long-wavelength region. When these were projected simultaneously but illuminated with two different broad bands of light, the natural color sensations were reproduced. Similar experiments with narrow bands of light (i.e., monochromatic lights) produced about two thirds of the possible colors. The effective colors depended only on the percent of the maximum (or average) of each light transmitted and not on their absolute intensities. It further depended on a random or Gaussian distribution of small patches of colors such as occur in the normal visual field.

8.5 Cortical Representation

The complex series of synapses of the visual pathway through the central nervous system is shown in Fig. 8.4. It should be noted that responses from either eye for a given area in the visual field eventually appear on (or are "projected onto") the occipital lobe of the cerebral cortex opposite to the half of the visual field containing the object. Further, the area of maximum acuity around the fovea occupies a major portion of the surface of the cortex.

The stimuli are not simply transmitted through the synapses. At various points in the midbrain, auxiliary fibers lead off to autonomic systems, such as the feedback loops controlling the iris, and to tear, blinking, and sudden

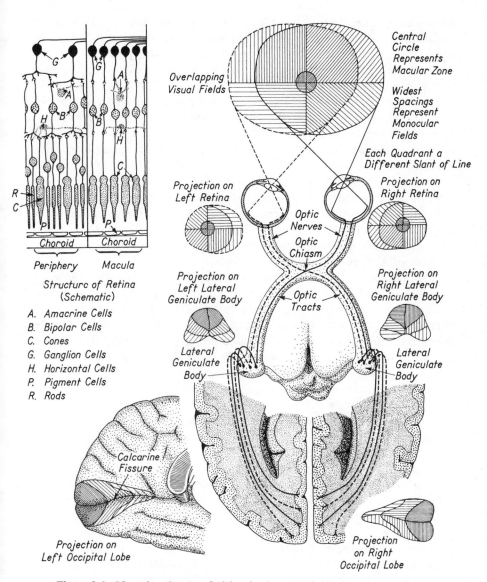

Figure 8.4   Neural pathways of vision in the central nervous system. Copyright *The CIBA Collection of Medical Illustrations*, Vol. 1, *The Nervous System*, by Frank H. Netter, M.D. (Summit, N.J.: CIBA Pharmaceutical Products, Inc., 1953).

withdrawal centers. Moreover, a great deal of data processing may occur at these synapses. For example, potentials at the retina follow a light blinking 1,000 times/sec, those in the midbrain barely follow 100 times/sec, whereas the cortical potentials can at most follow 10 per second. The potentials on the surface of the occipital lobe occur first locally and then spread over the entire cortex. Under the action of anesthesia, the local potentials do not spread as far. The exact role of these potentials or their relationship to conscious sensations is not yet known.

The retinal processing discussed in Sec. 8.4 is especially well developed in those organisms with little binocular vision. Animals with frontal eyes which have widely overlapping visual fields, such as humans and cats, appear to delay most signal processing until the visual cortex, when signals from both eyes may be compared.

In keeping with their use in binocular vision, most neurons in the visual cortex respond much more vigorously if stimulated by appropriate signals from two retinal fields, such as those generated when the animal views one object at a fixed distance. This distance varies from one cortical cell to another. Thus, the visual cortex is a complicated neural network which may be used for stereoscopic distance estimation.

This network is not inborn; it develops after birth. A wide range of experiments since the early 1960's have studied its development in newborn kittens. There appears to be a "sensitive" period, between 3 weeks and 3 months after birth, when kittens must make critical connections between the optic nerves and the visual cortex. If patterned visual experience is lacking in one eye, such connections do not develop and the animal is blind in the afflicted eye. However, if such experience is reintroduced within the sensitive period, and the animal is forced to use its inexperienced eye, these connections may be made correctly.

If the animal has an "incorrect" visual environment, incorrect connections are developed. Thus, animals can adjust to vertical misalignments caused by appropriate distorting goggles, and will be disoriented upon their removal. Similarly, if the kittens are reared in environments that restrict their visual experience to one orientation, they will develop cortical neural connections "tuned" almost exclusively to that orientation alone. For example, if they are reared with vertical lines, later they are more or less blind to horizontal lines.

Neural processing in the visual cortex thus may account for distance, size, shape, and orientation information. Color discrimination and analysis also may occur in part within the nervous system. Finally, cortical integration of other sense information (e.g., touch or hearing) to help interpret visual sensations is an additional neural aspect of vision.

8.6 Summary of Vision

Vision can be studied from many different points of view. In Chapter 2, the physical properties of light waves and optical systems necessary for vision were discussed. Likewise, the gross anatomy and histology of the vertebrate eye were described. These topics are all within the realm of definitive, quantitative knowledge unlikely to change in the near future. In Chapters 3 and 4, specialized uses of vision in navigation and homing of bees and birds were discussed. These uses depend critically on the actions of the central nervous system.

In this chapter the elements of color vision, including the older tricolor and antagonist theories and the elements of each that are supported by more recent experiments, were described. Visual information travels from the primary receptors, which can offer information on wavelength and intensity, to the brain, where the full range of visual effects, such as motion, shape, and distance, are perceived. The information must be processed in varied and complicated ways to account for such perceptions.

REFERENCES

Persons interested in any aspect of the visual process will find one or more articles in the following series useful. The first edition (in four volumes) was published in 1962; the six volumes of the second edition began appearing in 1969.

1. DAVSON, HUGH, ed., *The Eye* (New York, Academic Press, Inc.).

The following two books will provide a well-documented general summary of visual physiology.

2. DAVSON, HUGH, *The Physiology of the Eye*, 3rd ed. (New York: Academic Press, Inc., 1972), 643 pages.
3. BRINDLEY, G. S., *Physiology of the Retina and Visual Pathway*, 2nd ed. (Baltimore, Md.: The Williams & Wilkins Company, 1970), 315 pages.

The following are more specialized references to experiments described in the chapter.

4. HARTLINE, H. K., AND F. RATLIFF, "Inhibitory Interaction of Receptor Units in the Eye of *Limulus*," *J. Gen. Physiol.* **40**:357–376 (1957).
5. GRANIT, R., "A Physiological Theory of Colour Perception," *Nature* **151**:11–14 (1943).
6. WOLBARSHT, M. L., chairman, "Structure and Function of the Visual System," a symposium in *Fed. Proc.* **35**:23–67 (1976).
7. WAGNER, H. G., E. G. MacNICHOL, AND M. L. WOLBARSHT, "The Response Properties of Single Ganglion Cells in the Goldfish Retina," *J. Gen. Physiol.* **43**: Second Suppl. 45–62 (1960).
8. RATCLIFF, FLOYD, "Contour and Contrast," *Sci. Am.* **226** (June 1972):90–101.
9. LAND E. H., "Experiments in Color Vision," *Sci. Am.* **200** (May 1959):84–99.

10. BLAKEMORE, C., "Developmental Plasticity in the Cat's Visual Cortex," in *Sensory Physiology and Behavior*, Vol. 15 of the series *Advances in Behavioral Biology*, RACHEL GALUN, PETER HILLMAN, ITZHAK PARNAS, AND ROBERT WERMAN, eds. (New York: Plenum Press, 1975), pp. 5–9.

CHAPTER 9

Muscles

9.1 Introduction

A very general property of all living matter is the ability to alter its size or shape by contracting or expanding a given region of its body. In most of the higher animals, certain cells or groups of cells are specialized to contract or relax, thereby changing the position and shape of the animal. Other, similar groups of cells contract and relax to pump fluids (blood) through the animal, force food through the digestive tract, and so forth. Aggregates of these specialized contractile cells are called *muscle tissues*, or simply *muscles*. All other forms of protoplasm exhibit a contractility similar to that of muscles, but the latter are specialized to emphasize this property of contractility.

Muscles have been of interest to biophysicists for many years; their study will probably remain one of the fields of biophysical research for years to come. Most of the earlier studies on muscles were part of a larger field called *biomechanics*. This field was explored primarily by workers who, because of their backgrounds and training, labeled themselves physiologists and anatomists. Today, biomechanics research involves specialized topics such as body resonances, tissue elasticity, and fluid mechanics. These are part of biophysics (although not described in this text).

Starting some time in the 1920's, muscles were studied as biochemical complexes. At the same time, biophysicists related the heat changes that occurred in muscles to a mixture of chemical and mechanical effects. These studies markedly influenced the direction of biochemical research as a whole and still form part of the basis for current models of oxidative mechanisms in protoplasm.

A slight refinement in the abovementioned biochemical and thermal studies involves the use of extraction techniques. The muscles are ground up; certain compounds, for example, myosin, are extracted and purified; and then their properties are studied. It is believed that the nature of the contractile process should be related to the properties of the chemical constituents of muscles.

More recent advances in research on the contractile process in muscles have come about through the use of highly specialized physical instrumentation and by the introduction of the ideas and concepts of molecular structure and form. Thus, muscle studies are increasingly falling within the scope of biophysics and biophysical chemistry. For example, the enzyme reactions and the optical density changes in living muscle have been followed by using specially constructed spectrophotometers. Likewise, microelectrode techniques have made it possible to observe the magnitude and form of the electrical surface potentials, as well as the action potential spikes that precede contraction. Perhaps most important of all, a special physical tool, the electron microscope, has been used to extend the range of observation to smaller-size pieces of muscle than can be seen with the light microscope. The interpretation of electron micrographs of muscles dramatically altered the acceptable models of muscular contraction, at the same time emphasizing the need for further studies of protein structure before muscular contraction can be understood at a molecular level.

9.2   Muscle Structure

Muscles are found in all the more advanced animals, both invertebrate and vertebrate. All are transducers converting chemical energy into electrical energy, heat energy, and useful mechanical energy. Muscles appear in a variety of sizes and shapes; they differ in the forces they can exert and in their speed of action. In this chapter, only vertebrate muscles will be discussed.

Anatomically, muscles can be classified in many ways, in terms of function, of innervation, of body location, of embryological development, and of histology. The histologic classification is the most widely used and probably the least ambiguous. Histologically, one can distinguish two types of muscles in vertebrates: striated and smooth. Striated muscle, when viewed under the microscope, appears to have alternate dark and light bands distributed in a regular pattern across long fibers. Smooth muscle consists of shorter fibers with no striations.

Striated muscles form a large portion of the meat eaten by humans. If one examines a piece of steak, one notes there are large bundles or subdivisions of the muscle (Fig. 9.1(a)). The entire muscle is surrounded by a sheath of

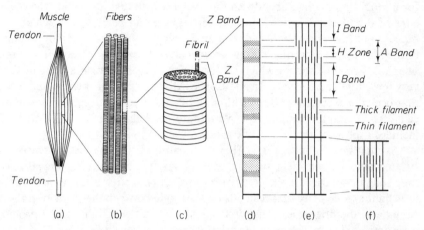

Figure 9.1   A muscle (a) consists of a large number of fibers, which in the light microscope show cross-striations (b). The single fiber is made up of numerous fibrils (c). At higher magnification these fibrils show alternating light and dark bands (d). The model (e), showing the regular arrangement of thick and thin filaments, is based on electron micrographs. During contraction, the thin filaments slide in between the thick ones, and the muscle fiber shortens (f). After K. Schmidt-Nielsen, *Animal Physiology* (Englewood Cliffs, N.J.: Prentice-Hall, Inc., reprinted by permission; 1970).

connective tissue. Between the large bundles comprising the muscle run connective tissues, blood vessels, and nerves. Each large bundle is then divided into smaller bundles, and each of these is finally subdivided into "muscle fibers" (Fig. 9.1(b)). The major portion of the striated muscle is made up of these fibers, 10–100 μm in diameter, and of lengths that reach 100 cm or more in the larger vertebrates. A piece of the fiber under high magnification would look something like that shown in Fig. 9.1(c).

The ends of the fibers of many striated muscles are attached to tendons. Throughout the length of the muscle fiber run still smaller fibers called *myofibrils*. These possess characteristic striations (Fig. 9.1(d)). Since the corresponding striations of adjacent myofibrils are lined up with one another, the entire muscle fiber appears striated. Besides the fibrils, a striated muscle fiber contains several other organelles and is surrounded by a special membrane called the *sarcolemma*. (The prefixes myo- and sacro- both are used widely to identify muscle and musclelike structures.) The organelles include small bodies associated with oxidative mechanisms known as mitochondria, as well as many nuclei. Thus, one may regard the striated muscle fiber as a single, polynuclear cell, but the entire concept of cell becomes less useful in this connection.

Three types of striated muscles are known: (1) the skeletal muscles, which form long, unbranched fibers with the nuclei distributed just inside the outer edge of the fiber; (2) special muscles of the face and head region, which are made up of branched fibers with cell nuclei located just inside the outer edge of the fiber; and (3) cardiac muscle, in which the nuclei are at the center of the fiber cross section and in which all the fibers branch to such an extent that very few ends can be found. In addition, cardiac muscle has intercalated disks that occur between the cell nuclei and divide the fibers into units resembling cells. This chapter emphasizes vertebrate skeletal muscles. Chapter 10 includes aspects of the action of cardiac muscle.

As mentioned above, Fig. 9.1(d) shows a number of bands present in all striated muscle fibers. The bands that stain dark are also birefringent; that is, they split unpolarized light into two beams. Any such substance also transmits light at a velocity that depends on the angle between the plane of polarization and the fiber axis. This birefringence is due to the lining up of protein macromolecules in the muscle striations as is discussed in Sec. 9.4. The birefringent bands are labeled A, for anisotropic; that is, index of refraction depends on direction of the incident light. By contrast, the less heavily stained bands have no polarizing properties. They are labeled I, for isotropic. In many ways, the *I bands* are harder to understand than the *A bands*, for protein molecules are oriented in both. In the center of the I band is a darker-staining membrane called the *Z line*. In the center of the A band is a lighter staining region called the *H zone*. As mentioned previously, the cell concept is not too helpful in discussing muscle fibers. However, a similar repeating unit is helpful to describe muscle structure and function at the "subcellular" level. Such a repeating unit, called a *sarcomere*, is chosen to run from one Z line to the next. A sarcomere may include no nuclei, or one, or even more than one; it is in no sense of the word a cell.

Modern electron microscope techniques permit the determination of still more details of the structure of the myofibrils. It is possible to make electron micrographs of the ultrastructure of the muscle without dispersing or homogenizing it in any way. For these studies, the muscle is first "fixed" to harden the protein elements. Then it is "stained" with a heavy metal to increase contrast in the electron microscope. Next, it is filled with, and embedded in, a plastic such as butyl methacrylate. Finally, it is cut into 10- to 20-nm sections. When these sections are examined in the electron microscope, most are cut at such angles to the myofibrils that they are useless for analysis, but a few will be either at right angles to the myofibrils or along the myofibril. A great deal of judgment is necessary to discard most of the sections as useless.

These studies show the existence of filaments of two types, thick and thin. The thick filaments are about 10 nm in diameter, about 2 μm (2,000 nm) long, and are composed almost entirely of the protein *myosin*. The thin filaments are about 5 nm in diameter and 1.5 μm long and are composed of the protein *actin*, with smaller amounts of two other proteins *tropomyosin* and *troponin*. These proteins are discussed in more detail in Sec. 9.4.

As shown in Fig. 9.1(e), the I bands consist of thin filaments joined by a membrane at their centers (the Z line). The H zone consists only of thick fibers, and the A band is a region of overlap between the thick and thin filaments. These are arranged in a regular array with a definite number of thin filaments surrounding a thick one, varying from two in the flight muscles of insects to six in some vertebrate muscles. Between the thick and thin filaments, there appears to be a series of cross bridges spaced about 5 or 6 nm apart.

When a muscle (or myofibril) contracts, the length of the A band remains constant. From Fig. 9.1(e) we see that the length of the A band is the length of the thick filament. This implies that the thick filaments do not change in length on contraction. Both the I band and the H zone are shortened on contraction. The decrease in length of both these regions is comparable. Therefore, as is shown in Figs. 9.1(e) and 9.1(f), the length of the thin filaments also must remain unchanged on contraction. The interpretation of the electron micrographs, then, is that the thin filaments slide in between the thick ones as the muscle contracts. This is discussed further in Sec. 9.4.

Electron microscope studies have shown that muscle fibers also contain a specialized endoplasmic reticulum (see Appendix E) called the *sarcoplasmic reticulum*. This includes *lateral tubules, transverse tubules*, and *cisternae*, all of which are illustrated in Fig. 9.2. Each fibril is completely surrounded by a lateral tubule. The transverse tubules run from the membrane through the Z bands of the muscle fiber to the lateral tubules, joining together with the latter to form structures called *triads* because of their appearance in transverse sections. Near the triads, the tubules become wider, to form the cisternae.

Although the transverse and lateral tubules come very close together, they are separate. The lateral ones are completely closed, whereas the transverse ones are open to the medium surrounding the sarcolemma. Thus, one may consider the transverse tubule as a part of the sarcolemma extending deep within the cell. As will be noted in the following sections, the sarcoplasmic reticulum plays an important role in muscle excitation and contraction. The tubules may, in addition, help maintain metabolic equilibrium throughout the muscle fiber.

Vertebrate muscles which are not striated are called *smooth*, because they are not made up of bundles of small groups of fibers. Smooth muscles, in

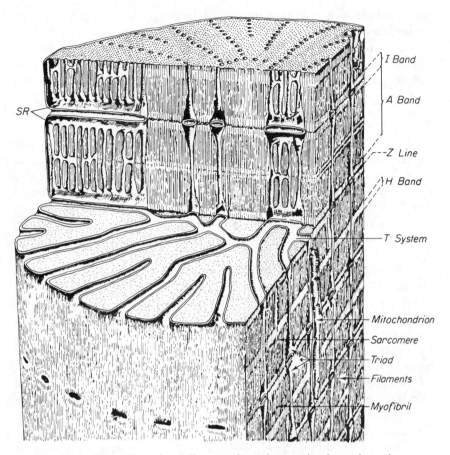

SR

I Band

A Band

Z Line

H Band

T System

Mitochondrion

Sarcomere

Triad

Filaments

Myofibril

Figure 9.2   Three-dimensional diagram of vertebrate striated muscle to show sarcoplasmic reticulum and muscle filaments in longitudinal sections with the transverse T system of tubules in cross sections and their interconnections (triads) at the Z regions. From K. R. Porter and C. Franzini-Armstrong, "The Sarcoplasmic Reticulum," Copyright © 1965 by Scientific American, Inc. (212(3):75). All rights reserved.

contrast to striated ones, consist of short spindle-shaped single-nucleated cells of isotropic material. The cells are usually 15–20 μm long, although some reach a length of 500 μm. A diagram of a typical smooth muscle cell is shown in Fig. 9.3. The maximum cell thickness at the center of the spindle is usually about 6 μm.

Smooth muscle cells differ from striated muscle sarcomeres in several other ways. For example, they lack both the oxygen-storing protein *myoglobin* (see Sec. 9.4) and a highly developed sarcoplasmic reticulum. Thus,

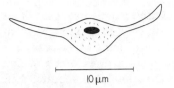

10 μm

Figure 9.3   Smooth muscle cell.

they must store oxygen and contract by different mechanisms. Indeed, while intact striated muscles rarely contract more than a small fraction of their original length, smooth muscles change their length manyfold. This large change is believed due to a slipping of one smooth muscle cell over another. In all cases of muscular contraction, little if any change of volume occurs.

Muscles are sometimes classified by criteria other than histological ones. In terms of function and innervation, muscles are separated into *voluntary* and *involuntary*. For an objective definition, those muscles under direct control of the frontal gyrus of the cerebral hemisphere might be called voluntary. By and large, striated muscles are voluntary and smooth muscles are involuntary, but this is not a hard and fast rule. Certain smooth muscles are under conscious voluntary control in some individuals and not in others. Likewise, very few individuals can voluntarily control all of their striated muscles.

Muscles may also be classified by their kinetic properties. In terms of speed of response, smooth muscles such as bladder and uterine muscles often take several seconds to contract. Striated muscle, in contrast, usually contracts rapidly, often reaching its maximum response in a few milliseconds. Within the same animal, faster muscles are usually paler, and slower ones are usually darker. The chicken is a particularly good example of this. The wing muscles work rapidly and are pale, whereas the slower leg muscles are dark. This color is more closely associated with the protein myoglobin than it is with the histological structure. In the next section, the kinetics of the contraction of striated skeletal muscles are described.

9.3   Physical Changes
During Muscular Contraction

A. CHANGES OF TENSION AND LENGTH

When a muscle is stimulated, it twitches. If the muscle is held at constant length, it develops a force, whereas if it is allowed to move, it contracts and may do work. Either type of response or any combination of the two is called a *twitch*. The two simplest situations to study are constant length (isometric) and constant force (isotonic). To eliminate the nervous control, it is possible to remove the muscle from the animal body or to cut the nerve fibers.

If one stimulates an excised muscle by means of an electrical shock (or a mechanical impulse, or heat, cold, and so on), a twitch occurs. If the stimuli are spaced a long time apart, the muscle relaxes to its original length between twitches, and, for isotonic contractions, a contraction curve is obtained as shown in Fig. 9.4(a). If the stimulus is repeated before relaxation occurs,

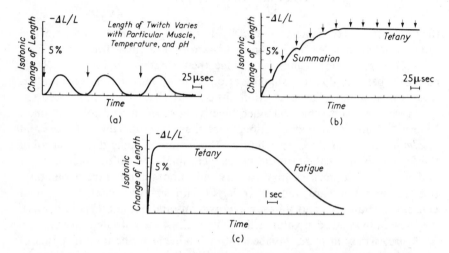

Figure 9.4 Curves of contraction. (a) Occasional stimulation shows twitches. Arrows indicate stimuli. (b) Frequent stimulation leads to summation. (c) Prolonged tetany leads to fatigue. Note the difference in the time scale as compared to (a) and (b). After S. Cooper and J. C. Eccles, *J. Physiol.* 69:377 (1930).

summation is observed as shown in Fig. 9.4(b). With still more rapidly repeated stimulation, a smooth contraction curve results such as shown in Fig. 9.4(c). The steady contraction is called *tetany*. All muscles will eventually fatigue and fail to contract, even though stimulated. This type of fatigue probably never occurs in the healthy intact animal, as the nervous system undergoes fatigue before the muscles do.

Curves illustrating the strength of isometric and isotonic contractions are shown in Figs. 9.5(a) and 9.5(b), in terms of effect of length on tension developed in isometric contraction and of load on shortening produced during isotonic contraction. Only in isotonic contraction is external work done. It is easy to show that the maximum work is done at half the maximum load for muscles for which the straight-line relationship of Fig. 9.5(a) is valid. The straight line can be described by

$$\Delta L = \Delta L_{max} \left(1 - \frac{F}{F_{max}} \right)$$

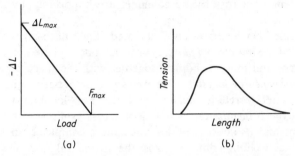

Figure 9.5 (a) Isotonic contraction. Change in length is plotted as a function of load for a muscle supporting a fixed load (isotonic). The straight line is an approximation only. (b) Isometric tension. Maximum tension developed is plotted as a function of length for a muscle held at various fixed lengths (isometric).

where ΔL is the contraction and F is the load. The work W done on the load is

$$W = F\Delta L_{\max}\left(1 - \frac{F}{F_{\max}}\right).$$

This work is a maximum when

$$\frac{dW}{dF} = 0 \quad \text{that is, when} \quad F = \tfrac{1}{2}F_{\max}$$

Striated muscles, in general, can develop large forces against a given load but even in tetany can contract only a small amount. In the vertebrate body, the skeletal muscles all develop far larger forces than the loads they move. However, the load moves more than the muscle contracts. This is accomplished by the lever action of the muscles and bones, with the joints serving as pivots. The mechanical advantage is considerably less than 1, that is, the force of the muscle is much greater than that of the load, and the muscle motion is much less than load motion.

To study muscular contraction, it would appear desirable to work with single muscle fibers. However, these are difficult to obtain and few people have succeeded in preparing them. Most experiments have been done on whole muscle.

B. INTERACTION WITH THE NERVOUS SYSTEM

In the intact animal, the muscle contracts following stimulation by the nervous system. The incoming impulses in the nerve fibers are called *electrical spike potentials*; similar spike potentials travel along the muscle fibers. The interactions of the nerve and muscle fibers and the magnitude and form of the electrical potentials across the sarcolemma are important physical characteristics of muscle.

Each muscle fiber is separately innervated. Each has at its end a special structure called the *muscle end plate*, near which one or more nerve fibers also end. The nerve and muscle endings, together with the space between them, is called the *myoneural junction*. When a spike potential reaches the nerve endings, the latter secrete a chemical activator acetylcholine, which is also important in the transmission of impulses across synapses between nerves. (Acetylcholine and its action are described more completely in Chapter 5.) The released acetylcholine diffuses across the myoneural junction (which is of the order of a few tenths of a micrometer) and stimulates the formation of a spike potential in the muscle fiber. The acetylcholine is rapidly destroyed by a protein catalyst, cholinesterase, present in the muscle end plate. Under certain conditions, the myoneural junction acts as a "computer," putting out a number of muscle spike potentials different from the number of incoming nerve spike potentials.

The muscle fiber membrane is polarized, just as is the axon membrane discussed in Chapter 5. An action spike potential, similar to that in nerves, is the first result of stimulation of a muscle fiber, whether the stimulus be the physiological one from the nervous system or an artificial one, that is, electrical, mechanical, or thermal. A typical muscle spike potential is shown in Fig. 9.6. The action potential differs from that in nerves only in the duration of the peak, which lasts much longer in muscle than in nerve.

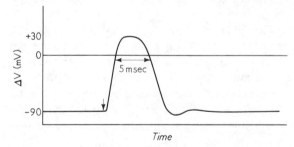

Figure 9.6 Spike potential of striated muscle. *V* is the potential difference inside, minus that outside the sarcolemma. The arrow indicates the application of stimulus. In cardiac muscle, the peak of the crest of the action potential lasts much longer.

Originally, the muscle potentials were recorded by means of *bipolar* or *differentiating electrodes*, which measured the potential difference between two neighboring spots on the muscle. These gave no possibility of measuring a resting or dc potential, nor any certainty of the size of the cellular potentials. These electrodes have been replaced by microelectrodes made by drawing out a capillary glass tube to a diameter of less than 1 μm. The tiny capillaries may be inserted through the wall of a single muscle fiber without damaging the fiber. With such probes, it is possible to measure both the resting potential and the action potential of skeletal muscle fibers. An additional difficulty is that the muscle fiber moves during contraction. Provision must be made to permit the microelectrode to move with the fiber. When this is done, consistent records can be obtained of the potentials across the sarcolemma of single muscle fibers. The striated muscle fiber typically both is very long and is composed of many myofibrils, all of which should contract (or twitch) simultaneously so that the entire fiber acts as a single unit. The spike potential is believed to travel along each transverse tubule through each Z band at the same time as it passes over the surface of the muscle fiber. From the transverse tubules, the excitation signal is apparently passed to the lateral tubules which are directly involved in initiating the twitch.

The spike potential always precedes contraction. After the crest of the spike has passed, the membrane potential starts to return to normal. At this time, the rate of heat production increases. A fraction of a millisecond later, there is a slight relaxation, and then the mechanical contraction starts. The method by which the spike potential "signals" the muscle fiber to begin the chemical changes necessary for a twitch is discussed in Sec. 9.4. The spike always precedes a twitch and the myofibrils usually all contract simultaneously. However, in pathological conditions spike potentials may occur without such contraction.

Within all skeletal muscles are sensing organs known as *proprioceptors* or *Pacinian corpuscles*. These continuously send back "reports" to the central nervous system on the state of contraction of the muscle. Thus, in any muscular motion, a complex process occurs, involving multiloop feedback systems. The nervous system signals the muscle to contract. As it does so, the muscle sends many reports, indicating its contraction to the central nervous system. These and similar proprioceptor reports from other muscles reach the central nervous system, where they are all "analyzed." As a result of this analysis, the original muscle is "instructed" or controlled to contract faster or slower so as to achieve the desired position. This process has appealed to servomechanism experts who have carried out quite detailed analyses of muscular contraction. Although such analyses can never supply new facts, they have made it possible to understand qualitatively the organizing principles of the muscle–nervous system relationship.

The problem of muscular fatigue also appears to involve the nervous system. A denervated muscle can be held in tetany by repeated stimulation until it tires. However, if the motor nerve causing a muscle to contract is stimulated, it can be shown that the myoneural junction fatigues before the muscle does. Similarly, if the entire normal animal is stimulated, it can be shown that fatigue sets in at the synapses in the central nervous system before the myoneural junction has fatigued.

C. HEAT PRODUCTION

Besides studying forces, work, and electrical changes, several biophysicists have followed a quite different approach, the measurement of the heat produced by resting and contracting muscles. Muscles produce extra heat when they are working; the extra heat accompanies the conversion of chemical energy to mechanical work. These heat measurements are based essentially on temperature measurements. They are difficult because the maximum temperature rise associated with a muscle twitch is only 0.003°C, and the heat is developed very rapidly. A. V. Hill refined his techniques to the point that he could resolve a few millionths of a degree change in a few milliseconds.

Hill's experiments showed that three different types of heat production occurred during muscular contraction. The first, called *resting heat*, is associated with metabolism in the resting muscle. The second type of heat production, *initial heat*, accompanies actual contraction and relaxation. The third general type is called *recovery heat*; it is liberated for 20–30 min following activity.

The resting heat is an indication of continuous metabolism in the muscle. It can be altered by stretching the muscle as well as by changes in ionic strength in the surrounding fluids. It is not a constant or simple quantity.

When a muscle contracts and then relaxes, the second type of heat production overrides the resting heat production. This initial heat consists of several components. While the muscle contracts, it develops a *maintenance heat*, which starts just after the spike potential passes and continues until relaxation. Some of this maintenance heat is actually produced before contraction occurs. There is, in addition, a *heat of shortening*. Under isotonic conditions when the muscle lengthens, a *heat of relaxation* is measured equal to the work done on the load.

These heat changes attracted the interest of many investigators. However, they are difficult to interpret. There is no simple relationship between the work done and the extra heat produced. The reasons for the rise in heat production before contraction and the dependence of resting heat on muscle length are still not understood. This basic lack of understanding emphasizes the incompleteness of current molecular models of muscular activity.

9.4 Muscle Chemistry

In the previous section, the various physical changes accompanying muscular contraction were presented. These all involve molecular changes and the conversion of chemical free energy to other forms of energy. (Free energy is discussed in Chapter 21.) Accordingly, it is appropriate to examine the chemical constituents of muscle. These include the types of molecules active during contraction and relaxation. The chemical transformations necessary for energy production are also indicated.

A. CHEMICAL COMPOSITION

There are more water molecules within the muscle, and indeed within the myofibril, than any other type of molecule. Present theories do not assign any specific role to these water molecules, aside from forming a medium through which the contractile molecules act and also through which the energy-carrying molecules diffuse. The various organelles within the muscle, for example, nuclei and mitochondria, have the same composition as those of other cells. Except for the contents of the transverse tubules, which is similar to the surrounding medium, the ionic concentration within muscle fibers resembles that within nerve fibers, as discussed in Chapter 5. However, relatively high concentrations of Ca^{2+} can occur in the cisternae, where this ion is presumably stored between contractions (see the end of the section).

Another component of striated muscle is the protein *myoglobin*. This red pigment, found only outside the myofibril, is similar to the hemoglobin of red blood cells except that myoglobin has about one fourth the molecular weight and only one iron atom per molecule (hemoglobin has four iron atoms per molecule). Myoglobin is generally believed to act as a storage for oxygen within the muscle fiber.

The myofibrils contain high concentrations of four proteins—myosin, actin, tropomyosin, and troponin—all of which are directly associated with muscular contraction and are usually not found elsewhere in the body. All four are members of a general class of proteins called *globulins*, when classified in terms of their solubilities. (Proteins are condensation polymers formed from small monomers known as *amino acids*. The structure of proteins, including those in muscle, is discussed more fully in Appendix D.) These proteins, like many globulins, can exist in either a globular (spherelike) form or a fibrillar form. Small changes in the ionic strength, pH, or temperature can convert some globulins reversibly from the fibrillar to the globular form. (In the fibrillar form, they are believed to be arranged in a helical structure described in Chapter 15.)

B. CONTRACTION

The striking physical changes that take place as these globulins shift from one form to the other first suggested that they might be the molecules actually responsible for contraction. Present evidence supports the conclusion that these four proteins form the contractile elements. However, the premise that they change from globular to fibrillar form appears to be fallacious. Rather, it appears that myosin and tropomyosin are always in the fibrillar form in intact muscles, and actin and troponin are always globular. They are formed into thin and thick filaments, visible only with the electron microscope, as shown in Figs. 9.1(e) and 9.1(f). These filaments are believed to develop the actual contractile forces.

As mentioned in Sec. 9.3, the thick filaments are composed primarily of myosin. The individual myosin molecules are thin fibers 145 nm long, with two small globular "heads" at one end. Within a thick filament, the myosin molecules are lined up with their heads oriented toward the closest end. The heads project out from the filament and are seen in detailed electron micrographs as fine cross bridges which appear to link the thick and thin filaments. Since the heads are oriented toward the ends of the filaments, there is a "bare zone," free of projections, in the center of the filament.

The thin filaments are composed of actin, tropomyosin, and troponin. Actin is present in largest amounts. It is a single-chain globular protein of 374 amino acid residues with a molecular weight of 42,000. Like the fibrillar myosin molecule, it is asymmetrical, with distinct "head and tail" or "front and back" regions. The actin molecules are arranged in the thin filament "head to tail" in two long strands; thus, the strand is polar, with a recognizable front and back. The two long actin strands form a "coiled coil" within a groove between the strands. The fibrillar tropomyosin molecules, each with a molecular weight of 35,000, form long, coiled coil strands themselves, one of which attaches to each long actin strand. One tropomyosin molecule extends over seven actins. One globular troponin molecular complex (three protein chains, total molecular weight 80,000) is attached near one end of each tropomyosin molecule. A detailed drawing of the thick and thin filaments is shown in Fig. 9.7.

Given the detailed three-dimensional representation of muscle filaments shown in Fig. 9.7, we can begin to explain the chemical events of muscle contraction. As mentioned in Sec. 9.3, the thin filaments slide past the thick filaments during contraction. Myosin head cross bridges cause that sliding by each attaching to an actin molecule, swiveling or turning in such a manner as to move the filaments relative to one another, and then detaching from the actin and reattaching to repeat the cycle. One cycle of attachment would cause a relative movement of about 10 nm.

C. ACTIVATION

The thick and thin filaments of the myofibril are quite large on an atomic scale. Yet, when a myofibril contracts and relaxes, it uses chemical energy that is derived from a much smaller molecule called *adenosine triphosphate*, or ATP. This small molecule is the source of immediately available chemical

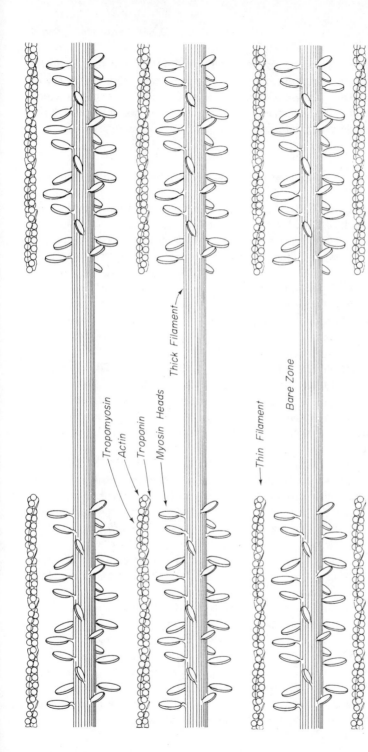

Figure 9.7 Thick and thin filaments interdigitate in an orderly array to form a muscle fiber. Two sets of thin filaments extend toward each other from adjacent Z lines. They lie between, and partially overlap, a set of thick filaments. The combination accounts for the striated appearance of muscle, as shown in Fig. 9.1. In this arrangement myosin heads of thick filaments act as cross bridges that make contact with actin molecules in the thin filaments. The heads are oriented in opposite directions on each side of the central bare zone. The muscle contracts when a nerve signal initiates a sequence of events that causes myosin heads to attach to actins, swivel, and then break contact, thus propelling the thin filaments past the thick ones and shortening the muscle. From J. M. Murray and A. Weber, "The Cooperative Action of Muscle Proteins." Copyright © 1974 by Scientific American, Inc. (230(2):60). All rights reserved.

energy for chemical syntheses, for muscular contraction, and for the active transport of ions and metabolites across cell membranes. A wide variety of systems within all cells can use ATP as a source of energy. One way in which this happens is that the molecule ATP is split into adenosine diphosphate, ADP, and inorganic phosphate, $\textcircled{P}$. Symbolically, one may write this as

$$ATP \; \rightleftharpoons \; ADP + \textcircled{P} + energy$$

(The structure and function of ATP is discussed in more detail in Appendix D.)

ATP binds to a myosin head cross bridge, activating it to a form that can bind to an actin molecule. When the ATP in this "active complex" is hydrolyzed, the energy released powers the swiveling of the cross bridge and a "rigor complex" results. This persists until a new ATP molecule can bind to the cross bridge to cause a new cycle of detachment, reattachment, and swiveling. (The stability of the rigor complex in the absence of ATP accounts for *rigor mortis*, the extreme muscular rigidity that occurs after death.) The exact molecular and atomic conformation of ATP and proteins in the active cross bridge, active complex, and rigor complex is a subject of continuing research.

There is yet another requirement for muscle contraction. The discussion above assumes that an activated cross bridge can attach to any actin and form an active complex. This is false. Rigor complexes can be formed in this manner, but the thin filament must itself first be activated before an active complex can be produced. This activation occurs when calcium ions attach to the troponin complex.

Thus, muscular contraction is initiated by the release of Ca^{2+} from the lateral tubules of the sacroplasmic reticulum in response to the action (or spike) potential transmitted to these tubules by the transverse tubular system. This Ca^{2+} is released from storage, presumably in the cisternae associated with the lateral tubules. The release of Ca^{2+} allows cross bridges to attach and contraction to occur.

The lateral vesicles also contain an active calcium pump. Thus, immediately following the rise of Ca^{2+} accompanying the spike potential, the excess Ca^{2+} starts to be removed, returning it to the cisternae. Unless another spike potential appears within a few milliseconds, the Ca^{2+} concentration at the myofibril falls to levels that allow relaxation to occur.

The exact mechanism of calcium's activating role is a subject of much debate, but the following seems clear. When there is no calcium present, troponin lies on the tropomyosin filament along the actin strands near the outer edge of the groove. When enough calcium ions bind to troponin, it changes conformation and tropomyosin moves deeper into the groove. This exposes a specific cross-bridge binding site on the actin.

Muscle chemistry is thus a complicated part of biochemistry. The biophysical aspects of this field are related to the processes that initiate and control contraction and the energetics of these processes.

9.5 Summary

Muscles are the contractile elements of animals. They act as transducers, converting chemical energy into mechanical energy and heat. Muscles in vertebrates can be classified according to function and to morphology. Of the various types, the striated muscles, usually associated with voluntary motion, have been studied in greatest detail. Their efficiency, the tensions developed at constant length, and the shortening produced with various loads have all been measured and are well known for many different muscles.

Each striated muscle consists of bundles of small groups of individual muscle fibers. The single, striated muscle fiber, about 10 μm in diameter, is surrounded by a single membrane electrically polarized in a fashion similar to that of a nerve fiber. The initial step in the contraction process is an action or spike potential, very similar to that of nerve fibers. This spike potential is normally initiated at the muscle end plate but can also be produced by the same types of stimuli which affect nerve fibers.

The striated muscle fiber contains many nuclei, mitochondria, and microsomes, as well as long myofibrils having the same striations as the muscle fiber. The myofibrils contain two types of filaments, which, in turn, are composed of helical fibers of the proteins myosin, actin, tropomyosin, and troponin. The two types of filaments appear to overlap slightly in electron micrographs of extended muscles; they intermesh more completely in similar electron micrographs of contracted muscles. The changes during contraction are brought about at the expense of chemical energy stored as ATP.

The energy of ATP is released when the latter is split into its components, ADP and phosphate. This splitting is catalyzed by enzymes called ATPases. The protein myosin is an ATPase. The molecular details of how the energy is transferred from ATP to mechanical contractions are not completely known, but calcium ion plays an important role. The release of calcium ion by the nerve impulse may well be the primary event in muscle contraction.

The basic physical parameters of the gross phenomena associated with muscular contraction are well known, and many of the chemical mechanisms are similar to those in other tissues. In contrast, the molecular and atomic description of muscular contraction is an active research area. The ideas involved demand a knowledge of active transport to understand the membrane action, enzyme kinetics to describe the synthesis and use of ATP, and protein structure to describe the filaments and their behavior during contraction. These and related concepts are presented in more detail in Chapters 15, 18, and 23 and in Appendix D.

REFERENCES

There are many books that deal only with the contraction of striated muscles. Most physiology, biochemistry, and anatomy texts have at least a chapter on this subject. The following list is neither complete nor exhaustive but contains a limited number of reviews and summary articles which should be especially useful to readers wishing to pursue this subject more thoroughly.

1. *Scientific American* continues a tradition of explaining topics of current research to the nonspecialist. Three articles relating to muscles:
   (a) MERTON, P. A., "How We Control the Contraction of Our Muscles," **226** (May 1972): 30–37.
   (b) MURRAY, J. M., AND A. WEBER, "The Cooperative Action of Muscle Proteins," **230** (February 1974): 58–71.
   (c) COHEN, C., "The Protein Switch of Muscle Contraction," **233** (November 1975): 36–45.

2. *Annual Review of Biophysics and Bioengineering*, Annual Reviews, Inc., Palo Alto, California, is a series well worth examining by every serious student of biophysics. Articles especially relevant to the current chapter include:
   (a) TONOMURA, Y., AND F. OOSAWA, "Molecular Mechanism of Contraction," **1**:159–190 (1972).
   (b) INESI, G., "Active Transport of Calcium Ion in Sarcoplasmic Membranes," **1**:191–210 (1972).
   (c) SQUIRE, J. M., "Muscle Filament Structure and Muscle Contraction," **4**:137–163 (1975).
   (d) MACLENNAN, D. H., AND P. C. HOLLAND, "Calcium Transport in Sarcoplasmic Reticulum," **4**:377–404 (1975).

3. HOAR, W. S., *General and Comparative Physiology*, 2nd ed. (Englewood Cliffs, N.J.: Prentice-Hall, Inc., 1975), 848 pages.

4. YOST, H. T., *Cellular Physiology* (Englewood Cliffs, N.J.: Prentice-Hall, Inc., 1972), 925 pages.

CHAPTER 10

Mechanical and Electrical Character of the Heartbeat

10.1 Role of the Vertebrate Circulatory System

All vertebrates possess a closed circulatory system. The blood that circulates through this system is a suspension of various types of single cells in a viscous solution containing, among other solutes, proteins, and inorganic salts. The blood is pumped; that is, it is forced to flow through the closed circulatory system. The organ which does the pumping is called the heart.

The circulatory system in vertebrates carries oxygen from special exchange organs (lungs or gills) to the other tissues. It also transports carbon dioxide from the tissues back to the lungs or gills. In some amphibians the skin also serves as an auxiliary gas exchanger. In any case, the blood flows through a special exchange organ in which very thin, moist walls separate the blood from the external environment.

The blood performs many other functions besides O_2 and CO_2 transport. Foods, metabolic wastes, and endocrine secretions are carried from their sites of production to target cells or organs. Likewise, blood contains antitoxins and phagocytic cells which help protect the organism from external invaders.

The kidney is a third major organ (besides heart and lungs) which interacts with the blood. It removes waste products and excess water from the blood. In this manner it maintains blood volume, ionic concentrations, osmolality, and, indirectly, arterial pressure as close as possible to neural-endocrine-selected set points.

The vertebrate circulatory system, then, is a major internal transportation line for chemical substances. The vessels into which the heart pumps blood are named *arteries*. These branch into smaller and smaller arteries; the smallest are called *arterioles*. The arterioles empty into *capillaries*. Here, most of the exchanges occur between the blood and the surrounding tissues. The capillaries join to form *venules*, which in turn join to form larger and larger *veins* leading back to the heart. This regular pattern is broken at several points, such as the kidneys, the pituitary, and the liver. For example, blood is brought to the liver both by the hepatic artery and by the portal vein, the latter coming from the intestines. Both the hepatic artery and the portal vein break up into capillaries in the liver; the blood leaves the liver via the hepatic vein.

The circulatory system is not completely closed, however. Some fluid leaves the capillaries, passing into the tissue spaces; it is then called *lymph*. The lymph filters back slowly through several nodes, finally entering the venous portion of the circulatory system. The circulatory system is continually in a dynamic steady state with water losses to kidneys, lungs, and exocrine organs (such as the salivary glands), as well as to the tissue spaces. Water is continually reabsorbed from the gut and also is formed by metabolic processes. As noted above, the primary regulatory organs for blood volume are the kidneys.

10.2 Blood Pressures and Velocities

Before the action of the heart is examined, the flow of the blood through the arteries and veins will be discussed briefly. The flow of the blood can be described in terms of its linear velocity (i.e., speed[1]) v and its pressure, p. The linear velocity is, in general, a function both of time and of the point in space at which it is measured. The pressure, p, is the force per unit area of the fluid. It is a scalar quantity; that is, p is independent of the orientation of the areas used to define it. The zero point for pressure is somewhat arbitrary. *Gauge pressure* is the difference between the absolute pressure and the atmospheric pressure. *Absolute pressure* is zero when there is no net external force on the system.

Pressure is a stress and has the dimensions of force per unit area. In the SI system it is measured in N/m^2. Instead of stress units, pressure is often

---

[1] Note that in this text we have used the customary words "velocity" or "linear velocity" rather than the more precise term, "speed."

measured in terms of the height of a column of liquid which it will support. Thus, it may be measured in terms of any convenient unit of height, such as meters of mercury or meters of water. The most frequent unit used in describing the circulatory system is mm Hg.

Besides pressure and velocity, another fundamental property of a fluid is its density, ρ. For all purposes in this chapter, the blood may be considered to be incompressible. Its density is approximately that of water. The percentage of the blood volume represented by formed cells (primarily erythrocytes) is called the *hematocrit*. Although diagnostically important, the hematocrit only affects the mechanical and electrical character of the heartbeat to the extent that it alters the apparent *viscosity* of the blood.

The variation of blood velocity, v, in a mammal is shown in Fig. 10.1.

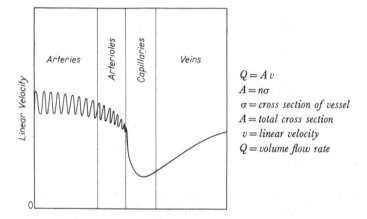

$$Q = A v$$
$$A = n\sigma$$
$$\sigma = cross\ section\ of\ vessel$$
$$A = total\ cross\ section$$
$$v = linear\ velocity$$
$$Q = volume\ flow\ rate$$

Figure 10.1 Linear velocity of the blood. Since the volume flow rate, Q, remains approximately constant throughout the circulatory system, a low linear velocity, v, means a large cross section A. In the capillaries, the vessel cross sections, σ, are small, but the number in parallel, n, is so large that A is greater in the capillaries than in the arterioles or veins. After C. H. Best and N. B. Taylor, *The Physiological Basis of Medical Practice*, 7th ed. (Baltimore, Md.: The Williams & Wilkins Company, 1961).

Although the arteries and veins are much larger than the capillaries, there are so many capillaries that the total cross-sectional area of the tubes open to the blood is much greater than in the larger vessels. Accordingly, the linear velocity of the blood in the capillaries is smaller than in the arteries and veins. The pulsations in the arteries are possible because the walls are elastic and stretch from the force of each heartbeat.

In a similar manner, one may diagrammatically represent the pressure variations. These are shown in Fig. 10.2. The maximum arterial pressure is called the *systolic pressure*, and the minimum arterial pressure is called the *diastolic pressure*. The pressure falls by the time the blood reaches the capillaries, and the pressure fluctuations are smoothed out. As the blood enters the venous system, the pressure is still lower. Just before the blood enters the heart, the gauge pressure is negative; because this pressure is very small, it is conveniently measured in mm water. In a normal adult human, the venous gauge pressure at the heart is about -40 mm H_2O. The postural position (e.g., standing, sitting, or supine) can alter not only the magnitude but also the sign of the venous gauge pressure at the heart.

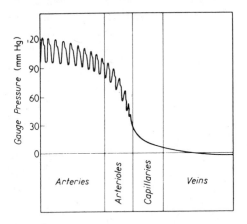

Figure 10.2 Variation of blood pressure at fixed time. Values shown are gauge pressure in a normal, adult human.

The arteries and veins have similar flow rates but very different pressures. Accordingly, both have about the same inside diameter (0.5–12.5 mm), but the arterial walls are thick and elastic, whereas the venous walls are very thin. The larger pressures in the arteries make reverse flow unlikely; valves limit reverse flow in the veins.

The capillaries are the location of most exchanges of gases, metabolites, and metabolic products. They are thin-walled and small in diameter. A red blood cell, 8 μm in diameter, distorts the shape of the capillary as it passes through. At the capillary walls, the excess gauge pressure, osmotic forces, and active transport combine to promote exchanges between the bloodstream and the surrounding tissues.

10.3   The Vertebrate Heart

In warm-blooded vertebrates, the heart keeps pumping for the entire life of the organism. If the heart stops even for a short time, the animal dies.[2] This

---

[2] During surgery, the heart beat may be stopped provided that circulation is maintained by an artificial pumping mechanism bypassing the heart and lungs.

continuous activity is regulated by both the nervous and the endocrine systems. However, even without these regulatory influences, the heart maintains its rhythmic beat. In cold-blooded vertebrates, the temperature also influences the heart rate. At close to freezing temperatures, their heart rate slows almost to zero.

The heart of the cold-blooded vertebrates is simpler than the mammalian heart. Most fishes and amphibians have a heart made up of a series of chambers, as shown in Fig. 10.3. The first, which receives the blood from the veins, is called the *sinus venosus*. It is the pacemaker and originates the heartbeat.

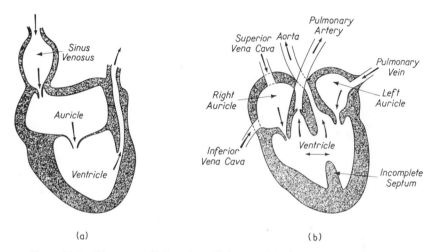

(a)   (b)

Figure 10.3   Diagrams of fish and reptile hearts. (a) Fish heart. The muscular walls develop successively higher pressures in the sinus venosus, auricle, and finally ventricle. (b) Reptile heart. Note the incomplete septum, allowing mixing of blood from both auricles within the ventricle.

The reptilian heart, also shown in Fig. 10.3, is more specialized. Instead of one auricle, there are two. One receives blood from the lungs only and the other from the remainder of the body. This system is more efficient in aerating the blood than is that of the amphibians and fishes. The sinus venosus does not exist as a separate chamber, but its homolog persists as a *sinoauricular* (s-a) *node* on the wall of the auricle serving the body proper.

The mammalian heart is illustrated in diagrammatic form in Fig. 10.4. It consists of four chambers: two auricles and two ventricles. The blood from all the body except the lungs enters the right auricle. It is forced from there into the right ventricle, then into the lungs and back to the left auricle. From there it is pumped into the left ventricle, and finally through the aorta to all arteries of the body except those going to the lungs. Thus, the blood in a complete circuit goes through the heart twice, once through the left side and

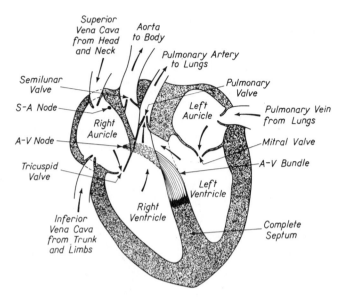

Figure 10.4   Diagram of the human heart. Arrows show direction of blood flow.

once through the right side. This system is highly efficient in supplying oxygen and removing carbon dioxide, for all the blood passes through the lungs on each trip around the circulatory system.

The walls of the heart consist primarily of muscle tissue. As in all other striated muscles, the fiber membranes are normally polarized, the inside being 90 mV negative relative to the outside. Just before contraction occurs, an action current passes over the membrane, reversing its polarity for a short period of time. The form and nature of these action currents are similar to those of nerve fibers discussed in Chapter 5. The large mass of fibers contracting simultaneously in the heart effectively acts as a large number of electric cells, all in parallel, and each with a high internal impedance. Although the net current from each fiber is small, the current from the entire muscle is appreciable, giving rise to measurable potential changes on the body surface. These potential changes are discussed further in Sec. 10.4, 10.6, and 10.7.

10.4   The Heart Sequence

A. OVERALL SEQUENCE

The mammalian heart pumps blood in a patterned sequence which repeats each beat. First the auricles contract, forcing blood through the *auriculo-ventricular* (a-v) *valves* into the ventricles. Then the ventricles contract. This shuts the a-v valves and opens the semilunar and pulmonary valves. As the ventricles continue to contract, blood is forced into the aorta and pulmonary arteries. Finally, as the ventricles relax, the semilunar and pulmonary valves close. The entire sequence is presented in more detail in Fig. 10.5, which shows, with a common time base: the auricular pressure, the ventricular pressure, the aortic pressure, and the ventricular volume for a human heart. Also, on the same base are shown the electrocardiograph (ekg) record and the sonograph record of a microphone placed against the chest. Heart pressures have been measured directly in both humans and other animals. The ventricular volumes have been found by X-ray and nuclear medical techniques.

Figure 10.5   Pressure sequences in the left side of the heart. The significance of the vertical lines is as follows: (1) the mitral valve closes; (2) the semilunar valve opens; (3) the systolic pressure reaches a maximum; (4) the semilunar valve closes; (5) the mitral valve opens; (6) end of heart sound; and (7) the auricle starts to contract. After C. H. Best and N. B. Taylor, *The Physiological Basis of Medical Practice*, 7th ed. (Baltimore, Md.: The Williams & Wilkins Company, 1961).

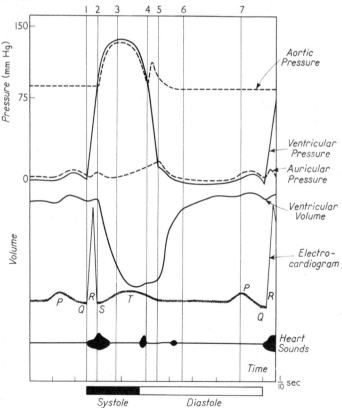

The diagram shows that the blood flows from the ventricle into the aorta during only a small part of the cycle. While this is happening, the ventricular volume falls to a minimum value, but the pressure remains close to its maximum. Likewise, the figure shows that the valves open and shut as the direction of the pressure difference across them changes.

B. ELECTRICAL EVENTS

The heart pulses rhythmically and with a definite sequence. The beat is initiated at the sinoauricular (s-a) node, shown in Fig. 10.4. The node acts in a fashion similar to an electronic multivibrator putting out an electrical pulse once per heart cycle (about 80 times per minute in humans). This pulse spreads in all directions as an electrochemical impulse over the surface of the auricle, causing the muscle fibers to contract. When two pulses reach the opposite side of the auricle from two directions, they annihilate each other, because the contracted muscle will not conduct another impulse.

Besides causing the auricle to contract, the electrochemical pulse, originating at the s-a node, also stimulates the auriculoventricular (a-v) node (see Fig. 10.4). This node, after a short time delay of about 0.1 sec or slightly less, puts out a new electrical pulse which is conducted down a special group of fibers called the *a-v bundle*, diagrammatically illustrated in Fig. 10.4. These special fibers pass down the septum and then divide into a *right branch* and a *left branch*, passing up along the outer side of the ventricles. These specially conducting fibers terminate at various points along the septum and the outer ventricular wall. While the wave of activity spreads from one myocardial fiber to the next, the conduction by the bundle fibers is faster, thereby tending to promote synchrony in the ventricular contraction. The bundle is sometimes called the *bundle of His*. Its conduction fibers are sometimes identified by the name *Purkinje*.

The s-a node resembles a free-running electronic multivibrator controlling a second multivibrator, the a-v node, which in turn controls a third multivibrator, the ventricle itself. Many factors suggest this analogy. The fundamental rate of the s-a node can be varied by two different sets of nerves which act to speed or slow the rate of firing of the s-a node. This is analogous to tuning either the resistance or the capacitance of a free-running multivibrator.

In some cases, the s-a node fails. Then the a-v node takes over control of the heart rate. The auricular contraction is no longer properly synchronized with the ventricular action, but this is by no means fatal. The a-v node behaves as an electrical multivibrator synchronized by pulses from the s-a node. When free running, it has a slower firing rate (about 50 beats/min in humans).

If the a-v node also fails, the heart does not stop and the individual does not die. Rather, the auricular and ventricular walls take over control directly. Their free-running rate is still slower (about 30 beats/min in humans). The ventricles and auricles are then completely independent in their times of contractions. On the average, the auricular beat then interferes with, rather than promotes, circulation.

The cardiac muscle fibers, like skeletal muscle and nerve fibers, have a resting potential around 90 mV, the outside being positive relative to the inside. As in skeletal muscle and nerve fibers, the action potentials are about 120 mV; that is, the outside is 30 mV negative relative to the inside at the peak of the spike. All three types of fibers are also similar in that the concentration of potassium ions is much higher within the cell than in the surrounding medium, whereas the sodium ion concentrations are just the reverse.

The cardiac muscle fibers differ markedly from skeletal muscle and nerve fibers in the kinetics of the recovery to the resting potential. In the largest mammalian nerve axons, this takes a fraction of a millisecond. In smaller nerve axons and skeletal muscle fibers, the recovery period is 2–5 msec. By contrast, some cardiac muscle fibers take as long as 200 msec to recover their resting potential. This period of time is comparable to the period of contraction of the ventricle.

A closely related property is the recovery of the normal low net permeability to potassium ions. When the resting potential of a voltage-clamped squid axon is suddenly decreased, the net permeability to potassium ions rises rapidly and then falls. The cardiac muscle cells, in contrast, do not recover their original impermeability to potassium until after the membrane potential returns to its original value.

Like nerve and skeletal muscle, cardiac muscle exhibits a "positive after potential," during which the resting potential is greater in magnitude, around 100 mV instead of 90 mV, the outside being positive relative to the inside. The after potential may last close to 500 msec before it is completely abolished. All experiments indicate that, except for time constants, and perhaps some absolute values, the electrical behavior of cardiac muscle is very similar to that of squid axons discussed in Chapter 5.

10.5   Ventricular Contractility and Energy

The contractile state of the ventricles as well as the work done by the heart have interested clinicians, biophysicists, and others concerned with the mechanical character of the heart beat. To some extent, the body can compensate for decreased contractile ability of the ventricles by increasing their filling. Thus, separate measures of the contractile status of the ventricular muscle and of the energy developed by the heart have been desired. The most direct measures of contractile status are obtained by placing strain gauges on the surface of the exposed heart and measuring the maximum velocity of change of the fiber lengths.

Even to interpret these data demands a model of some sort. The simplest, introduced originally by Hill, consists of a contractile unit, a series elastic element and a parallel elastic element as shown in Fig. 10.6. These two arms

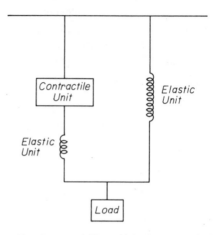

Figure 10.6   Simple contractile model of the ventricular muscle.

then support a load or mass. More complex models provide more precise descriptions, but none as yet developed has proved completely satisfactory.

During contraction of the ventricle, there is first an isovolumetric phase, as shown in Fig. 10.5, between lines 1 and 2. During this time the velocity of shortening of the contractile element V in the model of Fig. 10.6 must be equal in magnitude to the velocity of lengthening of the series elastic element. The latter has been shown to obey an exponential elastic law, so that one may write

$$\frac{d\sigma}{dl} \simeq k\sigma + c$$

where σ is the stress, l is the length, and k and c are constants. Thus, during the isovolumetric phase, it follows that

$$V = \frac{dl}{dt} \simeq \frac{d\sigma/dt}{k\sigma + c}$$

If one further assumes that the intraventricular gauge pressure, p, is a constant function of the stress, σ, then it is possible to rewrite the last equation as

$$V \simeq \frac{dp/dt}{kp}$$

If the latter is extrapolated to zero pressure, one finds that V_{max}, the maximum velocity of shortening of the contractile element, is

$$V_{max} \simeq \lim_{p \to 0} \left(\frac{dp/dt}{kp} \right)$$

This limit is independent of the filling of the ventricle and does give a measure of the state of health of the ventricular muscle. Contractility is often reported as V_{max}, as determined by the equation above. Measurement of p requires catheterization; the method is invasive but less traumatic than opening the chest and placing strain gauges on the heart wall.

In addition to the contractility, the energy associated with the heartbeat is also of interest. For this one needs to compute the kinetic and potential energy of the blood and to measure the heart output, Q (volume per unit time). A fluid such as the blood may possess both kinetic and potential energy. The kinetic energy per unit volume T is

$$T = \tfrac{1}{2}\rho V^2$$

The potential energy per unit volume Φ results from both the pressure on the fluid[3] and its height, h, above the earth. In physics texts, it is shown that, for an incompressible fluid,

$$\Phi = \rho g h + p$$

The total energy per unit volume, H, then is

$$H = p + \rho g h + \tfrac{1}{2}\rho v^2 \tag{10-1}$$

Bernoulli's equation states that H is a constant. It is true only for stream-line flow of nonviscous liquids. In general, the variation of H gives the change in energy per unit volume. The blood loses energy during each cycle in the capillaries. The heart, in pumping, increases the energy per unit volume of blood as the latter passes through the heart. Thus, the heart might be called a *chemicomechanical transducer*.

To apply Eq. (10-1) to the heartbeat, it is possible to drop the gravitational term, since relatively little change in height occurs in passing through the heart. Thus, the work per unit volume of blood is

$$H = \tfrac{1}{2}\rho v^2 + p$$

This is valid since both the kinetic and hydrostatic energies of the blood reaching the ventricle are very low. If q is the volume per stroke, then the work w for each ventricle for each heartbeat is

$$w = q\bar{H} = \tfrac{1}{2}\rho q \overline{v^2} + \bar{p}q$$

---

[3] Purists will no doubt object to calling p a form of potential energy per unit volume, but this is satisfactory for discussions of the circulatory system.

Of even greater interest is the power Π developed by the heart. To find this, one must replace q by the heart output Q and include the contribution of both ventricles. Note that Q will be the same for both ventricles, since the system, to a first approximation, is closed. Expressed analytically, this becomes

$$\Pi = \bar{p}_R Q + \bar{p}_L Q + \tfrac{1}{2}\rho\overline{v_R^2}Q + \tfrac{1}{2}\rho\overline{v_L^2}Q \qquad (10\text{-}2)$$

It is possible to introduce several additional approximations into Eq. (10-2) which aid in numeric calculation of values for Π. These are the following:

$$v_R \simeq v_L \qquad\qquad p_R \simeq \tfrac{1}{6}p_L$$

$$\overline{v^2} \simeq 3.5(\bar{v})^2 \qquad \bar{v} \simeq \frac{Q}{A}$$

where A is the average cross-sectional area of the aorta and the superscript bar represents an average over one heart cycle. Using these, Eq. (10-2) becomes

$$\Pi \simeq \tfrac{7}{6}\bar{p}_L Q + \frac{3.5\rho Q^3}{A^2} \qquad (10\text{-}2a)$$

It is instructive to compare this power for typical values for a human at rest and during exercise. A reasonable set of values are:

At rest	*Active*	*Both*
$\bar{p} = 100$ mm Hg	$\bar{p} = 100$ mm Hg	$A = 4.5$ cm²
$Q = 5.5\ \ell/\text{min}$	$Q = 35\ \ell/\text{min}$	$\rho = 1$ g/mℓ

Converting to SI units and substituting into Eq. (10-2a) gives

	At rest	*Active*	
$\tfrac{7}{6}\bar{p}_L Q$	1.5 W	10 W	"hydrostatic" power
$\dfrac{3.5\rho Q^3}{A^2}$	0.02 W	4 W	"kinetic" power
Π	1.5 W	14 W	total heart power

It should be noted that the kinetic power at rest is negligible. During exercise, both the heart rate and the stroke volume increase, making the kinetic power of the blood a significant component.

10.6 Electrocardiography

Every time the heart beats, electrical potential changes occur within it. These potentials spread to the surface of the body. Electrodes at almost any pair of points on the surface of the body will show potential differences related in time to the heartbeat. A record of these potential differences is called an *electrocardiogram*; the recording equipment is an *electrocardiograph*. The recording equipment and the records are often indicated by the abbreviations ekg or ecg.

Electrical changes at the surface of the heart were first demonstrated in 1856. Electrocardiography, the science of measuring the associated potentials, did not really develop until physical instrumentation made possible the detection of these small potentials. The first major advance was the application of the string galvanometer to electrocardiography in 1903. This was the work of Einthoven, whose ideas dominated the field for many years. Today, all electrocardiographs depend on the action of electronic amplifiers. In this field, as in many others, rapid advances have resulted from the widespread application of electronic techniques. The electrocardiogram is used in many clinical diagnoses of heart ailments. It is widely used because of its convenience and also because it can be obtained without surgical procedures or discomfort to the patient.

The electrocardiogram is a record of electrical potential differences at the surface of the body. The heart, however, is not the only source of potentials at the body surface; it is necessary to distinguish between those potentials due to the heart and those originating from other organs. Every muscle within the body undergoes potential changes as its fibers contract. The magnitude of the action potentials for all nerves and all muscle fibers is about 120 mV. The motion of any skeletal muscle can give rise to body surface potential differences comparable to the ekg potentials. To limit this source of distortion, the ekg is often recorded with the subject lying down.

The ekg potentials can be observed between almost any pair of points on the surface of the human body. If the two points are reasonably separated, the maximum potential difference observed is of the order of 1.0 mV. The ekg has the same period as the heart. Traditionally, three wires were attached to the subject, one to each arm and the third to the left leg. The ekg was then recorded between the members of each of the three resulting pairs of leads.

Whether the electrocardiogram is recorded between two points on the surface of the body or between one point and a neutral electrode, it has a form in most normal individuals somewhat similar to that shown in Fig. 10.7. A neutral electrode can be formed by immersing the subject in a tub of water and placing the electrode far from the body. Provided that low-resistance electrodes are used, the curve will usually have the general shape shown.

The various bumps on the ekg are called *waves*. The *P wave* occurs just before auricular contraction. The *QRS complex* is associated with the start of ventricular contraction, and the *T wave* occurs at the end of ventricular contraction. The amplitude of the ekg waves is shown in Table 10-1. In some cases a small *U wave* can be distinguished following the T wave, but it is

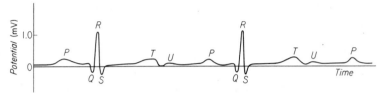

Figure 10.7 Typical ekg. The P wave precedes auricular contraction and the QRS complex is associated with ventricular contraction. Exact wave height depends on the lead used.

not usually as clear as in the idealized picture in Fig. 10.7. The U wave may represent ventricular relaxation.

Standard electrodes are placed on both arms and the left leg. In many ekg laboratories nine or more additional electrode placements are used. The ekg's are usually described in terms of *leads*, which means the potential difference between two points. This is confusing terminology because two wires, each ordinarily called a lead, are necessary for one ekg lead.

TABLE 10-1

NORMAL HUMAN ELECTROCARDIOGRAM PATTERNS

EKG interval	Amplitude (mV)	Duration (msec)	Relationship to heart cycle (Fig. 10.5)
P	0.1	8	Precedes auricular contraction by about 20 msec
P–Q	0.0	150–200	A–V delay time
Q	0.1	40–80	
R	1.0	40–80	Precedes ventricular contraction
S	0.1		
S–T	0.0	100–250	Ventricular ejection
T	0.1	100	Follows ventricular relaxation
T–P	0.0	300	Diastole

In ekg terminology, the potential differences in the three "standard" leads are numbered as

$$\text{lead I:} \quad V_I = V_L - V_R$$
$$\text{lead II:} \quad V_{II} = V_F - V_R \tag{10-3}$$
$$\text{lead III:} \quad V_{III} = V_F - V_L$$

where L, R, and F refer to the left arm, right arm, and foot, respectively, and the potentials with the three subscripts refer to the values between these points and a neutral electrode. Elementary algebra reduces these three equations to

$$V_{II} = V_{III} + V_I \tag{10-4}$$

That is, if any two of the three "standard" leads are measured, the third is thereby determined.

In the following sections of this chapter, the heart is approximated by an equivalent current dipole. On the cellular level, the heart cannot be regarded as a mere dipole. It was noted in the preceding discussion that at the start of every heartbeat, an electrical spike potential originates at the s-a node and spreads out in all directions over the auricle. Thereafter, the a-v node emits a pulse that causes spike potentials to travel down the fibers of the auriculo-ventricular bundle and its branches to initiate ventricular contraction. The spike potential travels down around the septum and then up the outer sides of the ventricles. In every region, the appearance of the spike potential is followed by a contraction. The spread of the spike potential over the ventricle takes about 60 msec. As it starts down the interventricular septum, the Q wave appears on the electrocardiogram recorded at the surface of the body. The R wave coincides roughly with the spike reaching the bottom (apex) of the heart and starting up the outer ventricular walls. The S wave appears as the spike potential reaches the top of the ventricle. Detailed physical models of the ekg based on observed potential changes in hearts have been simulated using observed tissue electrical characteristics. Space will not permit presenting these simulations here, but their understanding presumes a knowledge of the ideas presented in the next section.

10.7 Physics of Dipoles

Einthoven stated that if the three lead voltages given in Eq. (10-3) were represented as magnitudes directed along the sides of an equilateral triangle, all three could be represented as the projections of a single hypothetical "heart vector $\vec{H}$" on this triangle. As is seen in Fig. 10.8, this follows for any set of voltages.

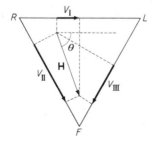

Figure 10.8 Einthoven's triangle. $\vec{H}$ represents the heart vector, whose magnitude is shown by omitting the vector sign. From the figure it can be shown that $V_I = H \cos \theta$; $V_{II} = H \cos (60° - \theta) = \frac{1}{2}H \cos \theta + \frac{1}{2}\sqrt{2}H \sin \theta$; $V_{III} = H \cos (120° - \theta) = -\frac{1}{2}H \cos \theta + \frac{1}{2}\sqrt{2}H \sin \theta$.

Although this procedure can be carried out for any three points, it has significance only if $\vec{H}$ indicates or is related to the electrical axis of the heart. The use of an equilateral triangle was based on the assumption that the three points chosen are electrically equidistant from the heart. If this is the case, one should find that

$$V_C = \tfrac{1}{3}(V_L + V_R + V_F) = 0 \qquad (10\text{-}5)$$

Although this is hard to test, because "neutral" electrodes are never truly neutral, the preceding condition is approximately satisfied. However, $\vec{H}$ remains at best an abstract construct.

To obtain three-dimensional information, a fourth electrode may be placed on the back or chest. Its voltage, relative to a neutral electrode, is designated by V_B. The ekg "lead" voltage V_{IV} is given by

$$V_{IV} = V_B - V_C$$

where V_C is an approximately neutral lead formed as above. V_{IV} tends to show up heart abnormalities in front or back of the midline of the heart, whereas the first three ekg leads tend to deemphasize this type of abnormality. Clinical ekg laboratories frequently use 12 leads to aid in diagnosis. The simulations referred to at the end of the previous section require 30 or more ekg leads.

To develop a better physical interpretation of the electrocardiogram, it is helpful to be familiar with electrical theory of a more advanced nature. This theory of current sources in a conducting medium is presented in this section. Those whose mathematical background does not include differential equations are advised to omit the remainder of this section and to accept certain statements in the next section.

The heart behaves as a group of current sources in a finite conducting medium. A current source is an emf whose internal resistance is much greater than the external load. Thus, the external current will remain constant no matter how the external load is varied. A current source is illustrated in Fig. 10.9. The tissues surrounding the heart are electrically similar and com-

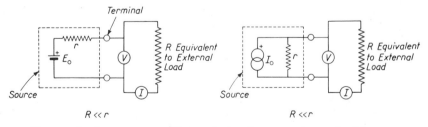

Figure 10.9 Current source. Two equivalent forms are shown. In either case, if $r \gg R$, the current source approximations can be made: $I \simeq I_0 \equiv E_0/r$ and $V \simeq RI_0$. Thus, V and I are determined by I_0 and the load.

paratively low in impedance. Because the heart muscle may be regarded as a group of current sources, the potential between any two external points will be the sum of the potentials due to each of the current sources acting independently.

The potential due to a group of current sources in an infinite conducting medium can be used to find an approximation to the currents produced in the body by the heart. For convenience, the two terminals in Fig. 10.9 will be treated as two sources, one positive and the other negative. Let the location of the ith current source be denoted by the vector distance $\vec{r}_i$ from the origin of the coordinate system as shown in Fig. 10.10. Then the current due to this

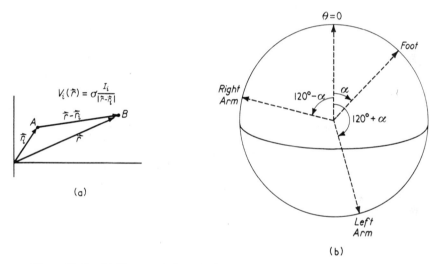

Figure 10.10   (a) Vector relationship for finding the potential, V_i, due to the current source, I_i, at A. (b) Geometrical relationship between heart dipole along $\theta = 0$ and arms and foot. This diagram is used in deriving the equations on page 187.

source, considered by itself, will spread throughout the medium, giving rise to a potential $V_i(r)$ at the point $\vec{r}$ from the origin. Because there are no net charges in the medium, the potential V_i must obey the Laplacian equation

$$\nabla^2 V_i = 0$$

(This is shown in electricity and magnetism texts.)

Because the tissues have a finite conductivity γ, there will be a current density $\vec{J}_i$ throughout the medium, originating from the ith current source, I_i

$$\vec{J}_i = \gamma \vec{\nabla} V_i$$

This is a special case of Ohm's law. The unique solution choosing $V = 0$ at infinite distance is

$$V(r) = \sum_{i=1}^{n} \frac{I_i}{\gamma|\vec{r} - \vec{r}_i|}$$

For large r this may be expanded in a series in $\vec{r}_i$; thereby one obtains

$$V(r) = \frac{1}{\gamma r}\sum I_i + \frac{1}{\gamma r^3}\left(\sum I_i\vec{r}_i\right)\cdot\vec{r} + \frac{1}{2\gamma r^5}\sum I_i[3(\vec{r}_i\cdot\vec{r})^2 - (\vec{r}\cdot\vec{r})(\vec{r}_i\cdot\vec{r}_i)] + \cdots$$

Because no net charge enters or leaves the heart, the first sum is zero. The second sum is a vector called the *dipole moment*, $\vec{p}$, defined by

$$\vec{p} = \sum_{i}^{n} I_i\vec{r}_i$$

A first approximation to the potential due to current sources in an infinite conducting medium is to replace them by an equivalent dipole, $\vec{p}$. The potential at r (referred to $V = 0$ at infinity) is

$$V(r) \simeq \frac{\vec{p}\cdot\vec{r}}{\gamma r^3}$$

The preceding expression was obtained for an infinite medium. If one restricts the heart to a sphere of radius R, a somewhat more complex expression is necessary. Consider the equivalent dipole $\vec{p}$ located at the center of a sphere and oriented along the $\theta = 0$ axis of the sphere. In this case

$$\vec{p}\cdot\vec{r} = pr\cos\theta$$

At small values of r, the potential must approach that of a dipole in an infinite medium:

$$V \to \frac{p\cos\theta}{\gamma r^2} \quad \text{as} \quad r \to 0$$

whereas at the surface, the radial current must be zero, so that

$$\frac{\partial V}{\partial r} = 0 \quad \text{at} \quad r = R$$

The unique solution to this approximation is

$$V = \frac{p\cos\theta}{\gamma}\left(\frac{1}{r^2} + \frac{2r}{R^3}\right)$$

which, at the surface of the sphere, reduces to

$$V(R) = \frac{3p\cos\theta}{\gamma R^2} \tag{10-6}$$

This is only an approximation but is useful in describing the electrocardiogram.

Equation (10-6) may be applied directly to the standard ekg leads. If the line to the foot makes an angle α with the heart vector, then the right arm is located at $\theta = \alpha + 120°$ and the left at $\theta = \alpha - 120°$, as shown in Fig. 10.10. Therefore, the three voltages, V_L, V_R, and V_F, should be

$$V_L = \frac{3p}{\gamma} \frac{\cos(\alpha - 120°)}{R^2}$$

$$V_R = \frac{3p}{\gamma} \frac{\cos(\alpha + 120°)}{R^2}$$

$$V_F = \frac{3p}{\gamma} \frac{\cos \alpha}{R^2}$$

The three lead voltages may be found by the appropriate differences, and the validity of Eqs. (10-4) and (10-5) can be noted. Thus, the Einthoven triangle is as valid as the spherical approximation with a dipole current source.

The representation of the heart as a single dipole and the torso as a homogenous conductor are at best a first crude approximation. For clinical purposes, ekg records are often determined using 12 leads. For simulation purposes, it has proved convenient to represent the ventricle as a group of 14 or more local dipoles, each with its own magnitude, orientation, and time sequence. As noted earlier, such simulations, which make use of detailed electrical theory to include the nonhomogeneities in the torso, are quite successful in predicting the body surface potentials of cardiac origin. Clinically, something simpler remains desirable. The scheme discussed in the following section has received widespread use.

10.8   Vector Electrocardiography

In attempts to increase the information obtained from the electrocardiogram, various schemes have been developed. The most successful, called *vector cardiography*, records the magnitude and spatial orientation of the equivalent heart dipole as a function of time. As has been pointed out, the physical relationship between the equivalent dipole and the cellular events in the heart is not in any way obvious. The abnormalities, producing a given change in the heart potentials, cannot be logically related to the change in many instances. In spite of the inability to logically interpret the vector electrocardiogram, it can still form a powerful diagnostic tool for clinical work. The equivalent dipole is referred to as the *heart vector*. The rationale behind these systems is presented in this section.

An integral part of vector electrocardiography is the equivalent dipole or heart vector. The discussion in the preceding section illustrated that a net dipole is the first approximation to any distribution of current sources whose net sum is zero. There are an infinite number of distributions which have the same vector dipole as a first approximation. It is in no way obvious that the heart should be well represented by the dipole approximation. Two different types of experiments, which tend to validate this approximation for the QRS complex, are discussed in the following paragraphs.

Let the dipole heart vector be denoted by $\vec{p}$. It is conventional to represent this as a sum of three vectors directed along the Cartesian axes. One may write

$$\vec{p} = p_x\vec{i} + p_y\vec{j} + p_z\vec{k}$$

where the subscripts refer to the scalar components of $\vec{p}$, and $\vec{i}, \vec{j}$, and $\vec{k}$ are Cartesian unit vectors. Then at any point on the periphery, the voltage V (relative to ground) may be written as a linear sum of the three components of the heart vector; that is,

$$V = \alpha p_x + \beta p_y + \eta p_z$$

In general, the three constants α, β, and η will depend on the location of the dipole, the location of the observation point, and the shape and electrical characteristics of the torso. The three quantities α, β, and η will be constant for the entire QRS complex if the heart can be represented as a dipole for that period of time.

If V is measured at four points, one may write four equations:

$$V_1 = \alpha_1 p_x + \beta_1 p_y + \eta_1 p_z$$
$$V_2 = \alpha_2 p_x + \beta_2 p_y + \eta_2 p_z$$
$$V_3 = \alpha_3 p_x + \beta_3 p_y + \eta_3 p_z$$
$$V_4 = \alpha_4 p_x + \beta_4 p_y + \eta_4 p_z$$

These may be regarded as four nonhomogeneous, linear equations in the three unknowns p_x, p_y, and p_z. For most sets of the (α, β, η)'s and the V's, there are no consistent solutions for p_x, p_y, and p_z. If, however, the heart vector is a good approximation, the fourth equation should be a linear combination of the first three. This, then, is a simple, unambiguous test of the dipole approximation.

Measurements on humans in which four pairs of wires are used (i.e., four independent leads) have shown that the QRS complex can be fitted very well by an equivalent dipole for a variety of different sets of points. The P and T waves of the ekg definitely cannot be described by the same dipole. Moreover, although the error in fitting V_4 with a linear combination of V_1, V_2, and V_3 is small, it is definitely larger than experimental error.

Another test of the dipole approximation is that of mirror images. For the central dipole in a sphere, discussed in Sec. 10.7, the equator of the sphere is a zero potential line. Any two surface points equidistant from but on opposite sides of the equator will have potentials that are equal in magnitude but opposite in sign. They are called *mirror points*. For the human torso (or indeed even for a cylinder with an eccentric dipole), the zero potential line is not an anatomically or geometrically obvious feature. Nonetheless, it could be located by finding mirror images if the dipole approximation is a good one. Experiments have revealed the existence of a mirror point for the QRS complex at any arbitrary point on the torso. This is also in accord with the dipole approximation. The data are good enough to show that the best mirror points are not perfect mirror points.

In practice, it is not possible to independently locate a single equivalent dipole and determine the time variations of its magnitude and direction. Instead, there are two alternatives to follow. One is to set up a system of leads that allows one to compute the magnitude and direction of the equivalent dipole without knowing its location. Various combinations of 5 to 16 electrodes have been devised which are claimed to be independent of exact body shape or the exact location of the heart. Such systems are called *orthogonal*. Although the vector dipole approximation is a poor one for the important P and T waves, nonetheless orthogonal lead systems are used throughout the heart cycle.

The second approach, rarely used but still in the spirit of vector cardiography, is to represent the heart as a dipole plus a quadrupole and higher moments if necessary, all located at the center of the heart. By using the added information of the location of the center of the heart, one can show that a quadrupole moment is necessary even to account for the QRS complex. Such information increases the diagnostic utility of the vector cardiograph; however, the added computational complexity, even with computer assistance, has dissuaded its widespread use.

The more physical approach of seeking to represent the heart as a collection of dipoles magnifies severalfold the electrodes necessary and manyfold the computational difficulty. Although in theory such an approach seems best, the use of techniques to estimate the characteristics of a multidipolar model has remained of research interest only. On the other hand, heuristic rules have been evolved to allow the use of the vector cardiograph for automated diagnosis.

10.9 Summary

The heart is a large mass of muscle that pumps blood through the vertebrate circulatory system. Its physical activity may be described in terms of the velocities and pressures acquired by the blood at various points of the circulatory system and also in terms of the power expended. The heart not only does work but also contains tissues that produce periodic beats in a fashion similar to that of a series of electronic multivibrators. The firing rate of the normal control element, the s-a node, can be increased or decreased both by the nervous system and by certain hormones.

Like the fibers of all striated muscle, the heart fibers are traversed by a spike potential before contraction. These spike potentials appear as current sources immersed in the surrounding fluid. The resulting body surface potentials are called electrocardiographic potentials. The heart may be simulated as a group of dipoles. For clinical purposes, a best equivalent dipole is often used; the systems based on this are called vector cardiography. Although a single dipole is a poor approximation from a physical perspective, the vector cardiogram has proved to be clinically useful, particularly for computer-based, automated interpretive systems.

REFERENCES

1. BEST, C. H., AND N. B. TAYLOR, *The Physiological Basis of Medical Practice*, 9th ed. (Baltimore, Md.: The Williams & Wilkins Company, 1973).

2. RAY, C. D., ed. *Medical Engineering* (Chicago, Year Book Medical Publishers, Inc., 1974), 1256 pages. Read Part III, Sec. 1, "Cardiovascular Systems" (Chaps. 18–27, pp. 219–321).

3a. BARNARD, A. C. L., I. M. DUCK, AND M. S. LYNN, "The Application of Electromagnetic Theory to Electrocardiology, I. Derivation of the Integral Equations," *Biophys. J.* 7:443–462 (1967).

3b. BARNARD, A. C. L., I. M. DUCK, M. S. LYNN, AND W. P. TIMLAKE, "The Application of Electromagnetic Theory to Electrocardiology, II. Numerical Solution of the Integral Equations," *Biophys. J.* 7:463–491 (1967).

3c. LYNN, M. S., A. C. L. BARNARD, J. H. HOLT, AND L. T. SHEFFIELD, "A Proposed Method for the Inverse Problem in Electrocardiology," *Biophys. J.* 7:925–945 (1967).

4. HOLT, J. H., JR., A. C. L. BARNARD, AND M. S. LYNN, with PEDER SVENDSEN (Part I only) and J. O. KRAMER, JR. (Part III only), "A Study of the Human Heart as a Multiple Dipole Electric Source: I. Normal Adult Male Subjects, II. Diagnosis and Quantitation of Left Ventricular Hypertrophy; and III. Diagnosis and Quantitation of Right Ventricular Hypertrophy," in *Circulation* 40:687–696, 697–710, 711–718 (1969).

5. GOLDMAN, M. J., *Principles of Clinical Electrocardiography*, 8th ed. (Los Altos, Calif.: Lange Medical Publications, 1973), 400 pages.

6. WARTAK, JOSEF, *Computers in Electrocardiography* (Springfield, Ill.: Charles C Thomas, Publisher, 1970), 250 pages.

7. PIPBERGER, H. V., J. CORNFIELD, AND R. A. DUNN, "Diagnosis of the Electro-cardiogram," in *Computer Diagnosis and Diagnostic Methods: the Proceedings of the 2nd Conference on the Diagnostic Process held at the University of Michigan*, J. A. Jacquez, ed. (Springfield, Ill.: Charles C Thomas, Publisher, 1972,) pp. 355–373.

8. PIPBERGER, H. V., R. A. DUNN, AND A. S. BERSON, "Computer Methods in Electrocardiography," *Ann. Rev. Biophys. Bioeng.* **4**:15–42 (1975).

DISCUSSION QUESTIONS—PART B

1. Action or spike potentials have been observed in the algae *Nitella* as well as in nerve and muscle. Describe the experiments necessary to demonstrate this biophysical phenomenon in *Nitella*. What is the implication of such behavior for the specialization necessary to conduct action potentials?

2. The Hodgkin–Huxley model presented in the text to describe the physio-chemical behavior of nerve axons was preceded by other models, while still other ones have attracted scientific interest in more recent times.

(a) What is the Lillie iron wire model? What was its historic importance?

(b) Cable theory models are generally believed to describe conduction between modes of Ranvier in myelinated nerves. Outline the significant features of cable theory. Detail the experimental evidence supporting the validity of this type of model.

3. Numerous anatomic, histochemical, and biophysical techniques have been used to determine nerve tracts and the roles of the various cell clusters and nuclei within the central nervous system. Describe a few of those techniques and examples of the structural and functional knowledge unraveled by the specific techniques you describe.

4. Present in detail the criteria used for sleep staging. What features make these hard to automate?

5. Deafness is a serious human handicap. Hearing aids attempt to correct this through amplification and other means. Describe the most common causes of deafness and the type(s) of hearing aids, if any, useful for each cause. Include the biophysical principles behind each type of aid.

6. Quite advanced mathematical models have been developed to describe the hydrodynamic responses of the cochlea to incident acoustic energy. Describe one of these in detail, including its mathematical development.

7. The control of the opening of the iris has been a favorite subject for the application of engineering analysis. Present one or more of these studies, including its experimental apparatus, mathematical analysis, and the scientific value of the study.

8. Describe the embryology of the vertebrate eye.

9. Contrast the optic nervous tract of humans and *Limulus*.

10. The muscles of the mammalian gastrointestinal tract inhibit a basic electrical rhythm (ber) whose frequency varies from one major region to the next. Describe the experiments necessary to demonstrate the ber and to relate it to peristalsis.

11. Calcium is believed to play a critical role in muscular contraction. Accordingly, it is important for the vertebrate to maintain a constant plasma calcium level. Normal humans do this in spite of enormous variations of calcium in the diet. Discuss the various mechanisms used to accomplish this control.

12. Various physical models have been proposed for muscular contraction. Explain the differences between these models and the evidence supporting each.

13. The electrical surface potentials due to the heart can be described as a sum of dipole, quadrupole, octupole, and higher terms. Develop mathematical expressions that relate the quadrupole and octupole terms to the elementary current sources.

14. Electrocardiography has been regarded as a major area in which automation could be used to assist the diagnostician (also see Chapter 30). Describe one vendor-supplied system for acquisition, analysis, and reporting of ekg data.

15. A number of attempts at automatic monitoring of the critically ill patient have used apparatus that computed the cardiac output from the shape of the pressure wave in the aorta or a major artery. Develop the theory on which these measurements were based. What are its limitations?

16. In Chapter 9, the proteins actin and myosin were said to occur primarily in muscles. They are also found in platelets, blood constituents important in clotting. Are platelets muscles? How do they use these contractile proteins? Are actin and myosin found elsewhere in mammals?

PART C

PHYSICAL BIOLOGY

Introduction to Part C

This section of the text deals with the physical properties of cells and groups of cells as revealed by various biophysical studies. The first chapter of Part C, Chapter 11, describes events produced by ionizing radiation in tissues and cells. This topic is continued in Part D, Chapter 17, which deals with the molecular effects of ionizing radiation. In Chapter 11, emphasis has been placed on the use of ionizing radiation to study the fundamental properties of biological systems.

Not all radiation is ionizing. In Chapters 12 and 13, the physical properties of cells and of groups of cells revealed by the use of nonionizing electromagnetic and ultrasonic radiation are discussed. Chapter 13 also describes the effects of high-intensity ultrasound.

Chapter 14 deals with the use of X-ray diffraction techniques to study biologically important molecules and structures. The last chapter of Part C is a description of the physicochemical properties of virus particles. Such particles lie between biological cells and molecules in their complexity and in their physical and chemical properties.

CHAPTER 11

Ionizing Radiation in Tissue

11.1 Introduction

Radiation has always been a significant factor in the environment of living creatures. Natural sources have included such diverse phenomena as cosmic rays, terrestrial activity, and even ingested radioisotopes. The slow genetic variation of species so crucial to Darwinian evolution may be at least partially assigned to those historical causes. In our society, the man-made sources of radiation now appear to contribute a component that is comparable to the environmental dose. This fraction could increase with the proliferation of nuclear power reactors and the general improvement of the worldwide level of medical care. In addition, the threat of military and terrorist usage of atomic or fusion weapons has caused great concern over acute doses of radiation to a general population. It is important, therefore, to understand the quantitative somatic and genetic results of a given exposure. In this chapter, events at the cellular and higher levels are discussed; molecular considerations are deferred to Chapter 17.

11.2   Quantitation of Dose

One of the fundamental concepts in the consideration of radiation effects is that of radiation dose. The primary dosimetric unit is the rad, defined as the absorption of 100 ergs/g of absorbing material. The assumption is that a given amount of energy deposited in a small mass is more significant than the identical amount of energy in a large mass. For example, 500 ergs delivered to 1 g of matter results in a radiation dose of 5 rad; yet if given to 10 g, a dose of only 0.5 rad is obtained. By "diluting" the ionization energy into enough mass, its gross effects can be made arbitrarily small. Such a macroscopic description of the energy deposition is not sufficient to categorize biological damage, however. This is due to the fact that, at a cellular level, some particles can generate ionization at a relatively more intense rate than others, yet have the same resultant rad dose. Large molecules in the cell nucleus are particularly liable to this sort of very localized damage.

It is thus necessary to define an effective unit of radiation dose whereby the spatial rate of kinetic energy deposition, dK/dx, is explicitly taken into account. A pure number, called the *quality factor* (QF), is made approximately proportional to the mean value of dK/dx. Typical quality factors are given in Table 11-1. These values, although generally agreed upon, are somewhat arbitrary for intensely ionizing particles. Low-energy electrons and photons are considered a standard radiation type and are assigned a QF of unity on that basis.

TABLE 11-1

LIST OF ASSIGNED QUALITY FACTORS

Particle	Quality factors
Photon	1
$e^{\pm}$	1
p, n	10
Alpha (^{4}He)	20

The biologically effective radiation dose in units of rad equivalent mammal (rem) is then computed by means of the equality

$$\text{effective dose (rem)} = \text{QF} \cdot \text{dose (rad)} \qquad (11\text{-}1)$$

Equation (11-1), while used in practice to calculate the effective radiation dose, neglects the spatial distribution of the radiation field. If this can be determined, the rad dose should be multiplied by a distribution factor as well as a quality factor to yield an effective dose.

The rad dose is directly measured by the amount of energy deposited in a given mass of absorber. The calculated rem value is used in the various state and federal limitations applicable to both radiation workers and the general public. Some of these are listed in Sec. 11.4. Before discussing such maximum permissible doses, the contemporary environmental and medical human radiation doses are presented.

11.3   Contemporary Dose Levels

A. BACKGROUND DOSES

The background whole body human radiation dose in a contemporary American sea-level location is slightly in excess of 100 mrem/year. This total is, roughly speaking, made up of contributions from cosmic, terrestrial, and ingested sources in the proportions: 1:2:1. Of these three components, the first two will typically increase in a mountainous locale so as to raise the sum to perhaps 150 mrem at the 1.6-km elevation of Denver, Colorado. At some seashore locations, however, particularly in Brazil, Australia, and India, there are radioactive beach sands (monazite) which contain thorium 232. In these instances the residents can have annual doses approaching several rem with no apparent detrimental effects.

Ingested radioactivity is caused by a multitude of radionuclide sources, but the largest dose contribution is due to ^{40}K, a primeval constituent of the earth's crust with a physical half-life of 1.3×10^9 years (see Chapter 29 for a discussion of half-life). Table 11-2 gives a list of nuclei incorporated in the

TABLE 11-2

DOSE RATES DUE TO EXTERNAL AND INTERNAL RADIATION
FROM NATURAL SOURCES IN "NORMAL" AREAS

Source of radiation	Dose rate (mrem/year)		
	Gonads	Bone	Bone marrow
External			
Cosmic			
Ionized component	28	28	28
Neutrons	0.7	0.7	0.7
Terrestrial	50	50	50
Internal			
^{40}K	20	15	15
^{87}Rb	0.3	<0.3	<0.3
^{14}C	0.7	1.6	1.6

Source: United Nations Scientific Committee on the Effects of Atomic Radiation. Twenty-first Session, Suppl. #14 (A/6314), 1966.

tissue of living humans and their various contributions. Carbon 14 is of secondary importance in this context, although its use as a tracer is of enormous significance in many fields such as nuclear medicine (see Chapter 29). This nuclide is generated by primary cosmic-ray neutrons high in the earth's atmosphere and has a half-life of approximately 5,700 years. Natural production of ^{14}C has been relatively constant over historical periods, which permits the dating of once-living biological materials.

B. MEDICAL DOSES

Medical sources of radiation are of increasing concern in most developed countries. It is estimated that the average yearly whole body medical dose in the United States is approaching 70 mrem, a figure comparable to the background value. A large fraction of the medical contribution is directly attributable to diagnostic radiographic procedures in medicine and dentistry. Table 11-3 gives some human skin and gonadal X-ray doses obtained under optimal shielding conditions. The maximum photon energy is given by E_x in kiloelectron volts, where an electron volt (eV) is 1.6×10^{-19} J. Notice that the fluoroscopic techniques are prolific sources of radiation, with rates approaching several rem per minute. In a strictly demographic sense, the human radiation exposures due to nuclear medicine and therapeutic radiology procedures are minimal compared to those due to diagnostic X-ray sources.

TABLE 11-3

SOME RADIATION DOSES DUE TO DIAGNOSTIC X-RAY PROCEDURES[a]

Site	E_x (keV)	$I \Delta t$ (mA·sec)[b]	Gonadal dose	Skin dose
Chest	100	5	0.00004 rem	0.009 rem
Chest	90	3	0.00002 rem	0.006 rem
Pelvis	120	70	0.07 rem	1.10 rem
Pelvis	85	250	Not given	3.00 rem
Lower bowel	90	Conventional fluoroscope	0.060 (♂) to 0.56 rem/min (♀)	2.40 rem/min
Lower bowel	90	Image fluoroscope	0.011 (♂) to 0.14 rem/min (♀)	0.48 rem/min

[a] Also see refs. 7 and 8 at the end of this chapter.

[b] I is the X-ray tube current; Δt is the time it is on.

A very important consideration is that the 70 mrem medical total refers to an average person; any given patient or volunteer must be individually monitored during irradiation to give exact figures for dose values. A film, ionization chamber, or thermoluminescent dosimeter (TLD) is appropriate in such circumstances. The TLD may also be placed in certain body cavities to record the absorbed dose at depth. Alternatively, skin readings and depth-dose tables may be used to extrapolate from external values. It is these extrapolations which are important if one is to estimate radiation damage at depth in a specific example. Table 11-3 or a local equivalent should be used only as a prior estimation of the skin dose.

11.4   Annual Dose Limits

Radiation workers are required to wear monitoring devices. They are limited by the National Council on Radiation Protection and Measurements (N.C.R.P.) to a whole body dose of 5.0 rem annually. Extremity limits are allowed to be 75 rem/year. These values are measured at the skin surface, so internal doses can significantly differ from those registered by the detector. In the case of a typical diagnostic X-ray beam, the skin dose is considerably higher than that at several centimeters' depth, because of the rapid decrease of radiation intensity in tissue. This is shown in Fig. 11.1. Note that the dose can achieve a maximum beneath the skin for higher-energy photons, such as

Figure 11.1   Percentage dose as a function of depth in water (a soft-tissue equivalent). The values shown are obtained along the central axis of the beam. Each curve is labeled with the maximum photon energy in the beam. After A. Casarett, *Radiation Biology* (Englewood Cliffs, N.J.: Prentice-Hall, Inc., 1968). (Original source: C. F. Behrens, *Atomic Medicine*, 3rd ed., Williams & Wilkins, Co., Baltimore, Md.; 1959, p. 401.)

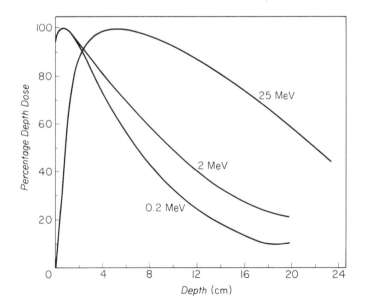

^{60}Co gamma rays (1.25 MeV) or X rays having a maximum energy of 25 MeV. The use of high-energy radiation for deep tumor therapy is based on these results.

Members of the general public are limited to a theoretical whole-body dose of 0.5 rem annually. An identical value is allowed the fetus being carried by a radiation worker.[1] The limit is usually meaningful only for radiation shielding and protection considerations. It should be emphasized that all listed annual limits are, by definition, in excess of the sum of environmental and medical exposures. Thus, as a patient, a member of the public or a radiation worker may receive any amount of radiation consistent with what may be considered good medical practice. The medicolegal aspects of a clinical study may involve radiation dose, but no specific maximum values are available from regulatory agencies. Such limits may be forthcoming.

11.5   Cellular Considerations

The reactions of single cells to ionizing radiation have been extensively studied. Experiments with both photon and charged beams have demonstrated that the nucleus of the cell is much more sensitive to a given rem dose than is the cytoplasm. Since the replication apparatus of a cell is primarily contained within the nucleus, the acute sensitivity of this region can be easily understood. In the following two subsections, the evidence for nuclear effects and their quantitative descriptions are considered.

A. NUCLEAR SENSITIVITY

Changes in cellular chromosomes and chromatids are typically found after irradiation. (An overview of DNA replication, covering both mitosis and meiosis, is given in Appendix E. The reader unfamiliar with these chromosomal changes should refer to that segment of the text before continuing.) These alterations are most readily described structurally in terms of new breaks and unions in the DNA molecules.

Radiation-induced DNA aberrations may lead to either somatic or genetic defects, depending upon the type of cell under consideration. The amount and configuration of the nuclear DNA allows one to make general predictions regarding the physical appearance of these various defects. Figure 11.2 shows the mammalian cell cycle, where M is the temporal duration of division, G1 and G2 are gap intervals, and S is the time during which the amount of DNA is actually doubled (replicated). Irradiation may occur during any of these periods. Chromosomal changes can be induced during the G1, S, and G2 (interphase) segments of the cycle, whereas chromatid effects will be produced during the M phase. These alterations will range from the very simple (e.g., a

---

[1] The 9-month pregnancy interval is appropriate in these cases, however.

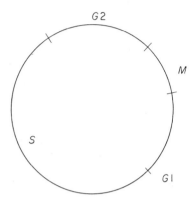

Figure 11.2 Mammalian cell cycle represented as a circle. Length of any arc is proportional to the time required to complete that part of the cycle. Interphase is the sum of G1, S, and G2. The period of DNA synthesis is labeled S. Cell division occurs during the M phase.

single break) to the complex geometrical configurations that result from several disruptions and subsequent new unions of DNA material. Figure 11.3 gives a graphical presentation of chromosome and chromatid configurations possible after exposure to radiation.

The defects diagrammed in Fig. 11.3 have been observed microscopically with the use of chromosomal dyes during the M phase of the cell cycle. One interesting discovery is that the actual time of transition from any point on the cycle to the next M phase is often delayed by radiation. Hence, the interval from G1 to M may be extended for such a long period of time that some fraction of the cells never enters into mitosis (or meiosis). These aberrants may, however, continue to function in a relatively normal metabolic fashion and can grow in cytoplasmic size. The result is that a certain fraction of the irradiated cells will become much larger than normal. These are the *giant cells*. They are frequently seen in tumors undergoing radiation and are often taken as evidence of a tissue's exposure to an ionizing beam. Production of giant cells is an indirect indication of the sensitivity of the nucleus to radiation. Other such indications are known from early work in radiobiology.

Differential sensitivity studies had indicated that certain cellular types were relatively more sensitive to ionizing radiation. Cell sensitivity here means the likelihood of cellular attrition in a given period of time after irradiation. This information led to the formulation of the empirical law of Bergonie and Tribondeau (1906). Their law states that cells having any of the following properties would be unusually susceptible to damage by ionizing particles: (1) high rate of mitosis; (2) cellular division throughout a major portion of the animal's lifetime; and (3) lack of specialization in a cellular developmental sequence.

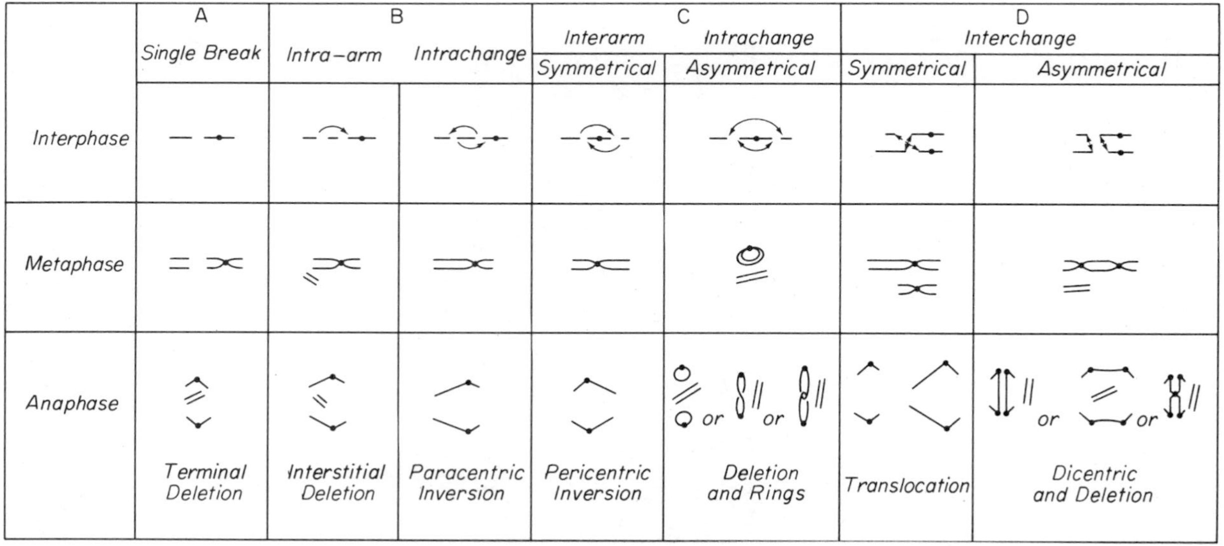

Figure 11.3 table — Selection of one and two break chromosomal aberrations

	A	B		C		D	
	Single Break	Intra-arm	Intrachange	Interarm (Symmetrical)	Intrachange (Asymmetrical)	Interchange (Symmetrical)	(Asymmetrical)
Interphase							
Metaphase							
Anaphase							
	Terminal Deletion	Interstitial Deletion	Paracentric Inversion	Pericentric Inversion	Deletion and Rings	Translocation	Dicentric and Deletion

Figure 11.3 (a) Selection of one and two break chromosomal aberrations. Types B, C, and D require two DNA breaks. Columnar chromosomes are not present during interphase. A very schematic representation of the interphase aberrations is included. Metaphase and anaphase drawings show the appearance of chromosomal aberrations. After Alison P. Casarett; *Radiation Biology*, © 1968, p. 10. Reprinted by permission of Prentice-Hall, Inc., Englewood Cliffs, New Jersey.

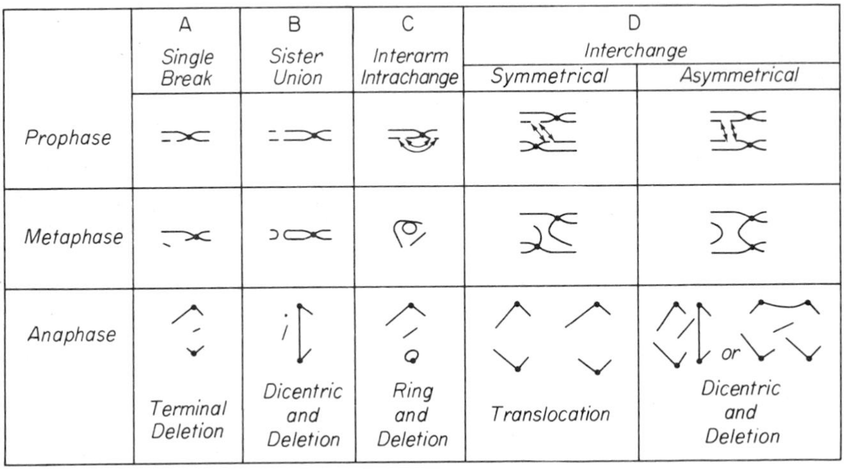

| | A | B | C | D | |
| | Single Break | Sister Union | Interarm Intrachange | Interchange | |
				Symmetrical	Asymmetrical
Prophase					
Metaphase					
Anaphase					
	Terminal Deletion	Dicentric and Deletion	Ring and Deletion	Translocation	Dicentric and Deletion

Figure 11.3 (b) Selection of chromatid aberrations. Here, the radiation lesion usually occurs after DNA replication, so that only one of two daughter chromatids may be affected, as in case A. Cases B, C, and D, however, involve two generally nonhomologous DNA breaks. The result of the various aberrations as the *M* interval progresses (metaphase and anaphase) is indicated. After Alison P. Casarett, *Radiation Biology*, © 1968, p. 105. Reprinted by permission of Prentice-Hall, Inc., Englewood Cliffs, New Jersey.

Attributes (1) and (2) refer to the total number of divisions in the cellular history. Therefore, those cells dividing relatively more often have a greater chance to manifest damage due to irradiation. An epithelium cell in the intestinal lining of the skin would be an example of a cellular type satisfying criteria (1) and (2). A central nervous system (CNS) cell would not satisfy either. Thus, the pathologist can give a relatively higher sensitivity index to epithelium relative to cells of the CNS. The hypothesis of Bergonie and Tribondeau is valid in this comparison, since the mammalian epithelial radiation syndrome can be seen at doses of less than 100 rem, while CNS radiation effects require doses of 10^3 rem or more.

In the preceding discussion, no distinction was made between the deaths of parent and daughter cells. From experiments done on other mammalian cell types, one expects that very few parent cells would die upon receiving even 10^4 rem. A more likely hypothesis is that these cell lines expire in the first generation after irradiation.

In the example given, both the epithelial and CNS cells are quite specialized, so that neither satisfies the third criterion of Bergonie and Tribondeau. This last is based on the statement that certain genetic information, repressed in the parent cell, must be manifest at a future time. If this is so, radiation may interfere with or prevent the expression of the correct cellular properties in the various daughter cells. As an example of a cell sequence, consider the production of red blood cells (RBC's). The erythroblast is the most primitive form in the RBC sequence. It satisfies all three Bergonie and Tribondeau criteria and is found to be extremely radiosensitive. By contrast, mature RBC's, which satisfy none of the criteria, are quite resistant to radiation.

These three criteria are not necessarily applicable in all cases; many exceptions, such as the human lymphocyte, are known. What must be emphasized, however, is that the hypothesis directly follows from the need for coherent transmission of genetic information at each DNA synthesis. Those cells which do not divide or specialize through division will not be sensitive to radiation effects. Moreover, the cells of a particular organism will not exhibit the same sensitivity throughout the individual's lifespan. The embryo and infant human have dividing CNS neurons. An irradiation during certain critical points in embryonic life (days 9 through 42 after fertilization in the human) may have disastrous consequences, while the same number of rem delivered to an adult will have no observable effects.

Both chromosomal aberrations and cell sensitivity implicate the nucleus as the major site of radiation damage. Historic experiments on nuclear irradiation were performed by Zirkle, Bloom, and their co-workers, who used narrow, highly collimated beams of protons and ultraviolet (UV) photons on single cells. They exposed single cells of different types to radiations using beam cross sections approximately 8 μm in diameter. The apparatus was arranged so that the area exposed could be located simultaneously with an optical microscope, and also so that the cell could be followed after irradiation. Experimental results were recorded on a motion picture film, with intervals of several seconds between pictures.

At low doses, no cytological changes were observed when the beam passed only through the cytoplasm. However, when the same types of cells were irradiated with the proton or UV beam passing through the nucleus, the process of mitosis was often altered. If irradiation occurred during the resting phase, when distinct chromosomes cannot be observed, a variety of abnormal effects were produced during the next mitosis, including broken chromosomes, pairs of chromosomes stuck together, and uneven division of chromosomes. During mitosis, when one particular region of the chromosome, the centromere, was hit by as few as a dozen protons, the chromosome no longer lined up with the others. Eventually, it was forced into one of the two daughter cells, forming either an auxiliary nucleus or a lobe of the existing one. Higher doses were needed on any other part of the chromosomes to alter mitosis, although these doses were small compared to those necessary to produce damage when used on the cytoplasm.

The nonmitotic cellular changes observed are much less pronounced. In extreme cases of high doses to single cells, the cell membrane is damaged. Most of the subcellular structures, such as mitochondria, neurofibrils, and myofibrils, remain unaltered at doses that eventually lead to cellular death. Thus, as indicated by the indirect experiments cited previously, it would appear that irradiation of the nucleus is the crucial stimulus leading to cell destruction.

B. QUANTITATION OF CELLULAR DEATH

Poisson statistics (see Sec. 2.4.A) have been used to model the cell's nuclear sensitivity. For this analysis the typical cell is assumed to consist of r targets ($r \geq 1$). Assuming also that a single hit destroys a target, the probability that one of these targets survives a single dose of radiation, D, in rem, is given by

$$\text{Prob (one survival)} = \exp(-SD) \qquad (11\text{-}2)$$

where S is called a *sensitivity coefficient* (rem^{-1}). If *all* r targets are hit, the associated probability is

$$\text{Prob } (r \text{ hits}) = [1 - \exp(-SD)]^r \qquad (11\text{-}3)$$

This event (all cellular targets destroyed) is assumed to lead to cellular death. Thus, the likelihood of cell survival is

$$\text{Prob (cell survival)} = 1 - \text{Prob } (r \text{ hits}) \qquad (11\text{-}3a)$$

This should be the surviving fraction of cells and may thus be rewritten as

$$\frac{N}{N_0} = 1 - [1 - \exp(-SD)]^r \qquad (11\text{-}4)$$

where N_0 is the initial number of cells.

Equation (11-4) is illustrated graphically in Fig. 11.4. Notice the shoulder

Figure 11.4   Idealized cell survival curve. Destruction occurs if $D \geq D_1$, the threshold dose. The ratio of N/N_0 is termed the *survival fraction*. The linear segment of the graph is described by Eq. (11-4b) and is characteristic of many actual survival curves.

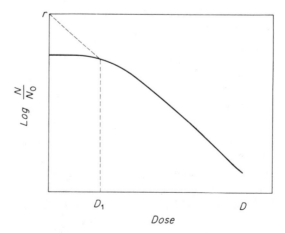

Dose

or "threshold" effect at low dose. For radiation doses less than a critical threshold value D_1, N is essentially equal to N_0, the original number of cells. If the exponential is expanded in the large dose region, where $\exp(-SD)$ is small compared to 1,

$$[1 - \exp(-SD)]^r \simeq 1 - r\exp(-SD)$$

Then Eq. (11-4) becomes

$$\frac{N}{N_0} \simeq 1 - [1 - r\exp(-SD)] = r\exp(-SD) \qquad (11\text{-}4a)$$

This may also be reexpressed as

$$\log_e\left(\frac{N}{N_0}\right) = \log_e r - SD$$
$$(11\text{-}4b)$$

Equation (11-4b) represents the linear part of Fig. 11.4. The target cross section (S) is the negative slope of this line. Its intercept on the ordinate defines r, the number of intracellular targets. One can also interpret differential cell sensitivity in terms of S and the threshold dose, D_1.

Up to this point, a single exposure of variable but controlled amount has been implicitly assumed. If there are multiple doses, but the time interval between doses is too short, the effects of the second dose will fall along the linear portion of the $\log_e(N/N_0)$ survival curve (i.e., it will be tantamount to a larger single dose). The threshold can recur in a serial irradiation experiment if a sufficient period of time is permitted between exposures. The appearance of similar survival curves after such an interim period as shown in Fig. 11.5, is sometimes used to indicate complete cellular recovery between exposures.

The effect of various environmental factors can also be evaluated using cell survival experiments. If the multitarget model given above is employed, these effects can be quantitated by S and r. A typical phenomenon is the increased sensitivity to radiation if the partial pressure of O_2 is increased. This is shown in Fig. 11.6. The difference in the survival curves demonstrates the significance of the O_2-associated reactions following irradiation. This effect of oxygen as well as that of certain other synergistic agents can be interpreted as evidence for the role of free radicals, such as $OH\cdot$ and $H\cdot$, as the primary instigators of cellular effects. This view is elaborated in Chapter 17. Alternatively, oxygen and the synergists may simply be involved with the direct chemical products of the radiation and somehow increase their effectiveness. A great deal of work is continuing in this area; both hypotheses may be valid.

In addition to models that assume single hits in multiple targets, radiobiologists use ones that involve multiple hits in either single or multiple targets. Curves predicted by multiple-hit models are similar to the curve illustrated in Fig. 11.4 for a single-hit model. Indeed, multiple-hit models are

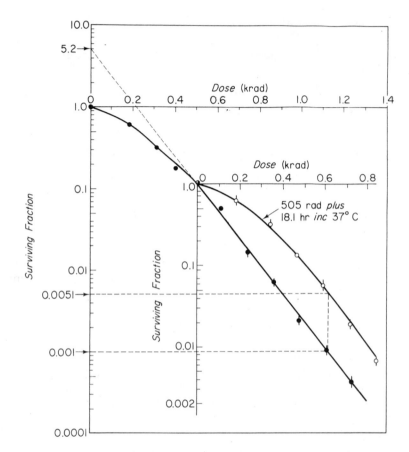

Figure 11.5 Actual cell survival curves with repetition of exposure regime. The second survival curve (insert) is a replica of the first. This establishes that the Chinese hamster cells originally exposed and surviving were not significantly changed in their radiation sensitivity to 55-keV X rays. After M. Elkind and H. Sutton, *Radiation Res.* **13**:556–593 (1960).

experimentally indistinguishable from single-hit models. Nevertheless, values estimated for parameters in any one of these models can be used for data reduction as described in the next paragraph. Such phenomenological descriptions are frequently used in biophysics; see, for example, the discussion of the Hodgkin–Huxley model in Chapter 5.

Parameters such as r and S for the single-hit model are used in comparing cell types and in measuring the effects of external variables. Estimates of the cross-sectional target areas, based on values of r and S, appear relatively consistent using different kinds of radiation on the same cell type. This is described in the reference by Andrews.

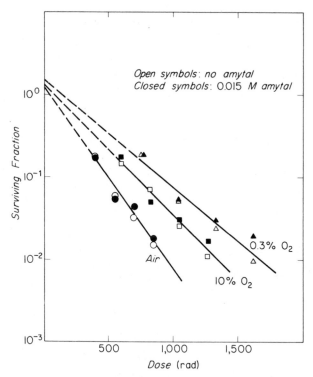

Figure 11.6   Oxygen effect with X rays. As the partial pressure of oxygen increases, the survival curves decrease more rapidly with radiation dose. This effect does not change if an agent (0.015 M amytal) is used to reduce the oxygen consumption rate of these ascites tumor cells. X rays of 250-keV maximum energy were used. After B. Cullen and T. B. Constable, "Independence of the Radiobiological Oxygen Constant K," *Intern. J. Radiation Biol.* **29**:348 (1976).

11.6   Macro Biological Effects

A. CANCER INDUCTION IN HUMANS

Cellular death has been described as an essentially nuclear phenomenon with little, if any, cytoplasmic involvement. The question arises as to the future course of those cells irradiated but *not* destroyed. From the earliest days of radiology, it was realized that malignancies could result after irradiation with X rays. This effect can be attributed to changes in DNA coding of cellular information. Generally, it is found that the probability of such effects in humans is linearly proportional to the dose received, the typical constant of proportionality being between one and two excess cases of neoplasm per 10^6 persons/year/rem. Figure 11.7 shows data for human lung carcinoma subsequent to occupational exposure of American uranium miners.

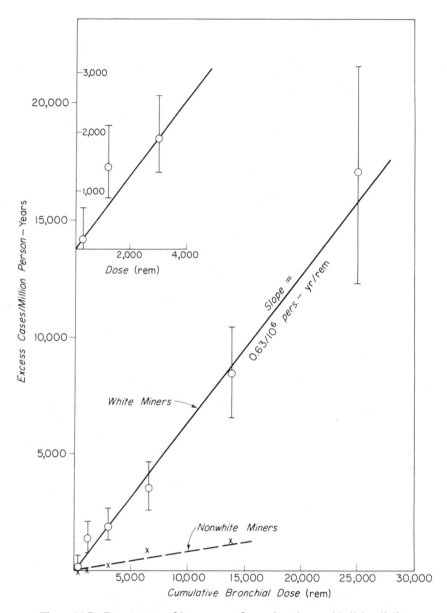

Figure 11.7 Excess cases of lung cancer for various lung epithelial radiation doses. The patient (worker) population included both white and nonwhite miners. Insert contains the low dose interval for white miners and establishes a linearity hypothesis (nonthreshold effect). After *The Effects on Populations of Exposure to Low Levels of Ionizing Radiation* (Washington, D.C.: National Academy of Sciences–National Research Council, 1972).

The linear relationship appears a reasonable fit to the data down to the lowest doses recorded (insert). The slope of excess cases per year vs. dose gives a probability of 0.63 excess case of lung carcinoma per 10^6 people/rem. The number is from the BEIR report of the National Academy of Sciences (NAS). There is, however, a controversy as to whether the graph is linear as the dose approaches zero. If one were to accept the concepts of threshold and recovery which are implicit in the multitarget model, a nonlinear relationship should, in fact, occur for small radiation exposures. The data are necessarily poor in the low-dose region. Thus, there is little hope of resolving the issue at this time. However, a conservative approach to radiation protection implies acceptance of the assumption of linearity for all doses. Table 11-4, compiled from the data of the BEIR report, gives the slope of various linearized cancer induction curves. Values tend to cluster around unity in the standard units used.

TABLE 11-4

Carcinoma (source)	Prob(excess cases/10^6 people/yr/rem)
Lung (uranium ore)	0.63
Bone (^{224}Ra)	0.96
Bone (^{226}Ra)	0.11
Lung (fluorspar)	1.96
Lung (weapons) Hiroshima and Nagasaki	1.56
Leukemia (weapons) Hiroshima	1.7 (assuming a QF = 5 for neutrons)
Nagasaki	1.0

The NAS Committee also estimated that although the excess cases of leukemia were between 1 and 2/10^6 people/yr/rem for adults, some 2 to 3 times this number of excess cases were observed for children. Other specific differential sensitivities may also be present in human populations. The relative constancy in the probability of induction of malignancy regardless of these details is quite impressive and can be used in calculations of the risks to a potentially exposed population. Such considerations arise in evaluations of both working and living conditions; that is, the nuclear reactor worker as well as the adjacent urban householder needs to know the likelihood of neoplasm induction in quantitative terms.

B. GENETIC CHANGES

In addition to the somatic cell mutations, the surviving viable germ cells may be changed by the irradiation. This effect was long suspected by early workers and first demonstrated by Müller at the University of Texas. His experiments concerned the fruit fly (*Drosophila*), which exhibits various genetic mutations after X irradiation. These specific changes are not different from those occurring spontaneously in the same species. This implies that radiation is an effective way of scrambling the DNA code, not a unique way.

Quantitative evaluation of the frequencies of mammalian genetic mutation following irradiation has been accomplished by Russell and his co-workers at Oak Ridge using the mouse as a model. These "Megamouse" experiments have involved recording seven externally observable parameters in the test animal (e.g., ear shape, coat color, and coat uniformity). Since the parent generation is of well-known genetic configuration, any unexpected results in the next generation are attributable to a genetic change of the parent's chromosomal makeup. Following the subsequent generations verifies that the change was indeed genetic. Oak Ridge data have revealed that the probability of recessive[2] genetic mutations in the mouse is approximately $2.5 \times 10^{-8}/$ gene/generation/rem. Geneticists had previously determined that the human exhibits each generation between 5 and 50 spontaneous recessive genetic mutations per 10^7 genes. Thus, a radiation dose of 20–200 rem/generation would cause a number of genetic changes equal to the spontaneous rate in humans. The calculation is necessarily based upon an extrapolation from mouse to human and may also be criticized in that, since there may be many unnoticed "silent" mutations, the mutation probability given is probably a low estimate. Nonetheless, most authorities prefer to quote the range of values 20–200 rem as the human genetic doubling dose. Note that this dose refers to a time period of one generation, i.e., approximately 20–30 years.

Direct data on human susceptibility to genetic changes caused by radiation are inconclusive at present. No obvious alterations have been documented among the descendants of the survivors of the nuclear weapons used on Hiroshima and Nagasaki. Nor are any unusual genetic problems associated with those people living in the monazite beach sand areas of various continents. The detection of recessive mutations is made difficult in these cases by the small number of generations over which the geneticist may track the various descendants. Likewise, the effects of such slow genetic changes are extremely difficult to separate from various concurrent environmental variables. No control group is usually available; the incidence of specific anomalies is known to be dependent upon race, nationality, and other factors. One can guess, however, that most spontaneous mutations are not due to environmental radiation since the present dose per generation is only one tenth that of the minimal genetic doubling dose of 20 rem.

There would appear to be little advantage in increasing human radiation exposure. The additional mutations would probably be nonviable and at best

---

[2] Dominant mutations are quite rare—probably only one tenth as likely as recessive.

a social burden. The minimization of extraneous radiation dose would appear to be a most profitable directive for this reason. Such considerations would, of course, also have to involve the cost of this protection in both societal standards of living and safety. This is a vital ecological issue.

11.7 Summary

Ionizing radiation has been seen to act through nuclear DNA transformations. These can lead to lowered cell viability and various pathological syndromes in the organism. While the environmental and medical doses appear small, they are not trivial compared to those associated with neoplasm induction and genetic changes. The presence of oxygen appears to enhance cellular sensitivity to radiation. A target-hit theory, based on Poisson statistics, can be used to quantitate the enhancement due to O_2. The vital significance of radiation in both energy generation and diagnostic medicine will continue to require that society carefully understand the overall cost of any environmental or medical innovations.

REFERENCES

1. MINER, R. W., ed., "Ionizing Radiation and the Cell" (monograph), *Ann. N. Y. Acad. Sci.* **59**:467–664 (1955).
   (a) BLOOM, WILLIAM, R. E. ZIRKLE, AND R. B. URETZ, "Irradiation of Parts of Individual Cells. III. Effects of Chromosomal and Extrachromosomal Irradiation on Chromosome Movements," pp. 503–513.
   (b) PATT, H. M., "Factors in the Radiosensitivity of Mammalian Cells," pp. 649–664.
2. HOLLAENDER, ALEXANDER, ed., *Radiation Biology*, Vol. 1, *High Energy Radiation* (New York: McGraw–Hill Book Company, 1954). (In two parts.)
   (a) MÜLLER, H. J., "The Nature of the Genetic Effects Produced by Radiation," Part I, pp. 351–473.
   (b) MÜLLER, H. J., "The Manner of Production of Mutations by Radiation," Part I, pp. 475–626.
   (c) BLOOM, WILLIAM, AND M. A. BLOOM, "Histological Changes After Irradiation," Part II, pp. 1091–1143.
3. ZIRKLE, R. E., "Partial-Cell Irradiation," in *Advan. Biol. Med. Phys.* **5**:103–146 (1957).
4. CASARETT, ALISON, *Radiation Biology* (Englewood Cliffs, N.J.: Prentice-Hall, Inc., 1968), 368 pages.

5. Advisory Committee on the Biological Effects of Ionizing Radiations, Division of Medical Sciences, *The Effects on Populations of Exposure to Low Levels of Ionizing Radiation* (Washington, D.C.: National Academy of Sciences–National Research Council, 1972), 217 pages. (This is known as the BEIR Report.)

6. ANDREWS, H. L., *Radiation Biophysics*, 2nd ed. (Englewood Cliffs, N.J.: Prentice–Hall, Inc., 1974), 314 pages.

7. ANTOKU, S., and W. J. RUSSELL, "Dose to the Active Bone Marrow, Gonads, and Skin from Roentgenography and Fluoroscopy," *Radiology* 101:669–678 (1971).

8. ARDRAN, G. M., AND H. E. CROOKS, "Gonad Radiation Dose from Diagnostic Procedures," *Brit. J. Radiol.* 30:295–297 (1957).

9. RUSSELL, W. L., "Genetic Hazards of Radiation," *Am. Phil. Soc. Proc.* 107:11–17 (1963).

10. MARSHALL, M., J. GIBSON, and P. HOLT, "An Analysis of the Target Theory of Lea with Modern Data," *Int. J. Radiation Biol.* 18:127–138 (1970).

11. SHAPIRO, JACOB, *Radiation Protection, A Guide for Scientists and Physicians* (Cambridge, Mass.: Harvard University Press, 1972), 339 pages.

12. YUHAS, J. M., R. W. TENNANT AND J. D. REGAN, eds., *Biology of Radiation Carcinogenesis* (New York: Raven Press, 1976), 347 pages.

CHAPTER 12

Biological Effects
of Electromagnetism

12.1   Introduction

The entire electromagnetic (em) spectrum encompasses radiation with frequencies from 0 to over 10^{19} Hz. This is conventionally divided into the regions named in Table 12-1, but the boundaries of each region are ill defined. The biological effects of the highest-frequency regions, X and γ rays, are discussed in Chapters 11 and 17, while visual aspects of em radiation are considered in Chapters 2, 8, and 19. There are, however, many other biophysical studies on the effects of electromagnetic fields, some of which have been grouped together in this chapter.

First among the concepts discussed in this chapter is biological impedance; knowledge of its variation with frequency is crucial in understanding low-frequency studies. A quantitative evaluation of the possible hazards of low-frequency irradiation is imperative in a society such as ours, which relies on above-ground power lines and electrocommunications. Next, the more common problem of whole-body exposure to microwave radiation is considered. Television stations, radar installations, and kitchen cooking devices all radiate energy in the microwave region. Included with the discussion of

TABLE 12-1

ELECTROMAGNETIC RADIATION

Frequency (Hz)	Wavelength (m)	Name(s)
$0.0–5.5 \times 10^5$	$\infty \to 5.5 \times 10^2$	Long wavelength
$0.55–1.5 \times 10^6$	$5.5–2.0 \times 10^2$	AM broadcast
$0.15–1.5 \times 10^7$	$20.0–2.0 \times 10^1$	Short wave
$0.15–6.0 \times 10^8$	$20.0–0.5$	TV, FM, UHF
$0.06–30.0 \times 10^{10}$	$50.0–0.1 \times 10^{-2}$	Radar, microwave
$0.15–25.0 \times 10^{11}$	$20.0–0.12 \times 10^{-3}$	Heat, far infrared
$0.25–40.0 \times 10^{13}$	$120.0–0.75 \times 10^{-6}$	Near infrared
$4.0–8.0 \times 10^{14}$	$7.5–3.5 \times 10^{-7}$	Visible light
$0.08–30.0 \times 10^{16}$	$35.0–0.1 \times 10^{-8}$	Ultraviolet
$0.3–30.0 \times 10^{18}$	$10.0–0.1 \times 10^{-10}$	X rays ($\sim$1–100 keV)
$0.3–30.0 \times 10^{20}$	$10.0–0.1 \times 10^{-12}$	γ rays (0.1–10 MeV)

microwave effects is a brief review of various far-infrared diathermy machines.

The chapter ends with a presentation of some of the uses of visible and ultraviolet light in cellular, tissue, and whole-body animal experimentation. The analyses employed are very similar to those applied to the biological evaluations of experiments using ionizing radiation. These have led to medical applications of high-intensity light sources such as lasers. The latter are also reviewed.

12.2 Electrical Impedance

Electromagnetic forces in a material medium are generally described by means of the spatial and temporal variation of the electric, $\vec{E}$, and magnetic, $\vec{B}$, fields. Table B-1 (Appendix B) lists several of the fundamental relationships between these fields and the other parameters that describe the motion of the electric charges. The quantities $\vec{D}$ and $\vec{H}$ are the corresponding fields generated by fixed charges and currents, respectively; they do *not* include the effects of the specific medium such as a cell suspension. In an isotropic material, one may obtain a simple multiplicative relationship between the actual fields and those generated by the sources:

$$\vec{E} = \left(\frac{1}{\epsilon}\right)\vec{D} \tag{12-1a}$$

$$\vec{B} = \mu\vec{H} \tag{12-1b}$$

where ϵ is the dielectric constant or permittivity and μ is the magnetic permeability. The description of a biological specimen's electromagnetic behavior may thus be completely described by the variation of ϵ and μ with field frequency and the specific tissue conditions such as temperature and pressure. Most of the experimental work has concerned the electrical behavior of tissue since the observed magnetic permeability of biological specimens is usually within 1 part in 10^4 of μ_0, the permeability of free space. It must be remembered, however, that a time-varying magnetic field induces an electric field. Thus, although static $\vec{B}$ fields are not believed to cause any biological responses, the effects of changing $\vec{B}$ fields may be observed.

The most general way to relate current I and voltage V in a biological circuit element is through the concept of electrical impedance, Z, defined by

$$Z = \frac{V}{I} \qquad (12\text{-}2)$$

Impedance, whose units are ohms, may be represented by a complex variable since that notation is used for V and I, which generally will not be in phase with one another. The conventional formulation of this concept is

$$Z = R + jX$$

where j is $\sqrt{-1}$, R is the resistance, and X is the reactance. The latter term is nonzero if either capacitive or inductive elements are present in the circuit element. In biological media, capacitive components are readily observed, with the cellular membrane being an example. The electrical description of a particular medium can be reduced to impedance per unit length of unit cross-sectional area through

$$z = \frac{E}{J} \qquad (12\text{-}2a)$$

where J is the current density and z is the specific impedance. In this context, one uses

$$z = \rho + jx$$

where ρ is the resistivity and x is the reactance per unit length of a unit cross-sectional area. Notice that the dimensions of the sample are no longer present in the specific impedance so that it is an appropriate characterization of a material. Techniques of cellular impedance measurement are described in the next section.

12.3   Biological Impedance Measurements

The membranes of most cells act as insulators at low frequencies. Thus, if one suspends cells in a saline solution, most of the current will flow around the cells. The electrical current must then follow a longer path than in the absence of the cells. In terms of impedance, the resistance (or resistivity; see Appendix B) will be higher for the suspension than for the pure suspending medium. The difference in these two resistivities can be used to find the volume of the cell.

For tissues, likewise, the low-frequency resistivity (ρ) is a measure of the free space between the cells. The resistivity is about the same for skeletal muscle, liver, and cardiac muscle. Typical values are about 900 ohm·cm, dropping slowly as the frequency increases from 10 Hz to 1,000 Hz. Blood, with fewer formed elements, has a much lower resistivity, whereas fatty tissue and bones have higher resistivities in the low-frequency region.

Between 1 and 100 kHz the specific impedance of a suspension of single cells drops quite sharply to a lower value. For cells of simple geometry in a saline solution, one can solve exactly the equations describing the flow of current. A first approximation is to assume that the cell is a spherical homogeneous conductor surrounded by a nonconducting (lipid) layer as shown in Fig. 12.1. This fits the impedance data in a qualitative fashion. The

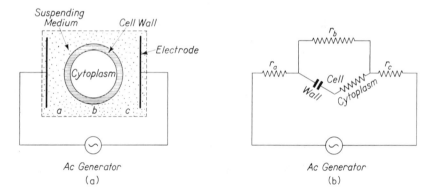

Figure 12.1 (a) Diagrammatic representation of a single cell in an electric field. The medium resistance is lower in regions *a* and *c* since the cell is a poor conductor relative to the medium. (b) Equivalent lumped electrical parameters for the physical arrangement of part (a). The cell wall is represented by a capacitor. A better approximation would include a leakage resistor in parallel with the capacitor. Resistors r_a, r_b and r_c represent media in regions *a*, *b*, and *c* of part (a).

lipid layer acts as a capacitor in series with the cell interior. The impedance of a capacitor (i.e., the lipid shell) is given by

$$Z_c = \frac{j}{\omega C}$$

where ω, the angular frequency, equals $2\pi\nu$. At low frequencies this impedance is very high because of the small value of ω in the denominator. Hence, little current can enter the cell, which then appears to be an insulator. As ω increases, Z_c becomes negligible and current can readily pass into the cell interior. At such increased frequencies, the impedance of the suspension will be less than it is at lower frequencies.

The experimental data can be fitted more quantitatively by allowing the cell wall to be a leaky capacitor, that is, to be equivalent to a high resistance in parallel with a capacitor. Using the actual cell shape instead of spheres also improves the application of the theory to the experiment. This last condition is difficult to implement except for cells with very simple geometrical forms.

The impedance of the cytoplasm can be represented as a resistivity, expressed in ohm·cm. Properties of the cell wall are expressed as an areal capacity in microfarads per square centimeter ($\mu F/cm^2$), and an areal resistance in ohm·cm². At one time, it was believed that these cell constants could be interpreted to find the effective thickness of the cell membrane, but this hope has not yet been realized. Perhaps the most impressive aspect of these data is their similarity from one cell type to another. Plant cells, animal cells, nerve axons, and egg cells all overlap in their electrical parameters.

The internal or protoplasmic resistivity varies from 10 to 30,000 ohm·cm, with 300 being common for most mammalian cells. The areal capacitance varies from 0.1 to 3 $\mu F/cm^2$. Few values lie outside the range 0.8 to 1.1; however, one low measurement of 0.01 $\mu F/cm^2$ has been obtained for a frog nerve. Values for the leakage areal resistance of the cell membrane vary from 25 to 10,000 ohm·cm² or higher. Nerve and muscle measurements have yielded both extremes.

Similar considerations apply to the electrical characteristics of whole tissues. Their impedance is hard to separate in terms of cellular parameters but may be represented as a lumped resistivity and capacitance.

If a medium is placed between the charge-bearing surfaces of a capacitor, the capacitance is multiplied by the value of the material's dielectric constant ϵ. A selection of dielectric constant data from the experiments of Schwan and his co-workers is given in Fig. 12.2. Notice that both axes utilize logarithmic scales in the figure and that ϵ_0 is the dielectric constant of a vacuum. The muscle data shown are typical of all living tissues. Such graphs of ϵ vs. ν are generally characterized by three rather gradual declines, each being several decades of frequency wide. These α, β, and γ regions are also seen in the behavior of tissue resistivity, as is shown in Fig. 12.3. Since the three transition regions in the resistivity curve occur at the same frequency ranges as those in the dielectric constant results, identical mechanisms may be involved in both sets of data.

It is generally assumed that the α, β, and γ regions are each associated with specific electrical relaxation processes. These are interactions between the $\vec{E}$ vector associated with the radiation and appropriate physical entities in the tissue or other sample. A characteristic time describes each relaxation, the

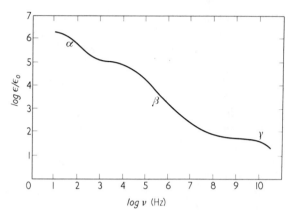

Figure 12.2   Frequency dependence of the dielectric constant of muscle. There are three fundamental regions labeled with Greek letters, indicating three different types of relaxation. After H. P. Schwan and C. F. Kay, *Ann. N.Y. Acad. Sci.* **65**:1007 (1957).

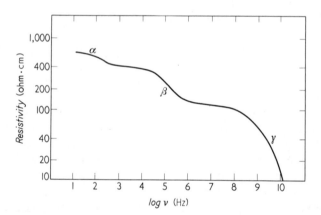

Figure 12.3   Resistivity of muscle as a function of frequency. Notice the similarity of the relaxation regions for Figs. 12.2 and 12.3. After H. P. Schwan and C. F. Kay, *Ann. N.Y. Acad. Sci.* **65**:1007 (1957).

numerical value being the inverse of the critical frequency at which ϵ/ϵ_0 is halfway between two relatively constant magnitudes. At this frequency, the absorption of energy from the radiation field is also a maximum. Since the speed of the electromagnetic wave depends on ϵ, it may be shown that absorption implies a complex dielectric constant. The data plotted in Fig. 12.2 only refer to ϵ, the real part of the complex dielectric constant ϵ', given by

$$\epsilon' = \epsilon + j\epsilon_I$$

where ϵ_I is the absorbed energy per cycle. A graph of ϵ_I vs. ν reveals a resonancelike behavior at the critical frequency. In the case of tissue samples, a distribution of such resonance frequencies is found experimentally. Some fine structure, the δ region occurring between the β and γ regions, is also seen.

Schwan has proposed specific mechanisms for each of the three relaxations. The α region is not clearly understood but is thought to correspond to excited cellular membranes and large intracellular structures. A characteristic frequency around 100 Hz is found in most tissues. The β interval, centered at 10^6 Hz, is probably due to the charging and discharging of a heterogeneous mixture of dielectrics inside the cell. This is sometimes referred to as the Maxwell–Wagner effect. The γ region, near 10 GHz, is considered a consequence of the rotation of electric dipole moments (Debye effect) of various biological molecules, primarily water and proteins. For example, the water dipole has an intrinsic frequency of 20 GHz, a value well within the γ band.

The γ band may physically result from molecules becoming polarized in an electric field. In an alternating field, the molecule must reverse this polarization each half-cycle. For relatively small permanent dipoles, the charge distribution actually rotates, whereas in the case of induced moments and compounds with higher molecular weights, a redistribution of charge occurs. As in any relaxation phenomenon, a monotonic increase in the frequency of the field will eventually lead to a situation in which the rotation or redistribution is unable to occur quickly enough and the effect ceases to be important. Some dielectric data have been used to indicate still higher relaxation frequencies, although these are only a matter of conjecture at the present time.

The frequency dependence of ϵ and ρ of most soft tissues is similar. Exceptions to this general pattern are found in bone and fatty materials. In the former case, the high content of calcium phosphate crystals causes increased impedance at lower frequency values. Fatty tissue is very different because lipid materials are excellent electrical insulators. Tissues of this nature show a much higher resistivity and a much lower dielectric constant than do those with more water. The general shape of the resistivity and dielectric curves is similar to that for muscle. Brain tissue has more fatlike material than does muscle. At lower frequencies, its resistivity is close to that of fatty tissues. However, this resistivity falls rapidly as the frequency is raised from 1 to 10 MHz. Its value above this frequency range is close to that of the nonfatty tissues. The dielectric constant of brain tissue is within the range of the watery tissues at all frequencies.

In concluding this section, it should be pointed out that the electrical impedance of biological cells supports the picture of a cell consisting of an electrically conducting cytoplasm surrounded by a poorly conducting lipid membrane with a high dielectric constant. These electrical data are interesting in themselves as physical properties of the cells, but have not been related in detail to the differences between cells. The classification of ρ and ϵ data into three relaxation regions is significant in that no sharp resonances are observed.

12.4   The Attenuation of Electromagnetic Radiation

Maxwell's equations predict the existence of electromagnetic waves which are propagated through free space at a speed given by

$$c = \frac{1}{(\mu_0\epsilon_0)^{1/2}} = 3 \times 10^8 \text{ m/sec}$$

In a tissue medium, this value is reduced to

$$c_{\text{med}} = \frac{1}{(\mu_0\epsilon)^{1/2}}$$

where the biological approximation

$$\mu \simeq \mu_0$$

has been used explicitly. The amount of radiant energy passing through a unit area per unit time is referred to as the *radiation intensity*, I. (This definition of intensity is identical to that in Sec. 1.2.)

Because the propagation of photons in a thickness of material x is a Poisson process, the attenuation of electromagnetic radiation is described by an exponential decay law. Historically, this law has been cast into various forms, depending upon the specific conditions in the medium. For example, the *Lambert equation* refers to constant molecular concentrations:

$$I = I_0 \times 10^{-Kx} \tag{12-3}$$

where K is the extinction coefficient (see Chapter 26 for a discussion of Lambert's and Beer's laws). The use of the base 10 in lieu of the base of the natural logarithms follows chemists' tradition. The exponential form

$$I = I_0 \exp(-ax)$$

where a is $2.3K$, is more common in the physics literature.

If the photon beam is not monoenergetic, one defines a differential intensity per wavelength interval:

$$dI = i(\lambda)\, d\lambda$$

and restructures the attenuation law as

$$I(x) = \int_{\lambda_{min}}^{\lambda_{max}} i(\lambda) \times 10^{-K(\lambda)x}\, d\lambda \tag{12-4}$$

Notice that an integral must now be evaluated to determine the intensity at any given depth x. If the molecular concentrations are not constant, a still more complicated expression is needed.

A photon may be scattered by the target molecule instead of absorbed. Ionizing photons such as X rays decrease in energy when scattered and thus can be separated from the incident beam by the detector. Non-ionizing photons will be confused with the incident beam unless directional discriminators (e.g., collimators) are used.

The following four sections concern specific frequency intervals of electromagnetic phenomena. All are presently under intensive investigation for environmental and biological reasons. The examples presented are illustrative and not exhaustive.

12.5   Low-Frequency Effects

The study of low-frequency electromagnetic radiation has been closely tied to the development of the electrical power industry. One example of obvious environmental interest is that in which the individual is inadvertently made a part of an electrical circuit (i.e., the arm-to-arm or arm-to-leg impedance is placed between two different voltages). At 60 Hz, a 1-mA current is generally discernible by a majority of the population. If the current is raised by an order of magnitude, the "let-go" threshold is often exceeded and the subject cannot release a hand-held electrode. Ventricular fibrillation has its onset with currents in the range 50–100 mA, approximately two orders of magnitude above the limits of detectability. Dangers of electrocution are maximized if the individual has reduced skin impedance (e.g., has wet hands or stands in a pool of water). The internal environment of the human body, presumably a replica of the ocean out of which life originated, is an excellent conductor of electrical current.

Fibrillation can be halted by an electrical current when it is correctly applied to the myocardium. However, not all electrocutions are due to the incoherent beating of a fibrillating heart. Certain electric shock victims have been observed to recover when given artificial respiration alone. Since such mechanical manipulations are thought to be ineffective in defibrillation, this implies that these victims' only problem was cessation of breathing, probably due to an interference in the respiratory centers of the brain. Other consequences of tissue current flow include such diverse phenomena as burns, due to resistive heating, and thrombi, produced by local vascular changes. The electroshocked patient often exhibits convulsive behavior patterns similar to an epileptic seizure.

In contrast to circuit potentials, lower-frequency electromagnetic radiation is rather ineffective in eliciting a biological response, for two reasons. The electric component cannot penetrate to any appreciable depth into the specimen, because of the free ions present in the body fluids. These make the inside of the living organism behave like a conductor of electrical charge, with the result that the internal environment is shielded by the surface movement of charges. In addition, the magnetic component of the radiation can penetrate tissue, but the permeability of this medium is much like that of free space, so no magnetic polarization effects are possible. The hazard of continued exposure to variable magnetic fields is somewhat more uncertain, however. Kouwenhoven and his co-workers followed personnel chronically exposed to alternating radiation fields in the extremely low frequency range. Some 11 power linemen were examined extensively for clinical effects caused by 60-Hz radiation. No detectable changes were seen, even though those individuals worked at close range to 100-kV lines carrying currents of hundreds of amperes.

Several biological effects are known to be induced by time- or space-varying magnetic fields. One is the phosphene effect, wherein the observer actually "sees" a vivid light display if an ambient magnetic field is suddenly changed in intensity. An exact explanation is not yet available, although the head of the observer appears to be the crucial locus. It has been shown that certain animals have learned to use spatially varying magnetic fields as a compass indicator. The homing pigeon (see Chapter 4) is among these. It is not clear as to whether the motion-induced time derivative ($d\vec{B}/dt$) or the $\vec{B}$ field itself is the primary significant physical input. As indicated above, sensing of a constant magnetic vector is difficult to explain, whereas the detection of the electromotive force caused by a temporal variation of $\vec{B}$ seems much more feasible. Part of the confusion as to the avian stimulus is rooted in the unknown nature of the birds' detection mechanism. Jamming has been accomplished using constant fields having magnitudes comparable to the earth's, but the implications of these results are not clear.

12.6 Microwave Radiation

This important band encompasses the range from 300 MHz to perhaps 300 GHz and includes the intensively investigated heating and diathermy frequency of 2.45 GHz. As in the discussion of extremely low frequency radiation given in the preceding paragraphs, the primary concern has been the effect of various ambient levels of radiation intensity. To place the topic in context, it should be recalled that the metabolic output of an adult human being is approximately 5 mW/cm². It would then be anticipated that intensity values appreciably above this might be difficult to withstand, particularly in those regions where blood circulation is rather minimal. One example of such a system is the lens of the eye.

The results of Carpenter and his co-workers on the irradiation of the rabbit eye are given in Table 12-2. The percentage of opacities is included as

TABLE 12-2

IRRADIATION OF THE RABBIT EYE

($\nu = 2.45$ GHz)

$I(mW/cm^2)$	$\Delta t\ (hr)$	*Percent positive results*
50	20	0
80	20–24	11
100	18–32	40
120	13–20	80

the right-hand column. These authors comment on the complete lack of human cases in the medical literature. At least one clinician, however, has observed marked opacities in several radar technologists. One difficulty in any such interpretation is that the cataracts discovered under controlled microwave exposure conditions do not appear to be different in any way from those which occur spontaneously. Thus, a control group is mandatory. A review by Appleton involving some 1,500 workers is in agreement with the hypothesis that the incidence of lens opacities is not demonstrably higher in the industrial group. Although the results in Table 12-2 indicate microwave levels leading to cataract in rabbits, it is important to point out that this animal model is of uncertain sensitivity relative to man. The critical intensities appear to be in excess of 50 mW/cm²; that is, they are elevated compared to typical metabolic levels in a large mammal. The present environmental limit in the United States is 10 mW/cm², a value that has been assumed to be approximately one order of magnitude below the threshold for observable effects.

The provision of heat within living tissue is of great clinical significance. The diathermy region largely corresponds to the microwave band. Figure 12.4 gives the penetration (a^{-1}) in centimeters for fat and wet tissues. Notice that the presence of H_2O in the tissue increases its attenuation by approximately 1 order of magnitude. This absorption is the basis of the microwave oven and was explored in 1940 by the Japanese armed forces as a death ray. One can realize, by extrapolation from Fig. 12.4, that the depth of penetration becomes minimal as the frequency approaches 10^{13} Hz, the lower limit of the infrared radiation band. In that case, in any tissue the heating is almost purely a skin effect.

Because of metabolic and other data, the American limit on ambient microwave radiation is set at 10 mW/cm². The comparable Russian and Eastern European value is only 10 μW/cm², an increased safety margin of

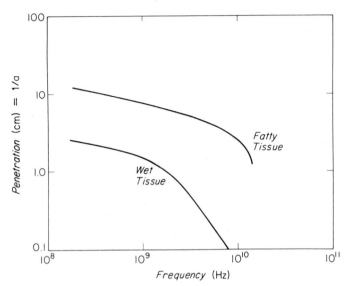

Figure 12.4 Penetration (a^{-1}) of microwave energy in fatty and wet tissue. Both axes are logarithmic. As the frequency rises, the increase in the linear attenuation coefficient (a) causes most photon absorption to be superficial. After H. P. Schwan, *Therapeutic Heat*, S. Licht, ed. (New Haven, Conn.: E. Licht, 1958).

some 3 orders of magnitude. The reason for this discrepancy is the Russian belief that subtle nervous system changes occur after chronic irradiation at levels comparable to the metabolic rate. The basic evidence for this effect is the disruption of conditioned reflexes in animal models. Their exposure regimes usually extended from periods of weeks to several months. Such chronic irradiations are assumed not to induce appreciable heating of nervous tissue due to the low power levels involved. Some additional information has been obtained that indicates an amplitude decrease in rabbit eeg patterns after acute exposure to 5–10 mW/cm² of microwave energy. Similar changes can occur spontaneously in control animals, however, and are not specific for microwaves. It would seem that more experimental data are required to establish the Russian claims regarding the dangers present at the higher U.S. limit. The interested reader should consult the Czerski reference for a review of this literature.

Temperature changes are not in themselves detrimental to the function of mammalian tissues. The microwave danger is that the change in temperature is of sufficient magnitude that the final tissue temperature is elevated beyond a destructive end point. This limit is on the order of 45°C for the central nervous system and 50°C for cardiac function. Thus, absolute power levels are not meaningful unless the organism is known to be in a normal temperature state prior to the experiment. Several authors have, in fact, considered warming hypothermal animals by means of high-intensity microwave sources.

12.7   Laser Irradiation

One of the important aspects of patient diathermic treatment is the actual size of the region affected by the radiation field. For example, an ultrasonic source has a wavelength of approximately 1 mm at MHz frequencies as compared to the 12-cm (2.45-GHz) microwave beam. Consequently, the ultrasonic treatment may be applied to a much more restricted volume. This topic is explored further in Chapter 13. As the electromagnetic source's frequency is raised to the near-visible and visible, however, the corresponding wavelength is on the order of 10^{-4} cm, or 1.0 μm. It would then appear possible to apply radiant energy to objects of this size using optical frequencies. Various laser sources have been utilized in diverse applications in biological media because of their focusing capability. Selected examples of this strategy are given below; the laser principle is considered in Chapter 27.

Pulsed ruby lasers have been used to repair retinal tears; the photocoagulation process takes about 1.0 msec. With argon lasers, which are also used, about 100 msec is needed. After the iris is dilated, a laser focal spot on the order of 100 μm in diameter is imaged on the loose tissue. The actual "weld" occurs in the pigmented epithelium, since the first layer of the retina is transparent at optical frequencies (see Chapter 2). The mechanism is essentially thermal, with threshold values being known from animal models (rabbits) and clinical studies. Energy densities between 10 and 100 J/cm² are commonly used. A small scar is left at the laser focal spot. The green light of the argon laser is more heavily absorbed in the pigmented tissue and somewhat sharper foci are possible with 50 μm being a nominal limit. Since the fovea is approximately 10^3 μm across, the improvement in spot size using 488 and 514 nm argon radiation can be significant in treatment of the patient's problem.

Vascular lesions are a second pathological condition that has been successfully treated by laser irradiation. One example is the neovasculature arising in the retina of diabetics (diabetic retinopathy). The new capillaries are closed by coagulation, using a short pulse of argon laser light, with the subsequent scar being about 50 μm across. A second problem of this type is the set of disfiguring skin conditions often associated with a congenital lack of an epidermal layer over the vascular bed. These areas appear reddish-purple because of the proximity of the venous blood to the skin surface. By repeated application of ruby (694.3 nm) or argon sources, the skin area can be sufficiently scarred to produce a bleached effect. Some local vascular destruction is also accomplished. The literature contains clinical studies in which over 100 patients have been laser-treated for such port wine lesions. Success rates appear to approach 50 percent. Alternative forms of treatment are generally difficult and have more traumatic consequences.

A cauterizing instrument such as the focused laser beam could have wide application in general surgical procedures. The organ or tissue to be excised can be cut free at the same time that the vascular input is sealed so as to minimize subsequent bleeding. Major organs have been removed using laser surgery, and relatively few complications are reported. Some clinicians have also demonstrated dramatic improvements in skin cancers using very intense pulsed sources. As one example, a skin melanoma can be completely excised by focusing a high-power instrument into areas on the order of 1 cm² in size. The effect is again one of intense heating, with perhaps 10^3 J being delivered to the lesion. Because motion is possible only in a retrograde direction (toward the laser source), the resultant "plume" of cellular debris may be several centimeters high. It must be captured in a disposable container to prevent tumor cell contamination of the surgical suite. Reaction pressures on the order of 50 atm have been measured under these conditions!

It is interesting to compare the temperature increments achieved in typical laser and X-ray irradiation therapies. Since 1 J is 0.24 gram-calorie, and 1 gram-calorie will raise the temperature of 1 g of tissue by approximately 1°C, 10^3 J of laser radiation delivered to 1 g of tissue will raise the mean temperature several hundred degrees Celsius. An X-ray dose of 100 rad is only 10^{-3} J/g, with a corresponding temperature change of a fraction of a millidegree. Because of the rapidity with which laser energy is delivered (1 μsec is not uncommon), heat conduction away from the impact site is not possible, even in the case of metallic targets. Thus, the energy must be absorbed locally, with little chance of transfer beyond the focal point.

As a specific illustration of the use of localized heating, let us consider the general subject of laser microirradiation. Figure 12.5 shows the result when a laser beam is first focused and then selectively attenuated so as to produce a very small effective beam spot. The transverse dimensions of the human red blood cell lesions generated in this manner ranged from 0.25 to 1.4 μm. The capability exists only in lasers in which the energy distribution across the beam is a simple function (e.g., a Gaussian curve). Moreover, since the biological effect is produced at a given radiation level, the actual area affected may be made almost arbitrarily small by lowering the overall intensity.

Most microirradiation experiments have studied individual cells such as protozoans, protists, and single cells in tissue culture. The relative biological roles of the nucleus and cytoplasm are the primary objectives of this research. A secondary interest has been *chemotaxis*, a process by which a chemical affects nearby cellular motion. For example, if a *Euglena* is irradiated by ruby laser light, the high absorption of the chlorophyll contained therein causes a very rapid thermal death of the protist. Neighboring *Euglena* are then found to move away from the site of the irradiation, presumably in

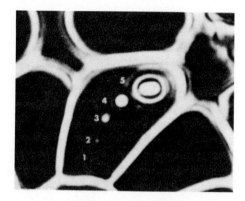

Figure 12.5   Human red blood cell lesions produced by visible-light laser illumination. The laser output was selectively attenuated for each irradiation. Filter transmission and hole diameter were as follows:

Lesion	Filter Transmission (%)	Lesion Size (μm)
1	0.48	0.25
2	0.62	0.40
3	0.79	0.50
4	1.0	0.90
5	1.15	1.4

Source: After M. W. Berns, *Biological Microirradiation, Classical and Laser Sources* (Englewood Cliffs, N.J.: Prentice-Hall, Inc., reprinted by permission; 1974).

response to some molecular form(s) released by the dead cell. Because it can specifically open certain parts of a cell or even an organelle of interest, the laser beam appears to be of great significance in investigations of this sort.

12.8   Ultraviolet Radiation

Ultraviolet (UV) radiation has a maximum wavelength just below that of visible light (360 nm) and extends down to low-energy X rays (100 nm). By this definition, ionizing processes are excluded from consideration (see Chapter 11). The biological importance of ultraviolet radiation is difficult to over-estimate. The solar electromagnetic spectrum contains a significant amount of energy in this band interval. Many biologists feel that ultraviolet radiation at the earth's surface was responsible for generation of the first organic compounds in the primitive lakes and tidal basins. The transparency of the early atmosphere to UV light was a consequence of the absence of ozone (O_3). These original organic syntheses are presumed to have started with methane, water, ammonia, and other simple compounds and progressed toward the various amino acids. Miller and Urey have demonstrated the feasibility of this scheme *in vitro*.

Proteins and the nucleic acids are also very strong absorbers of UV light. The absorption is so characteristic that it may be used to accurately determine the presence of various specific macromolecules. The continued existence of the ozone shield is thus essential to the persistence of present-day life forms. Both nitrogen oxides and fluorocarbons have been shown to act as catalysts for O_3 depletion. The ecological impact of all technological products must be carefully evaluated with regard to the production of these and similar chemical species. One need not, however, look only at these dangers on a geological time scale; much more immediate problems can occur after exposure to UV radiation.

Ultraviolet light is a well-known skin carcinogen—probably due to its strong absorption by the nucleic acids (see Appendix D). The generation of two types of cancer, squamous-cell and basal-cell carcinomas, is not uncommon in relatively fair-skinned people exposed to intense UV light. Among the hazardous sources are the tropical sun and the mercury arc lamp. Shielding is advisable under such working conditions. Certain anthropologists have developed a theory of human skin pigmentation largely based on these considerations. Those populations living in equatorial regions are presumed to develop more melanin granules per square centimeter of skin to protect against the overhead solar source. The amount of melanin cannot become arbitrarily large, however, since the penetration of some UV radiation is necessary for the activation of vitamin D and bone development. Thus, each regional subspecies presumably has an appropriate level of pigmented shielding, pertinent at its particular latitude.

Human skin cells are not the only cells sensitive to ultraviolet irradiation. Techniques of cellular experimentation have developed along with the various laboratory UV sources. The methods are similar to those employed in the ionization experiments described in Chapter 11. Generally, the results are expressed in a similar fashion (e.g., cell survival curves and mitotic inhibition). The advantage of a UV source is that optical focusing methods may be used instead of divergent beams of X or γ rays. The precise localization of the radiation impact point may be quite crucial in determining a biological mechanism. It is for this reason that interest continues in the ultraviolet irradiation of cellular samples.

12.9   Summary

Electromagnetic energy has been extensively used as a tool for investigating living tissue. Work on cellular impedance has identified at least three possible mechanisms for tissue–radiation interaction, each of these being associated with a given frequency interval. As the frequency increases, the responsible structure becomes progressively smaller—commensurate with the wavelength. Cellular models resulting from this picture are in reasonable agreement with other cytological data.

Environmental electromagnetic effects are a topic of great concern to health specialists as well as to the public. The most important problem is the hazard of electrocution due to 60-Hz power lines. Current flow through the heart muscle is the critical parameter in such considerations. Power-line radiation itself does not appear to have measurable consequences. Heating effects induced by microwave sources, however, are quite capable of producing thermal lesions. Regions of poor cooling capacity (e.g., the lens of the eye) are quite sensitive in this regard. At higher frequencies, ultraviolet solar radiation is a well-known carcinogen. Preservation of the earth's ozone layer is imperative if we are to be protected from this danger.

Laser sources, primarily at visible frequencies, have been applied to medical problems. Retinal tears and disfiguring skin pathologies have been the primary conditions treated. Excision of lesions is a growing area of interest. The self-cauterizing nature of the laser beam as well as its small focal spot size are important in surgical protocols.

REFERENCES

Impedance

1. SCHWAN, H. P., "Electrical Properties of Tissue and Cell Suspensions," in *Advan. Biol. Med. Phys.* **5**:147–209 (1957).

Low-Frequency Electromagnetic Attenuation

2. LLAURADO, J. G., A. SANCES, JR., AND J. H. BATTOCLETTI, *Biologic and Clinical Effects of Low-Frequency Magnetic and Electric Fields* (Springfield, Ill.: Charles C Thomas, Publisher, 1974), 345 pages.
3. KOUWENHOVEN, W. B., et al., "Medical Evaluation of Man Working in AC Electrical Fields," *IEEE Trans. Power App. Systems* **PAS-86**:506–511 (1967).

Relaxation Processes in General (see Sec. 12.3)

4. CZERLINSKI, G. H., *Chemical Relaxation: An Introduction to Theory and Application of Stepwise Perturbation* (New York: Marcel Dekker, Inc., 1966), 314 pages.

Microwaves

5. CZERSKI, P., et al., *Biologic Effects and Health Hazards of Microwave Radiation*, Proceedings of an International Symposium, Warsaw, Poland, October 15–18, 1973 (Warsaw, Poland: Polish Medical Publishers, 1974), 350 pages.

(This reference also contains a good review of impedance measurements by H. P. Schwan, pp. 152–159.)

6. APPLETON, B., AND G. MCCROSSAN, "Microwave Lens Effects in Humans," *Arch. Ophthalmol.* **88**:259–262 (1972).

Lasers

7. GOLDMAN, L., AND R. J. ROCKWELL, JR., *Lasers in Medicine* (New York: Gordon and Breach, Science Publishers, Inc., 1971), 385 pages.

8. WOLBARSHT, M. L., ed., *Laser Applications in Medicine and Biology*, Vol. 1 (New York: Plenum Press, 1971).

9. BERNS, M. W., *Biological Microirradiation, Classical and Laser Sources* (Englewood Cliffs, N.J.: Prentice-Hall, Inc., 1974), 152 pages. (A review of both charged particle and photon microbeams. An excellent list of references is included.)

Ultraviolet

10. JAGGER, J., *Introduction to Research in Ultraviolet Photobiology* (Englewood Cliffs, N.J.: Prentice–Hall, Inc., 1967), 164 pages.

CHAPTER 13

Sonic Irradiation

13.1   Introduction

The human sense of hearing, discussed in Chapters 1 and 7, involves audible mechanical vibrations that have a certain limited range of frequencies. However, from a physical or even biological point of view, there is nothing unique about vibrations within the limits of human hearing. This chapter uses the adjective *sonic* to refer to transmitted mechanical vibrations not involved with hearing. The frequency ranges covered include the sound waves audible to humans (roughly 30 Hz to 20 kHz) as well as higher- and lower-frequency regions called, respectively, ultrasonic and infrasonic.

The general properties of acoustic waves are not dependent on hearing. Accordingly, the characteristics of these waves introduced in Chapters 1 and 4 can be applied directly to sonic vibrations. A more general overview of acoustic theory is found in Appendix A. That theory is used here in the quantitative description of sonic radiation.

This chapter is concerned with the medical and biological effects of sonic irradiation and some of their applications. The most significant medical application of sonic radiation has been the noninvasive determination of

233

anatomic structures using echoes reflected from various organs and tissues. This technique, often called *echosonography*, is discussed in Chapter 28.

The growth of echosonography has increased the interest in the biological effects of sonic irradiation. These include effects due to repetitive exposures both at low levels of irradiation characteristic of echosonography and at higher levels. The studies at the higher levels indicate the maximum power levels that can be used with acceptable risks to the subject's safety.

There are numerous other factors that provide a continued stimulus for biophysical studies involving sonic irradiation. Some of these are concerned with searches for novel medical applications in such diverse areas as neurosurgery and cancer therapy. Others include an interest in the products formed by cell fracture. More basic to biophysics is the use of sonic vibrations to describe and quantitate the mechanical properties of biological cells and tissues as a function of frequency.

Several different biophysical aspects of sonic irradiation of biological cells and tissues are included in the current chapter. Sections 13.2 and 13.3 are concerned with low-amplitude effects. The first of these sections is an overview of pertinent aspects of physical descriptions of sonic irradiation, specifically wavelengths associated with sonic waves in tissues and measurement of the absorption of sonic energy. The application of these concepts to biological systems is presented in Sec. 13.3.

The remainder of the chapter is concerned with effects that occur at higher intensities. Section 13.4 describes the physical phenomenon known as cavitation, which can occur at sufficiently high intensities in a liquid. The following section reviews the destructive effects of cavitation. Depending on the goals of the study, these destructive results may be desired ends or unwanted side effects.

Even in the absence of cavitation, irradiation at sufficiently high intensities can lead to tissue damage due to heating and other, less clearly understood phenomena. These effects, which have been proposed as alternatives to surgery, form the basis for Sec. 13.6. Little direct application has been made of such destructive effects; they have been studied primarily to set limits for the intensity levels used in echosonography.

13.2   Sound and Ultrasound

As discussed in Chapter 1 and in Appendix A, material media possess elastic properties such that a mechanical disturbance will be propagated (transmitted) through a medium with a characteristic wave speed, c. Such disturbances in fluids can be conveniently described as a pressure, p, wave and the concomitant (vector) particle velocity, $\vec{v}$, wave, or in terms of the transmitted intensity, I, that is, energy per unit cross-sectional area per unit time. Simple wave motions occur at one frequency, ν; more complicated waves may be analyzed as a distribution of amplitude or intensity per unit frequency band using techniques such as the fast Fourier transform described in Appendix C. In the following discussions p and $\vec{v}$ are regarded as either Fourier transforms or vibrations at a single frequency.

Small-amplitude waves are transmitted through a medium in a fashion such that the energy in each small frequency interval can be considered independently of all other intervals. (This is described in mathematical physics by the statement that acoustic waves obey the linear wave equation. See Appendix A.) These waves will undergo diffraction and will exhibit interference phenomena. The critical parameter in diffraction and interference is the wavelength, λ, which can be computed by the relationship

$$\lambda = \frac{c}{\nu} \tag{13-1}$$

For soft tissue, c has a value close to that of water; for rough estimations one can use 1.5 km/sec. For studies on biological structures, it is sometimes desired to have λ small compared to organ sizes; if one desires a value of 1 mm, one should use a frequency of 1.5 MHz. However, for studies described in the next section, frequencies as low as a few hertz and higher than 30 MHz have been used. One parameter used to characterize tissues undergoing sonic irradiation is called the *acoustic impedance*, Z. This is defined by

$$Z = \frac{p}{v} \tag{13-2}$$

where the scalar v is the magnitude of the vector $\vec{v}$. In general, p and v will have a phase difference that can be represented by regarding both of them, and hence Z, as complex variables.

The impedance Z depends, among other things, on the spatial wave shape. The value of Z for a plane wave is called the *characteristic impedance*, z. In the absence of absorption, z will be a real number given by

$$z = \rho c \tag{13-3}$$

where ρ is the density of the medium in which the wave is traveling. The fractions of a plane wave transmitted and reflected at the interface between two materials (tissues) is determined by the geometry of the interface and the values of z in the two materials. For large changes of z, most of the energy is reflected, whereas for small changes, most is transmitted across the interface. Thus, in describing the intensity of the sonic radiation in a living system, it is important to know the tissue values of the characteristic impedance.

All media absorb some of the transmitted energy. In the presence of absorption there will be an imaginary part (acoustic reactance) of the acoustic impedance, Z, as well as a real part (acoustic resistance). Many experiments determine a complex z. Nonetheless, it is probably easier to discuss and analyze the attenuation coefficient, α, which is conceptually similar to the linear attenuation coefficient discussed in Chapter 11. Here, α is defined implicitly by the following equation, which describes the acoustic pressure p in a plane wave moving in the $+x$ direction:

$$ p = p_0 \exp\left(-\alpha x\right) \exp\left[j\omega\left(t - \frac{x}{c} \right) \right] \tag{13-4} $$

where p_0 is the amplitude, x is the distance, j is $\sqrt{-1}$, t is the time, and ω is the angular frequency ($2\pi\nu$). Equation (13-4) is an approximation that is valid for small-amplitude waves.

The attenuation coefficient, α, is dependent on the chemical and physical nature of the medium, the temperature, and the frequency. For a pure, simple substance it can be shown theoretically that the ratio α/ν^2 is independent of frequency. This has been experimentally verified for pure water and for water solutions with simple ions over a wide range of frequencies. Accordingly, this ratio (α/ν^2) is used to characterize the absorption of biological tissues and colloidal suspensions, even though α/ν^2 varies with frequency for these more complicated media.

13.3   Sonic Characteristics of Biological Materials

Acoustic impedance and absorption have been measured as a function of frequency for a variety of tissues and biological fluids. These measurements are interpreted using the terminology and formalism introduced in the preceding section. As noted there, the wavelengths involved are about 0.1 mm at 15 MHz and are larger at lower frequencies. While this is small compared to the size of biological organs, it is comparable to, or larger than, the dimensions of most biological cells. Accordingly, absorption of acoustic energy usually involves groups of cells or the bulk properties of cells in suspension.

This bulk behavior contrasts with the absorption of ionizing radiation, which corresponds to much shorter wavelengths and can thus affect a single molecule or even a small site on a large macromolecule (see Chapter 11). Most absorption studies with nonionizing electromagnetic radiation (see Chapter 12) take place at the molecular level. Ultimately it is desired to assign acoustic absorption to molecular processes; however, these always involve bulk properties of the tissue or suspension.

A. ACOUSTIC IMPEDANCE

The easiest parameter to measure is the characteristic impedance, z, which would describe the transmission of a plane wave if no absorption occurred. Table 13-1 shows values of the characteristic impedance for a variety of tissues. It can be seen that many of these are close to the value for water. However, the value for bone is manyfold greater and that for lung very much smaller, since it includes appreciable amounts of air.

TABLE 13-1
CHARACTERISTIC IMPEDANCES

Tissue or medium	c (km/sec)	ρ (g/mℓ)	z (or ρc)
Muscle	1.58	1.06	1.66
Liver	1.57	1.06	1.65
Kidney	1.56	1.06	1.64
Brain	1.52	1.03	1.57
Fat	1.48	.93	1.37
Bone	3.36	2.32	7.80
Water	1.43	1.00	1.43
Air	0.34	1.25×10^{-3}	4.2×10^{-4}

This variation of impedance is very important in considering transmission across the interface between bone and soft tissue and at air–tissue interfaces. Specifically, it can be shown that the intensity I_t transmitted into a medium with characteristic impedance z_2 from an incident beam with intensity I_i in a medium with characteristic impedance z_1 is given by

$$I_t = \frac{4|z_1 z_2|}{(|z_1| + |z_2|)^2} I_i \tag{13-5}$$

If the two impedances are equal, all the energy will be transmitted. However, if one is tenfold greater than the other, Eq. (13-5) indicates that only 40 percent of the incident beam will be transmitted, the remainder being reflected or absorbed at the interface.

Between tissue and air, there is an approximate difference in the characteristic impedance of a factor of 3×10^{-4}. Thus, a negligible portion of the incident energy will be transmitted across a tissue–air interface. This, in turn, indicates that effective sonic irradiation of tissues even at nondestructive levels requires a fluid coupling medium between the sound generator and the tissue.

B. ABSORPTION COEFFICIENTS

In addition to a characteristic impedance, an absorption coefficient or excess of the ratio α/ν^2 is used to describe the behavior of biological tissues. Figure 13.1 includes a curve which shows a typical relationship between the absorption and the frequency. To interpret such a curve, one must study

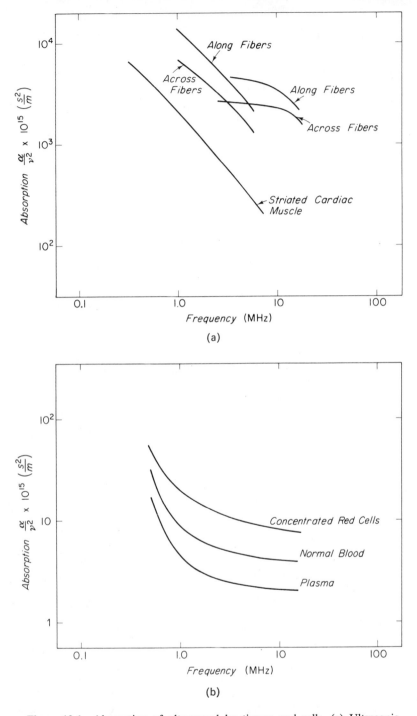

Figure 13.1 Absorption of ultrasound by tissues and cells. (a) Ultrasonic absorption spectra for muscle. The various curves indicate results obtained by different authors. (b) Ultrasonic absorption of normal beef blood and its components as measured at 15°C. After M. Hussey, *Diagnostic Ultrasound: An Introduction to the Interactions Between Ultrasound and Biological Tissues* (Glasgow: Blackie & Son Ltd., 1975).

simple solutions and cell fractions. These studies aim at understanding the biophysical mechanisms involved in the absorption of sonic energy. It was hoped that such studies would aid in modeling the mechanical form of biological systems. By and large this strategy has not contributed heavily to current concepts of biomechanics. On the other hand, the growing interest in echosonography has led to continued studies quantitating sonic absorption.

The simplest materials in which to study sonic absorption are water and dilute solutions of inorganic ions. Here it is found that the absorption, expressed as α/ν^2, is independent of frequency essentially from 0 Hz up to at least 100 MHz. Since over 60 percent of the weight of a human is water, this absorption might be quite important for studies of sonic irradiation of biological systems. The absorption of water has a value that decreases with temperature, having a value of about 2×10^{-14} sec^2/m at room temperature. This small value means that at 1 MHz, the decay factor α has a value of 2×10^{-2} per m; that is, the intensity in a plane wave would decrease by a factor of e in 50 m. Thus, for echosonography or diathermy, losses due to absorption by pure water are not significant under reasonable geometric constraints.

An additional effect seen when α/ν^2 is measured as a function of frequency for more complicated substances is called *relaxation*. The experimental observation is a decrease in the attenuation coefficient at higher frequencies, with the characteristic shape indicated in curve b of Fig. 13.1. Relaxations occur because the absorbers are unable to follow the motion of the very high frequency vibrations.

A single relaxation is described analytically by an equation with three constants, a, b, and τ, such that

$$\frac{\alpha}{\nu^2} = a + \frac{b}{1 + \nu^2\tau^2}$$

where τ, the *relaxation time*, is a characteristic of the absorber. Similar relationships are obeyed whenever electromagnetic or sonic radiation are absorbed (see Chapter 12). However, the physical interpretation of τ is different for these two types of irradiation.

Sonic relaxations are observed for solutions of inorganic ions. At low frequencies, these solutions have absorption coefficients appreciably greater than pure water. This is particularly true for bivalent and higher-valency ions. Thus, while univalent ions as Na^+ and Cl^- have only minimal effects, 0.1 M $MgSO_4$ has a value of α/ν^2 at low frequencies several hundredfold greater than pure water. Other bivalent ions have comparable effects. The absorption is apparently due to changes in the hydration and association of the ions as the acoustic pressure changes. This effect shows relaxations in the range 0.1–10 MHz, which agrees well in relative frequency dependencies and absolute magnitudes with theoretical predictions.

In all cases aqueous solutions of simple molecules show higher absorptions than pure water, and relaxations at high frequencies, due to the inability of the solutes to change their associations during the time interval of going from pressure minimum to pressure maximum. Aqueous solutions of more complex molecules also exhibit an excess absorption over that of pure water. The relaxations for these solutions may involve decreases of several thousand-fold in the excess absorption as the frequency is raised from 1 MHz to 100 MHz. This excess absorption is believed to be due to changes in the hydration shells of the macromolecules as a result of the acoustic pressure changes. However, the relaxation curves showing excess α/ν^2 as a function of ν are complicated and cannot be interpreted as being due to a single mechanism.

When solutions of single cells such as erythrocytes are sonically irradiated, additional absorption mechanisms become important. These include friction between the cell membrane and the surrounding liquid which is added to the absorption due to hemoglobin. The data in Fig. 13.2 show the

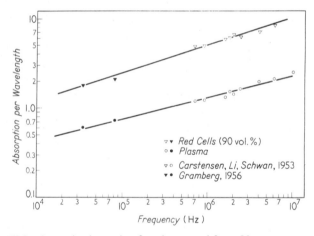

Figure 13.2 Acoustic absorption for plasma and for a 90 percent suspension of red blood cells.

absorption for plasma and a 90 percent suspension of red blood cells. The shape of this curve, although qualitatively what one would expect, quantitatively disagrees with all proposed models.

Similar data for a variety of tissues show still greater absorptions of sonic energy. Several mechanisms have been proposed, all of which undoubtedly contribute to the observed behavior. These include, in addition to the effects mentioned in the preceding paragraphs, the relative motion of the various cellular organelles, such as the cell nuclei, mitochondria, and ribosomes. (See

Appendix E for a discussion of these terms.) Figure 13.3 shows the results of a study of the absorption of particulate fractions of liver.

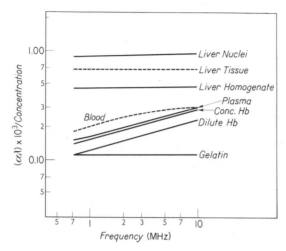

Figure 13.3   Absorption of particulate fractions of liver.

C. APPLICATIONS

Irradiation is used clinically to heat selected local regions of an individual, an effect that is called *diathermy*. Both electromagnetic (see Chapter 12) and sonic irradiation are used for this purpose, with certain notable differences. A variety of disorders is reported to be alleviated by diathermy; these techniques are standard ones in physical medicine and physical therapy.

The energy absorbed is proportional to α, which, in turn, at low frequencies is proportional to ν^2. Accordingly, at very low frequencies, a minimal amount of energy would be absorbed. Thus, sonic diathermy usually involves higher frequencies, in the 1- to 15-MHz region. Still higher frequencies prove less useful, since relaxation processes reduce the specific absorptions produced by the biological systems. In the useful range, quite narrow beams can be employed, allowing application of excess heat to selected organs, skeletal joints, and parts of muscles.

It is found that at the surface of bones, not only is a major fraction of the energy reflected, but also an excess absorption occurs very locally. This is believed to be due in part to the conversion of energy from the usual acoustic (compressional) wave, where the particle motion occurs in the same direction as the wave propagation, to transverse shear waves, where the particles move right angles to the propagation direction. Transverse waves are rapidly absorbed, thereby creating added heating.

For diseases of the joints, bursae, and the surfaces of the bones, this local excess heating may prove quite beneficial. Sometimes, hormonal preparations such as cortisol are injected into the joint, and then sonic irradiation is used to augment the hormonal effects. The effects can be synergistic; that is, the combination may be significantly better than the anticipated sum of the effects of each alone.

Thus, absorption of sonic energy can be used to characterize tissues and indicate the inadequacy of all mechanical models. Absorption also has clinically useful effects associated with diathermy. A combination of data from absorption and characteristic impedance measurements enables the prediction of the results of echosonographic procedures. However, if the intensity of the ultrasonic irradiation (or the time, or the product of the intensity times time) becomes too high, damage can occur. Such limitations make it necessary to observe suitable precautions, particularly when irradiating bony regions, since local spots of intense heating can occur. In addition, at higher intensities other phenomena can occur, which are examined in the following sections.

13.4   Cavitation

As the intensity (power per unit area) of the irradiating sonic beam is increased, a number of effects are observed which are not linearly, or even simply, related to the acoustic pressure. Among these are a rectification of the alternating pressures and velocities, resulting in streaming or continuous flow. The pattern involves flow away from the transducer in the center of the beam and, since the fluid must return, back toward the transducer at the edges of the beam. Such flows can produce considerable mixing and eddies, as well as carrying parts of a solution being irradiated to lower-intensity parts of the sonic field.

A related effect at higher intensities is the sharpening of the wave fronts. This occurs because the portions of the wave at higher acoustic pressures travel with a greater speed and tend to catch up with slower parts. The net result is to generate repetitive steep pressure rises, followed by slower pressure falls. In the limit, this is called a *shock wave*. In the frequency domain, this is described as generating higher harmonics.

Both of the preceding nonlinear effects can be observed during the sonic irradiation of tissues at higher intensities; both may contribute to the biological results obtained. However, another phenomenon, called *cavitation*, is probably more important than these or other nonlinear, high-intensity sonic effects; its biological actions are the basis for the section that follows.

Cavitation occurs because the local pressure in a sound wave drops to sufficiently low values that the liquid essentially tears apart. The acoustic pressure, p, is defined in Chapter 1 and Appendix A by the relationship

$$p = P - P_0 \tag{13-6}$$

where P is the local pressure and P_0 is the average local pressure. The absolute pressure P in a gas cannot be negative. Therefore, the acoustic pressure amplitude must always be less than the atmospheric pressure.

In contrast, liquids have a definite volume and can sustain negative pressures or tensions. Pressure amplitudes of hundreds of atmospheres can be generated in water. However, when the pressure becomes low enough (i.e., the tension becomes great enough), the liquid will fracture. The liquid breaks by forming small, more-or-less spherical holes called *cavities*.

The tension necessary to produce cavitation measures, in some sense, the cohesiveness present in the liquid. Various physical theories have attempted to predict the negative pressure at which pure, "clean" water will fracture. The most straightforward of these predicts about $-15,000$ atm. This theory essentially calculates the work to pull two planes apart. A somewhat more sophisticated theory employs both the spherical nature of the holes and another general semiempirical theory called absolute rate theory (see Chapter 21). This combination, named *nucleation theory*, predicts cavitation thresholds of about $-1,500$ atm.

No one has ever reached this value. The lowest threshold reported for cavitation (i.e., liquid rupture) in water is -350 atm. This is usually explained by dirt on the sides of the vessel, or suspended in the medium, which, even in triply distilled water, limits the tension the liquid can withstand before rupturing. For dirty liquids, saturated with gas and tiny gas nuclei, cavitation may occur at positive pressures. Indeed, cellular disruption due to cavitation is sometimes found with acoustic pressure amplitudes of only 0.1 atm, that is, at positive pressures of 0.9 atm.

Cavitation can be observed by various techniques. Intense cavitation is readily visible to the naked eye. A hissing sound is often heard. A probe microphone out of the main sound field will pick up noise radiated by the cavities. If the acoustic pressure is measured as a function of the energy applied to the transducer used to generate the ultrasonic field, a curve is obtained such as that sketched for p in Fig. 13.4. As the point of cavitation is reached, the curve bends over because the liquid tears rather than sustaining higher pressures. As the liquid fills with cavities, its density decreases; this permits the particle velocity to increase linearly with the square root of the applied energy to values well above the threshold for cavitation.

Cellular fracture and harmonic distortion of the pressure wave are observed at acoustic pressures lower than those shown by the break in the curve in Fig. 13.4, labeled cavitation. However, both cellular fracture and

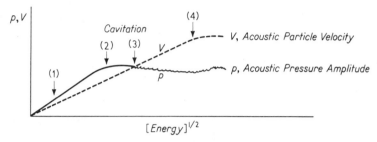

Figure 13.4 Variation of acoustic pressure and particle velocity with energy supplied to transducer. The arrows indicate points at which cavitation occurs based on (1) cellular disruption; (2) break in the pressure curve; (3) generation of noise; and (4) break in the particle velocity curve.

harmonic distortion can be suppressed by increasing the atmospheric pressure, a change that would interfere only with effects due to cavitation. These indicate that cavitation is occurring, at least to a limited degree, before the break in the curve. Thus, different tests for cavitation lead to different thresholds for cavitation.

Care must be taken to distinguish cavitation from heating. When an intense ultrasonic field is generated in a small body of liquid, acoustic energy must be dissipated as heat. If no method is provided to remove this excess heat energy, the temperature will rise. Then all biological effects of ultrasound are masked by heating. Any practical exposure technique must involve either very short periods of time or some form of cooling.

The threshold values for cavitation are convenient to use for the divison between low intensity and high intensity. To convert these values, quoted previously as approximately 1 atm of pressure amplitude to SI units of intensity, one may use the relationship noted in Chapter 1 and Appendix A:

$$I = \frac{p^2}{2\rho c}$$

and the approximate equality

$$1 \text{ atmosphere} \simeq 10^5 \text{ N/m}^2$$

These combine to give a value for I of about 3 kW/m². An irradiating sonic beam is called high intensity if its maximum acoustic pressure amplitudes are of the order of 1 atm or greater or its maximum intensities are at least 1 kW/m². This division is quite arbitrary, however. For example, Hussey divides intensities into three ranges, low, middle, and high; there is no detailed agreement on such divisions.

There is, however, general agreement that in studies using any type of irradiating, it is desirable to monitor the dose. In high-intensity cavitating sound fields this process may become quite involved, since one should know the acoustic intensities and pressures, each as a function of frequency and position. In addition, other measures of the intensity of cavitation are desirable. Much of the literature is confused by experiments where only the frequency and power supplied to the transducer are recorded. This was sufficient to allow the experiments to be duplicated with the original equipment, but not to allow its repetition by others. In spite of this limitation, cavitation has been used in a number of biophysical applications, which are discussed in the next section.

13.5   Cavitation Damage to Biological Cells

Cavitating sonic fields are used to fracture biological cells and to clean laboratory equipment. The cell disruption is used to obtain suspensions of subcellular particles, to extract enzymes, and to study the mechanical characteristics of single cells. Cavitation is harder to demonstrate within intact tissues and organs. Accordingly, most of this section deals with effects observed with suspensions of cells.

Destruction of biological cells in suspension is observed when cavitation occurs. It had been suspected that some other parameter of the acoustic field or perhaps heating associated with cavitation was responsible for the cellular destruction. For example, it was believed that the large acoustic pressures, or perhaps the local particle velocities and accelerations, were in some way responsible for the observed cellular rupture. All of these properties are, if anything, enhanced when cavitation is suppressed, either by increasing the applied pressure or by vacuum degassing. In contrast, cellular destruction disappears or is greatly diminished when cavitation is suppressed.

To prove that local heating is not the active agent when cavitation occurs near biological cells demanded additional experiments. If local heating were important, the rate of cellular disruption should be strongly temperature dependent. Experiments with bacteria, red blood cells, and protozoans have shown that quite the reverse is true—that the rate of cell breakage is independent of the temperature from 0 to 30°C. Another indication is to compare red blood cells heated to boiling with those agitated by a cavitating sonic field. The heat-destoyed cells break into small spherical pieces, each containing some methemoglobin.[1] However, the red blood cells which are "sonicated" form hemoglobin-free ghosts and cellular debris; the hemoglobin itself is slowly converted to the met form.

Another action of cavitation is to break molecular bonds. For instance, H_2O molecules are fractured into $H\cdot$, $OH\cdot$, and HO_2^- radicals and H_2O_2 is generated. Similarly, when a solution of NaCl is exposed to a cavitating

---

[1] Methemoglobin is an altered form of hemoglobin in which the iron in the prophyrin groups is in the ferric rather than the ferrous state (Fig. 18.11(b)).

sonic field, atomic Cl is produced. The Cl is apparently formed directly, for much more is present than could be formed by the H_2O_2 produced by the cavitation. Either free Cl atoms or H_2O_2 molecules could destroy biological cells. Quantitative studies, however, indicate that the concentrations of H_2O_2 and Cl are several orders of magnitude too small to account for the cellular disruption.

Thus, a large number of experiments with many types of cells, including blood cells, protozoans, bacteria, and algae, have all confirmed that the cells are broken by mechanical rupture as a direct result of cavitation. Electron microscope studies of cells exposed to sonic fields indicate that they are torn mechanically. Some typical electron micrographs are shown in Fig. 13.5. This tearing could occur in any of several closely related fashions. In the presence of an expanding and collapsing bubble (i.e., cavity), there will be very violent motions close to the bubble and relatively weak ones several diameters away. Thus, a part of the cell wall near the bubble will execute large displacements relative to the rest of the cell wall. The resulting shearing strains could easily rip the cell wall. A model of this action is shown in Fig. 13.6. Near the collapsing cavities, there is also an extremely vigorous stirring type of turbulence. The cell walls might be broken by the shearing stresses set up by this turbulence.

Similar mechanical rupture of cell walls of many types, such as those of amebas, yeast, and white blood cells, can be produced by rapid local shearing stresses. For instance, a micromanipulator needle can be slowly inserted into and then removed from these cells without damaging them. However, a rapid jab will permanently destroy the cell wall. By analogy, one might suspect that not only the shearing stresses produced by cavitation but also the rapidity of their application are important.

Thus, cavitation may tear cellular structures by a combined effect of local shearing stress, local turbulence, and rapidly applied shearing action. When, and only when, cavitation occurs, these effects are observed. The rate of destruction can be altered by a number of physical factors. Anything tending to decrease or suppress cavitation tends to protect the cells. Increases in the viscosity and in the wetting ability of the suspending medium have been found to raise the sonic pressure necessary for cavitation. Thus, red blood cells in isotonic saline are ruptured at lower sonic pressures than are necessary in whole blood.

The threshold for cavitation always depends on where the greatest sonic pressures in the field occur. For nonfocused sound fields, these maxima occur at or near the transducer surface. Experiments have shown that over a wide frequency range, from perhaps 250 Hz to 10 MHz, the threshold is unchanged. By contrast, in focused sound fields where cavitation occurs away from the surface of the transducer, the sonic pressure necessary to produce cavitation increases rapidly as the frequency is raised.

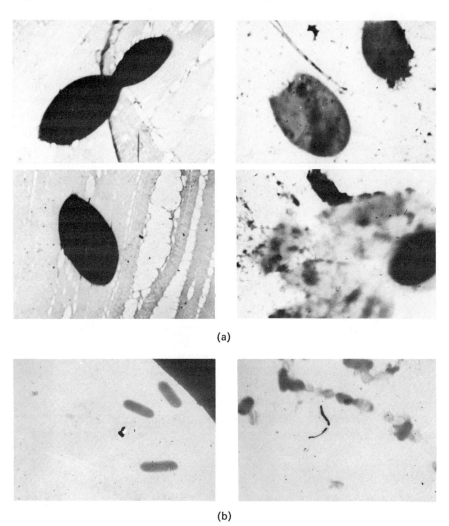

(a)

(b)

Figure 13.5 (a) Electron photomicrographs of *Saccharomyces cerevisiae*; unexposed cells illustrated on the left. Cells exposed to a 9 kHz-magneto-striction oscillator, illustrated on the right, show fragmented cells, some in which the end has been broken, and some intact cells exhibiting marked irregularity in density. (b) Electron photomicrographs of *Escherichia coli*, strain *B*. Unexposed cells illustrated on the left. Cells exposed to a sound field, illustrated on the right, show increased debris and fragmented cells. After H. Kinsloe, E. Ackerman, and J. J. Reid, "Exposure of Microorganisms to Measured Sound Fields," *J. Bacteriol.* **68**:373 (1954).

The threshold for cavitation in the body of the liquid is controlled, at a given frequency, by the existence of submicroscopic pockets of gas or vapor

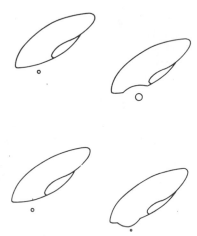

Figure 13.6 The distortions of a cell wall which might be caused by an oscillating bubble near the cell. After E. Ackerman, "Pressure Thresholds for Biologically Active Cavitation," *J. Appl. Phys.* **24**:1371 (1953).

called *nuclei*. These grow larger when the pressure is decreased and shrink when the pressure is increased. If the amplitude of the pressure changes during the sonic cycle is sufficiently great, the nucleus grows to such a large volume that it collapses violently when the pressure starts to increase. The product of the pressure amplitude times the period of the sonic wave determines the pressure amplitudes necessary for violent collapse of the cavities. At about 10 kHz, 1 atm of pressure amplitude is sufficient, whereas at 1 MHz, the threshold for cavitation in air-saturated water is about 30 atm.

Over very wide frequency ranges, the relative rates at which different types of biological cells are destroyed remain unaltered. These relative rates can be used to indicate the relative fragility of different cell types. To make this quantitative, one must study the time rate of cell destruction in a constant sound field. Studies of these rates show that as long as at least a few percent of the population remains undamaged, one may write

$$\frac{dN}{dt} = -RN \tag{13-7}$$

where N is the cell concentration, t is the time, and R is the *fragility*, a rate constant determining the probability of rupture per unit time. At lower fractions of the original population remaining undamaged, this equation is not always obeyed.

Sonic measurements of fragilities give rankings similar to other types of measurements. For example, the fragility of red blood cells can easily be measured by another method. This consists in suspending the cells in NaCl solutions of various concentrations. As the concentration is lowered, the red blood cells swell and eventually burst (lyse). The lowest NaCl concentration at which lysis does not occur is an osmotic measure of fragility. Both osmotic and sonic methods show that the fragility of human red blood cells increases as the cells age. The sonic measurements show this increase sooner than do the osmotic ones. Thus, the two do not measure exactly the same properties of the cell.

The role of the cell surface in determining *fragility* can be illustrated dramatically by the protozoan *Blepharisma*. These are pink ciliates somewhat similar to *Paramecia* in shape. When a culture of *Blepharisma* is exposed to suitable narcotics (e.g., morphine sulfate) the animals shed their pink outer walls, called pellicles, while retaining their shape, cilia, and so on. The relative fragility of the animals is doubled after they shed their pellicles. This indicates that although the pellicle does not control the shape of the animals, it does contribute to their ability to withstand mechanical disturbances.

Although at most frequencies the relative breakdown rates are the same, at certain frequencies, experiments with *Paramecia, Blepharisma,* and red blood cells indicate greatly increased sensitivity to sonically induced cavitation. These frequencies have been interpreted as cellular resonances induced by cavitation. From a knowledge of these resonances, it is possible to estimate the elastic properties of the cell walls.

These optimum frequencies at which cells are particularly sensitive to a cavitating sound field are characteristic of the particular species and strain. Table 13-2 shows the sizes and the characteristic frequencies of various strains

TABLE 13-2

OPTIMUM FREQUENCY VS. SIZE

Cell type	Maximum diameter (μm)	Minimum diameter (μm)	Optimum frequency (kHz)
P. caudatum	223	63	1.2
P. bursaria	118	51	1.7
P. aurelia G's	124	29	3.3
P. trichium	80	38	4.1
RBC–Amphiuma	45	10	16.5

of *Paramecia.* Other studies have shown optimum frequencies for red blood cells and for single-cell types of algae. It is easiest to interpret these as being due to mechanical resonances involving the cell surfaces. Similar surface resonances have been demonstrated and studied for air bubbles suspended in water and for rain drops.

One might wonder why surface modes of resonance are referred to as the basis for the characteristic frequencies for cellular disruption. More work has been done with pulsating bubbles in which the surface remains spherical than with bubble surface modes of vibration. However, all calculations show that any pulsating modes for biological cells (or any resonance dependent on the wavelength of sound in the suspending medium) should occur at much higher frequencies than those observed with cavitating sound fields. Accordingly, the characteristic frequencies seem to represent some other resonance of the biological cell, such as surface vibrations.

Two different physical models of spherical cells illustrated in Fig. 13.7

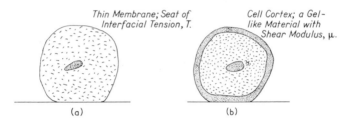

Figure 13.7   The interfacial-tension model (a) and the rigid-shell model (b). These two models of the surface of a biological cell are essentially different in that (a) presupposes that the cell wall lacks any rigidity or shear modulus, whereas, by contrast, (b) includes the rigidity of the outer cell layers but ignores any interfacial tension that may be present. Both disregard the contributions of the intracellular structure to the forces determining the cell shape.

predict surface resonances in the ranges observed. The first model approximates the biological cell by a spherical shell lacking any rigidity but possessing an interfacial tension. The cell is filled with liquid and surrounded by liquid. It makes no difference whether this interfacial tension is a true liquid–liquid interfacial tension, a liquid–membrane interfacial tension, or a surface-tension residual in a stretched membrane. Physically, all of these may exist at the cell boundary. Values of this interfacial tension T computed from static experiments ranged from 0.01 to 3.0 mN/m. The theory discussed here gives values of T from 0.03 to 15 mN/m for vertebrate red blood cells and ciliate protozoans.

The surface motions of this model are very similar to the resonant modes of a raindrop or of an air bubble in water. The raindrop and the bubble are different from cells, in having liquid on only one side of the boundary and also in possessing a much higher surface tension, around 75 mN/m. Nonetheless, the general forms of the motions are similar. These types have been photographed for oscillating bubbles and for liquid droplets.

The second model also treats the cell as a fluid-filled sphere with negligible internal viscosity. However, it assigns a rigidity to the cell cortex, or outer layers. In both models, the cells are considered, then, to be spheres filled with one ideal, incompressible liquid and surrounded by another. Clearly, no biological cell fits this description. Thus, both are at best approximations to the real biophysical system.

The rigid-shell model is very different from the interfacial-tension model, in terms of both its mechanical structure and its biochemical makeup. However, its predictions for distortions and resonances of biological cells are very similar to those of the interfacial-tension model. Indeed, there is no simple way to distinguish one from the other.

The rigidity of the cell cortex is negligible compared to steel, glass, or even wood. Rigidities are described by elastic moduli called *coefficients of rigidity* or *shear moduli*, which are about 10^7–10^9 N/m^2 for solid objects. All protein gels have much smaller, but nonetheless measurable, shear moduli in the range 10^2–10^4 N/m^2. Assuming gelatinous properties for the outer layers of the single cell leads to predicted resonant frequencies in the ranges observed for protozoans and erythrocytes.

Thus, exposure of suspensions of biological cells to cavitating sonic irradiation has been proved useful not only to obtain cell extracts but also to form estimates of the physical properties of the cell outer surfaces. The latter include fragility measurements and determination of characteristic frequencies at which the cells are destroyed at a maximal rate in a cavitating sound field. Similar results are not seen when tissues or intact organisms are exposed to sonic irradiation. This results from the absence of cavitation nuclei within the organisms. Nonetheless, other destructive effects are observed when high-intensity sonic irradiation is applied. These are discussed in the next section.

13.6   High-Intensity Irradiation

As noted in previous parts of this chapter, at all intensities sonic energy is absorbed by biological tissues. The rate of absorption varies linearly with the incident intensity, being determined by the linear attenuation coefficient, α. The latter is frequency dependent; for pure water α/ν^2 is independent of the frequency, whereas solutions and tissues exhibit relaxation effects.

If the incident intensity is sufficiently low, the heating rate will be less than the metabolic rate and no significant temperature rise occurs. It is possible under these circumstances that the added heat may result in increased circulation, which, in turn, may prove quite beneficial. However, unless intensities greater than 10 kW/m^2 are used, sonic irradiation has not been shown to produce any heating damage.

At still higher intensities, it is possible to destroy mammalian tissue and organs by sonic irradiation. The high intensities are most readily achieved by using a focused sonic beam. Such focusing is accomplished in fashions similar to those used to focus light. Moreover, acoustic focusing is often easier than optical, because all parts of the sonic source move in a fixed phase relationship with one another. (Similar optical behavior is illustrated by lasers, discussed further in Chapter 27.)

Sonic beams may be generated by ceramic transducers (made of materials such as barium titanate). These can be shaped into spherical or cylindrical sections to yield a focused sound field. It is also possible to use plastic lenses (Lucite, polystyrene) to focus the sonic beam on a desired region. In those fashions, very high intensity sonic irradiation can result from moderate power input to the transducers and relatively low intensities in the immediate neighborhood of the transducer.

Using focused sound fields it would then seem possible to destroy certain tissues without surgically removing them. This method has been suggested as an alternative to surgery and preliminary feasibility studies have been carried out. Two such studies are described in the following paragraphs. Although both appeared to offer much promise, neither has resulted in a practical application of the destructive effects of heating by high-intensity sonic irradiation.

In the first of these studies Cerino and his colleagues showed that under suitable conditions, intense sonic irradiation could destroy cancerous tumors transplanted into the bone of a rabbit limb. In the controls, these tumors grew and eventually would destroy the animal. The destruction of the cancer was apparently related to the high temperatures produced by the sonic beam focused on the affected bone. Since bone has a greater value of the attenuation coefficient α, and since added conversion to heat occurs at the interface between the bone and the soft tissue, this method was particularly effective for raising the temperature of selected portions of a given bone.

Although all the cells, normal and tumor, were killed in the irradiated part of the bone, the latter retained its structural integrity. The treated animals moved in the same fashion as completely normal rabbits. The temperature rise in the surrounding soft tissues was less than in the bone. Accordingly, if the tumor had broken through the bone and into the soft tissue, it could not be destroyed by sonic irradiation at the intensities used.

The human implications of these experiments are quite limited. Surgeons would by choice remove a limb rather than hope to destroy the tumor by sonic irradiation. However, it has been suggested that such methods might be used on tumors in individuals who for some reason refuse surgery.

Greater interest centered at one time on the high-intensity sonic destruction of nervous tissue. This was actually tried on a number of patients as an alternative to the then popular prefrontal lobotomy as a treatment for mental disturbance, Parkinson's disease, and other disorders. To achieve sufficiently fine focusing, four focused beams were used, each directed through its own hole in the skull at the same small volume of tissue. Since the wavelength of sound at 1 MHz is about 1.5 mm in tissue, it should be possible to achieve a volume of maximum intensity of the order of 1–2 mm^3, that is, 1–2 $\mu\ell$.

High-intensity sonic destruction of nervous tissues has several advantages over conventional neurosurgery. First, and most important, the dosage can be so adjusted that the blood vessels and supporting (glial) cells are undamaged when the neurons are destroyed. Second, it is possible to eliminate neurons in a small volume deep within the brain while leaving the surface and outer layers of the brain unaltered.

The action of destroying the neurons but leaving the surrounding cells undamaged is not clearly understood. However, a number of possible explanations can easily be eliminated. For example, the action on the neurons is not due to heating. Experiments at different temperatures showed the same results. Moreover, intermittent exposures produced the same destruction as continuous exposures of the same total exposure time. Likewise, the neuro-surgical effects do not depend critically upon the frequency of the applied signal. Thus, they are not a resonant type of phenomenon. Static pressures of a magnitude comparable to those occurring during the positive pressure of the acoustic cycle do not produce any damage.

The absence of other effects suggests cavitation as a possible cause of neuron destruction. Traditionally, the most reliable test for cavitation has been to apply an excess static pressure. If the effects observed are due to cavitation, these should be decreased when an excess pressure is applied. This is, indeed, the case for the damage to neurons; at any ultrasonic pressure levels, the destruction is much less when a static excess pressure of about 13 atm is applied. However, neuronal destruction is still observed when the acoustic pressure amplitude is 6 atm, whereas the static or average pressure is 13 atm. While making cavitation less likely as a cause for neuronal destruction, these observations do not rule it out completely, because cavitation is observed in particulate suspensions at positive pressures.

From the preceding paragraphs it would appear that high-intensity sonic irradiation was an ideal tool for deleting selected portions of the central nervous system. Yet it is rarely used today. This disuse results from a number of negative features. One is that highly focused sound fields often have secondary lobes or regions where the sonic intensities are also high, albeit not as high as at the principal focus. These secondary lobes, which depend on the shape of the head and on the exact locations of the holes in the skull, may produce undesired neuronal damage.

An additional disadvantage is the surgeon cannot see or touch the region being destroyed. In a bone tumor this is not critical, but in a brain a small undesired displacement might prove tragic. Finally, this general alternative lost favor as lobotomies became less popular. Chemical therapy is preferred today for the control both of mental disturbances and of tremor-inducing diseases such as Parkinsonism.

There have been several attempts to uncover other destructive effects not directly attributable to heat or cavitation. For example, Goldman and Lepeschkin demonstrated cellular injury to algae and rotifers which was apparently independent of cavitation or heating. The existence of such destructive results has increased the care given to establishing the safety of the intensity levels and durations of exposure used for echosonography.

13.7 Summary

The term "sonic" is used to describe mechanical vibrations whose effects are unrelated to the phenomenon of hearing. Biological materials have been irradiated with sonic beams for a number of purposes. Low-intensity studies, whose results are expressed in terms of an acoustic impedance or an attenuation coefficient, are used to study the dynamic mechanical properties of biological materials. The increased utilization of echosonography has led to renewed interest in the results of irradiating biological specimens with low-intensity sonic beams.

At higher intensities other phenomena become important. These include heating, which varies linearly with the intensity and which is used to advantage for diathermy. Other nonlinear effects are associated with cavitation, that is, with the tearing apart of the liquid due to the extremely low pressures that occur during the acoustic cycle. Cavitation results in rupturing suspensions of single biological cells; cavitating sonic fields are used to prepare suspensions of subcellular particles and to extract enzymes. Other applications include determination of cellular fragilities and cell surface resonances.

At very high intensities, still other destructive results have been obtained. Thus, heat damage to bony structures has been used to destroy bone tumors in experimental animals. Other less clearly understood effects include the selective destruction of neurons in the central nervous system. This has been tested as an alternative to surgery, but the disadvantages of sonic irradiation outweigh the advantages.

REFERENCES

1. Hussey, M., *Diagnostic Ultrasound: An Introduction to the Interactions Between Ultrasound and Biological Tissues* (Glasgow: Blackie & Son Ltd., 1975), 254 pages.
2. Skudrzyk, E., *The Foundations of Acoustics: Basic Mathematics and Basic Acoustics* (New York: Springer–Verlag, 1971), 790 pages.

3. GOOBERMAN, G. L., *Ultrasonics, Theory and Applications* (London: The English Universities Press Ltd., 1968), 210 pages.

4. HUETER, T. F., AND R. H. BOLT, *Sonics: Techniques for the Use of Sound and Ultrasound in Engineering and Science* (New York: John Wiley & Sons, Inc., 1955), 456 pages.

5. GOLDMAN, D. E., AND T. F. HUETER, "Tabular Data of the Velocity and Absorption of High-Frequency Sound in Mammalian Tissues," *J. Acoust. Soc. Am.* **28**:35–37 (1956).

6. CARSTENSEN, E. L., AND H. P. SCHWAN, "Absorption of Sound Arising from the Presence of Intact Cells in Blood," *J. Acoust. Soc. Am.* **31**:185–189 (1959).

7. ACKERMAN, E., "Resonances of Biological Cells at Audible Frequencies," *Bull. Math. Biophys.* (renamed *J. Math. Biol.*) **13**:93–106 (1951); "An Extension of the Theory of Resonances of Biological Cells: I. Effects of Viscosity and Compressibility," **16**:141–150 (1954); "II. Cross-Section in a Plane Wave," **17**:35–40 (1955); "III. Relationship of Breakdown Curves and Mechanical Q," **19**:1–7 (1957).

8. FRY, W. J., "Intense Ultrasound in Investigations of the Central Nervous System," *Advan. Biol. Med. Phys.* **6**:282–348 (1958).

9. CERINO, L. E., E. ACKERMAN, AND J. M. JANES, "Effects of Ultrasound on Bone Tumors," *Prog. Clin. Cancer* **3**:19–30 (1967).

CHAPTER 14

X-Ray Analysis

14.1 X-Ray Diffraction

The physical behavior of molecules found in biological structures can be investigated from various points of view. One of the most fruitful of these has been an analysis of the atomic architecture as determined by X-ray diffraction patterns. The term "X ray" refers to a beam of photons (electromagnetic radiation) formed by bombarding a metal target with electrons. These X rays are shorter in wavelength than other electromagnetic radiation, referred to as visible and ultraviolet light. (A more complete description of the electromagnetic spectrum is found in Chapter 12.)

The method of X-ray diffraction is a relatively new one in physical chemistry. X rays were discovered by Roentgen just before the start of this century. Quite a bit of simple X-ray crystallography was done between 1912 and 1920. However, accurate measurements of X-ray wavelengths and the corresponding studies of crystal structure have been possible only since about 1920. These studies profoundly affected scientists' ideas of the physical world at many levels. Details of the periodic table, the exact values of the electronic charge and Avogadro's number, and the arrangement of atoms in crystals and electrons within atoms all have been based on X-ray measurements.

Although the diffraction of X rays by simple crystals such as NaCl had been studied for many years, the present interpretations of X-ray diffraction patterns of biologically interesting molecules were formulated since World War II. These were made possible by the factor that is basic to so much of biophysics, the development of suitable electronic techniques. The detailed interpretation of X-ray diffraction data from complex molecules is possible only with the use of electronic computers.

These studies of the diffraction of X-ray beams by biologically interesting molecules have influenced current ideas of the structure and action of almost all forms of biological compounds. The arrangements of the atoms within small molecules such as amino acids, purines, and sugars have been (or are being) determined. The chemical structural formula of certain antihistamines and the various isomers of vitamin A can best be investigated by their X-ray diffraction patterns. The helical structures of crystalline and fibrous proteins and of the genetic material DNA have been established from their diffraction of X rays.

The resolving power of an X-ray diffraction apparatus is much greater than that of a light microscope. In the light microscope, the limit of resolution is set by the wavelength of light employed. With X-ray diffraction patterns, no such restrictions exist. Using monochromatic X rays such as the Cu-$K_{\alpha 2}$ radiation, the wavelength is 0.154 nm, but interatomic distances can readily be measured with an error of less than 1 pm. This value can be compared with a theoretical limit of resolution of 2×10^2 nm for blue light.

One may ask why a crystal has to be used rather than a single molecule if the resolving power is indeed of the order of 1 pm whereas the covalent bond lengths average about 0.15 nm. Perhaps the most obvious answer is that it is impossible to hold one molecule in place. In addition, some of the X-ray photons will break molecular bonds. Because many molecules are present, breaking a few bonds does not have an appreciable effect on the average diffraction pattern. Perhaps the most important advantage of a crystal is that it restricts the scattered rays to a finite number of sharp, intense diffraction maxima.

One of the difficulties of X-ray diffraction studies is that one ends up with a photograph or graph with a number of spots of varying intensity, such as that shown in Fig. 14.1. The problem of reconstructing the crystal and the spatial arrangements of the molecules from these spots has a simple solution only for crystals of very simple molecules, such as NaCl or H_2O. For more complicated molecules a series of trial-and-error solutions is necessary. The analysis that follows is presented in the hope that those readers unfamiliar with this technique will acquire some idea of the problems involved.

Bragg showed that in treating X-ray diffraction by single crystals, one may regard the atoms as making up reflecting planes. A beam of X rays is shown incident on a single pair of such planes in Fig. 14.2 (although there will in

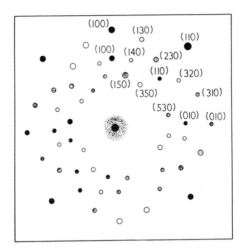

Figure 14.1 Laue pattern of NaCl. This has been redrawn from a photograph to emphasize the diffraction spots. The Miller indices of the corresponding planes have been labeled for some of the spots.

general be many planes for any given crystal). From this figure, one may see that there will be a maximum in the diffraction pattern if and only if

$$n\lambda = 2d \sin \theta \qquad (14\text{-}1)$$

where n is an integer, λ is the wavelength, and d and θ are as defined in Fig. 14.2. With monochromatic X rays, one set of planes, at most, will give a maximum for a given θ, and for any arbitrary θ there will probably be no maximum in the diffraction pattern. One might be produced by rotating the X-ray beam around the crystal or by rocking the crystal about an axis

Figure 14.2 X-ray diffraction. The dotted lines show perpendiculars to the wave front. For reinforcement, the rays reflected at planes 1 and 2 must travel distances that differ by a whole number of wavelengths. This will be fulfilled if $n\lambda = 2d \sin \theta$, where n is an integer. This is termed the Bragg relationship.

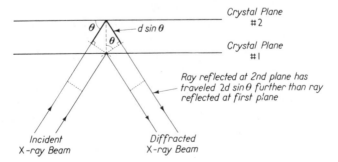

perpendicular to the plane of the paper. The latter alternative is more practical and is often employed. Another type of diffraction pattern, called a *Laue pattern*, avoids this problem of orientation by using an X-ray beam containing many wavelengths incident in a fixed direction. However, because λ will not be known for any particular spot, one cannot find the distance between reflecting planes from a Laue pattern.

Figure 14.3 shows several planes in a cubic crystal, each with a number of atoms per plane. These planes are numbered by Miller indices (hkl), which are described in Fig. 14.4. In Fig. 14.3, one may notice that the planes are

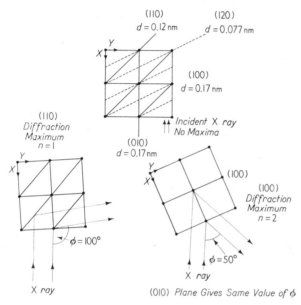

Figure 14.3   Diffraction of an X-ray beam. In working out angles, it is assumed that the X-ray wavelength λ is 0.152 nm and that the crystal had cubic symmetry with a lattice constant of 0.170 nm.

spaced at varying distances. By and large, as the Miller indices go up, the spacing d between adjacent planes decreases, and the number of maxima likewise decreases. For the example shown, there are two angles corresponding to $n = 1$ and $n = 2$ for the (010) and (100) planes. The (110) and (120) planes each have only one diffraction maximum, corresponding to $n = 1$. The maximum for the (120) plane essentially reflects the incident beam back on itself and could not be observed. None of the higher planes will exhibit diffraction maxima. However, planes not perpendicular to the xy plane, such as (101) and (011), will give maxima whose diffracted beams will not lie in the plane of the paper. Thus, the maxima will form a two-dimensional pattern such as that shown in Fig. 14.1. To show all the planes with monochromatic

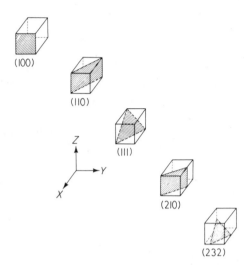

Figure 14.4 Miller indices (*hkl*) for some crystal planes illustrated for cubic crystals. The Miller indices are inversely proportional to the distance from the origin to the intersections with the crystal axes when these distances are expressed in terms of the lengths of the unit cell. The proportionality constant is so chosen that the Miller indices are the smallest possible whole numbers.

X rays, a number of different schemes have been developed, which lead to an easier interpretation of the Miller indices of the planes giving rise to a given maximum. For complicated molecules such as proteins and nucleic acids, it is necessary to use one of these schemes. The reader is referred to the references by Stout and Wilson for details of these techniques.

A sensitive test of an assumed atomic structure is to compare the relative intensities of the X-ray diffraction maxima observed with those computed from the model. The relative intensities are found by adding together the contributions of each atom, taking into account the phase differences due to the difference in path length to each atom. This is usually expressed by a crystal structure factor, F_{hkl}, for the beam perpendicular to the (*hkl*) planes. It can be found from

$$F_{hkl} = \sum_{n=1}^{N} f_n \, e^{2\pi j(hu_n + kv_n + lw_n)} \tag{14-2}$$

where N is the number of atoms in a unit cell of the crystal, n is a particular atom, f_n is the atomic structure factor for the nth atom as defined below, and u_n, v_n, w_n are coordinates of the nth atom expressed as fractions of the unit crystal lattice lengths.

The atomic structure factor f is defined by

$$f = \frac{\text{amplitude of the wave scattered by the atom}}{\text{amplitude of the wave scattered by an electron}} \tag{14-3}$$

In general, f depends on both the particular element (for example, Zn) and on the angle between the incident and the scattered beam; tables of f for various elements are available.

The object of this calculation is to obtain values for u, v, and w. Thus, one must measure the relative intensity, which will be proportional to $|F_{hkl}|^2$. In a typical experiment, these intensities are measured and the types of atoms present (and hence the values of f_n) are known. The initial assumed values of u_n, v_n, and w_n are used to compute theoretical values of $|F_{hkl}|^2$. It remains to adjust u_n, v_n, and w_n for each atom until the final structure agrees with both the chemical data and the X-ray diffraction pattern.

Complicated crystals or even simple crystals of larger molecules give rise to elaborate diffraction patterns. The number of points necessary to determine the crystal structure increases both as the number of atoms per molecule and the size of the unit cell increase. In order to obtain useful information from these complicated diffraction patterns, it is necessary to know the relative intensities of the various maxima as well as their direction.

In interpreting diffraction by large molecules, it is more convenient to deal with electron densities than with atomic positions. After the electron density has been mapped, the atoms may be located at the center of the density maxima. From the preceding paragraphs, it may be seen that the crystal structure factor F_{hkl} can be defined by an absolute value, $|F_{hkl}|$, and a phase angle, α_{hkl}, as

$$|F_{hkl}| = \frac{\text{amplitude of the wave scattered by all atoms in the unit cell}}{\text{amplitude of the wave scattered by an electron}}$$

α_{hkl} = phase difference between the wave scattered by the unit cell and the wave scattered by an electron at the origin

Adding the contribution of each electron, as before, and integrating over the unit cell,

$$F_{hkl} = \iiint \rho(u, v, w)e^{2\pi j(hu + kv + lw)} \, du \, dv \, dw \qquad (14\text{-}4)$$

where ρ, the electron density at u, v, w, is a real number. Readers familiar with Fourier series will recognize that Eq. (14-4) has the form of the coefficients of a Fourier series. Accordingly, one may invert it as

$$\rho(u, v, w) = \sum_h \sum_k \sum_l |F_{hkl}| \cos [2\pi(hu + kv + lw) + \alpha_{hkl}] \qquad (14\text{-}5)$$

Thus, if one can guess the values of α_{hkl} and can measure a sufficient number of intensities $|F_{hkl}|^2$, one can map the electron density ρ and hence locate all the atoms. This is called a *Fourier synthesis* (see Appendix C).

The problem of correctly guessing the phases in Eq. (14-5) has intrigued mathematically minded crystallographers. The general procedure is to guess phase values and then keep adjusting these to give sharper and sharper electron-density contours. Extensive use is made of fast Fourier transform (FFT) calculations, described in Appendix C. If one gets on the right track, these contours define atoms of the types known to be present at reasonable distances from other atoms to which they can be bonded. (The bonds can often be found by the methods of classical organic chemistry.) In the final analysis, phase guessing is very similar to working a crossword puzzle or solving a murder mystery; as in these, one finds, if successful, an answer that is no longer a guess.

Various schemes have been developed for the initial phase guessing. One of the most successful involves placing a heavy atom, such as I or Br, within the molecule.[1] The heavy atom diffracts more strongly than the others, so it may be located first. To do this, Eq. (14-2) is used, setting f_n to zero for all but the heavy atom. Once it is located, approximate values for many of the phases can be determined at once. With these as a starting point, one adjusts these phases and the others to give more and more sharply defined electron density contours. The final solution is an accurate determination of the structure.

For certain crystals, the unit cell is symmetric about the center, and it can therefore be shown that all the α_{hkl}'s have the value 0 or π. These are only two choices, but if 100 points were used, there are 2^{100}, or about 10^{30}, possible sets of guesses. By the use of heavy-atom substitution, this hopelessly large number may be reduced to a mere few billion. Protein and nucleic acid crystals are not generally symmetric about the center of the unit cell, so the problem is more difficult when using these molecules.

If $|F_o|$ and $|F_c|$, the absolute values of the observed and calculated diffraction amplitudes, are known, one can measure the goodness of fit using the residual R, a type of average relative error over all reflections:

$$R = \sum \frac{||F_o| - |F_c||}{\sum |F_o|} \qquad (14\text{-}6)$$

R values for trial structures are often in the range 0.4–0.5 ($R = 0.59$ for a random collection of atoms). Small molecules can be refined to an R of less than 0.1, and the structures of many biological macromolecules are known with an R of less than 0.2.

The entire adjustment of phase values and recomputing the F's and ρ is a lengthy, tedious process. With a pocket calculator and a protein crystal, this would take many human lifetimes. With electronic computers, it has been possible to find the details of the structural arrangements of the atoms within many smaller biological molecules (of molecular weight less than 2,000). The remainder of this chapter discusses the contributions of the

---

[1] This technique is useful only if the heavy atom does not alter the crystal structure; it is called *isomorphic replacement*.

method of X-ray diffraction to the determination of the structure of proteins and nucleic acids.

14.2   Protein Structure

In Chapter 9, it was mentioned that one class of proteins, the globulins, could exist in either a fiberlike or a globular state. Most proteins do not have these two alternatives but, rather, are only fibrous or only globular. X-ray diffraction studies have been applied to both types of protein structure, with varying degrees of success.

The problem of determining the structure of fibrous proteins is quite different from that of crystalline proteins. If a fiber were made up of small crystalline regions all lined up, then one should obtain spots similar to those from a single crystal. If these were slightly disoriented, the spots would become arcs. If the crystallites were completely randomly disordered, the spots would become circles. Early investigators took many pictures of X-ray diffraction patterns of fibrous materials, but no one understood the results.

Around 1930, Astbury studied many protein and nucleic acid fibers. He showed that the proteins gave rise to two types of X-ray diffraction patterns, called α and β. He recognized the α configuration as a folded or more dense structure, and the β pattern as being due to a stretched structure. To these, he assigned the forms

Astbury's α form   Astbury's β form

The double-bonded oxygen is slightly negative, whereas the hydrogen on the nitrogen is somewhat positive. Polypeptide chains were postulated to be held in place by the attraction between these two, called *hydrogen bonding*.

Pauling and Corey, in 1951, showed that if they drew a polypeptide chain with the known bond angles on a piece of paper and twisted it into a helix, the various turns could hydrogen-bond to one another. Astbury and others had tried helical models but always with an even number of amino acids per turn; Pauling realized that this was a mistake. Pauling and Corey demonstrated that a helix with 3.7 amino acid residues per turn, a diameter of 0.68 nm, and turns spaced 0.54 nm apart would be permitted by the observed bond angles. This is shown in Fig. 14.5 for a right-handed helix. Left-handed

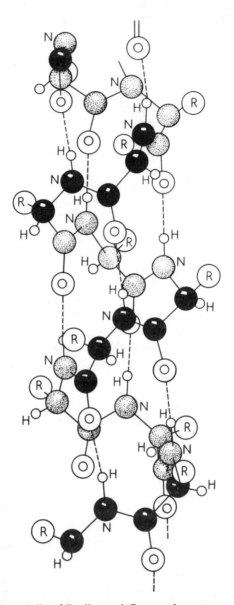

Figure 14.5 The α helix of Pauling and Corey redrawn to emphasize helical polypeptide chain. Dotted lines indicate hydrogen bonding of thé =O of one turn with the H—N of the turn below. This is a close-packed structure. The polypeptide helix has a radius of 0.34 nm and an interturn distance of 0.54 nm. There are 3.7 residues per turn. The R groups are much larger than shown; they extend as far as 1.0 nm.

α helices are also possible. However, the diffraction pattern for X rays due to such a helical structure was more complex than that for any of the simpler models.

Cochran, Crick, and Vand carried out a theoretical study of the predicted X-ray pattern for helical structures. They showed that it was necessary to orient the X-ray beam at an oblique angle to the fibers. In general, the helical structures lead to patterns similar to that diagrammed in Fig. 14.6. The

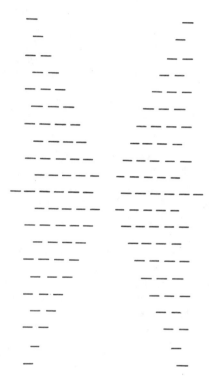

Figure 14.6 Diffraction pattern of a helix. Notice the clear area in the center of the pattern.

X-ray patterns of these helical structures can be described by Bessel functions of the distance from the center of the helix.

The α-helical structure of Pauling and Corey fits very well the diffraction patterns observed for synthetic polypeptides. It is generally accepted that many natural fibrous proteins occur as helices because their X-ray diffraction patterns are similar to that of Fig. 14.6. However, many unexplained spots are present. The data for the fibrous protein collagen can be fitted by a model involving a fiber made of three α helices twisted around each other in a helical fashion. The first direct demonstration of the α helix was in the globular protein myoglobin, to be discussed subsequently.

In addition to their α helix, Pauling and Corey made pleated sheet models of proteins similar to the β model of Astbury but did not restrict the peptide bonds to one plane. Both this and the α helix have retained the ideas of the α structure being compressed and the β structure stretched out, and also of hydrogen bonds being responsible for holding the shape of the protein fibers. They are superior to Astbury's earlier models in fitting known bond angles and in their agreement with the experimental results of X-ray diffraction studies.

Many attempts have been made to apply the general methods described in the previous section to crystals of globular proteins. Perhaps the most studied crystal has been that of the blood protein hemoglobin. However, the structure of a similar but simpler protein, myoglobin, was worked out to a resolution of about 0.6 nm before much progress was made with hemoglobin. Myoglobin is a red pigment similar to hemoglobin but occurring in muscle (see Chapter 9) rather than blood. It is believed to function by buffering the oxygen concentration within the muscle. Myoglobin has a molecular weight of about 16,000, very low for a typical protein. This corresponds to 153 amino acid residues, that is, about 1,200 atoms other than hydrogen in each myoglobin molecule. To locate all of these atoms in the molecule, one would need to measure the intensity and to guess the phases of perhaps 20,000 diffraction spots.

The results of analyzing and adjusting the phase for about 400 diffraction spots for myoglobin crystals, substituted with heavy atoms, showed that there were two myoglobin molecules per unit cell of the crystal and located the polypeptide chains and the iron-containing heme group within the myoglobin molecule. This procedure was then repeated to include 9,600 diffraction spots, which showed the electron density of the myoglobin molecule with a resolution of about 0.2 nm. This is not quite sufficient to indicate the separate atoms, but is adequate to confirm that the major part of the myoglobin molecule consists of right-handed α helices. To fit the single polypeptide chain of myoglobin into one globular molecule, it must be bent and twisted at various corners. Where this occurs, the α-helical form is lost usually for 3 or 4 amino acid residues. There is also one group of about 13–18 amino acid residues not in the form of an α helix.

Figure 14.7 shows a photograph of Kendrew's model of myoglobin, built to represent the structure that would give the 400 diffraction spots used. In addition, electron spin resonance measurements were used to locate the iron atoms in the heme group (see Chapter 26). However, the latter data were misinterpreted, so the heme group was tilted at the wrong angle. Figure 14.7(b), for comparison, shows the form of the polypeptide chain revealed by the 9,600-diffraction-spot study. Although not illustrated in the figure, all the chains are shown by the latter study to be hollow, cylindrical tubes of the form expected for helices. As noted above, the straight-chain portions are right-handed α helices. To obtain Figs. 14.7(a) and 14.7(b) it was necessary to use four different substitutions of heavy atoms to check the results and obtain suitable starting points for phase guessing. The model in Figs. 14.7(a) and 14.7(b) shows one continuous polypeptide chain, as demanded by chemical evidence.

Similar studies of hemoglobin (molecular weight about 65,000) have shown that it consists of four subunits, each with a heme group. These studies at a resolution of 0.55 nm showed that each subunit of hemoglobin is a continuous polypeptide chain folded around itself in a form very similar to that of the myoglobin molecule. There are two identical pairs of subunits in each molecule. These are shown in black and white in the model illustrated in Fig. 14.7(c), which summarizes the X-ray diffraction studies of Perutz and his co-workers.

The myoglobin and hemoglobin studies in the 1950's involved three new ideas not used in the 1920's in the X-ray determination of inorganic crystalline structure. The first was isomorphic replacement, the technique of the substitution of heavy atoms into the unit cell. The second involved the use of digital computers to adjust the phases until the electron density postulated and the diffraction spots observed were consistent with one another. The third innovation was the use of electron spin resonance to locate the iron atoms. These and other innovations, such as improved techniques for growing suitable protein crystals, have led to knowledge of over 50 complete protein peptide chain conformations. These conformations are not an end in themselves. Many questions about the relation between structure and function in proteins can be answered using X-ray analysis *if* we assume that the protein's structure in the crystal is the same or very similar to its structure in the cell and/or in the test tube. Although this is not always true, one sees very similar structures when a protein is crystallized under different conditions and even in different crystal forms. Also, many enzymes retain their catalytic activity in the crystalline form (see Chapter 18). This assumption is explored in more detail for the protein insulin in the reference by Blundell et al.

When studying enzymes using X-ray diffraction, one is interested not only in the structure of the protein alone, but also in the structure of the enzyme–substrate catalytic complex. Since these by their nature must be transient (see Chapter 18), X-ray crystallographers have looked at the much more stable complexes formed between enzymes and certain substrate analogs. Lysozyme was one of the first enzymes examined in this fashion; Fig. 14.8 shows a model of the enzyme–substrate complex based on X-ray structures of enzyme–substrate analog complexes. Close examination of the amino acid residues nearest the substrate has indicated a possible mechanism of action for this enzyme (see the reference by Phillips).

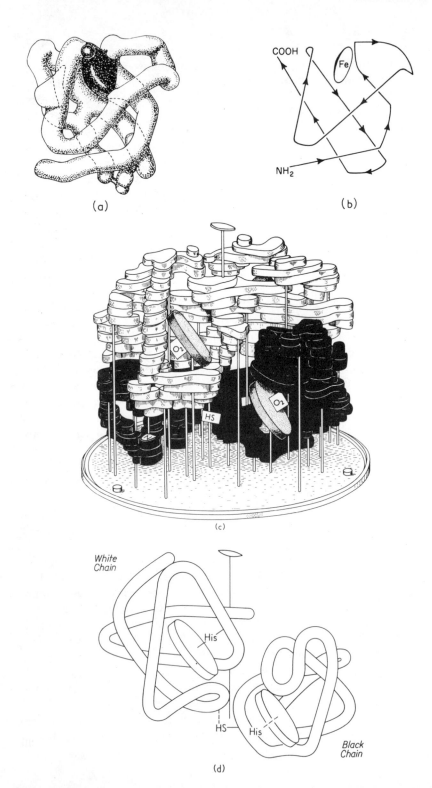

(a)

(b)

(c)

(d)

Figure 14.7 (Facing page)   Kendrew's model of myoglobin. (a) General shape
of the polypeptide chain at 0.6 nm resolution. The gray area is the heme
group. The round dark atom represents a heavy atom attached for iso-
morphous replacement. The tilt of the heme group is incorrect. (b) Course of
the polypeptide chain as determined by a three-dimensional Fourier synthesis
with a resolution of 0.2 nm. After J. C. Kendrew et al., *Nature* **185**:422 (1960).
(c) Perutz's model of hemoglobin at 0.55 nm resolution. The white units are
an identical pair, as are the two black units. Each unit is very similar to
myoglobin in the shape of the peptide chain. The heme groups are indicated
by gray disks. (d) Hemoglobin chain configuration in the two subunits facing
the observer. After M. F. Perutz et al., *Nature* **185**:416 (1960).

Finally, the question of why proteins fold as they do has received much
attention. It has been shown that, under certain conditions, some proteins
can be completely unfolded and yet will refold to their original shape once the
unfolding agent is removed. Biophysicists are attempting to correlate the
amino acid sequence of proteins to their shape, as shown by X-ray crystallography, to answer this very fundamental question.

14.3   Nucleic Acids

Deoxyribonucleic acid (DNA) is the genetic material in plants, animals,
protists, and many viruses (see Appendix D). Its central role in living systems
was not appreciated for many years. A major factor in increasing the significance assigned to DNA was the determination of its structure using X-ray
analysis by Crick and Watson in 1953. Their interpretation made use of the
theory referred to in the previous section for X-ray diffraction by helical
structures.

Crick and Watson showed from the X-ray diffraction patterns that DNA
consists of two antiparallel helices. The spiral is very large, having a diameter
of 1.8 nm and a spacing between turns of 3.4 nm. Thus, the helices have room
for other molecules, provided they are of the proper shape. On the basis of
chemical data and size considerations, Crick and Watson showed that the
helices are made up of –sugar → phosphate → sugar → phosphate– and so
forth, units. Between the two helices, as rungs along a spiral step ladder,
were strung pairs of hydrogen-bonded bases of the form purine—H—pyrimidine. These rungs are about 1.1 nm long. A piece of a pair of helices, also
called chains, is shown in Fig. 14.9.

Notice that the "top" carbon of the leftmost sugar–phosphate chain and
the "bottom" carbon in the rightmost chain are both the 5′ atoms in their
respective nucleotides (see Appendix D for the numbering system used).
Likewise, 3′ carbon atoms are found at the bottom of the left chain and the
top of the right. Thus, each chain in the DNA molecule has direction; if we

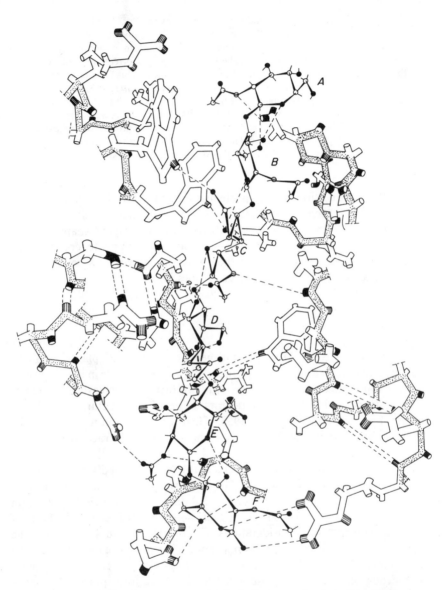

Figure 14.8 Atomic arrangement in the lysozyme molecule in the neighborhood of the active site with a hexa-N-acetylchitohexose shown bound to the enzyme. The main polypeptide chain is shown speckled, and NH and O atoms are indicated by line and full shading, respectively. Sugar residues A, B, and C are as observed in the binding of tri-N-acetylchitotriose (and β-N-acetylglucosamine for residue C). Residues D, E, and F occupy positions inferred from model building. It is suggested that the linkage hydrolyzed by the action of the enzyme is between residues D and E. After D. C. Phillips, *Proc. Nat. Acad. Sci.* (*U.S.*), 493 (1967).

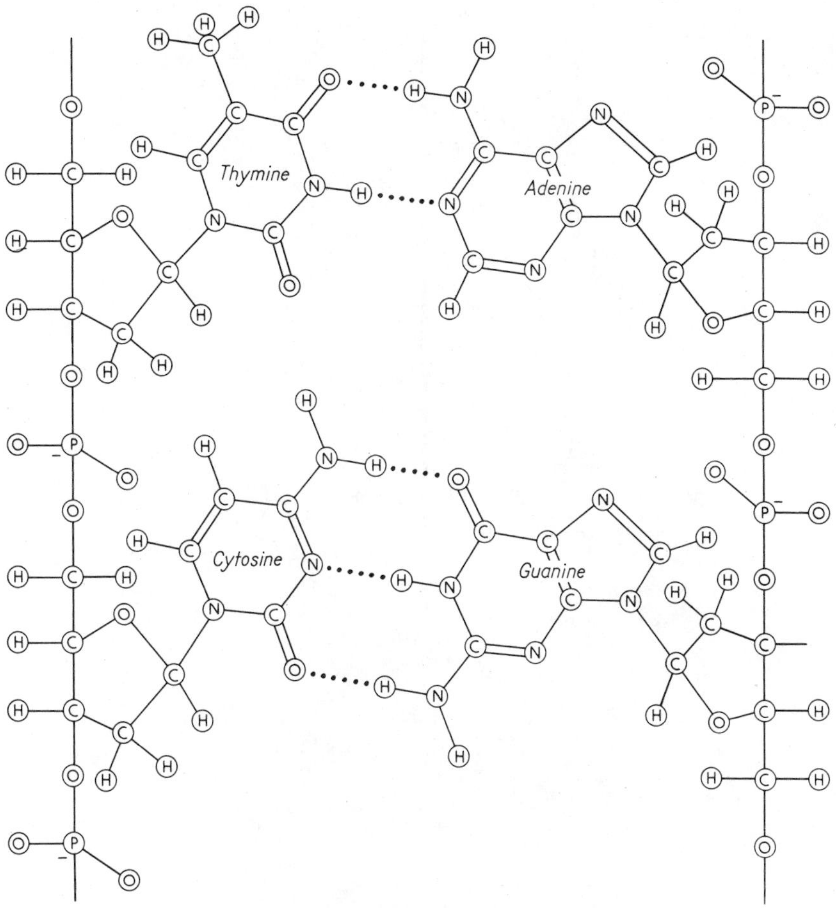

Figure 14.9 Double chain of DNA.

consider each to be an arrow →, the helix can be represented as a pair of arrows ⇆.

This type of unit is repeated into a long double chain. The entire double chain is then twisted to form the double helix shown in Fig. 14.10, with about 10 rungs per turn. The pairs of bases fit across the chain as rungs being supported in the middle by hydrogen bonds. It is necessary that each pair of bases fit very exactly. Measurements based on X-ray diffraction patterns of crystals of the purine and pyrimidine bases have shown that these do indeed fit, provided that adenine (A) is paired with thymine (T), and guanine (G) with cytosine (C). If this is the case, one should have the relative concentrations of organic bases in DNA related as

$$\frac{[A]}{[T]} = \frac{[G]}{[C]} = 1.0$$

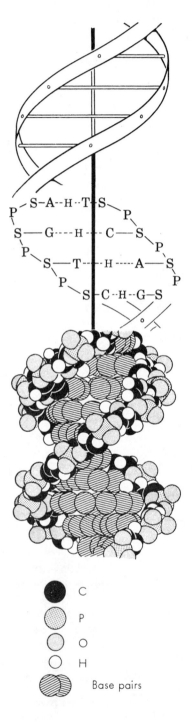

C

P

O

H

Base pairs

Figure 14.10   (Facing page) The helix of DNA, with three different ways of representing the molecular arrangement. Top, general picture of the double helix, with the phosphate–sugar combinations making up the outside spirals and the base pairs the crossbars; middle, a somewhat more detailed representation: phosphate (P), sugar (S), adenine (A), thymine (T), guanine (G), cytosine (C), and hydrogen (H); bottom, detailed structure showing how the space is filled with atoms: carbon (C), oxygen (O), hydrogen (H), phosphorus (P), and the base pairs. After C. P. Swanson, *The Cell* (Englewood Cliffs, N.J.: Prentice-Hall, Inc., 1960).

This relationship had been verified for all DNA and was one of the pieces of evidence used by Crick and Watson to construct their model.

Most ribonucleic acid (RNA) does not have the complementary base-pair ratios found in DNA. Indeed, it is far more heterogeneous than DNA and, within any one living cell, at least three main classes have been described. They are called transfer, messenger, and ribosomal RNA's; the significance of their names and their functions in the cell is described in Chapter 16.

The transfer RNA's are the smallest, with about 80 nucleotides each. Sequence analysis has shown that they have enough complementary base pairs to form intrachain hairpin folds, which allow double-helical segments. Figure 14.11(a) shows such a hairpin-fold model, termed a cloverleaf, for yeast alanine transfer RNA. X-ray analysis reveals that these folds are themselves folded compactly on each other (Fig. 14.11(b)).

The much larger messenger and ribosomal RNA's are similar in that there are intrachain double-helical sections separated by single-stranded sections (Fig. 14.12). Perhaps for this reason no crystals suitable for detailed X-ray analysis have been produced.

The helical structure of DNA and RNA is extremely important for their biological functions. The complementary nature of the two DNA strands allows each to serve as a template for the synthesis of a new DNA strand. Similarly, RNA is synthesized on a DNA template and proteins are synthesized according to an RNA template. Study of the structure of all nucleic acids is thus an active field; a current view of the actual mechanics of these syntheses is discussed in Chapter 16.

14.4   Summary

The X-ray study of the structure of living matter is a fascinating and growing field. It has made possible discoveries of the steric form of complex high polymers such as proteins and nucleic acids, both of which are responsible for many of the properties of all living systems. These studies, based on X-ray diffraction, have revealed both a complexity that was previously beyond imagination, and a simplicity and an ordering of atoms on a much

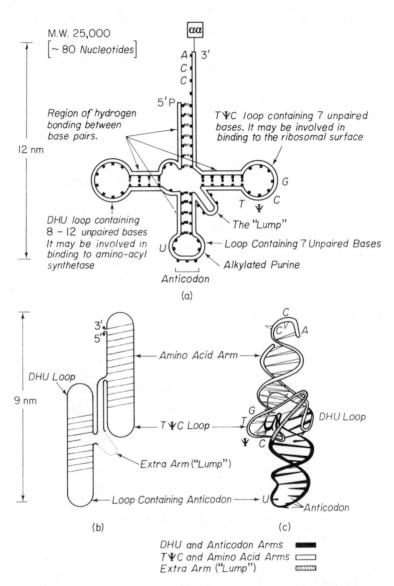

Figure 14.11   Diagram (a) of an aminoacyl *t*RNA molecule showing the clover-leaf convention of layout, and schematic diagram (b) and drawing (c) of proposed tertiary structure. Adapted from M. Levitt, *Nature* **224**:759 (1969).

larger scale than had been previously suspected. Analyses at the molecular level have contributed many of the major steps taken in recent years to the understanding of living matter. The advances discussed in this chapter are

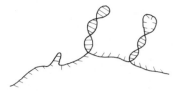

Figure 14.12   Schematic folding of an RNA chain, showing several double-helical regions held together by hydrogen bonds.

not the result of X-ray diffraction studies alone; rather, many divergent approaches, including those of chemistry, physics, crystallography, biochemistry, genetics, and virology, have been synthesized to elucidate the structure of proteins and nucleic acids. The task is far from complete and one may anticipate continued development of these techniques.

REFERENCES

X-Ray Diffraction

1. STOUT, G. H., AND L. H. JENSEN, *X-Ray Structure Determination: A Practical Guide* (New York: Macmillan Publishing Co., Inc., 1968), 467 pages.
2. WILSON, H. R., *Diffraction of X-Rays by Proteins, Nucleic Acids and Viruses* (London: Edward Arnold Ltd., 1966), 137 pages.

Proteins

3. EISENBERG, D., "X-Ray Crystallography and Enzyme Structure," in *The Enzymes*, Vol. I, 3rd ed., P. D. BOYER, ed. (New York: Academic Press, Inc., 1970), pp. 1–89.
4. DICKERSON, R. E., "X-Ray Studies of Protein Mechanisms," *Ann. Rev. Biochem.* **41**:815–842 (1972).
5. BLUNDELL, T., G. DODSON, D. HODGKIN, AND D. MERCOLA, "Insulin: The Structure in the Crystal and Its Reflection in Chemistry and Biology," *Advan. Protein Chem.* **26**:279–402 (1972).

Nucleic Acids

6. PHILLIPS, D. C., "The Three-Dimensional Structure of an Enzyme Molecule," *Sci. Am.* **215** (November 1966):78–90.
7. WATSON, J. D., *Molecular Biology of the Gene*, 2nd ed. (New York: W. A. Benjamin, Inc., 1970), 662 pages. (Especially Chapters 9 and 10 (DNA) and 11 (RNA).)
8. WATSON, J. D., AND F. H. C. CRICK, "A Structure for Deoxyribose Nucleic Acid," *Nature* **171**:737–738 (1953).

CHAPTER 15

Viruses

15.1 Introduction

In the border zone between small living cells and large macromolecules there is a class of particles that have some of the characteristics of each. These are called *viruses*. They are disease-causing particles; one or more can infect almost every type of living cell: plant, animal, or protist. For historical reasons, viruses infecting bacteria are given the separate name *bacteriophages*, or *phages* for short. All viruses are extremely small, on the order of from 20 to 400 nm. They contain none of the organelles found in living cells, such as mitochondria or nuclei, and very few of the molecular constituents. Often they are composed of only an outer *protein* coat and an inner *nucleic acid* core (proteins and nucleic acids are described in Appendix D). However, viruses can form many different shapes from these simple building blocks. Table 15-1 lists the properties of a few viruses.

Viruses reproduce in a manner different from that of living cells. In the viral reproductive process, the virus first attaches to a host cell. Then the outer protein coat of the viral particle breaks down and the viral genetic material (nucleic acid) is released into the host's cytoplasm. This serves as a

TABLE 15-1

SOME PHYSICAL PROPERTIES OF VIRUS PARTICLES

Name	Approximate molecular weight	Host	Size and shape
E. coli phage F2 or R17	3.6×10^6	Bacteria	20 nm
Bushy stunt tomato virus (BSV)	6×10^6	Plant	30 nm
Poliomyelitis virus	6×10^6	Animal	
Tobacco mosaic virus (TMV)	4×10^7	Plant	15 nm — 300 nm
E. coli phage λ	6×10^7	Bacteria	100 nm
Influenza virus	2×10^8	Animal	80 nm
E. coli phages T2, T4, and T6	2.5×10^8	Bacteria	100 nm
Herpes virus	10^9	Animal	150 nm
Smallpox virus	4×10^9	Animal	250 nm

template to direct the synthesis of new viral components. The precursors of these components and the energy necessary for this synthesis usually come from the host. Thus, viruses are obligate parasites and are distinguished from even the smallest parasitic living cells in that they do not possess the capacity to independently construct their own proteins and nucleic acids. The end result of viral infection is usually death of the host cell and release of many new viral particles, which may in turn infect other cells. Alternatively, the viral particles can be destroyed prior to attachment by various physical, chemical, and biological means, such as immune responses in higher organisms.

Virus studies have appealed to persons wishing to apply physics and chemistry to biology, for a number of reasons. First and foremost is the fact that viruses are simpler and exhibit a greater regularity than any single-celled plant or animal. At the same time, virus particles do reproduce and mutate in a fashion quite analogous to the more complex living organisms of a cellular nature. Another major reason biophysicists have been involved in virus research is that complex physical tools are necessary to study viruses; techniques used include electron microscopy, ultracentrifugation, spectro-photometry, and ionizing radiation. Although one can certainly use any of these without a knowledge of physics, it is also true that people with an inclination toward physics tend to feel more comfortable using these study tools. A third reason, albeit less important, is that many phases of virus research have involved the complicated mathematical manipulations of data that appeal to certain physicists.

The existence of viruses, as well as many of their basic characteristics, however, were discovered by "pure" biologists. After it was established that bacteria and other microorganisms caused human (and animal) diseases, occasional cases were found in which no organisms of a microscopically visible size were associated with a disease. Eventually, it was discovered that diseases of this type even killed bacteria. The latter could be studied by conventional bacteriological techniques; the destructive agents were called bacteriophages. The T series bacteriophages, which act on the bacterial species *Escherichia coli*, have been used in many studies. The T phages have the advantage that work in one laboratory can be compared with that in another; these T phages have been a standard in virus research for many years. Figure 15.1 shows T phage attached to empty bacterial shells ("ghosts").

15.2   Physical Methods Used in Virus Studies

A number of different types of physical techniques have been used to study the nature and activity of virus particles. Several of these methods are discussed briefly in this section: electron microscopy, X-ray diffraction, ultracentrifugation, electrophoresis, and bombardment with ionizing radiation.

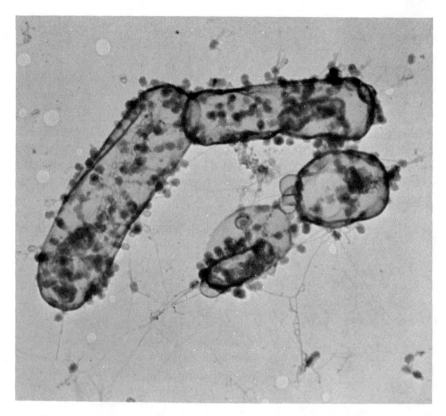

Figure 15.1   Electron micrograph of T2 phage particles attached to ghosts of
E. coli B. Note that many of the phage particles are attached to the bacterial
ghosts by their tails. After T. F. Anderson, *Am. Naturalist* **86**:91 (1952).

Electron microscopes are needed because virus particles are so very small.
A few of the largest viruses have maximum diameters of about 400 nm. The
smallest separation resolvable with a light microscope is about one-half of
this; thus, the largest viruses are barely visible in the light microscope. The
phages and most viruses are much smaller, as indicated in Table 15-1. They
cannot be seen with light microscopes.

Electron microscopes can resolve separations of 1 nm or slightly less.
Accordingly, suitable electron micrographs not only show the existence of the
viruses and phages as separate particles, but also allow one to observe the shape
and size of these particles. The major disadvantage of electron microscopy is
that the samples must be evacuated, and thus must be dried. The air–water
interface moves across the small particles as they dry; it then exerts tremen-
dous forces, tending to distort them.

One way of avoiding this interface effect is to replace the water with ethyl
alcohol and then with liquid CO_2. This latter substance has a *critical point*

(pressure and temperature at which the gas–liquid interface disappears) which can be approached under usual laboratory conditions. Thus, by going around its critical point, CO_2 can be removed with no interface forces distorting the specimen. Phage particles attached by their tails to *E. coli*, as shown in Fig. 15.1, were prepared by this method. The most common method of limiting surface-tension effects is by freeze-drying (sublimation).

Another difficulty in using the electron microscope arises from the fact that structures such as bacteria are so dense that it is not possible to observe the phage developing within the bacteria. This problem has been solved by embedding the bacteria in a suitable plastic and then cutting sufficiently thin sections. A stained section through an *E. coli* bacterium prepared in this fashion is shown in Fig. 15.2.

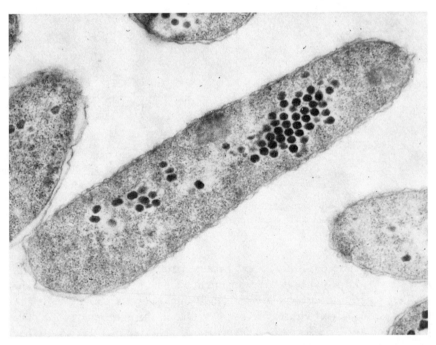

Figure 15.2 Electron micrograph of an *E. coli* bacterium infected with T2 bacteriophages. After E. R. Kellenberger, Laboratoire de Biophysique, Université de Genève, Switzerland.

As can be observed in Fig. 15.2, the phage particles are all very uniform. This is characteristic of many types of viruses. The extreme uniformity makes them similar to large molecules. Molecules can be crystallized, and so can several types of viruses. The historical example is a plant virus, tobacco mosaic virus (TMV), which infects tobacco leaves. Its crystallization led to an appreciation of the similarity of large molecules and virus particles. Virus crystals have been studied by the technique of X-ray diffraction (see Chapter 14). These studies demonstrated that the basic repeating unit of TMV consists of 49 protein subunits arranged around a 4-nm hollow cylinder in an 18- by 7-nm helix. The nucleic acid lies along the protein helix in its own 8- by 7-nm helix, as shown in Fig. 15.3.

Figure 15.3   Schematic representation of TMV structure derived largely from X-ray diffraction studies. TMV protein subunits (molecular weight, 17,500) are shown arranged to form about six helical turns whose pitch is 2.3 nm. There are 16⅓ protein units per turn and 49 units in the fundamental 6.9-nm repeat distance. The protein forms a cylinder with a hollow axial core of 4 nm diameter and with the maximum diameter of the protein about 18 nm. The RNA occurs between the subunits, forming a helix 8 nm in diameter with 49 nucleotides (3 per subunit) in each turn. It cannot be determined from X-ray studies whether TMV is a right- or a left-handed helix. After C. F. T. Mattern, *The Biochemistry of Viruses*, H. B. Levy, ed. (New York: Marcel Dekker, Inc., 1969).

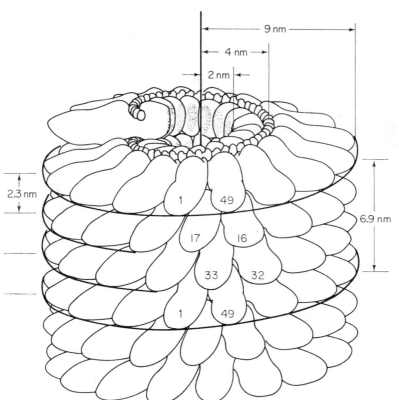

The concept that all viruses consisted of regularly packed, identical viral subunits originated in these X-ray studies of the rod-shaped TMV. Many viruses appear to be spheres, however, or have sphere-like portions attached to rods (see Table 15-1). There are five different arrangements of identical asymmetric protein units which could form a regular space-filling shell. X-ray diffraction studies on several "spherical" viruses, including bushy stunt tomato virus and poliomyelitis virus, have indicated an icosohedral symmetry, with integer multiples of 60 asymmetric subunits joined to form 20 triangular faces.

The density and uniformity of virus particle size can be determined with an instrument known as the *ultracentrifuge*, in which the suspension containing the particles is subjected to accelerations 10^4 or more times gravity, by rapidly rotating it about an axis. The tube containing the suspension is at an angle to the axis of rotation. Particles heavier than the suspending medium will tend to migrate "down" the tube. The analytical ultracentrifuge is equipped with optical systems to make it possible to observe the migration of particles in the high inertial field during rotation. By a series of calculations which are described in more detail in Chapter 16, it is possible to use ultracentrifuge data to determine the molecular weight of small particles and macromolecules, as well as to determine the uniformity of particle size and shape.

Another physical property of molecules is their rate of migration in an electrical field. This is called the *electrophoretic mobility*; it depends on the charge on the molecule, which is a function of the pH of the solution. Viruses, just as living cells, have a net negative charge at neutral pH and migrate to the anode. Electrophoretic studies have been used to demonstrate the uniformity of the virus particles, as well as changes in their charge as a function of pH. These studies, combined with ultracentrifuge and crystallization studies, have led to the picture of most viruses being uniform in particle weight, size, shape, and net charge. This contrasts with more complicated biological cells and structures all of which show distributions of weights, sizes, shapes, and charges.

A different approach to virus studies consists of bombarding a dried layer of virus particles with ionizing radiation. It is then possible to apply target theory (see Chapter 17) to the virus and determine a critical volume throughout which energy transfer may occur. Such measurements show that as many as 12 hits are necessary to destroy the infective properties of some viruses, whereas others are inactivated by the single-hit kinetics discussed in Chapter 17. These hits occur in a critical volume almost as big as the smaller viruses. For the larger viruses, the critical volume is much smaller than the particle size. In every case, this critical volume is about equal to the volume within the phage occupied by the viral nucleic acids mentioned earlier. Their properties are discussed further in the following section and in Appendix D.

15.3   Physical Biochemistry of Viruses

Most work on the chemistry and biology of viruses has centered on the *E. coli* phages. Not only are their hosts, *E. coli* bacteria, well characterized and easily grown in the laboratory, but these phages themselves are quite simple viruses. Before discussing their life cycle in detail, a brief look at the bacteriological techniques used in their assay is appropriate.

Petri dishes are partially filled with a gelatinous medium on which the bacteria can grow. Each dish is carefully sterilized. A suspension of bacteria, with or without phage particles, is poured into the dish and spread in a thin, uniform film over the surface of the gel. This is called a *plate*. The plate is then covered, and the bacteria are allowed to grow for 1 or more days. Bacteria are used at a concentration that would completely cover the plate in the absence of phage particles. If phages are present, clear areas develop on the surface of the plate. These result because each phage particle multiplies inside a bacterium until the cell wall is eventually ruptured. For every bacterium infected, as many as 300 new phage particles are sometimes produced. The new phage particles then enter other bacteria surrounding the original one, thereby producing a pattern that is characteristic of the particular phage. These clear spaces are called *plaques*. Figure 15.4 shows typical

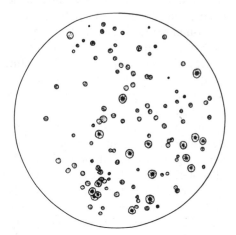

Figure 15.4   Phage plaques. This figure shows T2 plaques formed on *E. coli B*. The smaller plaques are wild type, whereas the larger ones are *r* mutants.

plaques for two strains of T2 bacteriophages infecting *E. coli*. The plaques are clear spots in a uniform layer of bacteria. Comparative counts show that the particles demonstrated by the plating technique are the same as those seen with the electron microscope.

As was mentioned in the introduction, proteins and nucleic acids are the only constituents of the simple viruses such as bacteriophages. Proteins form part of cell membranes and also part of all enzymes (i.e., substances controlling the rate of biological reactions). Nucleic acids are concerned with the transmission of genetic information and the synthesis of proteins. Two types of nucleic acids are known: DNA and RNA (see Chapter 14). DNA is associated with genetic information in plants, animals, protists, and certain viruses. RNA transmits genetic information in other viruses, and both types are associated with protein synthesis.

All viruses differ fundamentally from living cells in that they contain only one type of nucleic acid, that is, only DNA or RNA, but not both. As examples, bacteriophages such as R17 or F2 and the plant virus TMV contain only RNA, while the T phages and the animal virus herpes contain DNA. How the nucleic acid interacts with the host cell to produce new viruses has been the subject of much research, and the remainder of this chapter will be devoted to results obtained from studies of bacteriophages, especially the T phages.

The plating techniques and physical methods mentioned earlier have developed the picture of bacteriophage activity shown in Fig. 15.5. The T phages attach to the bacterial surface by their tails. This attachment is at first reversible, but then certain enzymes, presumably proteins on the tip of the tail, make it irreversible. Certain receptor sites appear necessary for phage attachment. If the nucleic acid is removed from the phage (which can be done in the case of the even-numbered T phages by osmotic shock) the phage particles attach to the bacterial surface exactly as if they were whole, but fail to reproduce. If an excess number of phage particles attack one bacterium, the cell undergoes "snap lysis"; that is, it breaks without reproducing phages. This also occurs when bacteriophages without nucleic acid are used.

After the complete bacteriophage attaches to the cell wall, it empties its nucleic acid content, but none (or almost none) of its protein, into the cell. The protein phage ghosts can be removed mechanically from outside the bacteria without interfering with phage reproduction. If phages are mixed with broken pieces of bacterial cell walls, they attach to these pieces, emptying their nucleic acid content out through the other side of the cell wall.

Once the phage nucleic acid is inside the bacterium, it alters the metabolic processes of the bacterial cell. In some cases, the cell may divide for several generations, carrying the phage with it in a latent form called a *prophage*. The cell is said to be in a *lysogenic* stage. Eventually, the prophage is induced to enter the active, *vegetative* stage. The T phages, in general, do not go through the lysogen stage, but enter the vegetative stage directly. In this stage, the bacterial cell starts manufacturing new proteins and nucleic acids typical of the phage. This period is called the induction period or the *eclipse*. At the end of a period of development, the nucleic acids are assembled. The proteins form into doughnutlike structures about the nucleic acids. These are

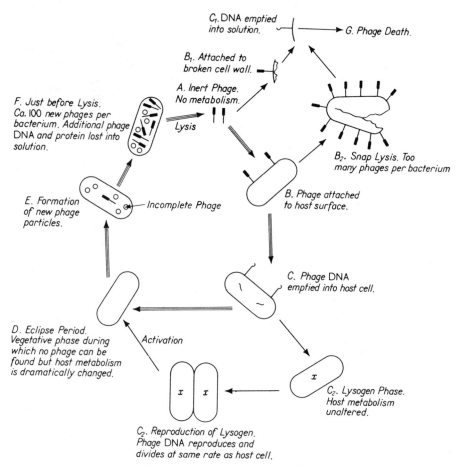

C_1. DNA emptied into solution.   G. Phage Death.

B_1. Attached to broken cell wall.

A. Inert Phage. No metabolism.

F. Just before Lysis. Ca. 100 new phages per bacterium. Additional phage DNA and protein lost into solution.

Lysis

E. Formation of new phage particles.

Incomplete Phage

B_2. Snap Lysis. Too many phages per bacterium

B. Phage attached to host surface.

C. Phage DNA emptied into host cell.

D. Eclipse Period. Vegetative phase during which no phage can be found but host metabolism is dramatically changed.

Activation

C_2. Lysogen Phase. Host metabolism unaltered.

C_2. Reproduction of Lysogen. Phage DNA reproduces and divides at same rate as host cell.

Figure 15.5 Life cycle of a lysogenic bacterial virus. We see that, after its chromosome enters a host cell, it sometimes immediately multiplies like a lytic virus and at other times becomes transformed into prophage. The lytic phase of its life cycle is identical to the complete life cycle of a lytic (nonlysogenic) virus. Lytic bacterial viruses are so called because their multiplication results in the rupture (lysis) of the bacteria.

then combined with other proteins to form whole phage particles. Eventually, the bacterium bursts. (This is called "lysis from within," in contrast to "snap lysis," which is lysis from without.)

The general character of many bacteria may be altered from without by two different processes, each of which bears some resemblance to the phage activity. The first way is by mating or conjugation. In this, two bacteria join together, some of the DNA from one passing into the other. The one receiving the DNA takes on new characteristics typical of the donor. These may include resistance to antibiotics, clone shape and size,[1] form of cellular wall, and metabolic nutrients required. Essentially the same results can be obtained by exposing the bacteria to high concentrations of DNA extracted from a strain having slightly different characteristics. The DNA molecules apparently pass through the cell membrane and alter the genetic properties of the cell.

Infection by bacteriophage is an extreme example of adding foreign DNA (the result of which is the acquisition of new properties which are fatal to the cell). During the formation of new phage particles, the nucleic acid threads appear to break and then recombine, not always with the same partners but always with partners of the same length. If a single cell is infected with several strains of the same type of phage, this *genetic recombination* can lead to new phages which have some properties of each of the parent strains. This makes it possible to study phage genetics. The experimental evidence for recombination does not necessarily imply that the nucleic acid thread actually breaks. Many other models of DNA replication also include the possibility of recombination.

It is also possible for a phage particle occasionally to change its characteristics, apparently spontaneously. The characteristics affected by such rare events include the size and shape of the plaque, the strains of bacteria it will infect, induction time, pH sensitivity, heat sensitivity, and shape and size as determined by the methods of Sec. 15.2. This spontaneous change is called a *mutation*. Once a mutation has occurred, descendants of the mutant phage will reproduce the new characteristics faithfully (until or unless another mutation occurs).

Thus, bacteriophages are similar to living organisms in that they reproduce, exhibit genetic recombination, and also undergo mutations. They differ from living cells in not metabolizing outside bacterial cells, in failing to show irritability outside cells, and in the simplicity and uniformity of the complete bacteriophage. Other viruses behave similarly to bacteriophages in most respects. The largest ones, such as influenza virus particles, show neither the simplicity nor the uniformity of bacteriophages. However, the general pattern of initial attachment, cellular entry, induction period, production of many replicas of the original virus particle, and eventual cellular destruction is common to all viruses.

15.4   Phage Genetics

The techniques of recombination between phage strains have been used to study the genetic fine structure of the *E. coli* phage T4. Different genetic characteristics of phage strains have been described by such factors as plaque

---

[1] A *clone* is an aggregate (colony) of bacteria which have grown from a single individual bacterium on the plate.

shape, strains of bacteria infected, rate of lysis, formation of lysogens, and details of the shape of the mature phage particles. If two strains of phage completely lack one common property due to mutations at different sites, it is comparatively easy to measure the occurrence of this property when the two strains are mixed. For example, if both lack the ability to form plaques on a given strain of bacteria, any plaques formed when a mixture is plated on this bacteria must be due to recombination. The probabilities of recombination to form viable phage particles when such a property is lacking in both parent strains is a measure of the distance between the locations of the two mutations along the DNA chain of the bacteriophage. The study of these distances has contributed much to a detailed knowledge of genetic structure and function.

The T4 bacteriophage has been used for these studies because it undergoes a particular type of mutation, labeled rII, which is easy to analyze for recombinations. The r-type mutations were originally characterized by their rapid lysis of *E. coli* strain *B*. Their genetic character is also shown by the types of plaques formed when plated with various strains of *E. coli*. The r-type plaque, as shown in Fig. 15.4, is larger than the usual *wild-type* plaque

TABLE 15-2

PLAQUE FORMS WHEN PHAGE STRAINS ARE
PLATED ON VARIOUS HOST STRAINS

Phage strain	E. coli strain		
	B	S	K
Wild type	Wild	Wild	Wild
rI	r	r	r
rII	r	Wild	(m)[a]
rIII	r	Wild	Wild

[a] The (m) means minute, turbid plaques; these are only occasionally formed when rII is plated with *E. coli K*.

and has sharper edges. Three different types of r mutants can be distinguished in terms of the plaques formed with different strains of *E. coli*, as described in Table 15-2. One may regard rII as a lethal mutation when the phage is grown on *E. coli K*.

The three types of r mutants can be considered to have one genetic character difference. In terminology applied to higher organisms, each type of r mutant of the T4 phage could be considered to have one gene altered. In this terminology, three different genes, I, II, and III, each lead to the same expression of genetic character, rapid lysis and r plaques, when the phage is grown on *E. coli B*. The rII strain is the most useful for studying (mapping) the fine structure of the gene (or genetic character), because these mutations are lethal on *E. coli K* but can be grown readily on *E. coli B*. If two *r*II-mutant strains of T4 phage are mixed and grown on *E. coli B* and then plated on *E. coli K*, any genetic recombination can be readily observed by the appearance of wild-type plaques. Thus, in a comparatively small number of experiments the frequency of recombination between the two mutants can be determined, even if it is as low as 1 in 10^7.

Mutations can be used to subdivide chromosomes into *cistrons*. A cistron is the smallest section of a chromosome which produces one specific cellular product. If we have several mutations which affect the same property, certain pairs of these will be found to allow normal (wild-type) offspring when they occur in the *trans* configuration, i.e. different members of a pair of homologous chromosomes. These mutation pairs are said to be *complementary* and to reside on different cistrons. Other pairs will be *noncomplementary* in the *trans* configuration. If noncomplementary mutations will allow normal offspring in the *cis* configuration (both mutations on the same chromosome) they are said to be in the same cistron.

In bacteriophage genetics, the phage nucleic acid is considered to be analogous to a chromosome. The wild-type phage is the analog of the normal chromosome. The cis configuration of mixed *r*II mutants and wild-type phage always leads to (normal) wild-type plaques (offspring). Thus, if two *r*II mutants are noncomplementary, they are on the same cistron. Complementarity can be checked in the trans configuration. Studies with more than 200 different *r*II mutants of T4 phages have shown that the "*r*II gene" consists of two cistrons. When phage strains with mutations in different cistrons are mixed, grown on *E. coli B*, and plated on *E. coli K*, many wild-type plaques are found. By way of contrast, wild-type plaques are rarely found when mutants in the same cistron are mixed. Thus, the study of T4 genetics of the *r*II mutants shows the existence of two cistrons, which together may be considered to make up the *r*II gene.

Recombinations within cistrons, although rare, do occur. Indeed, hundreds of distinct mutable sites have been found for each of the *r*II cistrons. The relative locations of a few are shown in Fig. 15.6. The distance between the closest distinct mutable sites approaches the distance between the nucleotide base pairs along the DNA chain (0.34 nm). Indeed, this and other evidence clearly demonstrates that a "point" mutation, one which changes only one such mutable site, alters one and only one nucleotide base pair.

The mapping of a large number of *r*II mutants in two cistrons in the T4 phage has altered the interpretation of the genetics of higher organisms. In particular, it has made untenable the idea that only a few mutations are possible per gene or even per cistron. Further, this mapping of the *r*II cistrons of T4 phage particles has supported the fundamental role of DNA in inheritance, including the possibility of recombination between almost every

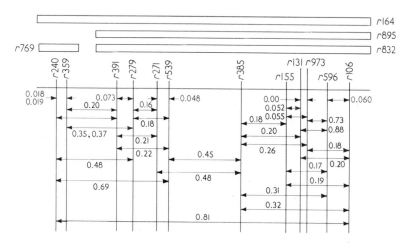

Figure 15.6   Map of the *r*164 region of the A cistron for *r*II mutations of T4 *E. coliphages*. The numbers along the horizontal lines give the recombination probabilities. The code *r*131, for example, means the 131st *r*II mutant isolated for T4. After S. Benzer, in *The Chemical Basis of Heredity*, W. D. McElroy and B. Glass, eds. (Baltimore, Md.: The Johns Hopkins Press, 1957).

DNA monomer (nucleotide pair) along the chain, and the possibility of a mutation involving only one such nucleotide pair. However, in Chapter 17, it is mentioned that the critical volume in which ionizations must occur in order to produce a mutation is about 180 nm³. Because this volume is much larger than the volume of one nucleotide base pair (~ 1 nm³), one may conclude that ionizations must occur near the DNA helix, but not necessarily in it, to produce mutations.

15.5 Summary

Viruses and bacteriophages lie between living and nonliving materials in terms of their size, structure, and behavior. Characteristic viruses infect all known living cells, usually causing the eventual death of the cell. Virus particles are too small to view with the light microscope. They are studied by conventional bacteriological techniques and by many complex physical techniques, including electron microscopy, ultracentrifugation, tracer analysis, and electrophoresis. A clearer understanding of the mode of action of viruses in general, and especially bacteriophages, has expanded knowledge of the cell surface, of the relationship of nucleic acids to metabolism, and most dramatically, of genetics.

REFERENCES

Viruses are responsible for serious diseases of man, plants, and animals. They have received much attention and are discussed in great detail in many books. Especially recommended for further reading are:

1. SMITH, K. M., AND M. A. LAUFFER, eds., *Advances in Virus Research* (New York: Academic Press, Inc.). (This appears annually; the first volume is dated 1953. Each volume contains at least one chapter which should be of interest to most biophysics students.)

2. WATSON, J. D., *Molecular Biology of the Gene*, 2nd ed. (New York: W. A. Benjamin, Inc., 1970), 662 pages. (Especially Chapters 7 and 15.)

3. MATTERN, C. F. T., "Virus Architecture as Determined by X-Ray Diffraction and Electron Microscopy," in *The Biochemistry of Viruses*, H. B. Levy, ed. (New York: Marcel Dekker, Inc., 1969).

4. BENZER, S., "The Elementary Units of Heredity," in *A Symposium on the Chemical Basis of Heredity*, W. D. McElroy and Bentley Glass, eds. (Baltimore, Md.: The Johns Hopkins Press, 1957), pp. 70–93.

5. LENNOX, E. S., "Genetic Fine-structure Analysis," *Rev. Mod. Phys.* 31:242–248 (1959).

6. DOUGLAS, J., *Bacteriophages* (London: Chapman & Hall Ltd., 1975), 136 pages.

DISCUSSION QUESTIONS—PART C

1. The concept of relative biological effectiveness (rbe) is used to quantitate the relative sensitivity of a specific cell type to a specific form of radiation. Define rbe. Can you relate this concept to the surviving fraction curves of Figs. 11.4 and 11.5?

2. What are the gross somatic effects of high doses of ionizing radiations in mammals? Relate these insofar as possible to the material of Chapter 11 and the laws of Bergonie and Tribondeau.

3. Using the data in the various references and other materials, determine the relative hazards of the American and Russian environmental microwave limits. Can you find evidence for any detrimental effects at the U.S. value?

4. Describe the construction of an ultraviolet or charged particle microbeam apparatus for the irradiation of small parts of living cells. Describe two experiments made possible by the apparatus you consider.

5. Derive the equations for the lines of flow of electric current through (and around) a spherical biological cell model which is suspended in a conducting medium, subjected to an electric field having plane symmetry at long distances from the cell. Sketch the lines of current flow in the various frequency regions.

6. W. Nyborg and his associates have emphasized the role of microstreaming near cavitating nuclei. Describe their theory and experimental results and the possible significance of this phenomenon in the disruption of single cells in cavitating ultrasonic fields.

7. The concept of a relaxation process has been cited as responsible for the gross features of the tissue impedance curves. Define this type of process mathematically and characterize it by means of a set of physically interpretable parameters. How many such variables do you need?

8. Increasing oxygen pressure (tension) inside a living cell may make the cell more sensitive to ionizing radiation. Review the evidence for this phenomenon and determine if there are any types of radiation which are an exception to the rule. Consider the concept of oxygen enhancement ratio in your discussion.

9. Discuss the history of the use of electrical shock in human therapy. Consider both muscular and neural applications. What effects are found as the current flow increases?

10. Thermal radiation (Stefan's law) has been applied to examine temperature distribution in the human body. What medical problems have been investigated in this fashion? Describe one apparatus and the statistical uncertainty of the determined temperature differences.

11. Mammalian joints emit acoustic radiation in both static and dynamic configurations. Explain how this occurs and see if any predictions regarding the condition of the joint might be made given the acoustical spectra.

12. The electron microscope has played an important role in many areas of biophysics, including viruses (Chapter 15), muscles (Chapter 9) and biological membranes (Chapter 24). Diagram the important components of an electron microscope and compare the path of an electron in it to the path of light in a conventional light microscope.

PART $\mathbf{D}$

MOLECULAR BIOLOGY

Introduction to Part D

In theory, all of biology could be explained in terms of molecular phenomena. Such descriptions appeal to the biophysicists as being in some way more fundamental; they involve the properties of far simpler systems than whole organisms. In this text, the earlier sections deal with properties of cells and groups of cells. In Part D, the molecular mechanisms are discussed.

Chapter 16 describes the molecular form of two important classes of biological molecules, the proteins and the nucleic acids. In Chapter 17, the interaction of these molecules and ionizing radiation is considered. One very important function of proteins is to control the rate of biological reactions. Proteins that are responsible for such catalytic action are called enzymes. Chapter 18 presents mathematical analyses of the kinetics of enzyme-catalyzed reactions.

Certain molecules owe their biological significance to their reactions with light. The roles of these photosensitive molecules in vision and in photosynthesis are described in the last two chapters of Part D.

CHAPTER 16

Proteins and Nucleic Acids

16.1 Introduction

Aspects of cellular metabolism are mentioned in several preceding chapters. For example, the transmission of nerve and muscle cell impulses discussed in Part B requires functioning nerve and muscle cells. The molecular biology examined in Chapter 14 and in the remainder of Part D has as background the living cell. This chapter discusses a few of the biophysical aspects of growth and reproduction of a cell, emphasizing the chain of events necessary for the synthesis of proteins and nucleic acids, two of the chief constituents of all living systems. The chapter assumes familiarity with the basic biochemical components of life (see Appendix D). To understand the experimental basis for much that is to follow, one must also understand the elements of ultra-centrifugation.

16.2 Ultracentrifugation

Sedimentation is the process by which solids (sediment) settle out of liquid (solutions). For example, silt is said to sediment out of pond water. The larger silt particles will settle more rapidly. If it were necessary to separate silt on the basis of size, one could stir the water, allow it to settle, and separate the supernatant fluid (smaller particles) from the precipitate (larger particles). As the sediment size decreases to the size of subcellular particles or macromolecules, sedimentation under gravity will be canceled by *diffusion*, the tendency for thermal motion to redistribute solute throughout a solvent (see Chapter 23). To separate these by sedimentation, one must accelerate the particle, using a force greater than gravity. Rapid centrifugal motion can induce such accelerations. A solution containing a mixture of particles may be placed in a container, called a centrifuge tube or bottle, and spun such that the largest particles sediment and smaller ones remain in solution. By pouring off the supernatant fluid and recentrifuging it, one may prepare several fractions of subcellular particles.

For finer analytic work, higher speeds and different techniques are necessary. The solution to be analyzed is layered on top of a second particulate free solution and the tube is placed in a rotor (rotating holder) of high-strength alloys of metals such as titanium. The rotor is then spun in an evacuated chamber. This *ultracentrifugation* can subject solutions to over 250,000 times the force of gravity. Figure 16.1 shows an analytic ultracentrifuge and illustrates how optical measurements can be made while a sample is rotating.

When a particle sediments at a constant velocity in such an ultracentrifuge, the centrifugal force balances the frictional resistance of the solvent. If the velocity of sedimentation, dx/dt, is measured in distance per second, the sedimentation coefficient, s, can be calculated as

$$s = \frac{dx/dt}{\omega^2 x}$$

where x is the distance from the center of the rotor and ω is the angular velocity in radians per second. As s value of 1×10^{-13} sec is termed 1 Svedberg unit (S), honoring the inventor of the ultracentrifuge. A subcellular particle such as a mitochondrion has a sedimentation coefficient of 10,000 S; values for proteins range between 200 and 1 S.

The sedimentation coefficient as derived from these *sedimentation-velocity* experiments is a function of the molecular weight, shape, and density of the particle. If a spherical shape is assumed, the molecular weight M can be calculated by

$$M = \frac{RTs}{D(1 - \bar{v}\rho)}$$

where R is the gas constant (8.31 J/mole/deg), T is the absolute temperature, $\bar{v}$ is the partial specific volume of the protein, ρ is the density of the solvent, and D is the diffusion coefficient of the protein.

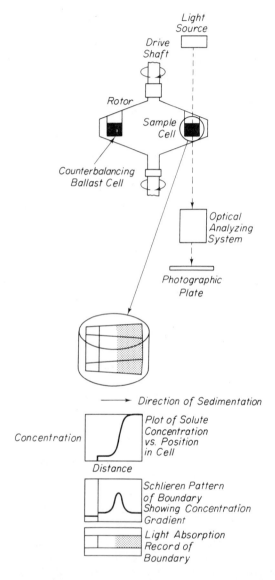

Figure 16.1 Principle of the ultracentrifuge, showing how optical measurements are made while sample is being centrifuged. From A. L. Lehninger, *Biochemistry*, 2nd ed. (New York: Worth Publishers, Inc., 1975, p. 175).

There are several other methods of ultracentrifugation besides sedimentation velocity; one of special interest in nucleic acid research uses a solvent of varying density upon which the solution to be analyzed is layered. This *density gradient* method increases resolution when separating molecules of similar size and shape but different density.

16.3 DNA Replication

The molecular constituents in a cell are either obtained directly from food or are synthesized from ingested precursors. These synthetic reactions are catalyzed by catalytic proteins called *enzymes* (see Chapter 18). Nucleic acid (usually DNA) is the genetic material in living cells and thus transmits to the cell all information necessary to growth and reproduction of that cell, including directions for synthesis of the relevant enzymes. Before investigating protein synthesis, one sould become familiar with some aspects of nucleic acid replication.

As is mentioned in Chapter 14, the structure of DNA was found to be a double-stranded helix with intrastrand hydrogen bonds between the nucleotide bases in each strand. The steric properties of the helix require that of the four bases found in DNA (adenosine (A), thymine (T), cytosine (C), and guanine (G); see Appendix D for their structures), an A on one chain can only pair with a T on the other chain, a C with a G, and vice versa. Thus, the two chains are complementary. It is fairly easy to imagine that DNA replication involves strand separation and the formation of complementary molecules on each of the free strands (Fig. 16.2).

This is termed *semiconservative replication*, since each daughter DNA helix contains one parental strand and one new strand. A classic experiment by Meselson and Stahl in 1958 provided support for this mechanism using mass tracer techniques (see Chapter 29). Bacteria were grown in a medium containing ^{15}N instead of ^{14}N, so that all the nitrogen bases in their DNA were labeled with the heavier isotope. They were then transferred to a light (^{14}N) medium. DNA was isolated from the bacteria and subjected to density gradient centrifugation at various times after this transfer.

The particular density gradient used, a cesium chloride solution ranging from 1.6 to 1.8 g/cm^3, will easily separate ^{15}N from ^{14}N DNA. Initially, all DNA is ^{15}N; after sufficient time for one DNA replication (one generation), the DNA has a density midway between that of ^{15}N and that of ^{14}N DNA. This may mean that all DNA molecules have one heavy and one light strand or that each daughter strand is a mixture of half new and half old material. After two generations, two bands are seen, one with the density of light DNA, the other the density of the hybrid DNA isolated after one generation. This is exactly as expected if DNA molecules undergo semiconservative replication.

This method of replication of genetic material from a doubly helical parent is found even in certain small viruses whose genetic material is not doubly helical DNA (see Chapter 15). Some, such as the bacteriophage Fl, have circular singly stranded DNA. This, in a host cell, serves as a template

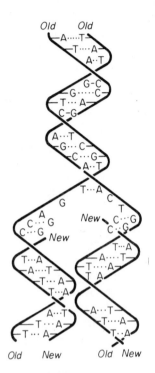

Figure 16.2 The replication of DNA. After C. P. Swanson, *The Cell*, 3rd ed., © 1969, p. 56. Reprinted by permission of Prentice-Hall, Inc., Englewood Cliffs, New Jersey.

for a complementary strand, and new viral DNA is synthesized from the double helix as explained above. Others, such as rheoviruses, have doubly helical RNA as the genetic material. This also replicates semiconservatively. Finally, in tobacco mosaic virus, bacteriophage F2, and others, the genetic material is singly stranded RNA, which is complemented inside the host cell. Thus, all genetic material is copied in essentially the same manner.

There are several enzymes involved in all these syntheses. In general, these enzymes have no effect on the structure of the new DNA (or RNA); this is completely determined by the parent molecule.

16.4 Transcription of RNA

The fact that the genetic material of even the simplest viruses can direct synthesis of more viral nucleic acids and proteins in a host cell emphasizes the importance of nucleic acids. Yet DNA itself does not serve as a template to order amino acids into proteins, for proteins can be constructed in the absence of DNA. The intermediate between DNA and protein has been shown to be RNA. All RNA is synthesized in close conjunction with DNA, for example, in the nucleus of cells. The amount of protein is directly related to the cellular content of RNA. In experiments using radioactive amino acid molecules and density gradient centrifugation, newly synthesized protein chains were found associated with RNA containing organelles called *ribosomes*.

If RNA is to transmit the information contained in DNA, it must be synthesized under the direction of the DNA. Its primary structure (see Appendix D) is similar to that of DNA, with the sugar ribose instead of deoxyribose and the base uracil instead of thymine. This leads to the hypothesis that a DNA strand may at some time act as a template for a complementary RNA, rather than DNA, strand. An appropriate enzyme, RNA polymerase, exists in virtually all cells.

This polymerase attaches to only one strand of the doubly stranded DNA. Indeed, it has been conclusively shown for certain viruses and assumed for higher organisms that only one strand of the two strands is transcribed. That is, there is only one RNA strand produced for each double-stranded DNA. Since, as described in Chapter 14, a DNA strand has polarity (one end terminated by a 5', the other by a 3' carbon atom), the RNA strand might be started at either one or the other end, or both. The direction of growth has been shown to be from the 5' end of the RNA to the 3' end; that is, it uses as a template a DNA strand in the 3' $\rightarrow$ 5' direction.

The DNA strand to be transcribed must carry at least one initiation marker or start signal to identify itself. Since the RNA chains produced by DNA transcription are of various fixed sizes and often much shorter than the original DNA strand, there must be several initiation and termination markers on the strand to be transcribed. These have been given the names promotors and terminators, respectively. Once again much of the knowledge of these sites is taken from studies with bacterial viruses; promotor and possibly terminator regions are frequently associated with AT-rich regions of DNA. However, not all such AT-rich regions are promotor regions.

Since the arrangement of these promotor and terminator regions governs the growth of every cell and hence the entire organism, there will be profound differences between different organisms at this level. This is a very complex subject; the interested reader is referred to the general references for an overview or to Chamberlain's article for a more detailed description of research in this field.

16.5 Translation of RNA into Protein

There are several lines of evidence, mentioned in Sec. 16.4, which point to RNA as the intermediate carrier of genetic information. The bulk of cellular RNA has a relatively long life in the cell, while we know that new and different proteins are being synthesized rapidly, within a time span of minutes. However, a small percentage of RNA is synthesized and broken down rapidly. Phage studies show that the RNA produced after phage infection has a base composition very similar to that of phage DNA but different from that of bacterial DNA or the rest of the bacterial RNA. In 1961 Jacob and Monod proposed the name *messenger RNA* (*m*RNA) for this species and hypothesized that it alone, of all the RNA present in the cell, carried genetic information from the DNA.

Proteins are quite different chemically from nucleic acids (see Appendix D), and their constituent amino acid side chains have no strong affinity for nucleotide bases. Thus, it is at first difficult to imagine how the genetic material in an *m*RNA molecule can be used to direct the amino acid sequence in a protein. Work started in 1950 by Zamecnik and colleagues demonstrated that translation was indeed a complicated process. Amino acids and *m*RNA were necessary, but so were many other components. A second type of RNA, *transfer RNA* (*t*RNA), whose structure is described in Chapter 14, was needed as a "connector." There are many different types of *t*RNA's in any one cell; each type will bind to a certain sequence of three *m*RNA bases, called a *codon*, and to one certain amino acid. The ribosomes, mentioned earlier, were the actual sites or factories for protein synthesis. Ribosomes are spherical particles composed of both protein and a third type of RNA, *ribosomal RNA* (*r*RNA). The major steps in protein synthesis are shown in Fig. 16.3.

Any amino acid, to be useful in protein synthesis, must first be attached to a *t*RNA molecule to form an *aminoacyl t*RNA molecule. This union is catalyzed by an aminoacyl *t*RNA synthetase enzyme. There is at least one such enzyme for each amino acid found in the proteins in each cell. This enzyme attaches the amino acid to a *t*RNA specific for that amino acid; again there is at least one such *t*RNA for every amino acid. In all bacteria one special molecule, *N*-formylmethionyl *t*RNA (fMet-*t*RNA), is necessary for the initiation of protein synthesis as described in the next paragraph.

Since ribosomes are the site of protein synthesis, they must coordinate the *m*RNA, aminoacyl *t*RNA's, and other factors involved in protein formation. A ribosome consists of two unequal subunits, which must disassociate before protein synthesis begins. The smaller subunit, with a sedimentation coefficient of 30 S in bacteria, binds both *m*RNA and a special initiation aminoacyl *t*RNA (in bacteria, fMet-*t*RNA); and this initiation complex reassociates with the larger, 50-S, subunit to form a functional "protein factory."

There are two sites on each ribosome which can bind *t*RNA. One, the A site, will bind aminoacyl *t*RNA. The other, the P site, will bind either the initiation aminoacyl *t*RNA or *t*RNA attached to peptides (*peptidyl tRNA*) but not ordinary aminoacyl *t*RNA. After initiation in bacteria there is an *m*RNA ribosome complex with fMet-*t*RNA in the P site and an empty A site.

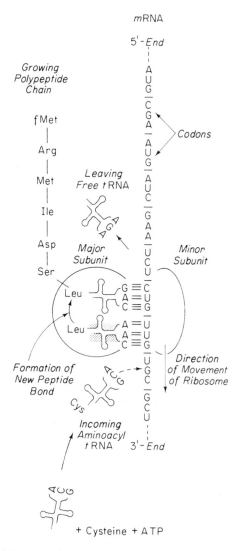

Figure 16.3 Major steps in protein synthesis. From A. L. Lehninger, *Biochemistry* (New York: Worth Publishers, Inc., 1970, p. 692).

A codon in the *m*RNA will specify which aminoacyl *t*RNA will next fill the A site.

The repeating cycle in protein synthesis begins: an aminoacyl *t*RNA is bound to the A site. In a reaction catalyzed by peptidyl transferase, an enzyme found in the large ribosomyl subunit, the amino group of the new aminoacyl *t*RNA displaces the *t*RNA attached to the initiation amino acid.

This gives a peptidyl *t*RNA in the A site and an empty P site. Then in a translocation reaction, the peptidyl *t*RNA is moved to the P site and, simultaneously, the ribosome moves along the *m*RNA. This allows a new codon to specify the aminoacyl *t*RNA permitted in the A site.

Referring back to Fig. 16.3, we see a leucyl *t*RNA (Leu-*t*RNA), which binds to the *m*RNA codon UUG. This is displacing the *t*RNA from a peptidyl *t*RNA, which also ends in Leu (this time binding to the *m*RNA codon CUG). After the formation of the new peptide bond, the growing peptide will have the amino acids *n*-formyl methionine (first), arginine, methionine, isoleucine, aspartate, serine, leucine, and leucine (most recent). After translocation, the codon CGU, which binds a *t*RNA specific for cysteine, will now be part of the A site, and thus the next amino acid to be added to the peptide chain will be cysteine.

Finally, the termination of polypeptide chains is signaled by one of three special termination codons, UAG, UAA, or UGA, in the *m*RNA. When these are encountered and protein release factors are present, the terminal *t*RNA is split from the nascent protein, which quickly assumes its final three-dimensional structure. The *m*RNA–ribosome complex dissociates and one cycle of protein synthesis is over.

This complicated description of translation is, at best, a great simplification of the actual process. Several essential cofactors and intermediates have been omitted. Interested readers are directed to the general references at the end of the chapter for more information.

16.6   The Genetic Code

One of the great questions facing molecular biologists and biophysicists in the 1960's was exactly how genetic information in nucleic acids was used to control protein synthesis. When it became apparent, from genetic experiments with mutants lacking one or more base pairs, that three bases in the RNA template specified each amino acid in the protein, the question reduced to: What is the genetic code? That is, which sequence(s) of nucleotide bases produced what amino acid?

As a first approach to this problem, in 1961 Nirenberg succeeded in synthesizing protein using a synthetic polyribonucleotide as a template. The synthetic template, containing only uracil (poly U), produced a protein containing only phenylalanine. Thus, UUU codes for phenylalanine. Similarly, CCC codes for proline and AAA codes for lysine.

In 1964 it became possible to measure the binding of *t*RNA molecules to ribosome–synthetic polyribonucleotide complexes. The synthetic template in these experiments could be quite small; even three nucleotides were sufficient. Many more codons were assigned in this fashion.

Finally, synthetic templates could be made with known sequences. A combination of all three of these methods permitted unequivocal assignment of the 64 possible codons, as shown in Table 16-1.

TABLE 16-1

GENETIC CODE

First position (5' end)	Second position				Third position (3' end)
	U	C	A	G	
U	Phe	Ser	Tyr	Cys	U
	Phe	Ser	Tyr	Cys	C
	Leu	Ser	Term[a]	Term	A
	Leu	Ser	Term	Trp	G
C	Leu	Pro	His	Arg	U
	Leu	Pro	His	Arg	C
	Leu	Pro	GluN	Arg	A
	Leu	Pro	GluN	Arg	G
A	Ileu	Thr	AspN	Ser	U
	Ileu	Thr	AspN	Ser	C
	Ileu	Thr	Lys	Arg	A
	Met	Thr	Lys	Arg	G
G	Val	Ala	Asp	Gly	U
	Val	Ala	Asp	Gly	C
	Val	Ala	Glu	Gly	A
	Val	Ala	Glu	Gly	G

[a] Chain terminating (formerly called "nonsense").

Source: After J. D. Watson, *The Molecular Biology of the Gene,* 2nd ed. (New York: W. A. Benjamin, Inc., 1970).

Notice that many amino acids are selected by more than one codon. For instance, phenylalanine is coded by UUC as well as UUU, and leucine is coded by UUA, UUG, CUU, CUC, CUA, and CUG. Some of this degeneracy is caused by multiple *t*RNA's, each recognizing a different codon but specific for the same amino acid. In addition, some *t*RNA's recognize more than one codon. When multiple codons can specify an amino acid, they do so with varying efficiencies. One is used preferentially in any given organism and others have much lower binding constants. Thus, the rate of synthesis of a given protein may be controlled in part by which codons code for its amino acid sequence.

16.7   Control of Protein Synthesis

In Sec. 16.5 a protein "factory" is described. All cells in an organism that synthesize protein have the same DNA and the same "factories"; accordingly, they might be expected to produce the same proteins. This is not the case;

every cell synthesizes varying types and amounts of proteins throughout its life cycle. Part of this variation is determined by the DNA and part is in response to the cellular environment. The mechanism of control of protein synthesis has been and is being investigated by many workers; the following brief discussion is only an introduction to this field; see the general references for more details.

Certain small molecules can be shown to influence the rate of synthesis of specific proteins. If the small molecules increase this rate, they are termed an *inducer* of the protein; if they decrease this rate, they are a *corepressor*. It is found that the variation in rate of synthesis of these proteins is related to the amount of *m*RNA present; thus, small molecules can somehow influence the translation of *m*RNA.

In Sec. 16.4 it is stated that the DNA region which initiates RNA translation is termed the promotor. A second region on the DNA, termed the *operator*, is present near those promotors for *m*RNA which codes for inducible or repressible proteins. If a small protein named the *repressor* binds to the operator, *m*RNA translation does not occur. Repressors can therefore control the synthesis of more than one protein; all that is necessary is that these proteins be transcribed by a single *m*RNA. Now the question is: How is repressor activity controlled by small molecules?

Repressors are always present for those proteins that can be induced or repressed. Those small molecules that act as corepressors bind to inactive repressors, and the resulting complex is an active repressor. Conversely, those small molecules that act as inducers must "turn off" active repressors, by binding to them to form an inactive complex.

Once an *m*RNA is synthesized, it must be transcribed. There are various controls possible in this process, also. All proteins coded along a given *m*RNA are not produced in similar numbers. Thus, ribosomes may attach to different initiation signals along the *m*RNA at different rates. Alternatively, since protein transcription occurs from the 5′ to the 3′ end of the *m*RNA (see Fig. 16.3), all ribosomes may attach to the 5′ end of the *m*RNA, but some may disengage as each chain termination codon is encountered. Finally, as mentioned in Sec. 16.6, the specific codon sequence may influence the rate of synthesis by requiring more or less efficient *t*RNA's.

Another factor influencing transcription is the rate of *m*RNA degradation. There are specific enzymes that will catalyze the hydrolysis of *m*RNA. The specific nucleotide sequence of an *m*RNA may help determine its probability of degradation and thus its synthetic lifetime in the cell.

16.8   Summary

DNA is replicated semiconservatively. The double-stranded helix partially disassociates and each single strand acts as a template for a new nucleotide chain. However, the genetic information in DNA must not only be duplicated; it must also be expressed in the form of proteins. The first process in this expression involves transcription of the DNA message into three types of RNA. These are called ribosomyl RNA (rRNA), transfer RNA (tRNA), and messenger RNA (mRNA). The mRNA alone contains the DNA's genetic information; this is translated into proteins on ribosomes composed of rRNA using tRNA molecules as adaptors. The entire process of replication, transcription, and translation is under complicated regulation and control, but can be simply related, as in the following scheme:

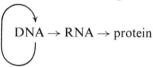

$$DNA \rightarrow RNA \rightarrow protein$$

REFERENCES

General

1. WATSON, J. D., *The Molecular Biology of the Gene*, 2nd ed. (New York: W. A. Benjamin, Inc., 1970), 662 pages. (Especially Chapters 9–14.)
2. LEHNINGER, A. L., *Biochemistry: The Molecular Basis of Cell Structure and Function*, 2nd ed. (New York: Worth Publishers, Inc., 1975), 1104 pages. (A lucid description of ultracentrifugation in Chapter 7, and protein and nucleic acid synthesis in Chapters 28–32.)

Specialized Articles

The following three papers were presented by their authors when they received the Nobel Prize in Physiology and Medicine for their pioneering work on protein synthesis.

3. DE DUVE, C., "Exploring Cells with a Centrifuge," *Science* **189**:186–194 (1975).
4. PALADE, G., "Intracellular Aspects of the Process of Protein Synthesis," *Science* **189**:347–358 (1975).
5. CLAUDE, A., "The Coming of Age of the Cell," *Science* **189**:433–435 (1975).

A specific review appears in:

6. CHAMBERLIN, M. J., "The Selectivity of Transcription," *Ann. Rev. Biochem.* **43**:721–775 (1974).

CHAPTER 17

Molecular Effects
of Ionizing Radiation

17.1 Introduction

The cellular and larger-scale effects of ionizing radiation presented in Chapter 11 occur as a result of molecular transformations produced by the radiation. Molecular changes are conveniently divided into two types: direct or local and indirect or nonlocal. The latter are probably the major cause of the effects of ionizing radiations on proteins and nucleic acids in solution and in the intact organism. However, direct changes are easier to describe and are important both in water and in dried films of proteins and nucleic acids.

In this chapter, the nature of the absorption of ionizing radiation in matter is briefly reviewed. This is followed by discussions of direct effects of ionizing radiation on water and on synthetic polymers. Before pursuing these further, the basic concepts of target theory are introduced. The preceding ideas are used in an analysis of the effects of ionizing radiation on dried films of proteins and nucleic acids. Finally, the biologically important effects of radiation on proteins and nucleic acids in solutions and in intact cells are examined.

Ionizing radiation consists of beams of particles that may be charged or uncharged. The exposed material is bombarded by the particles in the beam. According, the terms radiation, particle, and projectile are used interchangeably in this chapter.

17.2   Absorption of Radiation

A. CHARGED PARTICLES

Charged and uncharged radiations are absorbed by different mechanisms, although both produce ionization. Electrically charged particles, at least at moderate kinetic energies, interact with matter primarily by ionizing the outer electrons in the absorber's atoms. In this fashion, the charged particle of a given speed (v) loses kinetic energy (K) per unit distance (x) at a relatively continuous rate. This is approximated by

$$\frac{dK}{dx} \simeq -\frac{Z^2}{v^2} \cdot \rho_e \qquad (17\text{-}1)$$

where Z is the magnitude of the particle's charge in electron units and ρ_e is the absorber's electron density. In a biophysical context, dK/dx is referred to as the *linear energy transfer* (LET).

The operation of many charged particle detectors is based on Eq. (17-1). Except for positive and negative electrons, most charged particles have rather straight tracks through matter. The path ends at a predictable distance, called the *range* of the charged particle. Figure 17.1 shows a typical graph of

Figure 17.1   Number of monoenergetic charged particles as a function of distance into an absorber (integral range distribution curve). The dashed curve is the negative differential of the integral curve and represents the distribution of charged particle ranges.

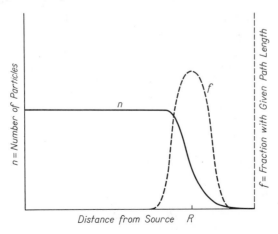

the number of charged particles versus the distance from the source. It also includes the bell-shaped distribution of particle ranges; the mode of that distribution is labeled R. A graph of LET versus distance is shown in Fig. 17.2. LET increases with distance (depth) since the particle's speed is con-

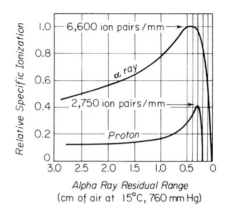

Alpha Ray Residual Range
(cm of air at 15°C, 760 mm Hg)

Figure 17.2 Relative magnitude of specific ionization versus distance into an air absorber. Specific ionization is the number of ion pairs per unit path length and is linearly related to dK/dx. Note the Bragg peaks at the range of the α and proton particles. After R. Evans, *The Atomic Nucleus* (New York: McGraw-Hill Book Company, 1955).

tinuously decreasing as it moves into the absorber. The maximum seen in LET, known as the *Bragg peak*, is characteristic of all charged particles.

B. UNCHARGED PARTICLES

Uncharged radiation is somewhat more difficult to detect since the almost continuous electronic excitation produced by charged particles does not occur. Instead, one must measure some discrete interaction between the neutral particle and the electrons or nuclei in the absorber. These interactions reduce the intensity of the neutral particle beam. An example is the *photoelectric effect*, whereby a photon (γ) may react with a bound K-shell electron and be absorbed with the release of energy. Symbolically,

$$\gamma + \text{atom} \longrightarrow \text{ion}^+ + e^- \tag{17-2}$$

In such single interactions, the neutral particle can be extinguished (photoelectric effect) or lose some or all of its kinetic energy. Thus, the neutral particle beam, which has an intensity I (number of neutral particles per unit area per unit time), is attenuated. This neutral particle attenuation can be described by a decrement in I per unit length, x:

$$dI = -\mu I \, dx \tag{17-3}$$

where μ is the linear attenuation coefficient,[1] a measure of attenuation probability per unit length. Equation (17-3) may be integrated to yield the exponential law:

$$I = I_0 \exp(-\mu x) \tag{17-4}$$

where I_0 is the value of I at the origin.

A beam of neutral radiation is defined by two specific attributes: particle direction and energy. If a given projectile within the beam has either (or both) of these properties changed, it is no longer in the beam, and thus the beam's intensity is decreased. Unlike a charged particle, a neutral particle does not have a precise range. Instead, beam penetration is quantitated by the distance in a given material that is required to reduce the intensity by a given factor. The usual reduction factor is 2; the corresponding distance is called the *half-value layer*. This parameter is mathematically similar to half-life in radioactive decay.

For photons with energies up to 10 MeV, two types of interactions, in addition to the photoelectric effect, are important. One of these, called the *Compton effect*, describes the inelastic scattering of photons from electrons. The other, called *pair production*, is the creation of an e^+ and an e^- pair from a photon:

$$\gamma + \text{nucleus} \longrightarrow e^+ + e^- + \text{nucleus} + \text{energy} \tag{17-5}$$

The nucleus is necessary to provide the electric field with which the photon will interact. Since the rest mass energy of each electron is 0.51 MeV, the energy of the photon must exceed 1.02 MeV for Eq. (17-5) to be energetically possible.

For photons at energies up to 10 MeV, the linear attenuation μ can be set equal to a sum of three mutually exclusive components:

$$\mu = \mu_{pe} + \mu_c + \mu_{pp} \tag{17-6}$$

μ_{pe}, μ_c, and μ_{pp} correspond to probabilities for the photoelectric effect, the Compton effect, and pair production, respectively. Each of these separate probabilities depends, among other things, on photon energy and absorber electronic composition. As the energy of the photon increases, other mechanisms involving the nucleus or even individual nucleons become important and there are more terms in the summation (Eq. (17-6)).

The mass attenuation coefficient is derived from μ by dividing by ρ, the mass density of the absorber. Figure 17.3 shows the mass attenuation coefficient as a function of photon energy ($E\gamma$) for various attenuators. There is a general tendency for an increase in μ as $E\gamma$ exceeds 10 MeV. This occurs in part due to the rise in μ_{pp}.

---

[1] The linear attenuation coefficient is symbolized as a in Chapter 12 and α in Chapter 13.

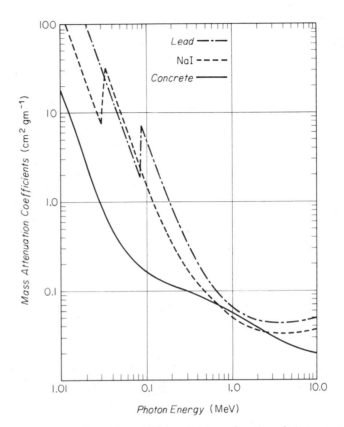

Figure 17.3   Mass attenuation coefficient (μ/ρ) as a function of photon energy. The general decrease in μ/ρ as E_γ increases is due to the E_γ^{-3} dependence of μ_{pe}. Sharp resonances in μ_{pe} do occur, however, at characteristic ionization energies. Ionization of K-shell electrons causes the peaks shown for NaI (30 keV) and Pb (90 keV). After H. Andrews, *Radiation Biophysics*, 2nd ed. (Englewood Cliffs, N.J.: Prentice-Hall, Inc., reprinted by permission; 1974).

The form of Eq. (17-6) is quite general, in that μ may be described as a sum of mutually exclusive processes (probabilities) for *any* neutral beam. For example, if that equation were written for neutrons with 10 MeV of kinetic energy (E_n) passing through tissuelike material, the dominant attenuation coefficients would correspond to *elastic scattering* from the protons and ^{12}C nuclei in the absorber. As E_n increases, various other reaction channels, for example,

$$n + {}^{12}_{6}\text{C} \rightarrow 3\alpha + n + \text{energy} \tag{17-7}$$

where α represents an alpha particle, would become available and contribute their terms to the total neutron attenuation coefficient. This added complexity at higher energies is similar to photon interactions.

The concept of linear energy transfer (LET) is applicable to both charged and uncharged radiation. In the latter case, a summation is carried out over the charged secondary particles produced by the neutral beam in the absorber. This allows assignment of the radiation damage on a particle-by-particle basis. The biological effect is related to the mean LET by the quality factor discussed in Chapter 11.

17.3   Direct Effects of Radiation on Water

The consequences of a series of molecular reactions describing the inter-action of radiation and H_2O are summarized by the formula

$$H_2O + \gamma \text{ (or other particle)} \rightarrow H\cdot + OH\cdot \qquad (17\text{-}8)$$

where the hydrogen ($H\cdot$) and hydroxyl ($OH\cdot$) free radicals have electronic forms $H\cdot$ and $\cdot\ddot{O}{:}H$, respectively. The dots in Eq. (17-8) symbolize the unpaired electron available in each species. The intermediate processes that lead to Eq. (17-8) are complicated. It is most probable that the first interac-tion is the production of $H_2O^+ + e^-$. The free electron then migrates through the medium to a second solvent molecule and converts it into H_2O^-. The decay of these ions, which occurs in less than 10^{-10} sec, leads to the free radicals of Eq. (17-8).

An unpaired electron is paradoxically very reactive and very long-lived, since it can persist via such reactions as

$$X\cdot + Y{:}Z \rightarrow X{:}Y + Z\cdot + \text{energy} \qquad (17\text{-}9a)$$

or

$$\rightarrow X{:}Z + Y\cdot + \text{energy} \qquad (17\text{-}9b)$$

where $Y{:}Z$ represents a covalent bond between any two atoms in a compound. In these reactions, excess kinetic energy is given to the final products. The original free radical ($X\cdot$) is replaced by a new form ($Z\cdot$ or $Y\cdot$) in this process.

A direct way to extinguish the unpaired electron is to combine two free radicals into a molecular compound. Some combinations of $H\cdot$ and $OH\cdot$ radicals can be used to confirm the occurrence of the reaction symbolized in Eq. (17-8). Thus, if $H\cdot$ and $OH\cdot$ were produced, one would expect that the reactions

$$H\cdot + H\cdot \rightarrow H_2\uparrow \qquad (17\text{-}10a)$$

and

$$OH\cdot + OH\cdot \rightarrow H_2O_2 \qquad (17\text{-}10b)$$

would occur in irradiated aqueous solution. Indeed, both hydrogen gas and hydrogen peroxide are detected. The volume of H_2 produced can, in fact, be used to quantitate the radiation dose. Figure 17.4 gives hydrogen yield as a

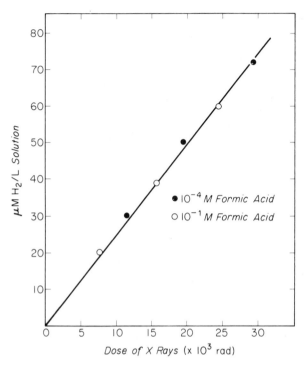

Figure 17.4 Indirect effect of radiation (X rays) on a formic acid solution. The yield of hydrogen gas is independent of solute concentration. After H. Fricke, E. Hart, and H. P. Smith, *J. Chem. Phys.* **6**:229–240, 1938. (Am. Inst. of Phys., Inc., New York.)

function of dose for two different solute concentrations; the latter parameter is not significant, which implies production of the H· directly from the water.

Alternatively, the presence of an unpaired electron gives rise to an unpaired magnetic moment, which can be detected by electron spin resonance (esr) methods (esr is discussed in more detail in Chapter 26). The presence of oxygen in the water will generally increase the yield of hydrogen peroxide. Figure 17.5 gives the yield as a function of dose for pure water and water with various solutes present. The presence of unpaired electrons tends to reduce the amount of H_2O_2 produced because of its various reactions with free radicals. Also, as shown in Fig. 17.5, a free-radical scavenger will decrease the yield of H_2O_2.

17.4   Radiation Effects on Synthetic Polymers

Ionizing radiation can also interact directly with most long-chain molecules. Although some of the specific examples chosen are not of biological signifi-

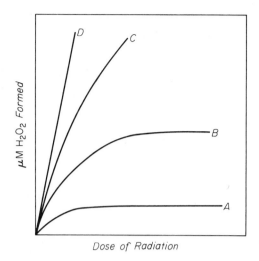

Figure 17.5 Schematic yield of hydrogen peroxide in various irradiated aqueous solutions. Cases are as follows: A, pure H_2O; B, H_2O with free radical scavenger solute; C, H_2O with dissolved O_2; D, H_2O with O_2 and oxidizable additives. After A. Casarett, *Radiation Biology* (Englewood Cliffs, N.J.: Prentice-Hall, Inc., 1968).

cance, the general results are applicable to biopolymers, such as proteins and DNA. As discussed in Chapters 14 and 16, a polymer is made up of a repeating unit (the monomer), which is duplicated again and again; the repeating units making up proteins are called *amino acids*, and those making up nucleic acids are called *nucleotides*. The chemical structures of both proteins and nucleic acids are presented in Chapter 14 and in Appendix D.

Both proteins and nucleic acids are complicated high polymers whose long chains consist of a mixture of different types of residues or monomers arranged in a definite pattern. It is easier to interpret the effects of ionizing radiations on simpler synthetic polymers, those made up of one or at most two types of monomers, than on the more complicated proteins and nucleic acids. A study of the radiation damage to these simpler polymers allows one to predict the types of effects to be expected when proteins and nucleic acids are exposed to ionizing radiation.

The effects of ionizing radiation on many different types of synthetic high polymers have been studied. Two used as examples in this section are polyethylene and polyisobutylene. They have the structural forms shown in Fig. 17.6. When the monomers (ethylene molecules) are added together to form polyethylene, it is possible to have carbon nuclei surrounded by only seven electrons, as shown in Fig. 17.7; these species are free radicals.

Free radicals are responsible for some of the effects of ionizing radiation

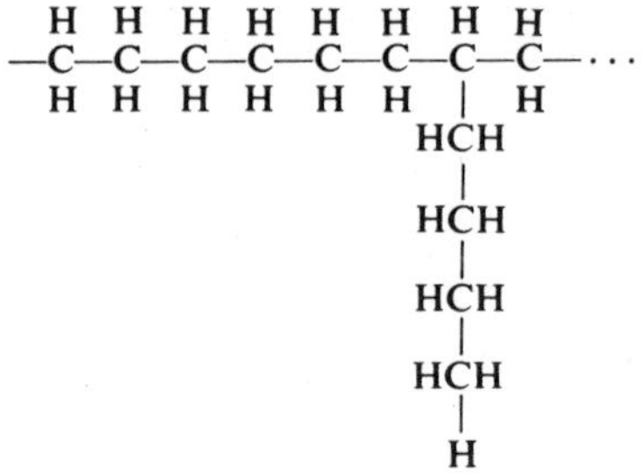

$$
\begin{array}{cccccccc}
H & H & H & H & H & H & H & H \\
-C-C-C-C-C-C-C-C- & \cdots \\
H & H & H & H & H & H & H & | & H
\end{array}
$$

HCH
|
HCH
|
HCH
|
HCH
|
H

(a) polyethylene

$$CH_2 = CH_2$$

(b) ethylene

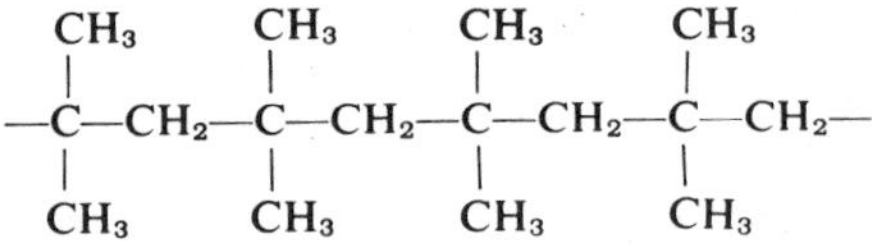

$$
\begin{array}{cccc}
CH_3 & CH_3 & CH_3 & CH_3 \\
| & | & | & | \\
-C-CH_2- & C-CH_2- & C-CH_2- & C-CH_2- \\
| & | & | & | \\
CH_3 & CH_3 & CH_3 & CH_3
\end{array}
$$

(c) polyisobutylene

$$
\begin{array}{c}
CH_3 \\
\diagdown \\
C=CH_2 \\
\diagup \\
CH_3
\end{array}
$$

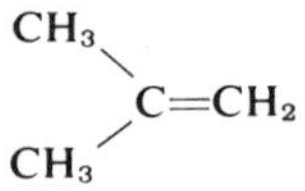

(d) isobutylene

Figure 17.6 Polyethylene (a) has branched side chains. It is an addition polymer of ethylene (b). Polyisobutylene (c) is an addition polymer formed from isobutylene (d).

on high polymers. In fact, the primary action of ionization is to knock an electron or proton away from the polymer, leaving the latter as a free radical. The extra energy imparted to the polymer in this fashion may result in a number of different changes, a few of which are discussed in the next paragraphs.

As a result of both direct and indirect damage, two major changes are found in synthetic high polymers. The first is called *cross-linking*, which means forming bonds between chains. It results in increased molecular weight, increased elastic moduli, increased transparency, and decreased solubility.

Figure 17.7 Free-radical formation during polymerization.

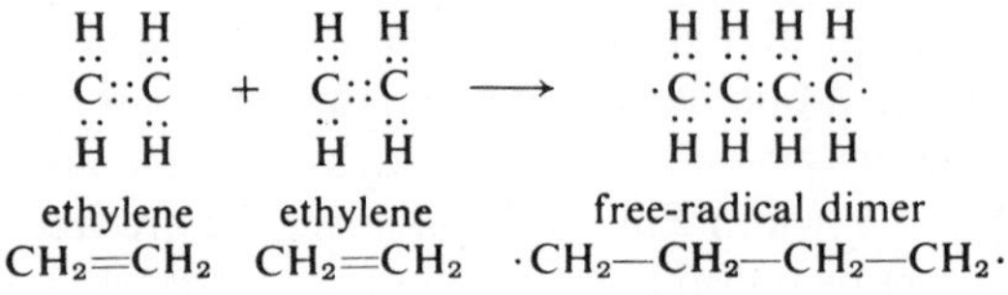

$$
\begin{array}{ccccc}
H\;\;H & & H\;\;H & & H\;H\;H\;H \\
\ddot{C}::\ddot{C} & + & \ddot{C}::\ddot{C} & \longrightarrow & \cdot C:C:C:C\cdot \\
\ddot{H}\;\;\ddot{H} & & \ddot{H}\;\;\ddot{H} & & H\;H\;H\;H
\end{array}
$$

ethylene   ethylene   free-radical dimer
$CH_2{=}CH_2$ $CH_2{=}CH_2$ $\cdot CH_2{-}CH_2{-}CH_2{-}CH_2\cdot$

The other change, called *scission*, consists of breaking bonds along chains. It is characterized by the opposite of the effects described for cross-linking. The results of cross-linking and scission are illustrated in Figs. 17.8 and 17.9, respectively.

In addition to cross-linking and scission, a small molecule such as H_2 or NH_3 is often eliminated. This third effect, *small-molecule elimination*, causes negligible changes in the molecular weight or physical properties of the polymers as compared to the changes due to scission and cross-linking. However, such small changes can completely alter the physiological actions of a protein.

Probably all high polymers undergo both scission and cross-linking when irradiated. However, in some, such as polyethylene (Fig. 17.8), physical

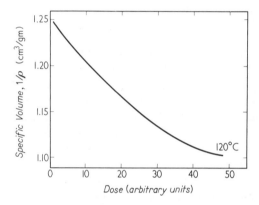

Figure 17.8   Cross-linking in polyethylene. An increase in density is one of the more obvious effects of radiation-induced cross-linking. This change is more pronounced above polyethylene's melting point (105°C).

characteristics are changed in a fashion that indicates that cross-linking is the predominant effect. By contrast, polyisobutylene (Fig. 17.9) shows only the effects associated with scission when it is exposed to ionizing radiation.

As indicated in Chapter 11, oxygen often enhances the large-scale effects of ionizing radiation. Polymer reactions are generally consistent with this picture as well. For example, if polyethylene is bombarded in the absence of oxygen, cross-linking occurs. If oxygen is present during the bombardment, considerably more cross-linking takes place. Likewise, the scission in polyisobutylene is greater if it is exposed to radiation damage in the presence of oxygen. This enhancement of the effects of bombardment is manifold larger than that expected from the ionization of the oxygen molecule by the radiation. The enhancement may result from the reactions of oxygen with free radicals and other less stable forms produced by the ionizations.

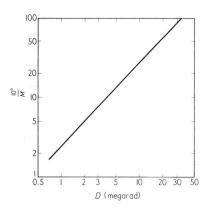

Figure 17.9 Scission in polyisobutylene. A decrease in the molecular weight of polyisobutylene is seen after irradiation.

In a molecule such as polymethylene, which has the same formula as polyethylene but has no side chains, there is no reason to suppose one carbon bond to be weaker than another. In other molecules, such as polyisobutylene, the carbon atoms attached to four other carbon atoms seem to be the weak links in the chain. In more complex polymers, there is usually one weakest bond per monomer. It appears that the extra energy imparted by the ionizing radiation is often carried to this spot. Thus, isooctane breaks preferentially at one particular monomer bond.

Another example of the transport of ionization energy can be found in solutions. Some molecules which fluoresce (see Chapter 27) because of the formation of free radicals will do so when the solution is irradiated, even if the ionization occurs in the solvent at a distance of 5 nm or more away. Similarly, most aromatic-ring compounds tend to stabilize free radicals and can protect polymers. This occurs either if the aromatic compounds are placed in solution with the polymer or if they are incorporated into the polymer, as in polystyrene.

The extra energy imparted by ionization may be transferred within molecules or between molecules. In some cases, this involves charge transfer within or between molecules. In others, less clearly understood, energy transfer occurs without charge transfer.

There are situations in which the extra energy imparted by the ionizing radiation is not transferred. For instance, color changes and electrical resistance changes in polyethylene, both of which are due to the presence of free radicals, may remain for weeks or even months after irradiation, provided that oxygen is excluded. In this case, there is no doubt that the extra energy is not transferred through the polymer. No general rule exists to predict why the energy is transferred in some cases and not in others.

In solutions of the more complex natural polymers, all of the radiation effects discussed in the preceding paragraphs occur. Dried protein films are simpler in that only direct effects are possible. Before discussing these films, we examine target theory in greater detail.

17.5   Target Theory

The basic concepts of target theory are introduced in Chapter 11. There, various models are considered which predict the cell's response in terms of the number of hits required in one or more targets by the particles of the ionizing radiation. From target theory, with suitable assumptions, one can calculate a critical or sensitive volume or cross section within the substance being bombarded. In this section, the physical correlates of this molecular application of target theory are considered.

If a large number (N) of molecules are exposed to a beam, the number (dN) of them destroyed by a small incremental radiation dose (dD) will be proportional to N. This leads to an exponential survival law of a form similar to Eq. (11-2):

$$N = N_0 \exp\left(\frac{-D}{D_A}\right) \qquad (17\text{-}11)$$

where D_A is the dose necessary to reduce N by a factor of e^{-1} and N_0 is the number of molecules present before irradiation. The dosimetric units of D and D_A are unimportant as long as they are the same. In the earlier literature, these were customarily expressed in units called roentgens (R), where a roentgen is defined as that amount of photon irradiation which releases a charge of 0.33 nC in 1 ml of dry air at standard temperature and pressure. However, as noted in Chapter 11, doses now are frequently measured in rem.

If this choice is made, the energy loss, L, per gram of target can be shown to be

$$L = \frac{D_A}{\text{QF}} \times 10^{-5} \text{ J/g} \qquad (17\text{-}12)$$

where QF is the quality factor, discussed in Chapter 11. If one assumes that one ionization (formation of one ion pair) within a critical volume V_C will destroy one target, it follows from the derivation of Eq. (17-12) that

$$V_c = \frac{E \cdot \text{QF}}{D_A \cdot \rho} \qquad (17\text{-}13)$$

where ρ is the density of the target and E is the energy required to form one ion pair. Using a typical value of 32 eV for E, and setting QF and ρ to 1 and 1 g/cm^3, respectively, Eq. (17-13) can be rewritten as

$$V_c = \frac{0.51}{D_A} \qquad (17\text{-}14)$$

where V_c is measured in μm^3.

Linear dimensions are extracted from the value of V_c by assuming a spherical (or other) shape and doing the appropriate geometrical calculation. This sort of model has been used to calculate critical volumes in viruses (see Chapter 15). Consideration must be given to the amount of ionization per unit volume, however, as the actual size of the calculated target depends on this important experimental parameter.

An alternative formulation of Eq. (17-11) makes explicit use of that parameter. First, one notes that the dose, D, is proportional to the volume density of ionization, ρ_I; that is,

$$D = \lambda\rho_I$$

where λ is a proportionality constant. Since D_A corresponds to one ionization per critical volume,

$$D_A = \frac{\lambda}{V_c}$$

Thus, $D/D_A = \rho_I V_c$ and hence

$$N = N_0 \exp\left(-\rho_I V_c\right) \tag{17-11a}$$

While this equation is compact, many experimental measurements yield a critical cross section, S_c, rather than a critical volume, V_c. This S_c can be thought of as the area of a plane through the target within which one "point" ionization would occur. These two parameters are related by the number of ionizations per unit distance along the path of a single particle. This number, the specific ionization, n_I, equals ρ_I divided by the ionized particles per unit area, which in turn, is equal to the time integral over the beam intensity. That is,

$$n_I = \frac{\rho_I}{\int I\, dt} \tag{17-15}$$

Thus

$$S_c = V_c \cdot n_I \tag{17-16}$$

Equation (17-16) indicates that S_c is directly proportional to n_I, with proportionality constant V_c. This is true for small n_I, but as the number of ionizations increases, S_c approaches a limiting value S_0, since the assumption of a single hit per target is no longer valid. Thus, as shown in Fig. 17.10, while the initial linear slope of S_c versus n_I is V_c, the complete curve may be better described by the saturation equation

$$S_c = S_0[1 - \exp\left(-n_I l\right)] \tag{17-17}$$

where l is the length (thickness) of the target being irradiated, that is, V_c/S_0.

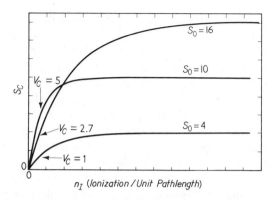

Figure 17.10   Computer-generated predictions of the critical cross section (S_c) as given by Eq. (17-17). The linear region, at relatively small linear ionization (n_I) values, follows from Eq. (17-16) or a power-series expansion of Eq. (17-17). As n_I increases, S_c approaches a saturation or limiting value, S_0.

To recapitulate, at sufficiently low beam intensities one may combine Eq. (17-11a), (17-15), and (17-16), to yield

$$N = N_0 \exp\left(-S_c \cdot \int I\, dt\right) \tag{17-18}$$

This equation is valid for all values of S_c if there exists some physical relationship (e.g., Eq. (17-17)), to describe the variation of cross section, S_c, with linear ionization density. This may be shown as follows. Consider the differential form of Eq. (17-18):

$$\frac{\Delta N}{N} = -S_c\, \Delta\left(\int I\, dt\right) \tag{17-19}$$

If we interpret S_c as an effective molecular cross section for beam interaction and the differential of the integral as the increment in integrated beam intensity, the product of those two gives the probability that an interaction will take place. The price one pays for such a simple relationship is the mathematical complexity of the S_c function. In general, S_c will depend on linear ionization density, beam type, and energy.

17.6   Inactivation of Dried Protein Films

The concepts of target theory are readily applied to studies of the irradiation of dried protein films. When these are subjected to ionizing radiation, the molecules are irreversibly altered. Because no solvent is present, the changes observed are of necessity direct ones in the protein itself. In most of the experiments described in this section, no attempt was made to exclude oxygen or determine its role, if any, in the final molecular alterations.

Protein films are usually tested for molecular changes after redissolving them in suitable media (usually buffered water). The physical and chemical properties of proteins can be measured by more sensitive methods than those used on synthetic polymers. However, in most cases, even though a change can be detected, it is not possible to determine whether cross-linking or scission has occurred within the protein molecule. The polypeptide chains making up the protein specimen may be initially so cross-linked that either scission or additional cross-linking can occur without altering the molecular weight. In other words, protein changes may be detected with a high sensitivity, but the tests yield little knowledge of the intramolecular alterations.

One criterion for physical changes in a protein is its solubility. Another is its isoelectric point, i.e., the pH of the medium at which the protein molecule will not migrate in an electrical field. If the protein is an enzyme, enzymic activity is a very sensitive indication of its physical and chemical condition (see Chapter 18). Finally, some proteins react with specific antibodies; in this case, the protein is an antigen. Its antigenicity may be altered after irradiation. Studies on a large number of dried protein films have shown all of the preceding changes.

The elimination of small molecules observed on irradiating synthetic high polymers can be readily demonstrated on irradiating proteins. When either the amino acid monomers or proteins are exposed to ionizing radiation, a number of small molecules are eliminated. These include NH_3, CO_2, and CO. The elimination of these smaller molecules represents a scission or a breaking of bonds. One amino acid that does not eliminate NH_3, CO_2, or CO during irradiation is cysteine, which contains a sulfhydryl group, —SH. In the absence of an oxidizing agent, irradiated cysteine will eliminate H_2S instead of other small molecules. In the presence of O_2 or other oxidizing agents, two cysteines may be oxidized and then may unite to form one cystine. In proteins, such a high fraction of the free —SH groups are oxidized by ionizing radiation that energy appears to be transferred preferentially to these sulfhydryl groups from other parts of the molecule. (Proteins apparently stabilize the free radicals in the cysteine residues so that little or no H_2S is released.)

It is possible to investigate energy transfer in dried protein films. Such studies are interpreted using the single-hit target theory discussed in Sec. 17.5. If one determines a damage or inactivation vs. dosage curve, a critical volume can be computed according to Eq. (17-11a). If energy transfer may occur throughout the protein molecule, this critical volume should be the entire molecule. It is also conceivable that energy transfer could occur in only part of the molecule, or, at the other extreme, energy transfer may take place between adjacent molecules. These conditions could lead to critical volumes that would be small or large, respectively, as compared to one protein molecule.

All three possibilities—critical volume equal to, smaller than, and larger than one molecule—have been observed by using dried films of different proteins. The case of the critical volume equal to the molecular volume has been observed for the enzymes DNase and invertase, and for many other proteins, as well as for one smaller molecule, penicillin, with a molecular weight of 600 daltons. A plot of typical experimental data is shown in Fig. 17.11. Although energy transfer throughout the molecule is a sufficient con-

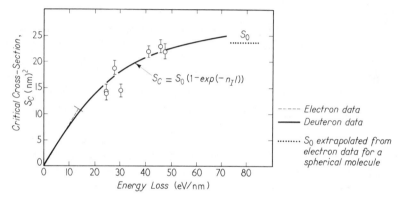

Figure 17.11   Critical cross section and limiting value for DNase. Explicit data are shown for deuteron results only. Rather than the related linear ionization density (n_I), energy loss per unit path length is plotted on the abscissa. After R. Setlow, *Ann. N.Y. Acad. Sci.* **59**:471 (1955).

dition for the equality of the molecular and critical volumes, it is not a necessary condition. It is possible that any of a variety of small changes in various parts of the molecule would all lead to the same conclusion of protein damage. Thus, finding a critical volume which equals a molecular volume suggests that energy transfer may take place throughout the entire molecule but does not prove it.

In contrast, a critical volume smaller than the molecular volume proves that energy transfer cannot occur throughout the entire molecule. This is the case for catalase, which has a critical volume, as determined by tests using enzymic activity, corresponding to one-half its molecular weight. Bovine serum albumin is a more extreme example. When a monolayer of this protein is irradiated, the critical volume determined by antigenic tests corresponds to one tenth of its molecular weight. This implies that severe damage to nine tenths of the molecule can leave the antigenic activity unaltered. Thus, both catalase and bovine serum albumin may be partially damaged without inactivating the remainder of the molecule.

Finally, some protein films are able to transfer excitation energy from one molecule to the next. Insulin films show a critical volume corresponding to a molecular weight of 23,000, whereas the molecular weight of the monomer is 6,000. This indicates that insulin in these films consists of four units, all linked together sufficiently tightly for energy transfer to occur between them. The digestive enzyme, trypsin, is similar in that it has a molecular weight of about 20,000, but its critical volume corresponds to a molecular weight of 34,000. Thus, energy apparently may be transferred from one molecule to the next.

If enzyme–substrate or enzyme–inhibitor complexes (see Chapter 18) are exposed as dried films to ionizing radiation, the enzymes can be inactivated. In these cases, it is found that one ionization anywhere within the complex is sufficient to inactivate the enzyme. Here, energy transfer occurs across the bonds holding the complex together.

Experiments similar to those with protein films can be done with dried viruses. As mentioned in Chapter 15, these have all shown that the critical volume for genetic changes is comparable to the volume indicated by the nucleic acid content. Moreover, they have shown the complexity of the virus, in that any gross criterion such as virus multiplication cannot be conveniently described in terms of Eq. (17-11). To use that equation, one may measure the rate of any genetic change (mutation) whatsoever, or else compare the rate of production of the same mutation (see Chapter 15).

In summary of this section, it should be noted that the inactivations of dried protein and virus films by bombardment with ionizing radiation have indicated several features of the physical properties of proteins and viruses. The similarities between proteins and synthetic high polymers are emphasized by the similarity in the types of radiation damage, whereas the greater complexity of the proteins is emphasized by the greater sensitivity of tests for changes in the proteins. Differences between proteins are demonstrated by the various ranges of energy transfer; in some, the excess energy may be transferred throughout the entire molecule, in others it is restricted to part of the molecule, and in a third group energy transfer could occur between molecules. Finally, the accepted role of nucleic acids in virus multiplication is in accord with the results obtained through the irradiation of virus particles.

17.7   Indirect Effects

The results of exposing protein solutions or intact cells to bombardment by ionizing radiation are more complicated than the direct effects discussed so far in this chapter for water, synthetic polymers, and dried protein films. All of these direct effects occur; however, *indirect effects* on proteins due to the direct formation of free radicals in the water seem more important. This is illustrated by many experiments that have confirmed that damage to both proteins and DNA in solution is much smaller either in the absence of oxygen or in the presence of a protective agent. The latter is more likely to react with the H_2O_2 and free radicals than are the proteins and nucleic acids. Protective agents of this type can also reduce damage due to direct effects of the ionizing radiations, because the extra energy may be transferred to the protective agent from the protein or nucleic acid. Among the most effective compounds of this nature are a group containing both —SH or —S—S— groups and amino groups —NH_2, such as cysteamine and cystamine. These compounds protect not only proteins and nucleic acids, but also synthetic condensation-type polymers such as polymethacrylic acid. Other protective agents with very different structures have also been used.

The net effect of the protective agent will be to reduce the critical volume far below the molecular volume. Various theories have been developed to explain the action of the protective agents on proteins as being due to stabilizing terminal —SH and —S—S— groups. However, because the same agents also protect synthetic polymers that do not contain sulfur, it appears unjustified to focus too much attention on one particular type of bond.

Studies have been conducted on the relative sensitivities of many amino acids, proteins, and nucleic acids to ionizing radiation. These have led to quite lengthy tables, but no one has succeeded in relating this information to protein structure. The relative sensitivities of different proteins can be indicated by a *G* value; this is the number of molecules altered per 100 eV deposited (of a fixed type of irradiation). Some typical values, shown in Table 17-1, indicate their widespread range.

These values indicate that as the complexity of the molecule increases, there is some tendency for the molecule to become less sensitive to ionizing radiation. Thus, adenine is more sensitive by itself than when combined with ribose to form the nucleoside adenosine. It, in turn, is more sensitive than the nucleotide adenosine monophosphate (adenylic acid). Complete nucleic acids are another order of magnitude less sensitive. However, the variation of sensitivity with size is not always observed. Some proteins are almost as sensitive as a typical small molecule, whereas others are far less sensitive.

17.8 Summary

Both charged and uncharged radiation are capable of altering intramolecular forms. In the case of synthetic high polymers, both formation of new bonds and the breaking of old bonds can occur after irradiation. Both of these effects are presumed to occur in biopolymers as well. Small molecules can

TABLE 17-1

G VALUES FOR BIOLOGICALLY ACTIVE MOLECULES[a]

Molecule	G
Protein enzymes	
Yeast alcohol dehydrogenase	3.4
Carboxypeptidase	0.55
Hexokinase	0.033
Catalase	0.009
Nucleic acids	
DNA	0.0039
RNA	0.0072
Nucleotides	
Adenylic acid (AMP)	0.161
Nucleosides	
Adenosine	0.196
Purine	
Adenine	0.676
Other small molecules	
NADH	1.7
Ethyl alcohol	5.9
Coenzyme A	9.2
Glutathione	10.7

[a] *G* value is the number of molecules reacting per 100 eV. All for X irradiation in aerated solutions.

also be eliminated from all types of high polymers. The presence of an aqueous medium can profoundly influence the results of irradiation due to the indirect effects of the generation of H· and OH· free radicals in the solvent. Target theory can be applied to establish sizes of various critical volumes or cross-sectional areas contained within the molecule. Radiation protection agents have been found to decrease the physical size of these target regions.

REFERENCES

1. JOHNS, H. E., AND J. R. CUNNINGHAM, *The Physics of Radiology*, 3rd ed. (Springfield, Ill.: Charles C Thomas, Publisher, 1969), 800 pages.
2. CASARETT, A. P., *Radiation Biology* (Englewood Cliffs, N.J.: Prentice-Hall, Inc., 1968), 368 pages.
3. ANDREWS, H. L., *Radiation Biophysics*, 2nd ed. (Englewood Cliffs, N.J.: Prentice-Hall, Inc., 1974), 314 pages.

CHAPTER 18

Enzyme Kinetics

18.1 Introduction

Modern physics has emphasized the description of the properties of matter and energy in terms of the interactions of fundamental particles. In a similar vein, molecular biology describes living systems in terms of their fundamental particles, molecules. This chapter is a discussion of an application of analytical methods to the dynamic behavior (kinetics) of a class of molecules called *enzymes.* Enzymes are biological catalysts of a primarily protein nature.[1] A *catalyst* is a chemical substance that changes a rate of reaction and, while it may be altered during the reaction, is in the same state after as it was before.

Many uncatalyzed systems exist in nonequilibrium conditions because the reaction rates to reach equilibrium are so slow. A catalyst speeds up the rate of attaining equilibrium but does not in itself alter the equilibrium. For example, glucose and oxygen may be thought of as being in equilibrium with carbon dioxide and water as shown in the equation

$$C_6H_{12}O_6 + 6O_2 \ \rightleftharpoons \ 6CO_2 + 6H_2O$$

[1] The chemical composition and structure of proteins are discussed in Chapter 14 and Appendix D.

However, glucose can remain in solution in the presence of oxygen for years without being significantly altered. Yet in the presence of certain catalysts glucose and oxygen react rapidly to form carbon dioxide and water while releasing energy. This oxidation of glucose is catalyzed by a series of enzymes found in almost all living cells. These enzymes not only promote equilibrium but also convert part of the energy released to forms available for other living processes.

Many biological reactions involve much more complex molecules than glucose. Others, involving smaller, simpler molecules, are easier to discuss. One such reaction, discussed in greater detail in Sec. 18.5, is the dissociation of hydrogen peroxide according to the scheme

$$2H_2O_2 \;\rightleftharpoons\; 2H_2O + O_2$$

Here, equilibrium also strongly favors the right-hand side of the equation. Nonetheless, hydrogen peroxide can be stored for years in a dark bottle in the absence of metallic ions. Many metal ions act as catalysts, accelerating the attainment of equilibrium. None of these is as effective as the protein catalyst *catalase.*

There are also examples of enzyme-catalyzed reactions whose equilibrium does not favor one side of the equation so strongly. A biologically important reaction of this type is the formation and destruction of carbonic acid according to the scheme

$$H_2CO_3 \;\underset{k_2}{\overset{k_1}{\rightleftharpoons}}\; H_2O + CO_2$$

In the absence of a catalyst, this reaction reaches equilibrium in a matter of minutes. However, this is too slow to remove the CO_2 from the bloodstream during its time in the vertebrate gill or lung. The rate of attainment of equilibrium is increased fourfold by an enzyme present in all red blood cells called *carbonic anhydrase.*

Some kinetic principles are briefly reviewed as follows: one principle states that if a molecule is unaffected by others when reacting, the rate of reaction (in units of change in concentration per unit time) is proportional to the first power of the concentration of the reactant. The proportionality constant, for example k_1 in the equation above, is called a *rate constant.* Thus, the rate of dissociation of H_2CO_3 is $k_1[H_2CO_3]$, where the brackets indicate concentration. A second principle, called the *law of mass action*, states that if more than one molecule must meet (collide, associate) to react, and all are free in solution, their rate of reaction must be proportional to the product of all their concentrations. For example, the rate of association of CO_2 and H_2O is $k_2[H_2O][CO_2]$. The k_2 parameter is another rate constant. Thus, the overall rate of change of H_2CO_3 concentration, $d[H_2CO_3]/dt$, is given by

$$\frac{d[H_2CO_3]}{dt} = -k_1[H_2CO_3] + k_2[H_2O][CO_2]$$

Note that k_1 and k_2 have different units; those of k_1 are time^{-1}, whereas those of k_2 are (concentration·time)$^{-1}$.

These kinetic principles assume that the rate constants are invariant. They should be if all conditions of reaction (e.g., temperature, reaction pathway) are the same. The addition of an enzyme alters the reaction so that new rate constants must be defined.

The mechanism of enzymic catalysis (how enzymes work) is one question approached through the study of enzyme kinetics; answering this type of question has proved attractive to biophysicists and physical chemists. Enzyme studies have become part of biophysics for several additional reasons. For instance, many reactions are observed by the use of physical instruments, such as recording spectrophotometers and paramagnetic resonance equipment. These have demanded a certain degree of training in physics as well as skill in electronics for their construction, maintenance, and data interpretation. A quite different reason for including enzyme studies in biophysics is that enzyme kinetics are necessary for a study of enzyme thermodynamics. Physics, to the extent it has any unifying factor, has emphasized the point of view that energy is the most fundamental, most significant quantity. This approach, expressed through thermodynamics as applied to enzyme systems, is presented in Chapters 21 and 22.

18.2   Enzymes

In this section, a minimal outline will be presented of many types of enzymes. These were first studied as a part of biochemical and physiological investigations. This section should be especially useful to readers with comparatively little biochemical background.

A complete classification of types of enzymes, in terms of the reactions they catalyze, can be found in many biochemistry texts. The functional classification described in the following four paragraphs, based on the International Enzyme Commission (EC) recommendations, has proved helpful in talking with biophysics students. Six major classes are described. Most of the classes are named by attaching the suffix *-ase* to a word describing the reaction catalyzed. The same procedure is used for the common name of many individual enzymes. When an enzyme catalyzes more than one type of reaction, or can use several different reactants, it is usually named on the basis of the most physiologically relevant reaction and reactants. The Enzyme Commission also gives each enzyme a longer systematic name, which completely identifies the reaction it catalyzes, and a unique, four-part EC classification number. The first three parts of this number identify the class, subclass, and sub-subclass to which the enzyme belongs, and the last numerical part is assigned in sequence as each new member of a sub-subclass is identified.

The members of class one are the *oxidoreductases*. They catalyze oxidation–reduction reactions and include the respiratory enzymes discussed in Sec. 18.5. For instance, catalase is EC 1.11.1.6 (i.e., the sixth member of the sub-subclass 1.11.1). Oxidoreductases have been studied extensively by spectroscopic methods, because many of them undergo changes in their absorption spectra during reaction. Many also have altered magnetic susceptibility during reactions (see Chapter 26).

Transferases belong to the second class of enzymes. These catalyze the transfer of a piece of one molecule to another. For instance, *transphosphorylases* (EC 2.7) catalyze reactions in which phosphate groups are changed from one molecule to another. If either the acceptor or donor of the phosphate is water (or phosphoric acid), the enzyme is a phosphatase (in class three) rather than a transphosphorylase. Other transferases may be defined in a similar manner; they include *transglycosidases*, *transpeptidases*, *transaminases*, *transmethylases*, and *transacylases*. The reaction catalyzed by each of these is contained in the name.

The third class consists of the *hydrolases*, which split molecules, adding H to one part and OH to the other. The kinetics of the hydrolases are discussed in Sec. 18.3. *Phosphatases* (EC 3.1.3) are hydrolases that catalyze the addition of water to an organic phosphate to form the corresponding alcohol and inorganic phosphate; symbolically, this may be represented as

$$R—O—PO_3^{2-} + H_2O \rightleftharpoons ROH + HPO_4^{2-}$$

In a like manner, *proteolytic* enzymes (EC 3.4) catalyze the splitting of peptide bonds, adding H and OH to the two split parts. Other hydrolases include: *glycosidases* (EC 3.2), which catalyze the splitting of complex sugars to simpler sugars with the addition of water; and *lipases* and *esterases* (EC 3.1.1), both of which catalyze the equilibrium between an ester and its hydrolyzed components. Acetyl cholinesterase (EC 3.1.1.7), introduced in Chapter 5, is an example of the second group.

Lyases make up class four. These are enzymes that catalyze addition of molecules to double bonds in other molecules. Class five, the *isomerases* or *mutases*, change the isomeric form of molecules. Finally, the sixth class is the *ligases*, those enzymes which form bonds between molecules with concurrent ATP cleavage.

Under each of the six major classifications just given, there are large numbers of enzymes. A major activity of biochemistry since about 1930 has been the discovery and characterization of most of these. Almost every reaction that occurs in a living organism, or just outside it, is catalyzed by a specific protein. Complicated reactions involve many enzymes. Specifically, in the oxidation of glucose referred to earlier, more than 50 different enzymes participate in a complicated pathway.

At the start of this chapter, an enzyme was defined as a biological catalyst of a primarily protein nature. The word "primarily" was inserted because many enzymes are covalently linked to carbohydrate or lipid material (i.e., they are glyco- or lipoproteins) or are associated with a small nonprotein molecule. In this latter case, the large, protein portion of the enzyme is referred to as the *apoenzyme*. The smaller part is called the *cofactor, coenzyme,* or *prosthetic group*. No sharp dividing line separates these three types of smaller groups. By and large, the word "cofactor" is used for a loosely bound, small inorganic ion. Coenzyme usually implies a larger group which separates easily from the enzyme, such as nicotinamide adenine dinucleotide (NAD), illustrated in Appendix D. Coenzymes separate and recombine with the various apoenzymes during normal physiological conditions. There may be many different types of apoenzymes for any given coenzyme. Although apoenzymes, like most other proteins, are irreversibly destroyed by prolonged boiling, the coenzymes in general are not so destroyed.

Prosthetic groups are small, nonprotein parts of enzymes which are attached so firmly that they cannot be removed easily without irreversibly altering the enzyme. Among the more frequently studied prosthetic groups are the hemes. A molecule consisting of a heme group and a protein is referred to as a *hemoprotein*. Examples include the carrier protein, hemoglobin, and the respiratory enzymes myoglobin, catalase, peroxidase, and about 20 different cytochromes. The chemical structure of the heme group is presented in Sec. 18.5, along with the reactions of two heme enzymes, catalase, and peroxidase.

It is interesting that the same types of respiratory enzymes (e.g., cytochromes) occur in almost every living cell from yeast to mammals (see Appendix E). A partial exception to this are the anaerobic bacteria. The similarities among the respiratory enzymes of various types of cells are much more impressive than their differences.

18.3   Michaelis–Menten Kinetics

A. HISTORICAL DERIVATION

One of the simplest formal chemical reactions involves the conversion of one molecular substance, S, into a second molecular product, P, with a rate constant, k. This can be symbolized as

$$S \xrightarrow{k} P \tag{18-1a}$$

In an enzyme-catalyzed reaction of this type, S would be called the *substrate*, or compound whose transformation is catalyzed by the enzyme.

The reverse arrow is omitted in this treatment, either because the equilibrium of the reaction greatly favors the product or because we start with a zero product concentration and look only at early time, before significant product accumulates. This also implies that the number and form of products is unimportant, thus increasing the number of enzyme reactions we can include with this model. We can also include reactions of the form

$$S + A \xrightarrow{k'} P \tag{18-1b}$$

if there is a large amount of A present compared to S, such that its concentration can be considered fixed throughout the reaction. In this case the constant k is equal to k' times this invariant concentration of A. For example, all hydrolysis reactions are of this type, where A is water.

Historically, the hydrolyases were the enzymes studied by Michaelis and Menten in 1913. Hydrolyases are an interesting class of enzymes to study because from many points of view they are very simple. Since many are extracellular enzymes, such as those of the digestive tract, it is relatively easy to obtain large quantities of these enzymes in an active form. Furthermore, most of them act independently rather than forming part of a complex chain or pathway.

Most hydrolase reactions have a negligible rate of product formation (or substrate depletion) in the absence of any enzyme. It is usual to speak of such rates as the *velocity*, V, of the reaction. When one adds a sufficient amount of enzyme, the substrate S is hydrolyzed with a measurable velocity, which decreases as time progresses because the S present decreases. If one assumes the simple mechanism shown in Eq. (18-1), the velocity is

$$V = \frac{-d[S]}{dt} = k[S] \tag{18-2}$$

However, under physiological enzyme and substrate concentrations, this

Figure 18.1 Concentration of the substrate S of a hydrolytic reaction as a function of time. This is actually observed.

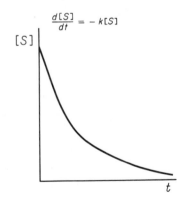

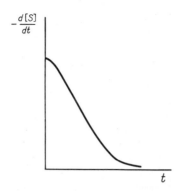

Figure 18.2 Rate of disappearance of S as computed from the curve in Fig. 18.1.

equation does not hold. The solid lines in Figs. 18.1 and 18.2 show what is observed in practice; k is no longer invariant and instead is proportional to the concentration of enzyme, as shown in Figs. 18.3 and 18.4.

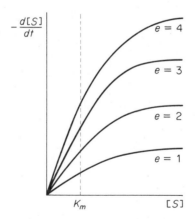

Figure 18.3 Rate at which a substrate S is hydrolyzed as a function of substrate concentration. Curves are plotted for four concentrations of the enzyme E. Note that all four curves reach half-maximum velocity at $[S] = K_M$.

Michaelis and Menten proposed an alternative reaction scheme to account for this change. They assumed that E and S formed an intermediate *enzyme–substrate complex*, $E \cdot S$. This complex was assumed to be in quasi-equilibrium

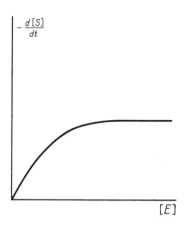

Figure 18.4 Rate at which a substrate S is hydrolyzed as a function of the concentration of the enzyme, E. In general, curves of this type are more difficult to obtain than those in Fig. 18.3.

with E and S, but could also break down to form product(s). The reaction would then follow the scheme

$$\overset{e-p}{E} + \overset{x}{S} \underset{k_2}{\overset{k_1}{\rightleftharpoons}} \overset{p}{E \cdot S}$$

$$\overset{p}{E \cdot S} \overset{k_3}{\longrightarrow} E + products$$

where the letters above the reactants will be used to denote concentrations, thereby avoiding the need for square brackets. Thus, e is the concentration of the total enzyme, p is the concentration of the enzyme bound with the substrate, and x is the substrate concentration. This situation is illustrated diagrammatically in Fig. 18.5. Although the shapes have no physical signific-

Figure 18.5 Diagrammatic representation of a hydrolytic reaction according to the Michaelis–Menten scheme. The shapes chosen here are purely for illustrative purposes and have no physical significance.

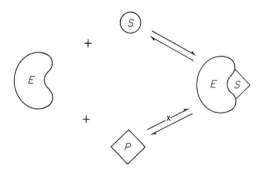

ance, the diagram illustrates the essentially cyclic nature of the enzyme process and also implies our belief that enzyme activity is dependent on the shapes of the reacting molecules. Sometimes enzyme catalysis has been compared to a key fitting into a lock. This is an oversimplification unless shapes are discussed in terms of the atomic configurations and of the electron orbitals in the substrate, intermediate complex and products.

B. DIFFERENTIAL EQUATIONS

The foregoing stoichiometric equations can be rewritten as differential rate equations. These are

$$\frac{dp}{dt} = k_1(e - p)x - k_2 p - k_3 p$$

$$\text{(18-3)}$$

$$\frac{dx}{dt} = -k_1(e - p)x + k_2 p$$

The velocity, V, at which substrate is consumed is

$$V = -\frac{dx}{dt} = k_1(e - p)x - k_2 p \qquad \text{(18-4)}$$

These three equations can agree qualitatively with the empirical observations presented in Figs. 18.1–18.4 provided that suitable values are chosen for the rate constants k_1, k_2, and k_3. Because, in general, V is the only measured quantity that must be fitted by these data, it is perhaps not too surprising that suitable values can be found. If p could be observed directly, this would permit a separation of the constants and add considerable strength to the theory. Normally, the rates of reaction are so rapid that this is not possible, but, as was mentioned in Sec. 18.2, often enzymes can react with alternative substrates. These substrates may form E·S complexes which are slow to dissociate to products; in fact, an intermediate complex of the hydrolase lysozyme with one such substrate has been analyzed using X-ray crystallography (see Fig. 15.8). Kinetic evidence that supports the Michaelis–Menten theory has been found using alternative substrates for several enzymes (see the book by Jencks).

The simultaneous differential equations (18-3) and (18-4) are nonlinear. In general, an analytical solution does not exist for these types of equations with arbitrary initial conditions. The equations can be solved by numerical computation, by analog computation, or by making suitable approximations. Within the current research laboratory, there is usually a continuous systems modeling program available on a digital computer system. Such modeling programs can numerically solve equations of this form and present the results in graphical or tabular form in less time and with less difficulty than any other method.

C. STEADY-STATE APPROXIMATION

The older methods, which solve Eq. (18-4) using a small number of reasonable approximations, have historic value and also aid in an intuitive appreciation of the implications of this equation. First, from Eqs. (18-3) and (18-4),

$$V = \frac{dp}{dt} + k_3 p$$

In all cases, p must be zero at the beginning of the reaction (t equals 0) and will return to zero at the end of the reaction. At the beginning of the reaction, dp/dt, the rate of change of p given by Eq. (18-3), must be positive; at the end it will be negative. Some place in between there will be a maximum value of p, designated p_{max} at which time dp/dt will become zero. At this time, the velocity V will become equal to $k_3 p_{max}$. Indeed, dp/dt can be ignored, or set to zero, whenever it is much smaller than the other terms with which it appears, even if it is not exactly zero. This is termed the *steady-state approximation*.

A second approximation, made when x_0 is very large with respect to e, is that p equals e. Thus,

$$V_{max} \simeq k_3 p \simeq k_3 e \tag{18-5}$$

V_{max} is called the *maximal velocity*.

If x_0 is not very large with respect to e, a few more algebraic manipulations are required. From Eq. (18-3) and the steady-state approximation,

$$0 = \frac{dp}{dt} = k_1 ex - k_1 px - k_2 p - k_3 p$$

$$p = \frac{k_1 ex}{k_1 x + k_2 + k_3} = \frac{ex}{x + (k_2 + k_3)/k_1}$$

If we define K_M, the Michaelis constant, as

$$K_M = \frac{k_2 + k_3}{k_1}$$

then

$$p = \frac{ex}{x + K_M} \tag{18-6}$$

Thus,

$$V = \frac{k_3 ex}{x + K_M} \tag{18-7}$$

and

$$k_3 et = (x_0 - x) + K_M \ln \frac{x_0}{x} \tag{18-8}$$

Figure 18.6 shows a comparison of functions of the values for x, p, and V as computed numerically from Eqs. (18-3) and (18-4) and as plotted from Eqs. (18-6) through (18-8). Note the excellent agreement.

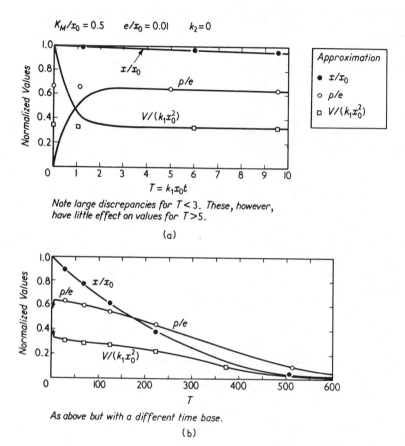

Figure 18.6 Exact and approximate solutions of Michaelis–Menten equations for hydrolase kinetics. Lines show exact values computed with an electronic digital computer. Points labeled show values predicted by the approximation.

Equation (18-7) shows that for x large compared to K_M, V will have the maximum value V_{max} given by Eq. (18-5). For x equal to K_M, V will be one half of V_{max}. This is sometimes used to find K_M. In any case, one may rewrite Eq. (18-7) as

$$V = \frac{V_{max}}{1 + K_M/x} \qquad (18\text{-}7a)$$

This form is particularly useful if the value of e is not known in absolute concentration units. Computer simulation confirms that Eq. (18-7a) agrees well with experimental data.

D. GRAPHICAL METHODS

Equation (18-7a) is not in a suitable form to determine graphically whether the reaction obeys these kinetics. The plot of V vs. x is a hyperbola; a linear plot with slope and intercept simply related to V_{max} and K_M would be more useful. By taking reciprocals in Eq. (18-7a), one obtains

$$\frac{1}{V} = \frac{1}{V_{max}}\left(1 + \frac{K_M}{x}\right) \tag{18-7b}$$

A graph of $1/V$ vs. $1/x$ is a straight line; it is known as a *Lineweaver–Burk plot*. Figure 18.7(a) illustrates in this fashion that sucrase does obey Michaelis–

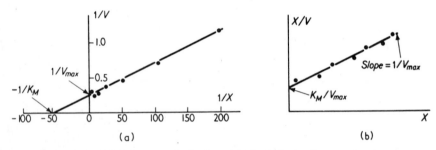

Figure 18.7   (a) Lineweaver–Burk plot of the hydrolysis of sucrose by sucrase. (b) Hanes–Woolf plot. After F. M. Huennekens, "Biological Reactions: Measurement and General Theory," in *Technique of Organic Chemistry*, Vol. 8, *Investigations of Rates and Mechanisms of Reactions*, S. L. Friess and A. Weissberger, eds. (New York: Interscience Publishers, Inc., 1953).

Menten kinetics. A better form is obtained by multiplying both sides of Eq. (18-7b) by x, giving

$$\frac{x}{V} = \frac{1}{V_{max}}(x + K_M) \tag{18-7c}$$

This equation also predicts a straight line; its graph is called a *Hanes–Woolf plot*. It is illustrated in Figure 18.7(b), where x/V is plotted against x. The most accurate points, obtained at large x, are spread out over a major part of the graph instead of being crowded near the axis. There are other possible linear plots (see Chapter 4 in the reference by Segel) and a computer simulation of data for a Michaelis–Menten enzyme, given normally distributed experimental error, has shown the Lineweaver–Burk method to be the least reliable graphical method to calculate K_M and V_{max} (see the reference by

Dowd and Riggs). However, granted the availability of a nonlinear parameter estimation routine on the laboratory computer system, the direct use of Eq. (18-7) seems preferable for quantitative research data.

The strongest argument for the Michaelis–Menten formulation is that it is the simplest theory that can be fitted to the reactions of most hydrolases. It also describes the reactions of several other types of enzymes. Its weakest point is that the intermediate complex has not been directly observed for most reactions; it could easily be an oversimplification. In spite of these uncertainties, this type of kinetics is the basis for many studies. Almost all enzyme kinetics are described in the language of intermediate complexes and Michaelis constants.

18.4 Action of Inhibitors

Many enzyme reactions have been studied in part through the use of inhibitors. Specific inhibitors are useful for determining the role of particular enzymes. Other inhibitors (such as para-chloromercuribenzoate, PCMB) are useful in determining the role of certain groups (e.g., sulfhydryl groups) in the enzyme activity. In this section, the action of inhibitors for systems obeying Michaelis–Menten kinetics will be analyzed.

Many inhibitors react with the enzyme in such a manner that the intermediate complex E·S cannot be formed. Inhibitors of this type are called *competitive inhibitors.* Often, their structure is similar to that of the normal substrate. For example, the enzyme succinic dehydrogenase catalyzes the removal of hydrogen from succinic acid. The enzyme is competitively inhibited by malonic acid. The structural formulas in Fig. 18.8 show the

succinic acid

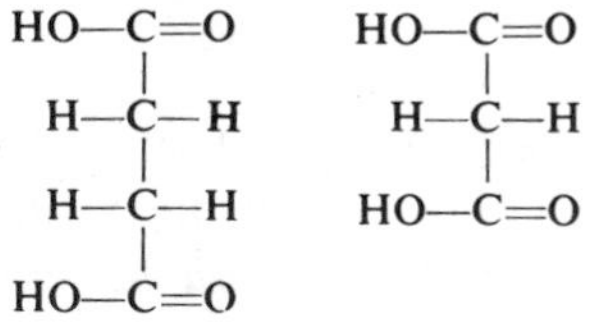

malonic acid

normal substrate

inhibitor

Figure 18.8 Similarity between structures of enzyme substrate and competitive inhibitor.

similarities of the normal substrate and its inhibitor *I*. The reactions involved in competitive inhibition are as follows:

$$E + S \rightleftharpoons ES \longrightarrow E + P$$
$$+$$
$$I$$
$$\Updownarrow$$
$$EI$$
(18-9)

A second group of inhibitors (*noncompetitive inhibitors*) do not interfere with the formation of the intermediate complex but block its hydrolysis or further reaction. The action of heavy metal ions on many enzyme systems is an example of this type of inhibition. The reactions are

$$E + S \rightleftharpoons ES \longrightarrow E + P$$
$$+ \qquad +$$
$$I \qquad I$$
$$\Updownarrow \qquad \Updownarrow$$
$$EI + S \rightleftharpoons EIS$$
(18-10)

The final simple group are *uncompetitive inhibitors* (Eq. (18-11). They react only with the ES complex. Uncompetitive inhibition may be rare in the simple systems we have been discussing, but it is frequently found in multi-reaction pathways.

$$E + S \rightleftharpoons ES \longrightarrow E + P$$
$$+$$
$$I$$
$$\Updownarrow$$
$$EIS$$
(18-11)

Although the inhibition schemes in Eqs. (18-9), (18-10), and (18-11) can all be represented easily using continuous modeling simulation programs, it is pictorially clear that noncompetitive inhibition is the most complicated.

Just as in the uninhibited case, it is instructive to analyze the actions of inhibitors using approximation methods. Only the competitive inhibitors are described in detail in the following. Either the normal substrate S or the inhibitor I may react with the enzyme E, but not both of them. If x, the concentration of S, is much greater than x', the concentration of I, the reaction must proceed as if I were not there. Thus, no matter how large the value of x', by choosing a sufficiently large value of x, it is possible to obtain the uninhibited maximum velocity.

The equilibria in Eq. (18-9) can be summarized by

$$\overset{e-p-p'}{E} + \overset{x'}{I} \underset{k_2'}{\overset{k_1'}{\rightleftharpoons}} \overset{p'}{E \cdot I}$$

$$\overset{e-p-p'}{E} + \overset{x}{S} \underset{k_2}{\overset{k_1}{\rightleftharpoons}} \overset{p}{E \cdot S}$$

$$\overset{p}{E \cdot S} \overset{k_3}{\longrightarrow} E + \text{products}$$

These may be rewritten as differential equations,

$$\frac{dp'}{dt} = k_1'(e - p - p')x' - k_2'p'$$

$$\frac{dx'}{dt} = -k'_1(e - p - p')x' + k'_2p'$$

$$\frac{dp}{dt} = k_1(e - p - p')x - (k_2 + k_3)p$$

$$\frac{dx}{dt} = -k_1(e - p - p')x + k_2p$$

$$V = -\frac{dx}{dt} = \frac{dp}{dt} + k_3p$$

As noted previously, these equations could be solved numerically using digital computer technology. However, the equations above also lend themselves to simplification by using the quasi-steady-state approximations,

$$\frac{dp}{dt} \simeq 0 \quad \text{and} \quad \frac{dp'}{dt} \simeq 0$$

Michaelis constants are defined as

$$K_M = \frac{k_2 + k_3}{k_1} \quad \text{and} \quad K_M' = \frac{k_2'}{k_1'}$$

The approximate equations can be solved to yield

$$p = \frac{ex}{x + K_M + x'(K_M/K_M')}$$

Solving this for V, inverting and multiplying by x, one finds that

$$\frac{x}{V} = \frac{1}{V_{\max}}\left[x + K_M\left(1 + \frac{x'}{K_M'}\right)\right]$$

The Hanes–Woolf plot would now be a series of lines of constant slope, intersecting the x/V axis at points depending on x'/K_M'. The Lineweaver-Burk plot is illustrated in Fig. 18.9.

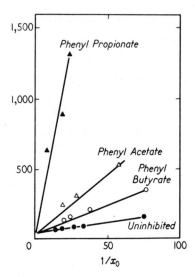

Figure 18.9 Competitive inhibition of the hydrolysis of carboxybenzoxy-glycyl-DL-phenylalanine by carboxypeptidase. The inhibitor concentrations were all 2×10^{-3} M.

The case of the noncompetitive inhibitor is algebraically so complicated that its derivation is left to the interested reader (see the reference by Segal). From Eq. (18-10), however, no matter what the relative concentrations of S and I, any trace of the inhibitor I will slow down the rate of hydrolysis of S. If one is willing to assume that

$$\frac{k_2''}{k_1''} = \frac{k_2 + k_3}{k_1} = K_M \quad \text{and also} \quad \frac{k_2'}{k_1'} = K_M' = \frac{k_2'''}{k_1'''}$$

one can show that

$$\frac{x}{V} = \frac{1}{V_{\max}} (x + K_M)\left(1 + \frac{x'}{K_M'}\right) \tag{18-12}$$

This equation indicates that the maximum velocity obtainable will be less than $V_{\max}$ even if only a trace of the inhibitor x' is present; it is illustrated in Fig. 18.10. Other types of inhibition, mixtures of these three types, have been found and studied. However, it does not seem fruitful to pursue their discussion here.

Inhibitors have been, and are, used widely to study enzymes and investigate enzymatic pathways. In several cases, as in helping unravel the pathways of the utilization of glucose, inhibitors have proved helpful in blocking the process at desired points. It has also been possible to find certain details of the active surface of the enzyme by observing inhibitor action. Of course,

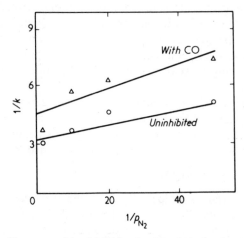

Figure 18.10   Noncompetitive inhibition by CO of N_2 fixation in *Azotabacter*. The data could also be interpreted as indicating that this reaction does not obey simple Michaelis–Menten kinetics. After F. M. Huennekens, "Biological Reactions: Measurement and General Theory," in *Technique of Organic Chemistry*, Vol. 8, *Investigations of Rates and Mechanisms of Reactions*, S. L. Friess and A. Weissberger, eds. (New York: Interscience Publishers, Inc., 1953).

inhibitors have been misinterpreted, especially when they have had more than one action.

In addition to specific substances which slow certain reactions there are others which tend to promote or accelerate specific enzyme-catalyzed reactions. Such activators in the most general sense can be thought of as noncompetitive inhibitors, where the ESI complex can dissociate to give product and, in fact, does so with a faster rate than the ES complex. In addition, studies of pH and ionic strength show that both of these, as well as temperature, can have major effects on the rate of catalysis. However, the kinetic effects of pH, and hence of hydrogen ions, are more complicated than those due to inhibitors of the types reviewed in the preceding paragraphs (see the references by Segal and Jencks).

18.5   Catalase and Peroxidase

The oxidoreductases catalyze the oxidation of many different substrates. Some are directly oxidized in one or two steps, whereas others follow a long pathway with many steps. The enzymes discussed in this section have been selected because of the reversible changes of their absorption spectra which occur during a catalyzed reaction. The changes in the absorption spectra of these enzymes can be related to the concentrations of intermediate compounds similar to E·S, postulated in the last two sections. Quantitative time studies of the absorption spectra during reactions have led to an understanding of the mechanism of action of these enzymes. In this chapter, the mechanism is emphasized, deferring most of the details of absorption spectrophotometry to Chapter 26.

Hydrogen peroxide, a byproduct of certain oxidation reactions, is very toxic to living systems. The first enzyme discussed here, catalase, catalyzes two types of reactions which remove this dangerous compound. The first of these, the *catalatic reaction*, is the destruction of hydrogen peroxide:

$$2H_2O_2 \longrightarrow 2H_2O + O_2$$

This is the oldest known biologically catalyzed reaction; its discovery was responsible for the name "catalase." The second, the *peroxidatic reaction*, is an oxidation of any of a variety of reduced substances. The overall reaction can be represented by

$$H_2O_2 + AH_2 \longrightarrow H_2O + A + H_2O$$

where AH_2 is the reduced substrate (hydrogen donor) and A is its oxidized form. This type of oxidation is catalyzed, with different kinetics, by both catalase and peroxidases.

Catalase occurs in many mammalian cells, including red blood cells and liver cells; it is also found in large amounts in certain bacteria. In one extreme case, 1–2 percent of the dry weight of the bacterial species *Micrococcus luteus* may be catalase. Whatever the source, all catalases contain the same type of prosthetic group, called a *heme*. This is a chelated iron compound containing the tetrapyrrole (porphyrin) ring structure, shown in Fig. 18.11(a). As mentioned in Sec. 18.2, many other enzymes also contain heme. The porphyrin ring also occurs in chlorophyll, discussed in Chapter 20.

The alternating single and double bonds give rise to a *resonance phenomenon*, in which several different structures have the same energy levels. For instance, in Fig. 18.11(b), if one exchanges all the single and the double bonds, and the dotted and solid bonds to Fe, one has an equally possible molecule. Quantum mechanics demands that one think of the molecule as existing not in either the first or second state, but as being partly in both. This leads to absorption bands in the visible and the near-ultraviolet regions of the spectrum. All the heme proteins have one or two absorption bands in the yellow–green region of the spectrum and a very strong absorption band in the blue–violet region. The latter, referred to as the *Soret band*, is shown for catalase in Fig. 18.12.

When catalase catalyzes either the peroxidatic or catalatic reactions, the absorption in the Soret band decreases during the reaction, indicating the formation of an enzyme complex with a different absorption spectrum. This

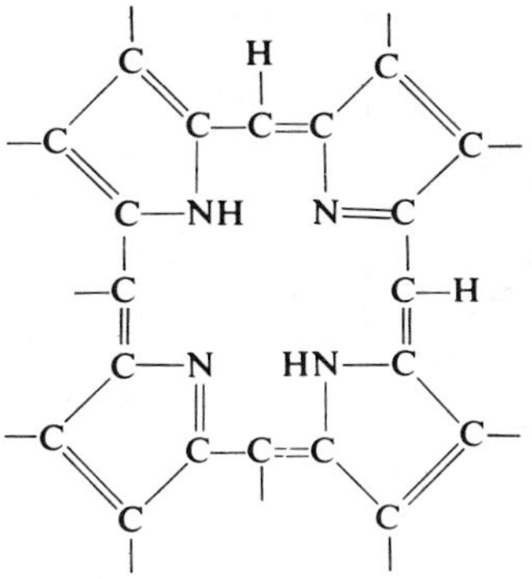

Figure 18.11 (a) Basic porphyrin structure is common to the prosthetic groups of hemoglobin, myoglobin, catalase, peroxidase, and cytochromes. It is also found in chlorophyll. Various groups are attached to the ten "dangling" bonds. (b) Ferroprotoporphyrin IX is the basic group of many of the heme proteins. Hemoglobin, myoglobin, and catalase contain this group. The other heme proteins contain either this porphyrin or ones derived from it by simple substitutions. The iron atom is in the ferrous state in both reduced and oxyhemoglobin and reduced and oxymyoglobin. In peroxidase, it is in the ferric state. Other heme proteins have their iron alternately reduced and oxidized during reactions.

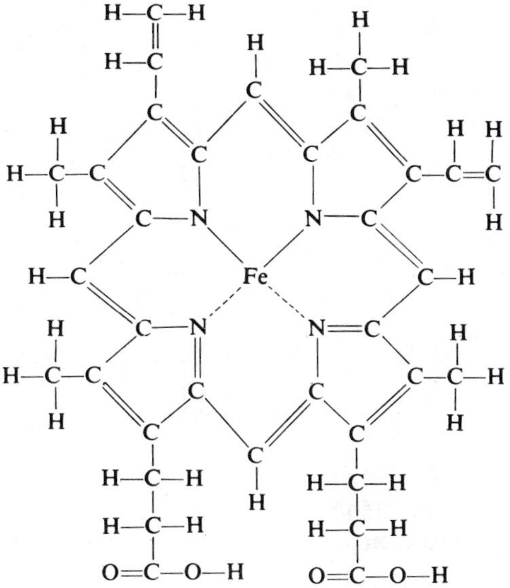

enzyme–peroxide complex (complex I) appears green. On prolonged standing in the presence of an excess of hydrogen peroxide, an entirely different, red complex appears. Both types of complexes are easier to study with alternative substrates, such as methyl hydrogen peroxide, CH_3OOH. Catalase will react with this compound to form both the green and red complexes, but it does not decompose this alkyl peroxide. Spectra for the complexes with methyl hydrogen peroxide are shown in Fig. 18.12.

Peroxidases, like catalases, are heme compounds and show similar spectral changes. They catalyze the peroxidatic but not the catalatic reaction. Peroxidases are widely distributed, being more abundant in plant cells than in animal cells. All can use the respiratory enzyme cytochrome c as a hydrogen donor. Cytochrome c is part of the cytochrome chain, which couples many oxidation chains to molecular oxygen (see Sec. 18.8). Thus, it seems possible

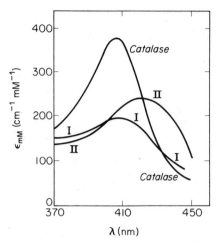

Figure 18.12 The spectra of catalase and its two complexes with methyl hydrogen peroxide. After B. Chance, *J. Biol. Chem.* **179**:1331 (1949).

that the peroxidases might function in normal respiration to use the enzymatically produced peroxides as oxidants. They are presented here as a slightly more complicated system to which the type of reasoning developed by Michaelis and Menten can be directly applied and the deductions tested spectrophotometrically.

Catalase and peroxidase reactions are convenient to study spectrophotometrically for another reason. Besides the possibility for rapid observations afforded by the spectral absorption changes of the enzyme, the peroxide concentrations can also be observed by measuring the absorption in the ultraviolet region. The spectra of peroxides do not show sharp bands but rather a curve that rises steadily as the wavelength is decreased from 300 nm to less than 200 nm. Because many proteins have a minimum in their absorption around 250–230 nm, this wavelength band has been used to measure the peroxide concentration. Finally, in the peroxidatic reactions, it is often possible to observe the oxidation of AH_2 to A in terms of spectrophotometric changes.

Such studies have shown that the peroxidatic reaction of catalase can be represented by the stoichiometric equations

$$\overset{e}{E} + \overset{x}{S} \underset{k_2}{\overset{k_1}{\rightleftharpoons}} \overset{p}{E \cdot S}$$

$$\overset{p}{E \cdot S} + \overset{a}{AH_2} \overset{k_3}{\longrightarrow} E + A + \text{products}$$

where E is catalase, S is peroxide, and AH_2 is the substance being oxidized.

As was done in the case of the hydrolase reactions discussed earlier, one may rewrite these as the differential equations

$$\frac{dp}{dt} = k_1(e - p)x - k_2 p - k_3 ap$$

$$\frac{dx}{dt} = -k_1(e - p)x + k_2 p \qquad\qquad (18\text{-}13)$$

$$\frac{da}{dt} = -k_3 ap$$

The algebra has become slightly more complicated because two substances are used up in the reaction instead of just one.

Peroxidases are even more complicated than catalases in that there are more active complexes formed between the substrate and the enzyme. Whereas only one complex (I) is enzymatically active in catalase, two are active in peroxidase reactions. Symbolically, one may represent the reactions as in the case of a single hydrogen donor such as reduced cytochrome c by the following:

$$\overset{e - p - p'}{E} + \overset{x}{S} \underset{k_2}{\overset{k_1}{\rightleftharpoons}} \overset{p}{E \cdot S_I}$$

$$\overset{p}{E \cdot S_I} + \overset{a}{AH} \underset{k_4}{\overset{k_3}{\rightleftharpoons}} \overset{p'}{E \cdot S_{II}} + A$$

$$\overset{p'}{E \cdot S_{II}} + \overset{a}{AH} \underset{k_6}{\overset{k_5}{\rightleftharpoons}} E + \text{products} + A$$

If we restrict our observations to situations where AH is present in excess and the "back reactions" are unobservable, these reactions may be represented by the following differential equations:

$$\frac{dp}{dt} = k_1(e - p - p')x - k_3 pa$$

$$\frac{dp'}{dt} = k_3 pa - k_5 p'a$$

$$\frac{dx}{dt} = -k_1(e - p - p')x \qquad\qquad (18\text{-}14)$$

$$\frac{da}{dt} = -k_3 pa - k_5 p'a$$

No exact solution in closed form exists for these nonlinear equations. However, it is possible to verify the reaction mechanisms shown in Eqs. (18-13) and (18-14) by a variety of experiments. This both supports the type of reasoning used in Michaelis–Menten kinetics and also suggests that digital computer simulation techniques involving nonlinear parameter estimation are more appropriate here than mathematical analysis.

Equations (18-13) and (18-14) can be simulated using any convenient programming system available. A particular system to allow the expression of the reactions in a form more natural for chemists has been developed by Garfinkel and is referenced at the end of the chapter. This system allows the direct entry of the stoichiometric equations. The computer then essentially develops, albeit in its own internal notation, equation sets such as (18-13) or (18-14). Thereafter, given values for the rate constants, initial concentrations, and times of adding further reactants and inhibitors, the program simulates the time courses of all the reactants, presenting the data in tabular and graphic formats.

In order to select best estimates for the rate constants, an iterative program is needed that simulates the reactants versus time for a variety of trial values for the rate constants. Based on the comparison of these results with the actual data, new sets of values are selected and tried. General computer programs also exist to perform these iterative trials and to select optimal estimates. Considerable skill, mathematical knowledge, and statistical sophistication are needed for the design of such estimation routines.

An added problem occurs as the number of equations in the reaction mechanism increases. It is found that the numeric integration programs tend to run very slowly or to diverge and fail completely. Such sets of equations are called *stiff* by numeric analysts. Intuitively, one can see the basis of this property by looking at Eq. (18-13). After p approaches its maximum, the right-hand side of the first equation in Eq. (18-13) will be almost zero. It represents the difference between the term

$$k_1(e - p)x$$

and the sum

$$k_2p + k_3pa$$

Very small errors in either term will contribute disproportionately to the difference and hence lead to large percentage errors in the derivative dp/dt. These, in turn, will lead to significant errors in p and then in a and x. This growth of errors can be limited by taking very small steps in the numeric integration. This seems wasteful since p, dx/dt, and da/dt are all changing slowly. Nonetheless, the usual techniques of numeric integration, such as the Runge–Kutta method or the various predictor–corrector methods, do not permit more rapid integration. A partial solution to this problem has been described in the reference by Gear.

Using approximation or more exact simulations, one can estimate the rate constants for the peroxidactic reactions of catalase and peroxide. The constant k_2 is very small; an upper limit is about $2 \times 10^{-4}\ sec^{-1}$. For k_1, some typical values are, using H_2O_2 for S, $2 \times 10^7\ (M \cdot sec)^{-1}$ for bacterial catalase, and $0.9 \times 10^7\ (M \cdot sec)^{-1}$ for horseradish peroxidase. If CH_3OOH and C_2H_5OOH are used for S, the value of k_1 is greatly reduced, being $0.9 \times$

10^6 and 1.0×10^4 $(M \cdot \text{sec})^{-1}$ respectively, for bacterial catalase. The values of k_3 for bacterial catalase depend on the hydrogen donor AH_2. Typical values are 171, 91, and 13 $(M \cdot \text{sec})^{-1}$ respectively, for formate, methanol, and ethanol. For horseradish peroxidase, using HNO_2 for AH, one can find that k_3 is 2×10^7 $(M \cdot \text{sec})^{-1}$ and k_5 is 2.4×10^5 $(M \cdot \text{sec})^{-1}$. These numbers are included here only to give the reader a feeling for the values involved.

The reaction of catalase and peroxidase have been discussed in comparative detail in this section. They have been emphasized because they are strong supporting evidence for the existence of intermediate complexes during enzyme reactions. Since it is possible to observe the concentrations of all the intermediates and reactants, and to vary these concentrations, it is possible to check that the reaction does obey the equations chosen. There is no evidence that the hydrolase reaction does not also follow the equations for the peroxidatic reaction of catalase (using H_2O instead of H_2O_2).

The spectra of catalase and peroxidase cannot be directly interpreted at the present time, other than giving the quantitative amounts of the various substances present. For additional information, such as the electronic state of the iron or the existence of other types of intermediates, one must turn to different lines of investigation. Some of these are discussed in Chapter 21 and still others in the sections on magnetic measurements in Chapter 26. Surprisingly, there is no spectral evidence for the existence of a compound $E \cdot S \cdot AH_2$. This is discussed further in Chapter 21.

18.6   Biological Oxidizing Agents

The previous section dealt with the enzymatically catalyzed oxidation of reduced compounds, using a peroxide such as HOOH as the oxygen donor. These reactions are convenient to study; they help to verify the physical existence of transient enzyme–substrate complexes. However, both peroxidase and catalase are believed to be unusual respiratory enzymes, in that the most frequent intracellular oxygen donor is the molecule O_2. The pathway from the reduced compound to the molecular oxygen is often a long one involving many catalyzed steps. Only the last step actually involves molecular oxygen, but many steps along the way are spoken of as oxidations.

In the more general chemical sense, *oxidation* is defined as removal of electrons from a molecule while *reduction* is the addition of electrons. Successive biological oxidations may move electrons in a series of compounds. Thus, in the oxidation of glucose, electrons are transported from the glucose molecule through numerous compounds to the molecular oxygen. This *electron transport* is usually implied by any series of biological oxidations.

Biological oxidations occur within a watery suspending medium, with an appreciable content of free hydrated protons, normally written as H^+. These can attach to an oxidized compound and will tend to do so if the resulting compound is stable. Similarly, if one removes an electron from a biological compound, an H^+ will usually dissociate, to leave the reduced compound neutral. Thus, removal or addition of an electron is equivalent, for biological compounds, to the removal or addition of a hydrogen atom. This fact is used to aid in balancing complicated oxidation–reduction reactions.

All the oxidations within the living cell serve two purposes. The first is to convert the chemical energy of the molecules being oxidized into a form useful to drive intracellular syntheses, muscular contractions, and active transport. The second is to produce heat to maintain the cellular temperature in an optimum range. Because all the energy-conversion processes are less than 100 percent efficient (see Chapter 21), some heat is always a by-product. The warm-blooded animals have internal systems to regulate the efficiency of energy conversion so as to maintain a more-or-less constant internal temperature. Some cold-blooded animals also tend to regulate their internal temperature but must vary their muscular activity to do so.

18.7 Glycolysis and the Citric Acid Cycle

Almost all biological organic molecules can be oxidized by some metabolic system. One ubiquitous molecule used as an energy source and an energy store by most biological cells is a six-carbon sugar (hexose), glucose. The metabolic oxidation of glucose can be conveniently divided into three parts: glycolysis, the citric acid cycle, and oxidative phosphorylation.

The overall results of the oxidation of glucose are its conversion to water and carbon dioxide. A major portion of the energy so released is used to drive the reaction

$$ADP + P_i \longrightarrow ATP$$

(The structure of these compounds is described in Appendix D.) The largest part of this conversion occurs during the oxidative phosphorylation. Because of its bioenergetic importance, oxidative phosphorylation is assigned a separate section in this chapter.

Glycolysis comprises the reactions that convert glucose to lactic acid. Since pyruvate is in equilibrium with lactic acid, it is also included in the glycolytic pathways. These pathways contain a large number of steps, some of which are summarized in Fig. 18.13. Each reaction in each direction is catalyzed by a separate enzyme. These enzymes occur in almost all cells, including human cells and yeast. Glycolysis occurs in the cell; its enzymes are essentially in solution in the cytoplasm.

During glycolysis two molecules of ATP are used, but four are formed. In addition, two molecules of the coenzyme NAD^+ (whose structure is shown in

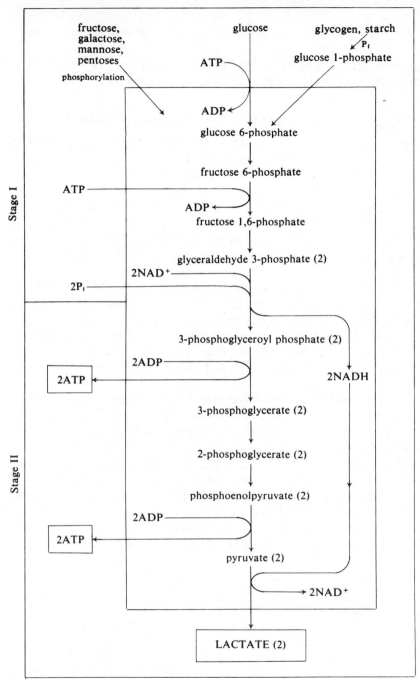

Figure 18.13 The two stages of glycolysis. Stage I: Collection of simple sugars and conversion into glyceraldehyde phosphate; input of priming ATP's. Stage II: Conversion of glyceraldehyde 3-phosphate into lactate, with conservation of energy as ATP. The *net* yield of Stage I plus Stage II is 2ATP per molecule of glucose. The outputs from the system are shown in boxes. The pathway from glucose 6-phosphate is identical in glycolysis and alcoholic fermentation. After A. L. Lehninger, *Biochemistry*, 2nd ed. (New York: Worth, 1975).

Appendix D) are reduced to NADH. Thus, using $\textcircled{P}$ for a high energy phosphate bond as found in ATP, one can summarize the net effect of glycolysis by

$$2H_2PO_4^- + glucose + 2NAD^+ \longrightarrow 2\,pyruvate + 2NADH + 2 \textcircled{P}$$

Although this converts only a small portion of the available energy to $\textcircled{P}$, it is used by anaerobic cells in human muscle and in yeast. (Yeast continue anaerobically to convert the pyruvate to ethanol.)

Provided that oxygen is present, the pyruvate reacts with another coenzyme, CoA, to yield acetyl CoA and carbon dioxide, simultaneously reducing another NAD^+ to NADH. The acetyl CoA then enters a cyclic pathway dependent on enzymes restricted to the mitochondria. This cyclic pathway, known by various names, one being the *citric acid cycle*, is diagrammed in Fig. 18.14.

For each acetyl CoA entering the cycle, two CO_2's are formed. Thus, all the carbons of glucose have been converted to CO_2. (Each glucose molecule oxidized causes two cycles to be completed.) In addition, the citric acid cycle forms one $\textcircled{P}$ (as GTP), reduces one flavoprotein (see Appendix D), and converts three NAD^+'s to NADH. Thus, the energy of the glucose molecule has been used anaerobically to form 4 $\textcircled{P}$'s, and to reduce 6 NAD^+'s and 2 flavoproteins.

The citric acid cycle is also involved in the metabolism of other molecules, including fatty acids and amino acids. Although no oxygen is used up to this point, the cycle stops in the absence of oxygen, owing to the accumulation of reduced flavoprotein. It is also possible for glucose to be metabolized by another chain of somewhat cyclic nature, the pentose phosphate shunt. This is less efficient than glycolysis and the citric acid cycle in that no $\textcircled{P}$ is formed. However, the glucose energy is converted to a different reduced coenzyme, NADPH. The latter, together with NADH and reduced flavoprotein, is oxidized by the oxidative phosphorylation chain. Both this chain and the citric acid enzymes are located in the mitochondria.

18.8 Oxidative Phosphorylation

Most of the energy originally within the glucose molecule is converted to ATP within the oxidative phosphorylation chain. Thus, while 2 $\textcircled{P}$'s per glucose molecule are generated in glycolysis, and 2 $\textcircled{P}$'s more during the two citric cycles per glucose molecule, a total of about 34 $\textcircled{P}$'s are formed during oxidative phosphorylation. Another feature of the oxidative phosphorylation chain is that oxidation and phosphorylation can be uncoupled, so that the 34 represents a maximum for the number of $\textcircled{P}$'s per glucose molecule.

The enzymes of the electron transport chain can be separated into three

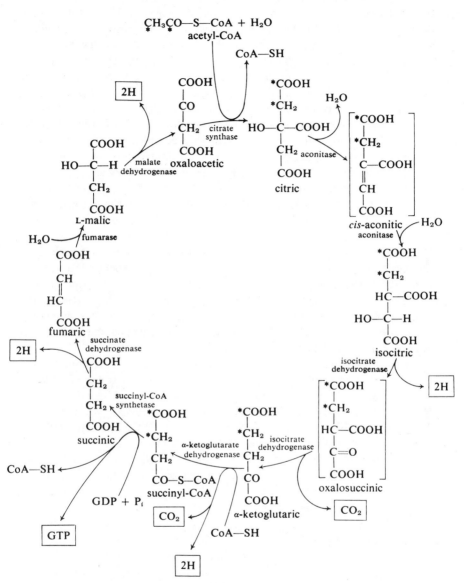

Figure 18.14 The citric acid cycle. The intermediates are shown as their free acids. The end products (two CO_2, four pairs of H atoms, and GTP) are in boxes. The carbon atoms entering as acetyl CoA are starred; their position in the cycle intermediates is given to the stage of succinyl CoA. The succinate formed, because of its symmetry, will yield fumarate, malate, and oxaloacetate, containing carbon from acetyl CoA in equal amounts in all positions, assuming no side reactions. After A. White et al., *Principles of Biochemistry*, 5th ed. (New York: McGraw-Hill Book Company, 1973).

classes. These are the pyridine-linked dehydrogenases, which contain the coenzyme NAD or NADP; the flavin-linked dehydrogenases, which contain the prosthetic group FAD; and the cytochromes, which contain a heme group. The structure of a heme group is shown in Fig. 18-10 and that of NAD and FAD in Appendix D. In intact mitochondria, whether from vertebrates or yeast, these enzymes react as shown in Fig. 18.15.

In the reactions of the chain in intact mitochondria, no spectroscopically observable intermediate complexes have been found. This is similar to the absence of spectrophotometrically detectable complex between the intermediate complex of peroxidase $E \cdot S_{II}$ and reduced cytochrome c. However, the various members of the chain do change spectroscopically from the reduced to the oxidized form.

If the structural integrity of the functional unit within the mitochondrion is not maintained, the reaction chain is altered. Kinetic experiments show that the types and the order of the enzymes involved in oxidation, as well as the active ones, are a function of their relatively fixed positions within the mitochondria. The reactions of the enzymes in intact mitochondria are also qualitatively different from those in mitochondria whose functional groups are disarranged. In damaged mitochondria, the energy liberated by the oxidation is converted to heat. By contrast, in intact mitochondria, the energy of oxidation may be converted to another form of chemical energy through oxidative phosphorylation.

The electron transport chain in intact mitochondria will react slowly in the absence of ADP and Ⓟ. The rate of oxidation is speeded manyfold when oxidative phosphorylation can occur. There are also other chemical substances, such as dinitrophenol, which will accelerate the reaction, although these do not conserve the chemical energy in a form useful to the cell. This indicates that the chain, in the intact mitochondria, can be inhibited at various points. For instance, in the absence of either phosphate or ADP, cytochrome c tends to accumulate in the reduced form and cytochrome a in the oxidized form. This inhibition can be accounted for with various models; however, it is difficult to find a real basis for distinguishing among them.

In order to compare the various models of oxidative phosphorylation with the experimental data, investigators have made considerable use of computer-based simulations. Such studies are often most pleasing to the investigator if they tend to indicate that the selected model is in accord with the data. However, the studies really generate the maximum of innovative information when they prove that a model that was accepted is untenable.

In studying glycolysis, citric acid cycle, and oxidative phosphorylation, the Garfinkel programs referred to earlier have proved to be particularly versatile. They have demonstrated that the views of glycolysis summarized in Fig. 18.13, although the product of intensive biochemical investigation, still encounter quantitative difficulties. These simulations have also been used in oxidative

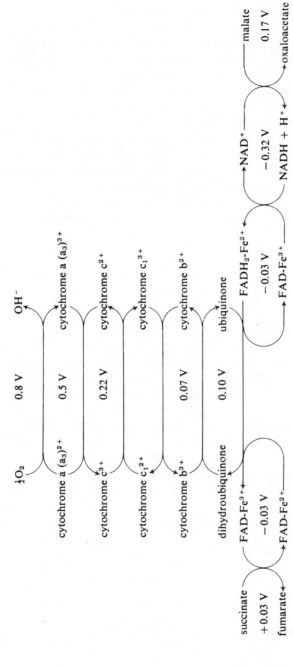

Figure 18.15 Apparent organization and electron flow in the mitochondrial electron-transport chains. The relative positions of cytochrome b and ubiquinone are not certain. In addition, electrons can flow to ubiquinone through a series of electron-transferring flavoproteins from fatty acyl thioesters of COA and from a flavoprotein, glycerol phosphate dehydrogenase.

phosphorylation research but have been less rewarding, in that they have failed to distinguish among several possible models.

For example, oxidative phosphorylation might not follow mass-law kinetics, owing to the steric restriction of the enzymes to the mitochondrial membrane. But such differences have not been demonstrated. Alternatively, phosphorylation might result at separate sites due to pumping charged molecules or electrons across the membrane. This would also allow different parts of the oxidative chain to be located at different places.

One of the limitations of the digital simulation models is the stiffness of the differential equations. This can be avoided to some extent by making approximations that the entire system of reactions is close to steady state. Under these conditions, a unique general set of rules was developed by King and Altman (see the references) for computing reaction rates. The use of this approximation can speed the simulations and aid in testing new models. However, the detailed mechanisms involved in oxidative phosphorylation remain unknown.

18.9 Summary of Enzyme Kinetics

Enzyme-kinetic studies apply to molecular biology the methods of mathematical analysis common to physics and physical chemistry. These strongly reinforce the view that many enzymes catalyze by entering the reaction forming intermediate complexes. In some cases, as with catalase and peroxidase, the intermediates have distinctive spectra which make it possible to follow the details of their formation and destruction. In other cases, as with the hydrolases, the intermediate complexes have been detected by inference from kinetic data. The action of enzyme inhibitors have also been analyzed mathematically. Inhibitors give some indication of the order of reaction of various enzymes in a chain and also of the mechanism involved.

Enzyme kinetics are best described in terms of nonlinear differential equations. Either these must be approximated and linearized, which was the historic approach, or these equations must be solved by numerical analysis. With the growth of digital computer technology, numerical solutions have become the method of choice. Schemes exist so that the differential equations are written from the chemical ones by the computer. Mathematicians have developed special methods to solve the stiff differential equations that describe enzyme kinetics.

Enzymes control the rate of most intracellular processes, such as biological oxidations. In biological oxidations, there are, in general, many steps between the original substance being metabolized (e.g., glucose) and the final products, such as CO_2 and H_2O. The oxidative steps include those which incorporate an atom of oxygen into the molecule, those which remove a hydrogen atom, and those which remove an electron.

Biological oxidations result in the formation of energy-carrier compounds such as ATP, which can move throughout the cell and supply energy to the various life processes, such as syntheses and muscular contraction. Much of the ATP in vertebrate cells and in yeast is formed by the cytochrome chain in the mitochondria. The characteristic spectra of some of the members of the chain have allowed their order of reaction to be determined. However, some of the steps in the synthesis of ATP from ADP are still unknown.

Enzyme kinetics and models of enzyme systems have appealed to biophysicists interested in spectrophotometry, in mathematical models, and in computer simulation. Enzyme kinetics also provides part of the basis for the applications of thermodynamics in biology. This is discussed further in Chapter 21.

REFERENCES

General References

1. WHITE, A., P. HANDLER, AND E. L. SMITH, *Principles of Biochemistry*, 5th ed. (New York: McGraw-Hill Book Company, 1973), 1295 pages. (Chapters 10–17.)
2. LEHNINGER, L., *Biochemistry: The Molecular Basis of Cell Structure and Function*, 2nd ed. (New York: Worth Publishers, Inc., 1975), 1104 pages. (Chapters 8, 9, and 16–19.)

Advanced References

The next three references are especially recommended for the serious student of enzymology. The first two are comprehensive one-volume works; the third is a multivolume series.

3. SEGEL, I. H., *Enzyme Kinetics: Behavior and Analysis of Rapid Equilibrium and Steady-State Enzyme Systems* (New York: John Wiley & Sons, Inc., 1975), 957 pages. (Chapters 1–4.)
4. JENCKS, W. P., *Catalysis in Chemistry and Enzymology* (New York: McGraw-Hill Book Company, 1969), 644 pages.
5. BOYER, P. D., ed., *The Enzymes*, 3rd. ed. (New York: Academic Press, Inc.; Vol. 1, *Structure and Control*, and Vol. 2, *Kinetics and Mechanism*, 1970; Vols. 11–13, on oxidation–reduction enzymes, 1975–1976.)

The next three references discuss mathematical methods useful in enzyme simulations.

6. GEAR, C. W., *Numerical Initial Value Problems in Ordinary Differential Equations* (Englewood Cliffs, N.J.: Prentice-Hall, Inc., 1971), 253 pages.

7. COLQUHOUN, D., *Lectures on Biostatistics* (London: Oxford University Press, 1971) 425 pages.
8. GARFINKEL, D., C. B. MARBACH, AND N. Z. SHAPIRO, "Stiff Differential Equations," *Ann. Rev. Biophys. Bioeng.* **6**:525–542 (1977).

The final group of references are a few representative journal articles.

9. DOWD, J. E., AND D. S. RIGGS, "A Comparison of Estimates of Michaelis–Menten Kinetic Constants from Various Linear Transformations," *J. Biol. Chem.* **240**:863–869 (1965).
10. GARFINKEL, D., "Simulation of the Krebs Cycle and Closely Related Metabolism in Perfused Rat Liver: I. Construction of a Model"; "II. Properties of the Model," *Computers Biomed. Res.* **4**:1–17, 18–42 (1971). (There are also three related articles in this volume co-authored by Garfinkel.)
11. KING, E. L., AND C. ALTMAN, "A Schematic Method of Deriving the Rate Laws for Enzyme-Catalyzed Reactions," *J. Phys. Chem.* **60**:1375–1378 (1956).
12. LAM, C. F., AND D. G. PRIEST, "Enzyme Kinetics, Systematic Generation of Valid King–Altman Patterns," *Biophys. J.* **12**:248–256 (1972).

CHAPTER 19

Molecular Basis of Vision

19.1 Vision and Photopigments

The phenomena of vision form the basis for Chapters 2 and 8. In these chapters, the eye is considered as an optical system which focuses images onto its retina. The retina is shown to act as a transducer converting the light to neural impulses. These, in turn, appear to be sorted and analyzed both within the retina and within the brain to give rise, finally, to the sensation of vision.

Part of the visual sensation consists in recognizing different colors. The color sensed is a function of the wavelength of the incident light; the complexity of this function is emphasized by the experiments described in Chapter 8. Nevertheless, any discrimination of different hues is due to the presence of receptors, which selectively absorb light energy in certain wavelength regions. The photosensitive pigments are altered by the photons absorbed. These changes take place on a molecular level and are far too small to be revealed by a histological method. This particular aspect of vision is, therefore, in the realm of molecular biology.

A variety of physiological and psychophysical experiments indicate that there are at least three pigments within the retina of those mammals with color vision. One of these pigments, *rhodopsin*, is found in the rods. It has been studied in detail; its structure and function form the basis for the next section. Other pigments are found in cones; the three cone pigments observed in humans are described in Sec. 19.3. The latter pigments are less well characterized than rhodopsin, yet offer the best-known explanation for color vision defects, discussed in Sec. 19.4.

The molecular basis for mammalian vision is far from being completely understood; this contrasts sharply with the knowledge of the geometrical optics of the eye. Invertebrate vision has been even less well studied on a molecular level. Although it is evident that many insects possess highly developed color senses which extend far into the ultraviolet, few of the photopigments responsible for insect vision have been isolated.

As stated earlier, molecular biophysics is regarded by some biophysicists as the most fundamental part of biophysics. From this point of view, the most significant aspects of the biophysics of vision are the least well understood.

19.2   Rhodopsin

It had been known for many years that pigments which might be associated with vision existed in the retina. In 1878, Kuhne observed the bleaching of one such pigment, rhodopsin, which he extracted with bile salts. Rhodopsin has also been called visual purple, because of its characteristic color. It is by far the most studied of the photosensitive pigments.

Today, rhodopsin has been purified and studied in many laboratories. In order to extract it, fresh, dark-adapted retinas are mashed in a dim red light. The mashed retinas are then subjected to differential centrifugation (see Chapter 16) until a fairly pure suspension of rods is obtained. The rods are treated with alum, which hardens them by making their proteins insoluble. Then the hardened rods are extracted exhaustively with buffers to remove all water-soluble material, after which they are dried. The rods are next extracted exhaustively with petroleum ether to remove all fat-soluble substances. This leaves insoluble particles that can be suspended only with suitable detergents. The particles containing the pigment rhodopsin are suspended in a 2 percent solution of digitonin, or in bile salts. The entire purification must be carried out in a deep red light in order to have an appreciable yield of rhodopsin.

Is the rhodopsin one ends up with anything like the original? Tests show that it has almost the same absorption spectrum, although at physiological pH the peak is shifted several nanometers (see Chapter 26). This similarity is taken as evidence that the purified rhodopsin is essentially the same as the initial rhodopsin.

Rhodopsin, with a molecular weight of approximately 34,000, is a conjugated complex of a particular protein, *opsin*, with a much smaller hydrocarbon chromophore (color producer), *retinal.* A covalently bound sugar oligosaccharide and a noncovalently bound lipid also form part of the complex.

If the retina extraction, mentioned previously, is performed in a bright light, no rhodopsin is present in the final suspension; only the protein, opsin, remains. Retinal is an aldehyde of the alcohol vitamin A_1 (retinol), shown in Fig. 19.1. The carbon atoms in the ring and along the chain are numbered for

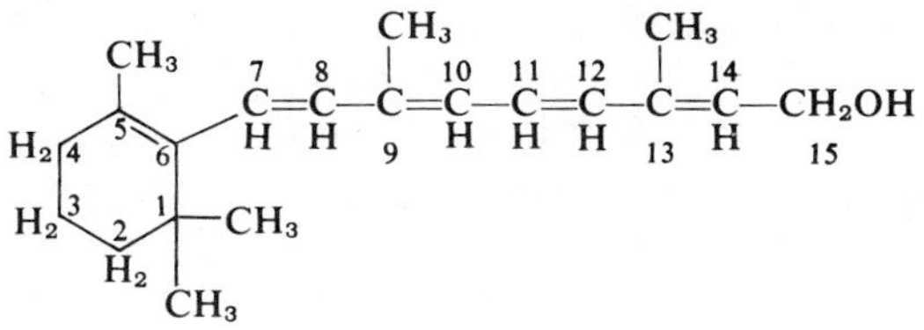

Figure 19.1 Structural formula of vitamin A_1 (retinol).

convenient reference. The only difference between retinol and retinal is the oxidation of the alcohol at C-15 (—CH_2OH) to an aldehyde (—CHO). Pigments with structures such as these are called *carotenoids,* because many early ones were found in carrots.

For every double bond along the vitamin A_1 chain, there are two spatial isomers. If one of the atoms on each carbon is hydrogen, one can represent the *cis* isomer as

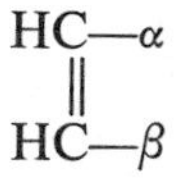

$$HC—\alpha$$
$$\|$$
$$HC—\beta$$

and similarly the *trans* isomer as

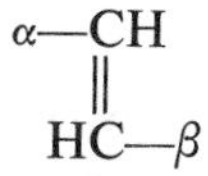

$$\alpha—CH$$
$$\|$$
$$HC—\beta$$

It is often possible to distinguish between *cis* and *trans* isomers by either chemical or X-ray techniques. However, sometimes this distinction is difficult to demonstrate. In the case of vitamin A_1 and retinal, there are 2^4, or 16, different isomers which can be drawn with pencil and paper. Only one of these retinal isomers is physiologically active.

Essentially, one may regard the visual process as the splitting of retinal from opsin and then the resynthesis of rhodopsin, as shown in Fig. 19.2. Unfortunately, this is a gross oversimplification, for if the usual form of vitamin A is converted to retinal, it does not react with opsin at all. This is because the common vitamin A is the all-*trans* isomer. If retinal is split from rhodopsin, it is also the all-*trans* isomer and will not recombine. Chemical

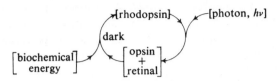

Figure 19.2   Simplified version of the visual cycle of rhodopsin.

methods can show that the active retinal is a mono-*cis* compound but cannot distinguish between the four possible isomers having the *cis* configuration at the 7-8, 9-10, 11-12, and 13-14 positions, respectively.   On the basis of a study of the actual dimensional configurations, Pauling predicted that the only four stable isomers should be the all-*trans*, 9-*cis*, 13-*cis*, and 9,13 di-*cis*. Of these, only the 9-*cis* isomer reacts with opsin, and, although it forms a photosensitive pigment, it does not form rhodopsin. Thus, retinal cannot be one of the more probable forms.

The other mono-*cis* isomers of retinal, 7-8 and 11-12, are hindered forms. This means that the interaction between different parts of the molecule twists the long side chain and the ring, distorting the normal planar form. Of these two hindered molecules, the isomer with the *cis* bond at the 11-12 position is less hindered and does indeed serve as the active form.

The plane projection of the 11-12 *cis* compound is illustrated in Fig. 19.3.

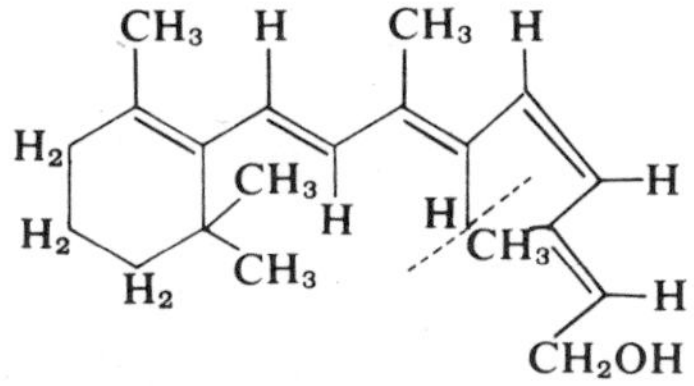

Figure 19.3   The 11-12 mono-*cis* isomer of vitamin A_1. The plane projection is distorted so that it does not show the steric hindrance. This is indicated by the dashed line between the methyl group at 13 and the hydrogen attached to position 10.

This compound possesses a large steric hindrance between 13-CH_3 and the 10-H. The conversion from the all-*trans* form of vitamin A aldehyde to 11-*cis* retinal occurs in the presence of iodine and light. It also occurs within the retina, where the conversion is enzymatically catalyzed.

Another enzyme within the retina maintains an equilibrium ratio between the aldehyde retinal and the alcohol retinol (vitamin A). This equilibrium is in the direction of much greater concentration of the alcohol. If, however, the aldehyde concentration is sufficiently low, the enzyme catalyzes the aldehyde production from the alcohol. Little is known about this oxidoreductase (an enzyme that oxidizes and reduces other compounds; see Chapter 18), except that it apparently requires a coenzyme, reduced nicotinamide adenine dinucleotide phosphate (NADPH), to donate hydrogens (or, in its oxidized form, $NADP^+$, to accept them).

The vitamin A produced by the action of the oxidoreductase and NADPH can pass through the rod membrane and into the blood. The vitamin A in the retina is in equilibrium with that in the blood under steady-state light conditions. The vitamin A in the blood is, in turn, maintained at a more-or-less constant concentration by the liver, which stores any excess.

The overall process is then a complicated cycle, which is shown in diagrammatic form in Fig. 19.4. In the dark, the cycle is stopped by all the opsin

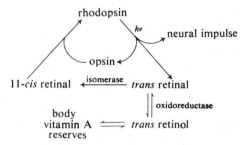

Figure 19.4   Visual cycle of rhodopsin in the retina.

being bound in the form of rhodopsin. In the light, an equilibrium must be established with a steady-state concentration of rhodopsin. The concentration may be very close to zero, but a small part of the opsin should always exist in the form of rhodopsin.

From the point of view of the oxidoreductase, one may regard the opsin as a trapping reagent which effectively shifts the equilibrium so that the alcohol (vitamin A) is converted to the aldehyde (retinal). The reaction of retinal with opsin is exothermic and goes spontaneously. From the point of view of the retinal, the opsin acts as an enzyme, converting the retinal with the help of the photon *hv* from the less probable *cis* form to the more probable all-*trans* form. Opsin will cause this transformation, even when the much more light-stable 9-*cis* or 9,13 di-*cis* retinals are bound to it. One may thus describe the protein opsin as a photoisomerase.

An additional feature of this reaction is that it tends to stabilize the reactants. Opsin and retinal are both relatively unstable. For instance, by changing the pH of an opsin solution from a neutral 7.0 to either 5.0 or 8.0, opsin is 50 percent irreversibly altered (denatured) in 1 hr. The *cis* retinal is relatively easy to isomerize to the all-*trans* form and is primarily converted to the alcohol (vitamin) form. In contrast to opsin and retinal, the compound rhodopsin presents remarkable stability, as indicated by the extraction procedure. It is stable over the pH range 3.9–9.6. It is easy to imagine that in vitamin A deficiency, the opsin might quickly degenerate. (Indeed, in this deficiency, rods do show rapid degeneration.) On addition of vitamin A to the diet, the opsin formed would be stabilized as rhodopsin. Hence, the rods (and cones?) could rebuild. This suggests that opsin is a type of adaptive enzyme. This mutual stabilizing may be typical of other adaptive enzymes.

Under certain conditions, the step from rhodopsin to *trans* retinal and opsin may be stopped at several intermediate points. The intermediate compounds identified in this manner in vertebrates are related as shown in Fig. 19.5. The numbers on top of the arrows are the very approximate half-times in seconds for the various reactions.

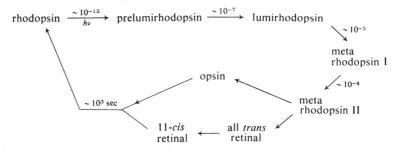

Figure 19.5 Intermediates in the transition from vertebrate rhodopsin to *trans* retinal and opsin. The small numbers above the arrows are the very approximate half-times of the reactions in seconds. After T. G. Ebrey and G. Honig, "Molecular Aspects of Photoreceptor Function," *Quart. Rev. Biophys.* **8**(2):140, 1975 (Cambridge Univ. Press, England).

The nature of the bond between the carotenoid retinal and the protein opsin has been investigated. The aldehyde group in retinal was shown to condense with the amino group of a lysine residue in opsin to form what is called a *Schiff's base*:

$$
\begin{array}{ccc}
\text{H} & & \text{H} \\
| & & | \\
R\!-\!C\!=\!O + H_2N\!-\!R' & \longrightarrow & R\!-\!C\!=\!N\!-\!R'
\end{array}
$$

The main spectral absorption band in free retinal is shifted to longer wavelengths when it is bound to opsin; this can be explained if the Schiff's base is protonated, to $R\!-\!CH\!=\!N^+H\!-\!R'$. This would allow a greater degree of delocalization in the electron system than is present in free retinal or in an unprotonated Schiff's base derivative.

There must be secondary points of attachment; it would otherwise be hard to explain why only the 11-*cis* isomer serves as the chromophore in visual pigments. The detailed reaction of the two compounds, still under investigation, may be implicated in the origin of the early receptor potential in vision (see Sec. 19.5).

19.3   Other Photopigments

Rods in all organisms contain rhodopsin, or a very similar pigment such as porphyropsin. Porphyropsin has a different aldehyde, 3,4-dehydroretinal, in place of retinal. 3,4-Dehydroretinal is the aldehyde of vitamin A_2, which differs from A_1 in having an extra double bond in the ring between 3-C and 4-C (see Fig. 19.1).

Cone pigments are more difficult to isolate. The best known cone pigment, iodopsin, was isolated from birds by Wald in 1937. An examination of iodopsin shows that it is very similar to rhodopsin. Iodopsin is a conjugated compound consisting of retinal and another protein, cone opsin or photopsin. (The rod protein is referred to in various places as rod opsin or scotopsin.) The rod opsin and the cone opsin can be extracted from a wide variety of animals, including most vertebrates. Although corresponding proteins from different organisms are distinguishable, they are very similar in structure and function.

Birds and reptiles have mostly cones in their retinas, and probably lack rods. The cones contain bright-colored oil globules between the inner and outer cone segments. These oils are all carotenoids. Wald extracted 1,600 bird eyes to isolate the oils shown in Table 19-1. Color vision is possible with just these carotenoid filters and iodopsin.

TABLE 19-1
AVIAN CAROTENOIDS

Carotenoid	Absorption maximum (nm)	Color
Astacin	497	Purplish red
Xanthophyll	463	Golden yellow
Galloxanthine	450	Greenish yellow

The three cone pigments responsible for human red, green, and blue color vision—erythrolabe, chlorolabe, and cyanolabe, respectively—had long been predicted, but were not characterized until the late 1950's. Rushton and colleagues used instruments known as reflection densitometers to measure the intensity of light reflected by the pigments of living eyes. Reflection densitometry of normal and color-blind subjects (see Sec. 19.4) allowed them to identify one green-sensitive pigment (chlorolabe) and one red-yellow-sensitive pigment (erythrolabe). Their experiments, although not set up to detect blue-sensitive pigments, could not rule out the possibility that some (or all) cones contained a mixture of pigments. In the 1960's Marks, MacNichol, and co-workers overcame severe technical difficulties to investigate the spectra of single-cone cells. Representative spectra for the three types of cones observed are shown in Fig. 19.6.

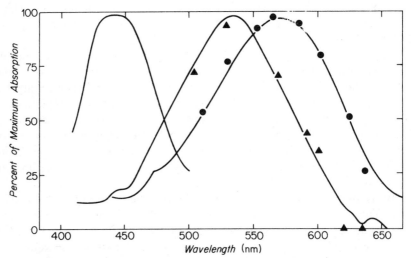

Figure 19.6   Three cone pigments of normal color vision absorb lights of different wavelengths as plotted here. The curves are the average spectral absorbance from single cones in excised eyes of humans or monkeys scaled to the same maximums. The measurements were made by E. F. MacNichol, Jr., and his colleagues at Johns Hopkins University. The triangles and circles represent green-sensitive and red-sensitive pigments in cones, respectively, as they were measured in the living human eye by H. D. Baker and W. A. H. Rushton at the University of Cambridge. The coincidence of the two sets of measurements demonstrates that single cones contain single pigments. From W. A. H. Rushton, "Visual Pigments and Color Blindness." Copyright © 1975 by Scientific American, Inc. (232(3):68). All rights reserved.

19.4   Color Vision Defects

In the past, one of the major factors considered in testing any theory of color vision was its ability to account for various types of color defects. People whose color vision is normal are called *trichromats*, since they need three colored lights to match all hues. There are some anomalous trichromats, who need abnormal proportions of red and green in their mixtures. *Protanomalous* trichromats require too much red; *deuteranomalous* trichromats require too much green.

Those persons needing only two colored lights to match hues are called *dichromats*. Four different types of dichromasy are known. Persons with two of these distinguish only blue and yellow. In this category, the *protanopes* identify red and blue–green colors as gray and have low luminosity sensitivity in the red. In contrast, the *deuteranopes* have a normal luminous sensitivity in the red, but identify greens and purple–reds as gray. The other two types of dichromats distinguish red and green but not blue. In this category, the *tritanopes* see purplish blue and greenish yellow as gray, whereas the *tetranopes* see all blue and yellow as gray.

Finally, those with no color distinction are called *monochromats*. Several types of monochromasy exist. In one type, called *cone blindness*, only rods are present in the retina. This type of monochromat shows a loss of acuity; they retain the scotopic luminosity curve only. This strongly supports the connection between the rods and the scotopic vision.

As mentioned in Sec. 19.3, Rushton and co-workers, using reflection densitometry, examined the eyes of persons with color vision defects. In the normal fovea (to exclude rod contributions) the densitometer can detect two cone pigments, erythrolabe and chlorolabe, but in the fovea of a dichromat, it detects only one. Protanopes have only chlorolabe, and deuteranopes have only erythrolabe in the red–green range. These single pigments are found to be identical to those found in the normal subjects.

As shown in Table 19-2, the majority of persons with defective color vision are not dichromats, but rather anomalous trichromats. It can be shown that these persons must have two pigments in the red–green range, yet at least one must be defective. Reflection densitometry cannot demonstrate this, however. For example, protanomalous trichromats, like protanopes, appear to have only chlorolabe. Apparently their anomalous erythrolabe, called protolabe, must have a spectrum very similar to chlorolabe. Likewise deuterolabe, the anomalous pigment of deuteranomalous trichromats, cannot be distinguished from erythrolabe.

Through a technique involving color matches of mixtures of polarized red and green light, Rushton and co-workers were able to characterize these anomalous pigments. Although these techniques have not yet been continued to the rarer tritanopes and tetranopes, who presumably have defective cyanolabe, it is very encouraging that in the course of not more than 20 years, questions on the nature of most defective color vision could be answered so definitively on a molecular level.

TABLE 19-2

PERCENTAGE OF HUMAN POPULATION
WITH COLOR DEFECTIVE VISION

Type of color defect	Percentage
Anomalous trichromatism	
Protanomaly	1.0
Deuteranomaly	4.6
Tritanomaly	0.0001
	5.6
Dichromatism	
Protanopia	1.2
Deuteranopia	1.4
Tritanopia	0.0001
	2.6
Monochromatism	0.003
Total	8.2

Source: From C. H. Best and N. B.
Taylor, *The Physiological Basis of Medical
Practice,* 7th ed. (Baltimore: Williams &
Wilkins, C., 1961). Originally from W. D.
Wright *J. Ophth. Soc. Am.* **42**:509 (1952).

19.5   Origin of the Early Receptor Potential

Many experiments indicate that one photon dissociates one retinal group from one rod opsin molecule. It is unknown how one or two such dissociations can give rise to the first neural spike or discharge in the visual pathway, the early receptor potential (see Chapter 8). This is one of the most fundamental questions that may be asked about the molecular changes in vision. One suggestion is that it is related to the retinal binding site on opsin. The binding site for retinal on opsin is a hydrophobic, apolar liquid environment. Yet within 100 μsec after light is absorbed by rhodopsin, before the formation of meta-rhodopsin II, the retinal has been transferred to a hydrophilic, polar, protein environment. This change in environment, and the charge separation that occurs with it, may be the critical step that produces the early receptor potential.

Also implicated in the production of the early receptor potential is a transmitter substance which acts upon the permeability channels of the plasma membrane. Rather than rhodopsin itself triggering the change, it would cause release of this substance (perhaps calcium ion), which would then diffuse to the permeability sites and interact with them to decrease the ionic permeability and generate the potential.

The information necessary to choose between these alternatives or devise others more consistent with the facts is missing. X-ray crystallography could be of benefit, but rhodopsin has yet to be crystallized. Much also could be learned by measuring the potential of intact single photoreceptors, yet their very small size makes these measurements extremely difficult. Thus, in common with hearing, olfaction, and taste, it is impossible to state precisely the method by which the neural impulse is started.

19.6 Summary

The molecular events in vision are still not completely known, but biophysicists and photochemists are approaching this goal. The photopigment responsible for rod vision, rhodopsin, has been extensively characterized. It is a very sensitive photopigment; indeed, bacteriorhodopsin, a closely related compound, is used as the primary light acceptor for photosynthesis in certain bacteria (see Chapter 20).

The photopigments responsible for color vision in humans have not been as well characterized. They have, however, provided a reasonable hypothesis to explain the most common types of color blindness. Finally, the most challenging problem remaining in the molecular aspects of vision is discovering the mechanism by which light energy captured by photopigments is transduced into an electrical signal.

REFERENCES

The general references listed in Chapter 8 all have sections or chapters dealing with photopigments and the molecular basis of vision. The following are more specific:

For a history of the earlier work by a pioneer in the field:

1. WALD, G., "The Biochemistry of Visual Excitation," in *Enzymes: Units of Biological Structure & Function*, O. H. Gaebler, ed. (New York: Academic Press, Inc., 1956), pp. 355–367.

The next three *Scientific American* articles span a decade of molecular photobiology:

2. MACNICHOL, E. F., JR., "Three-Pigment Color Vision," **211** (December 1964): 48–56.

3. HUBBARD, R., AND A. KROPF, "Molecular Isomers in Vision," **216** (June 1967): 64–76.

4. RUSHTON, W. A. H., "Visual Pigments and Color Blindness," **232** (March 1975):64–74.

The following two articles are at a higher level. The former is concerned with the binding site for retinal on opsin, the latter is a review with over 160 references on selected aspects of the molecular photobiology of vision; the journal it appears in is well worth examining by all serious students of biophysics.

5. KIMBEL, R. L., R. P. POINCELOT, AND E. W. ABRAHAMSON, "Chromophore Transfer from Lipid to Protein in Bovine Rhodopsin," *Biochemistry* **9**:1817–1820 (1970).

6. EBREY, T. G., AND B. HONIG, "Molecular Aspects of Photoreceptor Function," *Quart. Rev. Biophys.* **8**:129–184 (1975).

CHAPTER 20

Photosynthesis

20.1   Introduction

The surface of the earth continually receives radiant energy from the sun. This may be dissipated as heat or used to drive the syntheses of new molecules. These new molecules, in turn, can serve as sources of energy for later reactions and syntheses. The primary synthesis of new compounds driven by radiant energy is called *photosynthesis*. It is catalyzed by colored pigments, found in many plants. Photosynthesis occurs in all green plants, including all of the higher plants and some of the algae. In addition, many other protists carry out photosynthesis. The blue–green algae and bacteria of a variety of colors all photosynthesize. There is also a genus of mobile protists, called *Euglena*, which contain a green-pigmented organelle capable of catalyzing photosynthesis.

All living processes other than photosynthesis involve the degradation of chemical energy to heat energy. Eventually, all sources of chemical energy would be consumed and life on earth would stop if photosynthesis did not occur. This process of building up of the chemical energy available to living organisms is continuously driven by the sun. It can also be described in terms of entropy or information, as discussed in Chapters 21 and 25.

Photosynthesis is necessary for life on earth for another quite different reason. The entire chemistry of the surface of the earth has a net reducing property which, in the absence of photosynthesis, would blind all oxygen in the form of oxides. If this happened, protoplasm as we known it, which depends on oxidations to use chemical energy, would not be possible. However, photosynthesis produces sufficient molecular oxygen to control the oxygen in our atmosphere, raising it to an equilibrium value of about 20 percent. Thus, photosynthesis is necessary for living organisms, as they exist on the surface of the earth, both in supplying the necessary energy-rich organic compounds and also in producing the oxygen necessary to use the energy in these compounds.

The overall reaction occurring for photosynthesis in all organisms except bacteria is the fixation of CO_2 and water to form a sugar and molecular oxygen. This may be written symbolically as

$$6CO_2 + 6H_2O + 48h\nu \longrightarrow C_6H_{12}O_6 + 6O_2 \qquad (20\text{-}1)$$

In this formula, $h\nu$ represents a photon of visible light, and $C_6H_{12}O_6$, a hexose sugar.

Bacteria neither use water nor produce molecular oxygen in their photosyntheses. Instead, various species use compounds such as hydrogen sulfide (and produce sulfur) or isopropanol (and produce acetone). Van Niel, an early pioneer in photosynthesis, predicted that plant photosynthesis and bacterial photosynthesis are essentially similar:

$$\text{light} + 2H_2D + CO_2 \longrightarrow (CH_2O) + H_2O + 2D \qquad (20\text{-}2)$$

in which H_2D is a hydrogen donor.

This reaction can be generalized still further, for while CO_2 is the major hydrogen acceptor, other compounds, such as nitrogen or hydrogen ions, may serve that function as well. Thus, the completely general reaction for photosynthesis is

$$\text{light} + H_2D + A \longrightarrow H_2A + D \qquad (20\text{-}3)$$

Light energy will force the reduction of more electronegative compounds. The energy stored in these compounds can produce useful work for the organism.

Equation (20-3) is, of course, deceptive in its simplicity. Many steps and subprocesses occur at the molecular level for each acceptor–donor pair. Because of its importance, and because most work has been done on it, the remainder of this chapter will restrict discussion to Eq. (20-1), although it is expected that the same general steps do occur in the other variants as well. Research in photosynthesis has moved rapidly forward since about 1940, and there is no indication that the progress has stopped. Increasingly, this research has involved the tools and ideas of the biophysicist.

20.2   A Little Plant Histology

All green plants and *Euglena* contain organelles called *chloroplasts*. The chloroplasts can be removed from the cells by suitable fractionation procedures and, when resuspended in media containing the necessary additives, will catalyze photosynthesis at rates comparable to those in the intact plants. The chloroplasts contain the pigments primarily responsible for the green color. A wide variety of experiments indicate that photosynthesis in green plants and protists occurs only in the chloroplasts.

The size and shape of the chloroplasts vary quite widely. The most studied organism are two genuses of one-celled green algae called *Chlorella* and *Scenedesmus*. A diagram of a cross section through a *Chlorella* cell is shown in Fig. 20.1(a). *Chlorella* has only one cup-shaped chloroplast per cell. It differs in this respect from many other algae and most higher plants, all of which have many chloroplasts per cell. The chloroplasts of higher plants are shaped like a saucer with a diameter of 4–6 μm and a thickness of 0.5–1.0 μm.

Figure 20.1   (a) Three *Chlorella* cells. This diagram emphasizes that the single cup-shaped chloroplast occupies most of the cell. The pyrenoid is associated with starch and/or protein synthesis and/or storage. (b) Corn chloroplast. Sketch of an electron micrograph of a chloroplast from *Zea mays*. The dark regions are the grana. They are cylinders about 400 to 600 nm in diameter and 500 to 800 nm in height. After E. I. Rabinowitch, *Photosynthesis*, Vol. II, Part 2 (New York: Interscience Publishers, Inc., 1956), from Vatter, unpublished, unmodified.

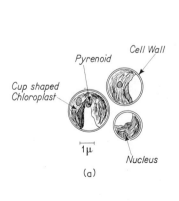

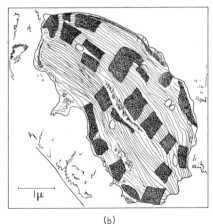

(b)

In nongreen plants, the pigmented organelles responsible for photosynthesis are called by other names, such as chromoplasts. A more general term used for both chloroplasts and chromoplasts is *plastid*. Plastids possess self-replicating properties; like mitochondria, they contain their own DNA and can code for some of their own proteins (see Chapter 14).

The algal cells have 1–50 plastids per cell. When these plastids first form, they are homogenous, but as they develop, structure appears. They are filled with smaller dark bodies called *grana*, which contain all the photosynthetic pigment. A simplified cross section of a corn plastid is shown in Fig. 20.1(b). In *Euglena*, the entire chloroplast is one granum. In most other organisms, there are 10–100 grana per plastid. All the chlorophyll, and presumably all the light-absorbing pigments of the plastid, are contained in the grana.

The grana can be isolated by breaking the plastids and then centrifuging the resulting suspension. Each granum contains large numbers of lipid, protein, and pigment molecules. These are highly ordered in anisotropic arrays.

Figure 20.2 shows the fine structure found within grana by electron

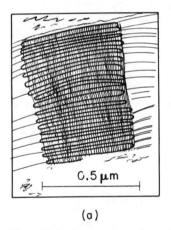

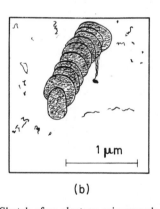

0.5 μm

1 μm

(a)   (b)

Figure 20.2   Green grana of corn leaves. (a) Sketch of an electron micrograph of a section of granum within an intact chloroplast of *Zea mays*. There are about 50 such grana per chloroplast. Each granum has about 15 parallel lamellae, which are about 40 nm thick and about 400 nm in diameter. (b) Sketch of an electron micrograph of a granum showing the thylakoid disks. After E. I. Rabinowitch, *Photosynthesis*, Vol. II, Part 2 (New York: Interscience Publishers, Inc., 1956), from Vatter, unpublished, modified.

microscopy. Lamellae are continuous throughout the chloroplast; inside grana they widen into objects called *thylakoid disks*, about 10 nm thick. A granum is composed of piles or stacks of these disks. This structure is very similar to that of the rods in vertebrate retinas.

Although apparently endless variations exist on the structures outlined in the preceding paragraphs, the general characteristics are common to all photosynthetic organisms, except bacteria; the absorbing pigments are oriented on a molecular level in small thylakoid disks. The disks are assembled or stacked up in larger ordered structures called grana, each surrounded by its own membrane. The grana, in turn, are ordered within the chloroplasts, each within its own membrane. The chloroplasts tend to be arranged in a random fashion in the cytoplasm of the cell, although chloroplasts of many cells are oriented in the light.

20.3   Basic Processes of Photosynthesis

The overall reaction of photosynthesis (Eq. (20-1)) consists of the conversion of CO_2 and water to carbohydrate at the expense of the energy contained in photons of visible light. As shown in Sec. 20.3.C, this requires 116 kcal/mole, a large amount of energy on a biological scale. The reaction is broken into three main processes: (1) the formation of free oxygen, (2) the reduction of NADP, and (3) the conversion of CO_2 into sugar. The first two of these processes require light energy, obtained from different photosystems. The third can be carried out in the dark.

A. OXYGEN PRODUCTION

In Eq. (20-1), CO_2 is used up and O_2 is produced in equal stoichiometries. It is therefore natural to guess that the O_2 might be produced from the CO_2. However, if some oxygen atoms in the CO_2 are ^{18}O, no such isotope is seen in the O_2. If the oxygen in the water is thus labeled, such a label does appear in the O_2. Thus, the oxygen produced in photosynthesis comes from water, not CO_2. This is also consistent with the more general equations of photosynthesis, Eqs. (20-2) and (20-3).

If water is split to produce free oxygen, hydrogen atoms remain. The production of free hydrogen is neither necessary nor desirable, however; instead, an electron and an H^+ ion can be transferred to suitable acceptors. The actual acceptors (carriers) in the intact plant are still not completely characterized.

In this process some light energy is used to synthesize molecules of adenosine triphosphate (ATP) from adenosine diphosphate (ADP) and inorganic phosphorous (P_i). This is termed *photosynthetic phosphorylation*, by analogy to oxidative phosphorylation, the series of oxidations that lead to the formation of ATP in mitochondria, as described in Chapter 18.

While oxidative phosphorylation results in the *reduction* of molecular oxygen to oxides, photosynthetic phosphorylation instead *oxidizes* "hydrogen oxide" (water) to produce molecular oxygen. Both involve a series of oxidations and reductions. The initial step may be regarded as the formation of two separate compounds which can serve as the oxidized [OH] and reduced [H] ends of the phosphorylating chain. The overall first process in photosynthesis can be summarized as

$$4h\nu + 2H_2O + 2ADP + 2P_i \longrightarrow 4e^- + O_2 + 4H^+ + 2ATP \quad (20\text{-}4)$$

B. THE REDUCTION OF NADP

The electrons produced at the end of the first process in photosynthesis cannot directly reduce CO_2 to CH_2O. Additional energy from light is required, and there is a second electron carrier chain to transfer this energy into a form where it can be chemically useful.

One compound that can store energy has already been mentioned: the ATP molecules produced in the first process can be hydrolyzed back to ADP and inorganic phosphate with the release of about 12.5 kcal/mole of energy. Some of the energy added in the second process can be cycled back into ATP as well. A second compound, reduced nicotinamide adenine dinucleotide phosphate (NADPH), is also used to store energy. This compound is quite easily oxidized and thus is a powerful reducing agent.

The overall reaction for the second process in photosynthesis is

$$4e^- + 4h\nu + 2NADP^+ + 2H^+ \longrightarrow 2NADPH \quad (20\text{-}5)$$

C. CO_2 CONVERSION

The fixation of CO_2 and its reduction to sugars is, in one sense, the central net result of photosynthesis. The simple sugars formed have the general formula $C_nH_{2n}O_n$, where n is in the range 3–7. These sugars may exist in either a straight chain or in one of several ring forms, as discussed in Appendix D. The six-carbon sugars and their polymers are the ones produced in largest amounts. The various hexoses are all stereoisomers, differing only in the relative locations of the —H and —OH groups. They can be converted from one form to another by suitable catalysts with very little expenditure of energy. Thus, if the cell forms or obtains one hexose, it can, with suitable enzymes, readily convert this hexose to other hexoses.

Forming the hexose from CO_2 and water requires energy. Specifically, it requires Gibbs free energy. (This is discussed more fully in Chapter 21.) In describing energy changes, it is customary to divide Eq. (20-1) by 6, giving

$$8h\nu + CO_2 + H_2O \longrightarrow \tfrac{1}{6}(CH_2O)_6 + O_2 \quad (20\text{-}6)$$

The value of the extra Gibbs free energy per mole, ΔG_0, necessary to drive this reaction to the right has been measured to be

$$\Delta G_0 = 116 \text{ kcal/mole} \quad (20\text{-}7)$$

This value may be compared with the energy of photons of red (680 nm) and violet (400 nm) light. These are

$$\Delta G_0 = 41 \text{ kcal/mole} \quad \text{(red photons)}$$
$$\Delta G_0 = 65 \text{ kcal/mole} \quad \text{(violet photons)} \tag{20-8}$$

If the process were 100 percent efficient, about 3 moles (einsteins) of red photons would be needed for each mole of CO_2 converted to hexose. As will be seen, 8 moles are, in fact, required, leading to an efficiency of about 38 percent.

There is no reason why all the energy for this process need come from photons. Indeed, under suitable conditions all living cells fix CO_2 and reduce the product to hexose. In other words, CO_2 conversion to hexose is not a unique property of photosynthetic cells. In most cells and tissues, this conversion takes place at the expense of metabolic energy. Photosynthetic tissues are distinguished by fixing CO_2 and converting the product to hexose, using the energy obtained by the absorption of photons of visible light to drive the reactions.

The net reaction for the conversion of CO_2 into sugar is

$$2H^+ + NADPH + 3ATP + CO_2 \xrightarrow{\text{enzymes}}$$

$$3ADP + 2NADP^+ + 3P_i + \tfrac{1}{6}(CHO)_6 + H_2O \tag{20-9}$$

Thus, the overall reaction for all processes in photosynthesis, reactions (20-4) + (20-5) + (20-9) + one additional ATP hydrolysis, equals reaction (20-6).

To recapitulate, the chloroplast catalyzes the splitting of water to form the reduced compounds necessary for CO_2 fixation and also effectively splits water to drive the photosynthetic phosphorylation chains. If the latter is limited by the ADP available, then the two uses of "split water" must keep in step. For if the CO_2 fixation goes faster, the ADP available will increase, thereby speeding up the rate of phosphorylation. Likewise, if phosphorylation proceeds more rapidly for a time, the ADP supply will be depleted and the rate of phosphorylation decreased. This is a simple example of the action of negative feedback on a chemical scale serving to keep two processes in step (see Chapter 6 for further examples).

One may then summarize the action of the chloroplast schematically as shown in Fig. 20.3. This shows H_2O, CO_2, and photons being used up, and H_2O, O_2 and hexose being formed. It emphasizes the three types of reactions catalyzed by the chloroplasts; the splitting of water, the fixation of carbon dioxide, and photosynthetic phosphorylation.

20.4   The Photosynthetic Pigments

The grana within the chloroplasts contain the pigments responsible for the absorption of light and its conversion into the forms useful for photosynthesis.

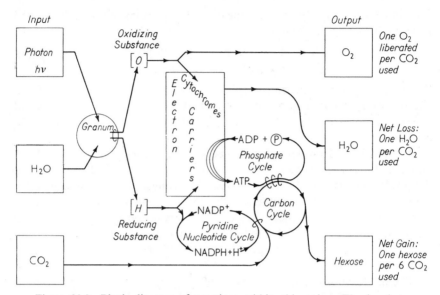

Figure 20.3 Block diagram of reactions within chloroplast. The brackets around the H and O indicate that these do not imply molecular or atomic hydrogen and oxygen, but rather reducing and oxidizing compounds. Considerable evidence indicates that there is a flavin mononucleotide intermediate between (H) and pyridine nucleotide. The circle for the carbon pathways and the square for the phosphorylating chain are purely diagrammatic. The carbon pathway is discussed in more detail in Sec. 20.6. There is not an instantaneous balance between CO_2 fixed and O_2 released. A negative feedback mechanism, controlling the rate of phosphorylation, assures that over a period of time the number of moles of O_2 released is equal to the number of moles of CO_2 fixed.

These pigments may be grouped in three classes: the chlorophylls, the carotenoids, and the phycobilins. All photosynthetic cells contain chlorophyll. Carotenoids are likewise found in grana from all photosynthetic cells. The phycobilins are found only in blue-green and red algae. In some fashion, all of these pigments act together so that light photons absorbed by any of them are equally effective for photosynthesis. (Some carotenoid pigments occur outside the chloroplasts and are completely ineffective for photosynthesis.)

A. STRUCTURE

All chlorophylls contain a porphyrin structure, similar to that of the heme groups discussed in Chapter 18. An atom of magnesium rather than iron is found in chlorophyll. Attached to the hydrophilic prophyrin ring is a long hydrocarbon side chain which is soluble in lipids. Thus, chlorophyll should be a detergent.

The lamellar microstructure of the chloroplast is due in part to this chemical detergency. The main body of the molecule can bind to protein and the tail to lipid, allowing long layers to form. This contributes to the lamellae so noticeable in electron micrographs (see Figs. 20.1 and 20.2). All green plants have a type of chlorophyll called chlorophyll *a* and most also have another, chlorophyll *b*. The two can be converted from one to another in extracts, but no such change has ever been demonstrated in whole chloroplasts. The structure of chlorophyll *a* and its absorption spectrum are shown in Figs. 20.4 and 20.5. The absorption spectrum shifts to the red on extraction from the whole cell.

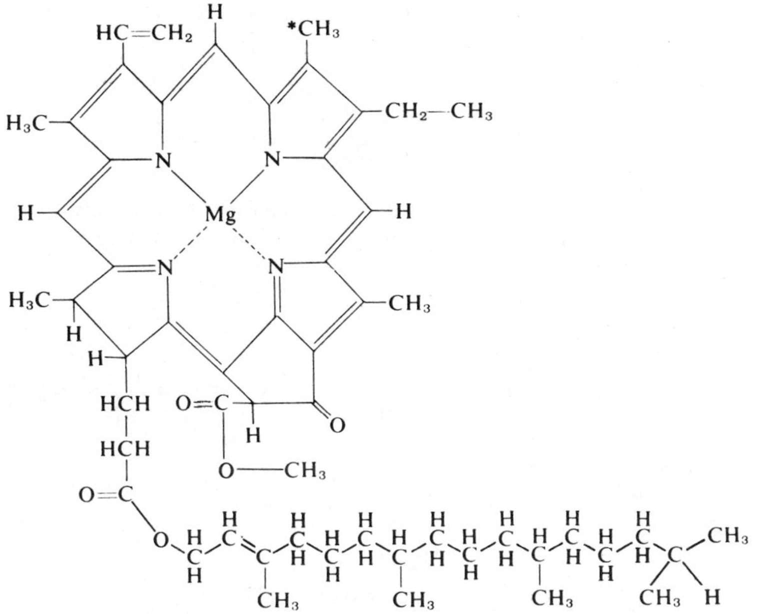

Figure 20.4 Structure of chlorophyll *a*. Chlorophyll *b* differs in having an aldehyde group   O   in place of the starred —CH₃.

Similar chlorophylls are found in blue–green algae and *Euglena*. Bacteriochlorophyll, the form found in photosynthetic bacteria, has a single bond substituted for one of the double bonds in the conjugated ring. All of these chlorophylls perform essentially the same function in photosynthesis.

Carotenoid pigments are involved in photobiology, not only in photosynthesis but also in vision. As described in Chapter 19, many vertebrate visual pigments involve a carotenoid derivative called retinal. In addition, the eyes of snakes and birds have carotenoid oil droplets which appear to act as filters. The general structure of carotenoids is shown in Fig. 20.6, while two absorption spectra are illustrated in Fig. 20.7. Carotenoid pigments can be

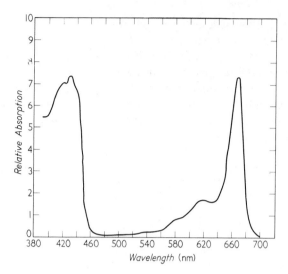

Figure 20.5   Relative extinction coefficient of chlorophyll *a* in methanol. After D. G. Harris and F. P. Zscheille, *Botan. Gaz.* **104**:515 (1943). Copyright 1943 by The University of Chicago.

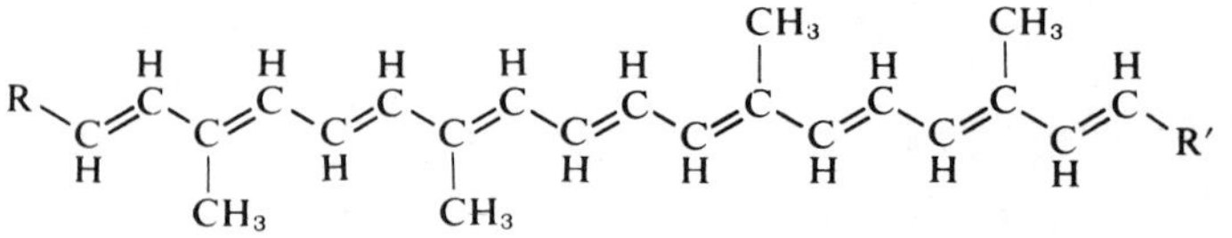

(a)

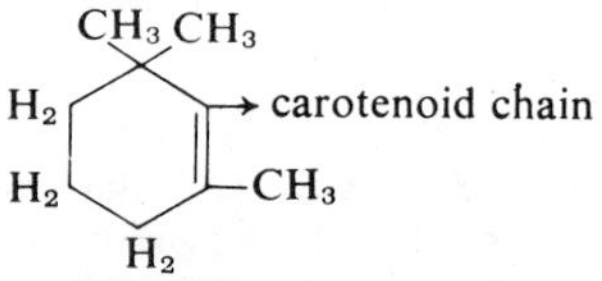

(b)

Figure 20.6   Carotenoid structure. Different carotenoids have different R and R′ groups and exist as various *cis* and *trans* isomers. The all-*trans* isomer is illustrated. For β-carotene, both R and R′ have the form shown in (b).

divided into two classes: carotenes, in which the R and R′ groups contain only carbon and hydrogen, and the carotenols, which are alcohols and ketones and thus contain oxygen as well.

B. FUNCTION

An action spectrum, that is, the relative yield of hexose at constant light intensity, is shown in Fig. 20.8. The most striking feature of this curve is its

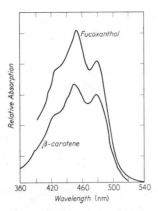

Figure 20.7 Absorption spectra of two carotenoids. These were both dissolved in hexane. The solvent alters the location of the maxima as well as the height of the curve. Different carotenoids in the same medium have different peaks. The two curves shown are not plotted on the same scale along the absorption axis. The *β*-carotene is after F. P. Zscheille, J. W. White, B. W. Beadle, and J. R. Roach, *Plant Physiol.* **17**:331 (1942). The fucoxanthol is after E. I. Rabinowitch, *Photosynthesis*, Vol. II, Part 1 (New York: Interscience Publishers, Inc., 1951), from Wald, unpublished, modified.

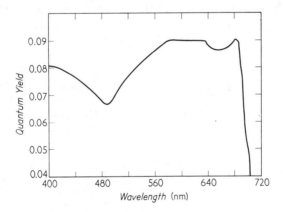

Figure 20.8 Action spectrum of *Chlorella*. Note the extremely flat curve as compared with those of Figs. 20.5 and 20.7. After R. Emerson and C. Lewis, "Dependence of the Quantum Yield of Photosynthesis on Wave Length of Light," *Am. J. Botany* **30**:165 (1943).

flatness as a function of wavelength. This confirms that light photons absorbed by pigments other than chlorophyll are effective in photosynthesis.

When extracted, chlorophylls fluoresce very strongly (see Chapter 27), and each type exhibits a characteristic fluorescence spectrum. Whole green cells also fluoresce, always exhibiting the chlorophyll *a* fluorescence spectrum.

This is true no matter what visible wavelength is used to illuminate the cells. Thus, all light energy absorbed that can be used in photosynthesis may be transferred to the chlorophyll *a* molecules. The chain of light-energy transfer among pigments is

$$h\nu \longrightarrow \text{carotenoids} \longrightarrow [\text{phycobilins}] \longrightarrow \text{chlorophyll } a \longleftarrow h\nu$$
$$\uparrow \qquad\qquad\qquad\qquad \uparrow \qquad\qquad (20\text{-}10)$$
$$h\nu \longrightarrow \text{chlorophyll } b$$

When phycobilins are absent, energy can be transferred directly from carotenoids to chlorophyll *a*.

While each pigment has its characteristic absorption and fluorescence spectrum, each may have various forms which show slightly shifted spectra; for example, although chlorophyll *a* has an absorption maximum near 660 nm (Fig. 20.5), spectral forms in which this maximum occurs from 670 to as high as 720 nm have been observed. This demonstrates the effect different environments can have on such spectra.

As was discussed in Sec. 20.3, there are two sites of light absorption in photosynthesis, one associated with O_2 formation, the other with NADPH formation. Each requires its own pigment system to capture light energy or to receive it from other pigments. Although these two pigment systems are similar, they differ in the amounts and types of their pigments, as shown in Table 20-1.

TABLE 20-1
PIGMENTS FOUND IN PHOTOSYNTHESIS

Process	Pigment type	Spectral maxima (nm)	
		Absorption	Fluorescence
O_2 formation (photosystem II)	Carotenes Chlorophyll *b* Chlorophyll *a*	660 670 678 685 690 705–720	670 680 687 695 700 710–715 720–735
NADPH formation (photosystem I)	Carotenols Chlorophyll *b* Chlorophyll *a*	660 670 678 685	670 680 687 695

Once the light energy has been transferred to chlorophyll a it must be used by the electron carriers discussed in Sec. 20.5. The pigment molecules which act as go-betweens in this conversion have not yet been fully characterized chemically (they may be specialized chlorophyll molecules), but they have been given names: the letter P (for pigment) and the number (in nm) of their absorption maximum. The transfer pigment used for NADPH formation is called P700; a similar one used for O_2 formation is called P680.

20.5 The Light Reactions

From the previous analysis of photosynthesis into three processes, it seems almost trivial that a part of the reaction could proceed with the light off, but historically this was not always understood. At high light intensities, the photosynthetic yield of carbohydrates (or oxygen) per incident photon is greater with a flashing light than with a constant light. Essentially, the light may be thought of as inducing stable intermediates up to saturation during the "on" period, which continue to drive the carbon cycle during the "off" (or dark) period.

The separation of the *light* and *dark reactions* was experimentally demonstrated in 1937 by Hill. He isolated chloroplasts and found that, although they could no longer convert CO_2 to hexose, they could still generate oxygen.

$$\text{light} + A + H_2O \xrightarrow{\text{chloroplasts}} AH_2 + \tfrac{1}{2}O_2 \qquad (20\text{-}11)$$

where A is a suitable electron (hydrogen) acceptor. This is now known as the *Hill reaction*. In 1950, Ochoa and Vishniac showed that the biological electron acceptor $NADP^+$ could be used in the Hill reaction.

$$\text{light} + NADP^+ + H_2O \xrightarrow{\text{chloroplasts}} NADPH + H^+ + \tfrac{1}{2}O_2 \quad (20\text{-}12)$$

Photosynthetic phosphorylation was independently demonstrated in isolated chloroplasts exposed to light. This also occurred in the absence of CO_2 consumption, and was shown to be the result of two separate processes, only one of which occurs in conjunction with O_2 formation. Thus, as was described in Sec. 20.3, there are at least two separate light reactions.

Experiments that measured the efficiency of light at different wavelengths to promote photosynthesis in different organisms, the action spectrum of the organism as shown in Fig. 20.8, had hinted earlier at two parts to the light reaction. This efficiency of light is fairly constant in most of the visible spectrum, yet it drops off in the red above 680 nm. However, if red 680-nm light is supplemented with bluer (< 600 nm) light, the two will add synergistically as if the red light were restored to full efficiency. Thus, two photosystems were envisioned: one to capture energy from long-wavelength light, called *photosystem I*, and a second for shorter wavelengths called *photosystem II*.

The latter was shown to be involved in oxygen production; photosynthetic bacteria, which do not evolve oxygen, contain only photosystem I.

Most intermediates in the light reaction have been identified, yet some are not well characterized. Early studies used electron spin resonance (esr), which is discussed in Chapter 26. Esr measures magnetic signals from unpaired electrons in excited states, produced by compounds called free radicals. Such magnetic studies with chloroplast materials indicate that free radicals form when light falls on the chloroplast. The very nature of the magnetic studies makes it impossible to tell just what compound(s) contain the unpaired electron. In contrast to most free radicals, these are of the minority class called *stable free radicals*, which can be maintained for a comparatively long time. (Most free radicals, for example OH·, are highly reactive and therefore cannot be kept as such except for very short periods.)

By repeating the magnetic measurements at 25°C and −150°C, it is possible to show that the free radicals build up in a fraction of a second at both temperatures. This suggests that free-radical production does not involve a separate chemical reaction. The unpaired electrons disappear much more rapidly at the higher temperature. This indicates that their disappearance is associated with a chemical reaction.

The available evidence on the nature of the light reactions, then, may be summarized as follows. Initially, in each reaction sequence, photons are absorbed by any of a number of pigment molecules, raising these to an excited level. All the different types of pigments are somehow coupled together so that the energy absorbed by any one of them may be transferred to any other one. This is indicated by the appearance of the chlorophyll *a* fluorescence spectrum in intact chloroplasts.

When the energy has been transferred to a pigment reaction center such as P700 shown in reaction sequence (20-13a) (there is estimated to be one such center for every 300 or so pigment molecules) it can enter into the first chemical reaction in photosynthesis. The excited electron can transfer to a lower level in an acceptor molecule, step (20-13b). Then a donor molecule can transfer an electron from its ground state to the lower ground state of the pigment molecule, step (20-13c).

$$
\begin{array}{lll}
\text{P700} + h\nu \longrightarrow \text{P700*} & \text{(a)} \\
\text{P700*} + \text{A} \longrightarrow \text{P700}^+ + \text{A}^- & \text{(b)} \\
\text{P700}^+ + \text{D} \longrightarrow \text{P700} + \text{D}^+ & \text{(c)} \\
\hline
\text{A} + \text{D} + h\nu \longrightarrow \text{A}^- + \text{D}^+ & \text{(d)}
\end{array} \quad (20\text{-}13)
$$

Equation (20-13d) is very similar to the general photosynthetic equation, (20-3). To generate Eq. (20-1), one must transport the separated charges in reaction (20-13d) to form (in most cases) O_2 from H_2O using D^+, and CH_2O from CO_2 using A^-.

The electron carriers used in this process are diagrammed against their oxidation potential in Fig. 20.9. Hill and Bendall of Cambridge University first proposed this scheme; it has been modified since by other workers. Symbols in parentheses refer to Fig. 20.9; compounds followed by question marks in that figure have not been fully characterized. Light striking photosystem II (P680) results in transfer of an electron from unknown donor (S) to an acceptor cytochrome C550. After donating four such electrons, this donor accepts four electrons from water and one molecule of oxygen is produced.

Electrons are transferred from reduced C550 to cytochrome b559, plasto-quinone (PQ), cytochrome f (cyt$_f$) (these compounds are analogs of mito-chondrial electron carriers; see Chapter 18), and plastocyanin (PC), a copper-containing protein. The step from the plastoquinone to cytochrome f provides enough energy to phosphorylate one ADP to ATP; a second ATP is produced earlier in the sequence.

Light energy striking photosystem I (P700) can promote transfer of an electron from plastocyanin to pigment P430. From there it is transferred to a ferredoxin-reducing substance (FRS), ferredoxin (FD) itself (an iron- and sulfur-containing protein), and the enzyme ferredoxin-NADP$^+$ reductase (FP). This last enzyme carries out the final electron transfer unique to photosynthesis, the two-electron reduction of NADP$^+$ to NADPH.

The electron transfers listed above are repeated for each of the four electrons, so eight light quanta are required for the production of one molecule of oxygen, two NADPH molecules, and eight ATP's. All these ATP's are coupled to the splitting of water, yet as mentioned earlier, photo-phosphorylation can happen without O_2 production. This is termed *cyclic phosphorylation*, and it occurs when electrons from reduced P430 are cycled back via cytochrome b564 to the electron carriers joining photosystem II to photosystem I (dotted line in Fig. 20.9). When this cyclic flow goes past plastoquinone, one ATP is produced.

Research in the field emphasizes the chemical identification of the less fully characterized pigments, electron carriers, and photophosphorylation intermediates. Biophysicists are concerned with the identification, through physical techniques, of structure–function relationships between these com-pounds at a molecular level. The light reactions may soon be as well under-stood as the related respiration reactions in mitochondria.

20.6   The Path of Carbon in Photosynthesis

The pathways followed by the carbon of the CO_2 during its fixation and conversion to hexose have been studied with radioactive carbon as a tracer. To do this, $^{14}CO_2$ is introduced into a suspension of photosynthesizing cells or isolated chloroplasts. If the reaction is stopped after a brief time, one may

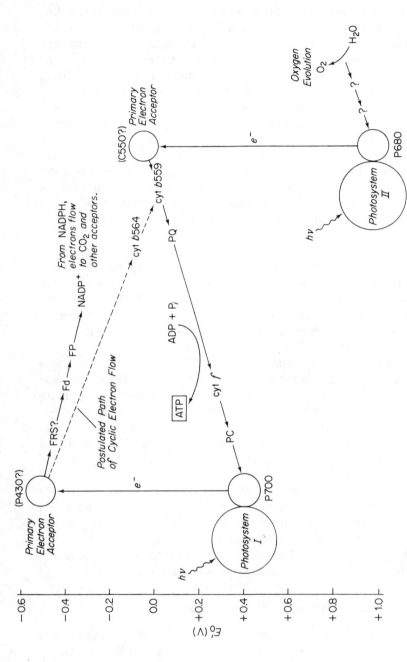

Figure 20.9 Flow of electrons from water to carbon dioxide proceeds against an electrochemical gradient of 1.2 V and requires two photochemical events, as described in the text. There is also a cyclic system of photosynthesis, in which electrons pass from P700 to X and then return through various electron carriers to P700; in this system only ATP is produced. After A. L. Lehninger, *Biochemistry*, 2nd ed., Worth Publishers, Inc., New York, 1975, p. 605.

384

determine from cellular extracts the compounds containing labeled carbon. Many similar experiments may be performed using different times of exposure to $^{14}CO_2$; from these data the relative amounts of the radioactivity in the various labeled compounds may be plotted as a function of time. Thus, it is possible to establish the order in which the compounds appear and hence their relationship to one another.

In order to separate the various labeled compounds, a type of paper chromatography is used. *Chromatography* is a very important tool in the separation of mixtures of compounds. In paper chromatography, a drop of solution containing the substances to be separated is spotted near one edge of a sheet of special filter paper, and the paper is allowed to dry. The edge is then immersed in a solvent and the various compounds in the mixture often migrate at different rates in the solvent and are thus separated. When the solvent has reached the other edge of the paper, the sheet is removed and the solvent is evaporated. This technique can be easily demonstrated by using it to separate the different colored pigments in water-soluble ink, using water as the solvent and strips of blank newsprint as the paper.

To use chromatography on the carbon compounds formed in photosynthesis, a more complicated two-dimensional system is necessary. The cellular extract is placed near one corner of the paper. A first solvent (often a phenol–water mixture) is used. Then the paper is dried, turned 90°, and the adjacent edge immersed in a second solvent.

The partly separated compounds migrate at different rates in the two different solvents and so are arranged in a two-dimensional array. The filter paper is dried. To find the radioactive-labeled compounds, the filter paper is placed against a sheet of X-ray film. The location of labeled compounds identifies them; the intensity of the darkening of the film shows the extent of labeling. This process is known as *autoradiography*, and is described in more detail in Chapter 29.

Because the initial compounds form in a fraction of a minute, very short exposures are necessary. Several of the steps known to occur from these studies are summarized in Fig. 20.10. Note that the final compound formed is a hexose sugar, fructose-6-phosphate. Enzymes exist within the chloroplast to change some of the fructose to glucose. Some of the glucose, in turn, is polymerized to starch, whereas the remainder is combined with fructose to form sucrose.

To confirm the scheme shown in Fig. 20.10, it was necessary to introduce the labeled CO_2 very rapidly into the mixture. Then the entire process had to be stopped in a matter of seconds. It was possible that the compounds found labeled might have reflected the method used to stop the photosynthetic reaction. However, plunging into boiling water, strong acid, and strong alkali all showed 3-phosphoglyceric acid as the first compound and confirmed the general scheme shown.

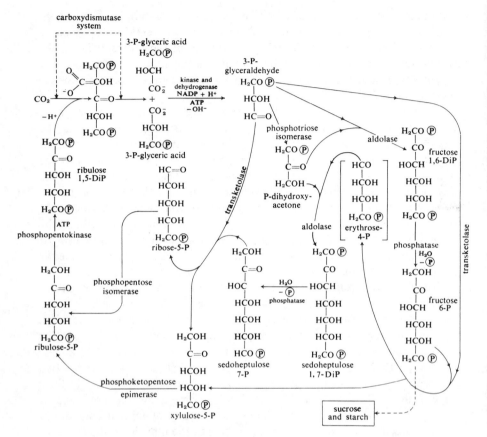

Figure 20.10 Some of the steps in photosynthesis. ℗ stands for phosphate in the diagram and P for phosphate in the names. Note that all compounds with five or more carbons are sugars and probably exist primarily in a ring isomer (see Fig. D. 4); straight-chain formulas are for convenience only. After J. A. Bassham and M. Calvin, *The Path of Carbon in Photosynthesis* (Englewood Cliffs, N.J.: Prentice-Hall, Inc., 1957).

The initial reaction of CO_2 with ribulose-1,5-diphosphate is interesting in that it is exothermic, liberating energy and apparently going even in the absence of any specific enzyme. NADPH is used to convert 3-phosphoglyceric acid to 3-phosphoglyceraldehyde. ATP is used at this step and in the formation of the ribulose-1,5-diphosphate.

Although the scheme shown in Fig. 20.10 may seem extremely complex, all evidence indicates that it is an oversimplification. Nonetheless, the general cyclic character and the need for NADPH and ATP seems established beyond question. The steps at which ATP and NADPH are used are the ones utilizing free energy, ultimately derived from the absorption of visible light. Each $NADP^+$ reduction requires about 8 kcal/mol. The cycle of Fig. 20.10 uses two NADPH's and three ATP's per molecule of CO_2. Thus, the total free energy required for this scheme is about 135 kcal/mole of CO_2, just a little higher than the 116 kcal/mole necessary to form hexose.

20.7 Summary

Photosynthesis is the trapping of free energy carried by photons of visible light, converting the energy into stable chemical forms. The process of photosynthesis makes possible life as it exists on earth, both by producing carbohydrates, the ultimate source of food energy for almost all organisms, and by liberating molecular oxygen into the atmosphere. Photosynthesis is catalyzed by all green plants, by the green protist *Euglena*, by the blue–green algae, and by a variety of pigmented bacteria.

In all the higher forms, photosynthesis is catalyzed by intracellular organelles called *chloroplasts*. Within the chloroplasts there are smaller organelles called *grana* which contain the pigments necessary for photosynthesis. For convenience the reactions can be divided into three parts: two light reactions or quantum conversions, which occur in the grana and lead (a) to the dissociation of H_2O and (b) to the formation of NADPH, and (c) the dark reactions, by which CO_2 is converted to carbohydrate.

In the light reactions, the incoming photon is first absorbed by any of a variety of pigments in the grana, including chlorophyll *a*, chlorophyll *b*, carotenoid pigments, and phycobilins. By a mechanism not clearly understood, the electronic excitation can be passed from one molecule to another. This process occurs with very high efficiency. Thus, energy must be transferred before much degradation occurs; accordingly, the transfer must be very rapid. The electronic excitation produces a charge separation, the resulting unpaired electrons being trapped or stored at certain sites where they may be regarded as stable free radicals. These then react to drive the phosphorylation chain of enzymes and the carbon cycle. The entire reaction has a very high efficiency; 30–40 percent of the photon energy absorbed is used to produce carbohydrate and oxygen.

The study of photosynthesis is in a state of transition. In the not too distant past, standard histological techniques and simple chemical procedures were used to reveal many of the basic characteristics of photosynthesis. More recently, highly specialized chemical and physical tools have become an essential part of photosynthetic studies. It appears that the outstanding advances of the future will involve the application of physical techniques such as the X-ray determination of molecular structure and arrangement.

REFERENCES

The number of books and articles on photosynthesis is very large. Owing to the rapid advances in this field, however, many of these become outdated very rapidly. This applies especially to the interpretation of the mechanism of photosynthesis on the molecular level. The following selections should be helpful to readers interested in more detailed discussions of photosynthesis than it was possible to include within the limits of this text.

1. BASSHAM, J. A., AND M. CALVIN, *The Path of Carbon in Photosynthesis* (Englewood Cliffs, N.J.: Prentice-Hall, Inc., 1957), 104 pages. (Extremely complete compendium replete with references.)

2. RABINOWITCH, E. I., AND G. GOVINDJEE, *Photosynthesis* (New York: John Wiley & Sons, Inc., 1969), 273 pages.

3. GOVINDJEE, G., AND R. GOVINDJEE, "The Absorption of Light in Photosynthesis," *Sci. Am.* **231** (*December 1974*):68–82.

DISCUSSION QUESTIONS—PART D

1. Several enzymes and cofactors are involved in DNA synthesis. What is the minimum system necessary for DNA synthesis *in vitro*? What additional factors may be important *in vivo*?

2. Describe repression and induction in the histidine synthetase pathway of *E. coli*. What are the chemical structures of the corepressors and inducers?

3. The repressor for β-galactosidase in *E. coli* has been isolated. What is known about its physical and chemical properties? Its amino acid sequence? Its three-dimensional structure?

4. Derive Eq. (18-11), the rate equation for noncompetitive inhibition. Describe one enzyme that follows this equation. What are the values of the kinetic constants for this enzyme?

5. What are the effects on enzyme kinetics of temperature, pH, and ionic strength?

6. Electron spin resonance, discussed in Chapter 26, was used to locate the Fe atoms in the myoglobin unit cell. Describe in detail how this was accomplished.

7. The enzymes in the electron transport chain in intact mitochondria have been most clearly demonstrated through the use of selective inhibitors. Describe how this can be accomplished, giving examples of specific useful inhibition.

8. Describe the metabolic and morphological changes that occur in *Euglena* when they are transferred from a well-lighted environment to a dark one.

9. Describe the current theories concerning the role of the carotenoid oil droplets in the vision of snakes and birds. What is the evidence supporting these theories?

10. What is the evidence that the red "eye spot" in *Euglena* actually functions as a visual receptor?

11. What is the evidence that P680 and P700 are specialized chlorophyll molecules? What is their concentration in a plastid? What special physical techniques are necessary to study a species in such low concentrations?

12. Review in detail the experiments that have led to the present ideas about the path of carbon in photosynthesis.

13. Describe the changes in the mechanical properties of several high polymers following exposure to ionizing radiations. Correlate these changes with other changes in the polymer.

14. Describe the difficulties of assigning a specific inactivation cross section to virus particles.

15. What are the detailed sequence of events necessary for protein synthesis? Include all known or postulated factors and cofactors in this synthesis.

PART $\mathbf{E}$

THERMODYNAMICS AND TRANSPORT SYSTEMS

Introduction to Part E

Thermodynamics and statistical mechanics are major branches of physics and physical chemistry. They emphasize the application of the concept of energy and its conservation law; these have proved extremely fruitful in modern physics as well as in classical physics and physical chemistry. Therefore, it is quite appropriate that the application of thermodynamics should be a basic part of biophysics.

The concepts of chemical thermodynamics which are important for biophysics are introduced in the first chapter in this section. These ideas are applied in two succeeding chapters to irreversible thermodynamics and to the diffusion of molecules through fluids and their transport through membranes. The preceding ideas are used in the next chapter to develop molecular models to explain the nature of the action of membranes. In the last chapter in this section, information theory is introduced; its relationship to thermodynamics and kinetic theory is emphasized.

Thermodynamics is a part of physics and is described best in the language of physics—mathematics. All the chapters in this section could be called mathematical biophysics (as could also parts of several other chapters). It is important that the mathematical developments be related to experiments. In Part E, the experimental applications are outlined, but the mathematical analyses are granted more space.

CHAPTER 21

Thermodynamics
and Biology

21.1 The Role of Thermodynamics in Biology

Thermodynamics is one of the major fields of classical physics. The concepts
of energy and of its changes of form are central to thermodynamics. Many
physicists and chemists have come to regard this approach as the most basic
and most important. Accordingly, they consider the application of thermodynamics to biological systems to be the core of biophysics.

Thermodynamics can be applied to various aspects of living systems. The
chapters in this part of the text illustrate some of these applications. They
include the behavior of enzyme systems, the transport of molecules against
chemical and electrical gradients, and the molecular bases of membrane
function. There is virtually no field of biological science to which the concepts
of thermodynamics cannot be applied.

Thermodynamics has long been recognized to be of prime importance to
biophysics. One of the most distinguished of the early biophysicists of the
current century, A. V. Hill, is best known for his heat measurement on muscles.
Other biophysicists have followed this path toward an understanding of life.

Thermodynamics is inherently a physical discipline whose theory is expressed most readily in mathematical terms. Because of this, and because the theory and applications of thermodynamics are often found in separate courses, many students receive a baccalaureate degree in physics or chemistry with little or no knowledge of thermodynamics. Accordingly, the development of thermodynamics is included in this chapter. No attempt has been made either to include rigorous proofs or to eliminate the fundamentally mathematical symbolism involved. Those terms and parts of thermodynamics of greatest application to biology are emphasized, particularly the concepts of energy, entropy, and Gibbs free energy. The last-mentioned concept is applied in this chapter to a discussion of chemical equilibria. This application is one of the important uses of thermodynamics in describing biological systems. The chapter is concluded by several sections describing the thermodynamic aspects of enzyme reactions. The last topic is continued into the following chapter, which deals with the thermodynamics of nonequilibrium systems.

21.2 The Laws of Thermodynamics

Thermodynamics is a study of the exchange of heat between bodies and the conversion of heat to and from other forms of energy. Energy is a concept constructed by humans to describe the external world; it is defined as the ability to do mechanical work, W. This, in turn, is defined as the product of a force F exerted times the distance s moved in the direction of the force. In the language of integral calculus, this last statement becomes

$$W = \int_1^2 \vec{F} \cdot \vec{ds}$$

where W is the work done by $\vec{F}$ in moving from position 1 to 2 and the arrows indicate vectors.

Mechanical energy may exist in two general forms, potential and kinetic. Potential energy includes elastic energy and gravitational energy. Sound or acoustic energy is a mixture of potential and kinetic energy. In a frictionless system, mechanical energy would always be conserved. Because friction occurs in all real macroscopic systems, mechanical energy is lost as heat. Moreover, mechanical energy sources are also known; thus, in a real system mechanical energy may be both generated and dissipated.

To physicists, the idea of conservation is a pleasing one. It was proposed to retain the concept of conservation of mechanical energy, even in the presence of friction and heat-driven machines, by including heat as a form of energy. Joule proved experimentally that for every unit of mechanical energy dissipated, a fixed number of heat energy units were generated. Moreover, if heat energy were used to operate a machine, the same ratio of energies was valid. By extending the concept of energy to include electric and magnetic energy, chemical energy, and finally mass energy, it has been possible to retain the conservation of energy as a fundamental law.

Another name for this fundamental law is the *first law of thermodynamics.* Symbolically, it may be written

$$dE = \delta Q - \delta W \qquad (21\text{-}1)$$

where E is the internal energy, Q is the heat put into the system, and W is the work done by the system. The symbol δ is used instead of d for differences, because neither δQ nor δW is an exact differential. A differential is the difference in a thermodynamic quantity when the system is changed from one equilibrium state to a neighboring equilibrium state. (The *states* are defined in thermodynamics by the pressure p, volume V, temperature T, and concentrations c_i.) In the expression above, dE is an exact differential because it depends only on the initial and final states, whereas δQ and δW will vary with the path between these two states. In fact, if one considers a heat engine going around a cycle, dE, dT, dp, dV, and dc will all be zero for a complete cycle or any integral number of cycles. In contrast, δW and δQ will increase with each complete cycle.

It is always preferable to work with exact differentials, if this is possible. Those who have studied differential equations will know that it is often possible to multiply by a suitable function, known as an *integrating factor*, to make a differential exact. This can be done for both δQ and δW. It is then possible to rewrite the first law using exact differentials only.

The differential of added heat δQ may be made exact by dividing by the absolute temperature. The resultant differential is dS, defined by

$$dS = \frac{\delta Q}{T} \qquad (21\text{-}2)$$

It is called the differential of *entropy*. The entropy S is interpreted in statistical mechanics as a measure of the disorder of the components of the system. In information theory, the entropy is a measure of the information to be gained by determining the locations, and so on, of all the parts of the system. From the point of view of thermodynamics, the importance of entropy is that it returns to its original value after a complete cycle; that is, dS is an exact differential.

Strictly speaking dS can be calculated from Eq. (21-2) only for reversible changes between equilibrium states. Most state changes, however, are irreversible in any real system (and biological ones are real). Under these conditions, the thermodynamic analysis becomes more complex. Discussion of the biological applications of irreversible thermodynamics is deferred to Chapter 22.

The differential of work done by the system δW may be represented as a sum of inexact differentials, each of which can then be made exact by suitable integrating factors. If the system is a gas, δW is particularly simple; dividing it by the pressure p gives the differential of volume dV. In other words,

$$\delta W = p\, dV$$

The gas system can do work δW only by expansion. For more complex systems, it is convenient to discuss the difference $\delta W'$, defined by

$$\delta W' = \delta W - p\, dV \qquad (21\text{-}3)$$

This work, other than expansion, may be elastic or mechanical, $\delta W'_M$; electromagnetic, $\delta W'_E$; or chemical, $\delta W'_C$. In each case, it is possible to find an expression similar to $p\, dV$.

For any elastic or mechanical type of work, one may always write for the work done by the system

$$\delta W'_M = -F\, d\xi$$

where F is the force exerted on the system and $d\xi$ is the displacement in the direction of the force. If many forces are present, $\delta W'_M$ is the sum of terms similar to the preceding equation; that is,

$$\delta W'_M = -\sum F_i\, d\xi_i \qquad (21\text{-}4)$$

where the summation includes all active forces.

If charges dq_i are added to the system, each will increase the potential energy by an amount ϵ_i per unit charge. Hence, one may write

$$\delta W'_E = -\sum \epsilon_i\, dq_i \qquad (21\text{-}5)$$

again summing over all altered charge distributions.

For biomolecular studies the most important term of this type is often the chemical work $\delta W'_C$ done by the system. By analogy with the electrical case above, it is convenient to assign a chemical potential energy per mole μ_i to each substance. If more than one phase exists, for example, intracellular and extracellular, then a different subscript i is required for each phase in which a molecular species is found. Letting n_i be the number of moles for substance i, it follows that

$$\delta W'_C = -\sum \mu_i\, dn_i \qquad (21\text{-}6)$$

where the summation is over all possible values of i.

The first law of thermodynamics may be rewritten by combining Eqs. (21-1) through (21-6) into the form

$$dE = T\, dS - p\, dV + \sum \mu_i\, dn_i + \sum F_i d\xi_i + \sum \epsilon_i\, dq_i \qquad (21\text{-}7)$$

The first summation in Eq. (21-7) is used implicitly in the discussion in this chapter dealing with chemical equilibria and with enzyme kinetics. The last sum is biologically important at membranes, whereas the next-to-last sum is significant in problems involving muscular contraction or the elastic properties of tissues. In discussions of enzyme activity, both of the last two sums may be set to zero.

There are two other laws of thermodynamics, both of which may be expressed in terms of the entropy. The *second law of thermodynamics* is concerned with the direction of time. In all of mechanics and in electricity and magnetism, there is nothing to distinguish the positive and negative directions of time. The second law of thermodynamics states essentially that the positive direction of time is that in which heat flows from a hot body to a cold body in an isolated system. When a given amount of heat δQ leaves a body at T_1 and flows to a colder body at T_2, the net change of entropy, for the system consisting of the two bodies,

$$dS = \frac{\delta Q}{T_2} - \frac{\delta Q}{T_1}$$

is greater than zero. The second law of thermodynamics states that in an isolated system, the entropy will be a maximum at equilibrium. The conditions for equilibrium then are

$$dS = 0$$
$$dE = 0$$
$$d^2S < 0 \quad \text{that is,} \quad S \text{ is a maximum}$$

Although it is mathematically useful to discuss isolated systems, they are as unreal as frictionless systems; no set of bodies is known which is completely thermodynamically isolated. In a real system, the entropy may decrease with time. Entropy changes in nonequilibrium systems are discussed in Chapter 22.

The *third law of thermodynamics* is more recent than the first and second laws. The third law is concerned with what happens to the entropy as the absolute temperature approaches zero. Equation (21-2) shows that the definition of dS has a factor of $1/T$. If the specific heat of a substance at constant volume (c_V) remained greater than zero as the absolute temperature approached zero, then the entropy change would approach minus infinity when a body was cooled toward absolute zero. The third law states that this is not true—that the entropy change remains finite. In other words, the third law states that c_V of every substance goes to zero at least as fast as the temperature in the neighborhood of absolute zero.

A somewhat stronger form of the third law states that the entropy of all single crystals is the same at absolute zero and may be conveniently chosen as zero. This means, for example, that if one measures the entropy changes for 2 moles of hydrogen and one of oxygen from absolute zero to room temperature and adds the entropy due to the formation of liquid water, the final sum should be identical to the entropy change from ice at $0°K$ to water at room temperature. This stronger version of the third law has been verified for every substance tested.

21.3   Other Thermodynamic Functions

Three thermodynamic functions, other than the entropy, are more satisfactory for discussing equilibrium conditions in nonisolated systems. One of these is the *enthalpy*, H, defined by

$$H = E + pV \tag{21-8}$$

From elementary calculus,

$$dH = dE + p\,dV + V\,dp$$

In an isobaric system (constant pressure), doing only $p\,dV$ work, since dp vanishes,

$$dH = \delta Q$$

For this reason, the enthalpy is sometimes called the *heat* or *heat function*, even though these names are misleading.

The other two functions are both called *free energy*; one or the other is designated by F in many texts. A less ambiguous approach is to use A for the *Helmholtz free energy*, defined by

$$A = E - TS \tag{21-9}$$

and G for the *Gibbs free energy*, defined by

$$G = H - TS \tag{21-10}$$

A little manipulation shows that, for the isolated system,

$$dA = -S\,dT - \delta W \tag{21-11}$$

and

$$dG = -S\,dT + V\,dp - \delta W' \tag{21-12}$$

If the system, instead of being isolated, is maintained at constant temperature, then

$$dA = -\delta W$$

At equilibrium, in this isothermal system, δW vanishes, and hence

$$dA = 0 \quad (A \text{ is a minimum})$$

If this system starts out other than at equilibrium, the Helmholtz free energy in excess of the equilibrium minimum value is the maximum work obtainable from the system.

Most biological changes occur with external restraints which maintain not only the temperature but also the pressure approximately constant. Under these conditions, Eq. (21-12) shows that

$$dG = -\delta W'\qquad(21\text{-}13)$$

and equilibrium corresponds to a minimum of G:

$$dG = 0\qquad(21\text{-}14)$$

If one starts with reactants in a nonequilibrium condition, the excess of the Gibbs free energy above this minimum is the work (other than $p\,dV$) available from the system.

In a system that is restricted by its surroundings to isobaric, isothermal conditions, and in which $\delta W'$ consists only of chemical work, one may write

$$dG = \sum \mu_i\, dn_i\qquad(21\text{-}15)$$

Under conditions of constant temperature and pressure, and assuming no mechanical or electrical changes, it can be shown that at equilibrium

$$\sum n_i\, d\mu_i = 0$$

Adding the last two equations allows one to integrate at once, obtaining

$$G = \sum n_i\mu_i$$

Advanced thermodynamics texts, including ones dealing with physical chemistry, show that the chemical potential may be expressed for an ideal solution as

$$\mu_i = RT\ln\left(\frac{c_i}{c_i^0}\right) + \mu_i^0(T)\qquad(21\text{-}16)$$

where μ_i^0 is the value of the chemical potential associated with the ith substance at its standard concentration c_i^0 and subject to the constant temperature (and pressure) selected.

A useful quantity for chemical thermodynamics is the partial molal Gibbs free energy, $\tilde{G}_i$. This is the Gibbs free energy per mole of i which the system possesses because of the presence of i. From Eq. (21-15) it is seen that this is just the chemical potential of i for an isothermal, isobaric system with no electrical or length changes. Thus, Eq. (21-16) can be expressed in the alternative form

$$\tilde{G}_i = \tilde{G}_i^0 + RT\ln\left(\frac{c_i}{c_i^0}\right)\qquad(21\text{-}17)$$

The term $\tilde{G}_i^0$ is the value of $\tilde{G}_i$ when the concentration is c_i^0. One may choose the latter as unit concentration and rewrite Eq. (21-17) as

$$\tilde{G}_i = \tilde{G}_i^0 + RT\ln c_i\qquad(21\text{-}17a)$$

It is important in (21-17a) to realize that c_i does not represent the concentration; rather, it is a dimensionless concentration ratio, because one cannot take logarithms of numbers with dimensions.

The term $\tilde{G}_i^0$ will depend on T, p, and the standard concentration of substance i. It may also depend on the concentrations of other molecular species. To uniquely define $\tilde{G}_i^0$, it is necessary to state all of these standard (or unit) concentrations. This group of standard concentrations is called the *standard state*. To recapitulate, the quantity $\tilde{G}_i^0$ is the Gibbs free energy per mole of substance i, due to the presence of substance i, when the system is in its standard state at absolute temperature T and pressure p.

The standard state need not be a real state. For instance, one might use 1 mole/ℓ for the concentration of catalase in the standard state. A concentration of 1 mmole/ℓ is a large one for catalase, and 1 mole/ℓ is physically unrealizable. In this case, $\tilde{G}_i^0$ means the value of the partial molal free energy obtained by extrapolating from infinite dilution to the standard state under the hypothesis that the solution acted as an ideal one. Even though this hypothesis is wrong, the term $\tilde{G}_i^0$ is a useful one for thermodynamic calculations.

From Eq. (21-10), one sees that it is possible to write

$$\tilde{G}_i^0 = \tilde{H}_i^0 - T\tilde{S}_i^0 \qquad (21\text{-}10a)$$

Changing the standard state by decreasing c_i^0 by a factor of 10^3 will decrease $\tilde{G}_i^0$ by an amount $RT \ln (10^3)$. Because the partial molal enthalpy, in the standard state $\tilde{H}_i^0$, is not dependent on c_i^0, this change in $\tilde{G}_i^0$ must be an increase in $\tilde{S}_i^0$ of $R \ln (10^3)$. Statistical mechanics interprets this increase as the equivalent of saying that 1 mole of substance i can be distributed 10^3 times more ways at 10^{-3} of the original concentration.

In order to assign meaning to $\tilde{S}_i^0$, it is necessary to have a zero for entropy. This is provided by the second statement of the third law of thermodynamics as the entropy of a simple single crystal of a single substance at $0°K$. By varying the standard-state concentration, it is possible to change the magnitude and even the sign of $\tilde{S}_i^0$. Thus, no physical significance can be attached to either the magnitude or the sign of $\tilde{S}_i^0$; it is a useful mathematical construction only.

21.4   Equilibrium

A. EQUILIBRIUM CONSTANTS

The concepts of thermodynamics, and particularly the Gibbs free energy, may be applied directly to the enzyme kinetic rates discussed in Chapter 18. In cases where all (or most) of the kinetic constants are known, equilibrium constants can be found in three fashions: directly, from the rate constants, and from thermodynamic arguments. All three types of values are found to be in good agreement. The equilibrium constant K for the reaction

$$B_1 + B_2 \underset{k_2}{\overset{k_1}{\rightleftharpoons}} C_1 + C_2 \tag{21-18}$$

is defined by the equilibrium value of the ratio

$$K = \frac{[C_1][C_2]}{[B_1][B_2]} \tag{21-19}$$

where the brackets indicate concentrations. Since it is a ratio, the constant K is independent of the concentrations of the reactants. In general, it will depend on temperature, dielectric constant of the suspending media, pH, ionic strength, and so forth.

The equilibrium constant K is directly related to the rate constants k_1 and k_2. By definition of the latter, the rate of formation of $[C_1]$ is given by

$$\frac{d[C_1]}{dt} = k_1[B_1][B_2] - k_2[C_1][C_2]$$

At equilibrium, this expression must vanish. Rearranging the resultant equation, one finds that

$$K = \frac{k_1}{k_2} \tag{21-20}$$

In the case of reactions involving equal numbers of molecules on both sides of the equation, the equilibrium constant K is dimensionless if the same units are used on both sides of the equation. Even if this is not convenient, it is still possible to regard K as dimensionless, provided that one regards the brackets as ratios of the concentrations to those in the standard state.

In general, the number of molecules on the two sides of the equation are unequal; one must either specify a standard state or treat K as a number with dimensions. Although, for most purposes, the second of these is a satisfactory procedure, it is necessary to use some artificial construct as the standard state to relate K to the Gibbs free energy G.

In the last section, it was noted that under isobaric, isothermal conditions, equilibrium occurred at a minimum for the Gibbs free energy G; that is, dG must vanish at equilibrium. To relate Gibbs free energy to K, it is necessary to obtain an expression for dG for a small disturbance of the equilibrium in Eq. (21-18). Symbolically, one may represent G for the reactants in Eq. (21-18) by

$$G = V([B_1]\tilde{G}_{B_1} + [B_2]\tilde{G}_{B_2} + [C_1]\tilde{G}_{C_1} + [C_2]\tilde{G}_{C_2}) \tag{21-21}$$

where V is the volume and where the partial molal free energy has been defined by Eq. (21-17a) as

$$\tilde{G}_{B_i} = \tilde{G}_{B_i}^0 + RT \ln [B_i]$$

Differentiating Eq. (21-21), using the definition of G in Eq. (21-17a), leads to the relationship

$$dG = \left(\frac{G}{V}\right) dV + V(\tilde{G}_{B_1}d[B_1] + \tilde{G}_{B_2}d[B_2] + \tilde{G}_{C_1}d[C_1] + \tilde{G}_{C_2}d[C_2])$$
$$+ RT(d[B_1] + d[B_2] + d[C_1] + d[C_2])\, V \qquad (21\text{-}22)$$

The various differentials are restricted by Eq. (21-18), so that, if the volume remains constant,

$$d[B_1] = d[B_2] = -d[C_1] = -d[C_2] = dx \qquad (21\text{-}23)$$

in which x is an arbitrary parameter expressing the amount that the reaction in Eq. (21-18) has progressed to the right. Substituting Eq. (21-23) into (21-22) and setting dV to zero, one arrives finally at the desired equation,

$$dG = V[\tilde{G}_{B_1} + \tilde{G}_{B_2} - \tilde{G}_{C_1} - \tilde{G}_{C_2}]\, dx \qquad (21\text{-}24)$$

for the change of Gibbs free energy for a small displacement of Eq. (21-18) from equilibrium.

For equilibrium, this last expression must vanish. Replacing the $\tilde{G}$'s by their expressions in Eq. (21-17a) and rearranging terms leads to

$$\tilde{G}_{B_1}^0 + \tilde{G}_{B_2}^0 - \tilde{G}_{C_1}^0 - \tilde{G}_{C_2}^0 = RT\,(\ln[C_1] + \ln[C_2] - \ln[B_1] - \ln[B_2])$$
$$(21\text{-}25)$$

The left-hand side of this equation is the difference in the partial molal free energies of Eq. (21-18) when all substances are in their standard state. This is usually denoted by ΔG^0, defined by

$$\Delta G^0 = \tilde{G}_{C_1}^0 + \tilde{G}_{C_2}^0 - \tilde{G}_{B_1}^0 - \tilde{G}_{B_2}^0 \qquad (21\text{-}26)$$

The right-hand side of Eq. (21-25) may be recognized as $RT \ln K$, thereby allowing one to write

$$\Delta G^0 = -RT \ln K$$

or

$$K = \exp\left(\frac{-\Delta G^0}{RT}\right) \qquad (21\text{-}27)$$

In words, Eq. (21-27) says that equilibrium will favor the side of a chemical equation that has the lowest total partial molal Gibbs free energy in the standard state; that is, the bigger the difference in the free energies of the standard states, the greater will be the tendency of the equilibrium to favor one side of the reaction equation. In cases where the value of ΔG^0 is not known, it can be determined by a graph of $\ln K$ plotted against $1/RT$. A graph of this nature is called an *Arrhenius plot* (see Fig. 21.3).

If there are unequal numbers of reacting molecules on the two sides of the stoichiometric reaction, the sum of concentration changes in Eq. (21-22) will not vanish. Accordingly, Eq. (21-27) must be modified to

$$K = \exp\left(\sum \nu_i - \sum \mu_i\right) \exp\left(\frac{-\Delta G^\circ}{RT}\right) \tag{21-27a}$$

where ν_i are the coefficients of the reactants on the left-hand side of the chemical equation and μ_i are those on the right-hand side. Note also that $\tilde{G}_i$ must be replaced by $\nu_i \tilde{G}_i^0$ or $\mu_i \tilde{G}_i^0$ as appropriate.

B. CATALASE

Equation (21-27a) is illustrated in the remainder of this section by computations of the hydrogen peroxide in equilibrium with air-saturated water without and then with the enzyme catalase (discussed in Chapter 18). The overall reaction being considered obeys the equation

$$2H_2O_2 \underset{k_2}{\overset{k_1}{\rightleftharpoons}} 2H_2O + O_2 \tag{21-28}$$

As standard states, one may choose 1 mole/ℓ for both hydrogen peroxide and oxygen, and 55 moles/ℓ for water. The free energies of these compounds, relative to gaseous molecular hydrogen and oxygen at normal temperature and pressure, may be found in the International Critical Tables. For this reaction

$$\Delta G^0 = 2\tilde{G}_{H_2O}^0 + \tilde{G}_{O_2}^0 - 2\tilde{G}_{H_2O_2}^0$$

Therefore, the following sum is needed:

$$
\begin{aligned}
2\tilde{G}_{H_2O}^0 &= 2(-56,560) &&= -113,120 \text{ kcal/mole} \\
\tilde{G}_{O_2} &&&= + \quad 3,904 \\
-2\tilde{G}_{H_2O_2}^0 &= -2(-31,470) &&= + \quad 62,940 \\
\hline
&&\Delta G^0 &\simeq - \quad 46,300
\end{aligned}
$$

All these numbers are for 15°C. Because there is one more molecule on the right than on the left in Eq. (21-28), Eq. (21-27a) becomes

$$K = \exp\left(\sum \nu_i - \sum \mu_i\right) \exp\left(\frac{-\Delta G^0}{RT}\right) \simeq \exp(-1)\exp\left(\frac{46,300}{600}\right) \simeq 10^{+33}$$

This extremely large equilibrium constant implies that at equilibrium it will be very difficult to detect any hydrogen peroxide. In fact, if one chooses practical values for the ratios of the oxygen and water concentrations to the standard state concentrations—1 and 0.24×10^{-3}, respectively—one may compute the equilibrium concentration of H_2O_2 as

$$[H_2O_2] = \left(\frac{[H_2]^2 \cdot [O_2]}{K}\right)^{1/2} = 5 \times 10^{-19}$$

This value is so low that it represents, on the average, less than 10 molecules of hydrogen peroxide per milliliter.

Anyone can go to a drug store and buy a bottle of hydrogen peroxide solution. This is a nonequilibrium solution. It will remain in this nonequilibrium condition in a dark bottle for many years providing that there are no catalysts present to speed the attainment of equilibrium. Many substances will accomplish this catalysis, including ferric, ferrous, cupric ions, and the iron-containing enzyme catalase. Per metal ion present, the most effective catalyst is the enzyme catalase. None of these catalysts alter either the value of ΔG^0 or K, but they do alter the rate at which equilibrium is attained.

It was shown in Chapter 18 that the reaction catalyzed by catalase could be represented as two successive reactions

$$\overset{e-p}{E} + \overset{x}{S} \underset{k_2}{\overset{k_1}{\rightleftharpoons}} \overset{p}{E \cdot S}$$

$$\overset{p}{E \cdot S} + \overset{x}{S} \underset{k_4}{\overset{k_3}{\rightleftharpoons}} E + 2H_2O + O_2$$

where E represents catalase, S represents hydrogen peroxide, and E·S is the intermediate complex. For each of these two reactions, there is an equilibrium constant defined as

$$K_1 = \frac{p}{(e-p)x} \quad \text{and} \quad K_2 = \frac{(e-p)[H_2O]^2[O_2]}{p \cdot x}$$

Both K's can be represented as quotients of rate constants

$$K_1 = \frac{k_1}{k_2} \quad \text{and} \quad K_2 = \frac{k_3}{k_4}$$

or as exponentials involving changes in free energy,

$$K_1 = \exp\left(1 - \frac{\Delta G_1^0}{RT}\right) \quad \text{and} \quad K_2 = \exp\left(-2 - \frac{\Delta G_2^0}{RT}\right)$$

Straightforward substitution confirms the logically required relationships:

$$K = K_1 K_2 \quad \text{and} \quad \Delta G^0 = \Delta G_1^0 + \Delta G_2^0$$

Since k_4 has not been determined directly, these equations cannot be used to check experimental consistency. However, they can be used to estimate k_2, k_4, and the equilibrium values of the intermediate complex p and the peroxide concentration x. The standard-state concentrations used are 1 mole/ℓ for all substances except water, which is 55 moles/ℓ. Then, selecting the approximate values,

$$k_1 = 2 \times 10^7 \text{ sec}^{-1} \qquad k_3 = 2 \times 10^7 \text{ sec}^{-1}$$
$$k_2 = 2 \times 10^{-4} \text{ sec}^{-1} \qquad e = 5 \times 10^{-6}$$

where all concentrations are dimensionless ratios, one may compute

$$K_1 \simeq 10^{11} \qquad\qquad x_{eq} \simeq 5 \times 10^{-19}$$
$$K_2 \simeq 10^{22} \qquad\qquad p_{eq} \simeq 2.5 \times 10^{-13}$$
$$k_4 \simeq 2 \times 10^{-15} \text{ sec}^{-1} \qquad \frac{p}{e} = 5 \times 10^{-8}$$

The extremely small values of x_{eq}, p_{eq}, and p/e indicate that it would be difficult, if not impossible, to observe them. It is nonetheless customary to assume that the reverse reactions do indeed occur. In other enzyme-catalyzed reactions, such reverse reactions are observed and do agree well with thermodynamic computations.

21.5   Collision Theory

Thermodynamics, as described so far in this chapter, deals with the gross, macroscopic properties of a system. It can, however, also be related to events on a molecular scale by kinetic theory. This and the next section of this chapter contain thermodynamic approaches to various aspects of enzyme systems. As a first step, some applications of thermodynamics to chemical reactions in general will be considered. Because gas-phase reactions are easiest to discuss, they are introduced first.

A. GAS PHASE REACTIONS

Naively, it might appear that energy changes during a reaction border on the trivial. If two molecular species are mixed together, one observes equilibrium and measures ΔG^0, or possibly if the equilibrium is hard to quantify, one could measure ΔH. These have been discussed in previous sections. The simplest assumption would be that every time two reactant molecules, say hydrogen and oxygen, approached each other they reacted, giving off energy.

The obvious fallacies of this are emphasized by the stable coexistence both of hydrogen and oxygen in the atmosphere and of glucose and oxygen in solutions. Reaction rates increase far more rapidly with temperature increases than do collision rates, and catalysts profoundly alter reaction rates, although leaving collision rates unaltered. A reasonable explanation of this was originally proposed by Arrhenius and is summarized in Figs. 21.1 and 21.2. Here it is assumed that a molecule of A with internal energy E_A approaches one of B with internal energy E_B. Their total energy will be given by

$$E_T = E_A + E_B$$

Note that there is a potential-energy barrier tending to keep A and B apart. Only if E_T is sufficiently great can they cross this barrier and react to form C.

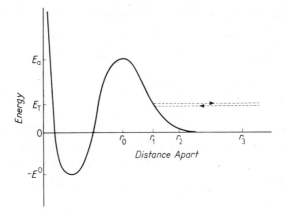

Figure 21.1 Molecular collision. The solid curve represents the potential energy of a molecule of A and a molecule of B when separated by a distance r. It is seen that their thermal energy E_T is less than the activation energy E_a. As a consequence, they cannot come close enough together to react. The dashed line represents their motion relative to one another.

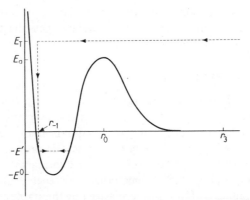

Figure 21.2 Molecular reaction. In contrast to the two molecules depicted in Fig. 21.1, the two here have a sufficient thermal energy E_T to come closer together than r_0. They react at r_{-1}, releasing the energy $E_T + E'$.

Quantum mechanical considerations allow tunneling through the potential barrier, but this process is unlikely at energies much less than E_a.

Figure 21.1 illustrates the case where E_T is not great enough so that the entire internal energy becomes potential at r_1. In Fig. 21.2, the total internal energy is great enough to allow them to react, falling into the potential well and giving off energy. For A and B far apart, E_T represents the thermal energy.

In the following discussions all energies are expressed on a per-mole basis. Statistical mechanics shows that the fraction of molecule pairs having a total energy E_T greater than the height of the barrier E_a is $\exp(-E_a/RT)$. It also allows for gases a computation of the collision frequency f. Thus, the preceding theory indicates for the reaction

$$A + B \xrightarrow{k} C$$

the rate constant should be given by

$$k = f \exp\left(\frac{-E_a}{RT}\right) \qquad (21\text{-}29)$$

If as a result of steric or other factors, not all reactants that approach as closely as the two in Fig. 21.2 actually react, Eq. (21-29) will predict too large a value for k. To avoid this, a probability factor α (less than 1) can be introduced, leading to

$$k = \alpha f \exp\left(\frac{-E_a}{RT}\right) \qquad (21\text{-}30)$$

A test of this theory is to plot the logarithm of k against $1/RT$. If a straight line results, one may calculate E_a even if α and f are unknown. Such lines are illustrated in Fig. 21.3. For gases, since f can be calculated from kinetic theory, one can compute α from Eq. (21-30). The constant E_a is called the *activation energy* or the *Arrhenius constant*. Note that it is not the same as the average energy E^0 obtained from the reactions. Referring again

Figure 21.3 Arrhenius plots. Curves expected for different reactions when the log of the rate constant is plotted against $1/RT$. On the basis of this plot, one determines a slope μ called the *Arrhenius constant*. As explained in the text, low values of μ for reactions in liquids (that is, curve 1) suggest that the reaction may be diffusion-controlled. However, high values of μ (curves 2 and 3) probably reflect the intrinsic properties of the reacting molecules.

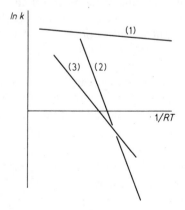

to Fig. 21.2, it is seen that E_a is the height of the barrier and E^0 the depth of the well.

According to kinetic theory for gases, f should show a temperature dependence but vary only slowly compared to $\exp(-E_a/RT)$. Thus, the linear nature of the curves in Fig. 21.3 seems to support collision theory. However, it is not at all clear that E_a should remain constant with temperature (except for reactions between single atoms). Moreover, arguments can be developed to indicate that Figs. 21.1 and 21.2 should have considered Gibbs free energy G rather than total internal energy E. To avoid this problem, it is customary to call the slope of a line in Fig. 21.3 the Arrhenius constant μ and to call the graph an *Arrhenius plot*.

B. REACTIONS IN LIQUIDS

The role of catalysts in general and enzymes in particular may be thought of as lowering the potential-energy barrier which keeps the molecules apart. It would appear desirable within a fluid to talk of a potential barrier of activation $\Delta G^{\ddagger}$. There remain, however, profound difficulties in applying collision theory to enzyme-controlled reactions. Some of these involve practical considerations of the limited temperature range over which the reaction can be studied. More basic are ones that deal with reactants in liquids. Although these questions are best avoided by the methodology discussed in Sec. 21.6, a brief discussion of reactions in liquids using collision theory may give a better intuitive appreciation of the motivation for absolute rate theory.

In a liquid as in a solid, there are equilibrium distances from one molecule to the next. The difference is that in a solid, long-range regular patterns are maintained for large numbers of molecules, whereas in a liquid the order is only local falling off in a few molecular diameters. Diffusion in a liquid occurs by a molecule moving abruptly from one quasi-stable position to the next. Each time two reacting molecules enter neighboring sites is defined as an encounter. During an encounter many collisions may occur, owing to vibrations and rotations about the quasi-stable site. The rate of encounter, f_e, can be readily computed and is proportional to the diffusion constant (see Chapter 23). In contrast, the collision rate f cannot be computed from existing kinetic theories of liquids.

Two extreme types of reactions can be considered. In the first, called *diffusion limited*, each encounter leads to a reaction. Anything lowering the diffusion constant will decrease the reaction rate. For this case one may write

$$k = f_e$$

In this case, the slope of the Arrhenius plot, μ, represents the temperature dependence of the diffusion constant.

The other extreme is one in which most encounters end by the molecules diffusing away from each other rather than reacting. Then, decreasing the diffusion constant decreases f_e but increases the length of each encounter. Accordingly, the collision rate f will be little changed and the reaction will be *diffusion independent*. The slope μ of the Arrhenius plot is then an intrinsic property of the reaction. Its interpretation by strict collision theory is difficult, however.

In spite of its limitations, μ is used to compare enzymes and to summarize temperature dependence of reactions. An equivalent term, Q_{10}, is often used. It is defined as the ratio of the rate constants measured at two temperatures 10°C apart. That is, symbolically defined,

$$Q_{10} = \frac{k_{t+10}}{k_t} \qquad (21\text{-}31)$$

where the subscripts indicate the temperature. It is possible to express the rate constant as proportional to $e^{-\mu/RT}$. Since both the numerator and the denominator of Eq. (21-31) would have the same proportionality constant, one may approximate that equation by

$$Q_{10} \simeq \exp\left(\frac{10\mu}{RT^2}\right) \quad \text{or} \quad \mu \simeq \frac{RT^2}{10}\ln Q_{10} \qquad (21\text{-}32)$$

For biological reactions Q_{10} is usually in the range 1.2–4, with the majority of reactions having values close to 2.0. By substituting in Eq. (21–32), one can show that these values correspond to ones for μ ranging from 3.4 to 25 kcal/mole, with the majority falling in the neighborhood of 12 or 13 kcal/mole.

If one plots the log of the diffusion constant for water against $1/RT$, an approximate straight line results with a slope of about 3 kcal/mole. Thus, most biological reactions are diffusion independent, but a few with the lowest values of Q_{10} might be diffusion limited.

21.6   Absolute Rate Theory

Collision theory, although an intuitive summary of the thermodynamics of reaction rates, has many limitations. An alternative theory, making use of quantum mechanical reasoning, albeit in a heuristic fashion, has been called *absolute rate theory*. To describe the latter, it is convenient to use the diagram in Fig. 21.4. This is more sophisticated than Figs. 21.1 and 21.2 in several regards. The ordinate is now Gibbs free energy, and the abscissa is a reaction coordinate that may be more complicated than a mere linear separation of centers of mass. There are now three regions. Far out on the reaction coordinate, the reactants are treated as separate molecules A and B, and in the potential well as a single molecule C. Near the top of the potential barrier, they are considered a single activated complex $A \cdot B^{\ddagger}$.

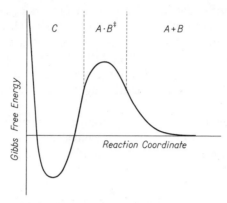

Figure 21.4   The absolute rate theory. The absolute rate theory postulates an activated complex A·B‡, which must be in equilibrium with A + B in order to apply this theory to the interpretation of k. The rate of crossing the barrier from A·B‡ to C is an absolute quantity if certain general assumptions are valid.

The complex, just as in Michaelis–Menten kinetics, controls the rate (see Chapter 18). However, the activated complex stays around for a shorter period of time, has never been directly observed by spectrophotometric or other physical changes, and breaks down to C at an absolute rate given by

$$\frac{1}{[A\cdot B^{\ddagger}]}\frac{d[A\cdot B^{\ddagger}]}{dt} = \frac{RT}{N_0 h}\exp\left(\frac{\Delta S_i}{R}\right) \qquad (21\text{-}33)$$

where R is the gas constant per mole, T is the absolute temperature, N_0 is Avogadro's number, and h is Planck's constant. The term ΔS_i represents the contribution of the entropy of A·B‡ due to motion along the reaction coordinate. The derivation of Eq. (21-33) is beyond the level of this text but is found in several references cited at the end of the chapter.

If the reaction is diffusion-independent, A·B‡ may be thought of as being in quasi-equilibrium with A and B. Using Eq. (21-27), this equilibrium may be expressed as

$$\frac{[A\cdot B^{\ddagger}]}{[A][B]} = \exp\left(\frac{-\Delta G^0}{RT}\right) \qquad (21\text{-}34)$$

where ΔG^0 is the height of the barrier in Fig. 21.4. It is convenient to define an activation free energy $\Delta G^{\ddagger}$ by

$$\Delta G^{\ddagger} = \Delta G^0 - T\,\Delta S_i$$

This relationship permits rearranging Eq. (21-34) as

$$[A\cdot B^{\ddagger}]\exp\left(\frac{\Delta S_i}{R}\right) = [A][B]\exp\left(\frac{-\Delta G^{\ddagger}}{RT}\right) \qquad (21\text{-}34a)$$

Next, since C is formed by, and only by, the breakdown of $A \cdot B^{\ddagger}$, one may write

$$\frac{d[C]}{dt} = -\frac{d[A \cdot B^{\ddagger}]}{dt} \tag{21-35}$$

Recalling the definition of k,

$$k = \frac{1}{[A][B]} \frac{d[C]}{dt}$$

Combining this with Eqs. (21-34a) and (21-35), and rearranging terms leads to the absolute-rate-theory formulation for k,

$$k = \frac{RT}{N_0 h} \exp\left(\frac{-\Delta G^{\ddagger}}{RT}\right) \tag{21-36}$$

It may also be convenient to use the definition of G in Eq. (21-10) to divide $\Delta G^{\ddagger}$ into an enthalpy and an entropy of activation. Then Eq. (21-36) may be rewritten

$$k = \frac{RT}{N_0 h} \exp\left(\frac{-\Delta H^{\ddagger}}{RT}\right) \exp\left(\frac{\Delta S^{\ddagger}}{R}\right) \tag{21-37}$$

The form of Eq. (21-37) argues for assigning μ to $\Delta H^{\ddagger}$, although one might also argue for an internal energy $\Delta E^{\ddagger}$. In any case, all the variables in Eq. (21-37) may be determined experimentally except $\Delta S^{\ddagger}$.

Provided that k refers to concentration ratios, Eq. (21-37) can be used for monomolecular, bimolecular, and polymolecular reactions. For any reaction involving more than one molecule, k and $\Delta S^{\ddagger}$ will depend on the standard state used. Changes both in bond structure and in number of particles can contribute to $\Delta S^{\ddagger}$, as can dielectric effects.

The activated complex should be clearly distinguished from the intermediate complex of enzyme reactions. This is illustrated in Fig. 21.5 for the decomposition of H_2O_2 catalyzed by catalase. The intermediate complex $E \cdot S$ is observable by spectrophotometric means. However, the two activated complexes have not been directly observed.

Absolute rate theory can be used to interpret a number of other kinetic phenomena. For example, the calculation of $\Delta H^{\ddagger}$ and $\Delta S^{\ddagger}$ allows two estimates of the number of bonds broken in enzyme and other protein denaturation. These two estimates agree reasonably well. It is also possible to interpret the values of $\Delta S^{\ddagger}$ in diffusion-independent reactions provided that one can separate the electrostatic effects. Both of these types of interpretation are discussed in references at the end of the chapter.

A summary of thermodynamics is included at the end of the following chapter. The topics in this chapter have dealt exclusively with thermodynamics of reversible systems. Absolute rate theory also contains the implicit assumption that such is the case. However, as noted in Sec. 21.3, even for catalase

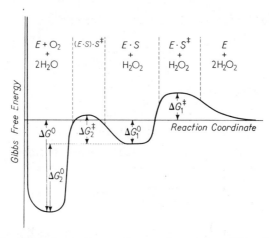

Figure 21.5 Absolute-rate-theory diagram for catalase. This diagram distinguishes the activation energies and the free-energy changes due to the reaction. It also emphasizes the difference between the intermediate complex $E \cdot S$ and the activated complexes $E \cdot S^{\ddagger}$ and $(E \cdot S) \cdot S^{\ddagger}$.

the reverse reaction cannot be demonstrated. Most biological systems are strictly both irreversible and nonequilibrium. Chapter 22 discusses special analyses and rules applicable to irreversible thermodynamics.

REFERENCES

1. DAVIDOVITZ, P., *Physics in Biology and Medicine* (Englewood Cliffs, N.J.: Prentice-Hall, Inc., 1975), 298 pages.

2. BREUER, H., *Physics for Life Science Students* (Englewood Cliffs, N.J.: Prentice-Hall, Inc., 1975), 453 pages.

3. MAHAN, B. H., *Elementary Chemical Thermodynamics* (New York: W. A. Benjamin, Inc., 1963), 155 pages.

4. KLOTZ, I., *Introduction to Chemical Thermodynamics* (New York: W. A. Benjamin, Inc., 1964), 244 pages.

5. LEHNINGER, A. L., *Bioenergetics: The Molecular Basis of Biological Energy Transformations* (New York: W. A. Benjamin, Inc., 1965), 258 pages.

6. GLASSTONE, S., K. J. LAIDLER, AND H. EYRING, *Theory of Rate Processes: The Kinetics of Chemical Reactions, Viscosity, Diffusion and Electro-chemical Phenomena* (New York: McGraw-Hill Book Company, 1941), 611 pages.

CHAPTER 22

Irreversible
Thermodynamics

22.1   Nonequilibrium and Irreversible Systems

Chapter 21 discusses applications of classical thermodynamics to biological systems. Such approaches have proved to be useful in describing energy utilization by living systems and the kinetics of enzyme reactions. In essence, the approach of the previous chapter restricted the considerations to equilibrium systems and to changes that occurred reversibly. Under those circumstances it proved useful to consider three laws of thermodynamics and certain special thermodynamic functions. The first law, conservation of energy, applies unaltered in nonequilibrium and other irreversible systems. Its discussion led to the introduction of the thermodynamic function called entropy, S, which, however, was defined only in terms of reversible changes between equilibrium states. Entropy, in turn, led to the second and third laws of thermodynamics. It also permits defining a number of other thermodynamic functions. Of these the Gibbs free energy, G, proves most useful for biological applications since it is a minimum for equilibrium systems subject to constant temperature, T, and constant pressure, P, constraints.

   If one examines any real biological system in greater detail, it becomes clear that it is a nonequilibrium system. The various carbohydrates, proteins, and lipids all coexist with molecular oxygen, whereas equilibrium would imply conversion to carbon dioxide and water. Electrical charges are separated across membranes and unequal concentrations are maintained across membranes in living cells. These are also nonequilibrium conditions. The living animal continually maintains body posture and position in a mechanical nonequilibrium state. Plants promote nonequilibrium by storing the energy in the sun's rays in the form of sugars. Thus, the study of living systems strongly suggests that a theory of nonequilibrium thermodynamics would be useful. Since changes from nonequilibrium states toward equilibrium are, in general, irreversible, this branch of study is often called *irreversible thermodynamics*.

This chapter presents the general theory of irreversible thermodynamics, the methodology of its application, and several physical and biophysical examples. Some of the more significant advances in biophysics during the 1960's resulted from the applications of irreversible thermodynamics to various membrane-related phenomena. These are included in Chapter 23, which deals with diffusion, permeability, and active transport.

To introduce irreversible thermodynamics, it is convenient to restate the second law in alternative forms. One formulation states that it is impossible to derive mechanical work from a single heat source by cooling it below the temperature of the surrounding objects. In other words, for any cyclic process returning to its original state, one can express this symbolically as

$$\oint_1 \delta W \leq 0 \qquad (22\text{-}1)$$

where the subscript 1 indicates that there is only one heat reservoir. (A sufficient, but not necessary, condition is that the entire cycle be isothermal.) The equality in Eq. (22-1) holds only if the process is reversible. Two examples follow which illustrate for idealized physical systems the basic application of Eq. (22-1) when irreversible processes are involved.

The first example considered is a cyclic process consisting of three parts, as illustrated in Fig. 22.1. The first leg from point 1 to point 2 is an irreversible adiabatic process (an *adiabatic process* is one which occurs with no transfer of heat energy). To calculate the entropy change involved, the cycle is completed by a reversible adiabatic change from point 2 to point 3 and then an isothermal one back to point 1. For these latter two steps it follows by definition that

$$T_3 = T_1 \quad \text{and} \quad S_3 = S_2$$

as illustrated in Fig. 22.1. There is a complete cycle, so it follows, since dE is an exact differential, that

$$\oint dE = 0 \qquad (22\text{-}2)$$

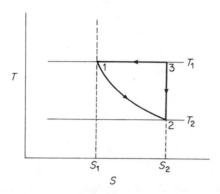

Figure 22.1   Adiabatic irreversible change. The system moves from point 1 to point 2 in the temperature (T) and entropy (S) plane along an irreversible adiabatic path. The cycle is completed by two reversible steps: first, an adiabatic one from point 2 to point 3 and then an isothermal one to return to point 1. After A. Katchalsky and P. F. Curran, in *Nonequilibrium Thermodynamics in Biophysics* (Cambridge, Mass.: Harvard University Press). Copyright © 1965 by the President and Fellows of Harvard College.

Hence, the first law of thermodynamics shows for this cycle that

$$\oint \delta Q = \oint \delta W \tag{22-3}$$

The second law, as stated in Eq. (22-1), can be applied to this cycle since the only heat reservoir is involved from 3 to 1. Hence,

$$\oint \delta W \leq 0$$

Therefore,

$$\oint \delta Q \leq 0 \tag{22-4}$$

But δQ vanishes except from 3 to 1, a reversible step; hence,

$$\oint \delta Q = T_1 \cdot \int_3^1 dS = T_1(S_1 - S_3) \leq 0 \tag{22-5}$$

Thus, the entropy will increase in an irreversible adiabatic process.

For an irreversible isothermal process, one can return to the starting point, as shown in Fig. 22.2 by a reversible isothermal process alone. Then Eqs. (22-1) through (22-5) will remain valid. Calling the point reached by the irreversible process 2, one can now write

$$\oint \delta Q = \int_1^2 \delta Q^{irr} + T \int_2^1 dS = \Delta Q^{irr} + T(S_1 - S_2) \leq 0$$

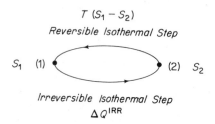

$$T\,(S_1 - S_2)$$
Reversible Isothermal Step

S_1 (1)   (2) S_2

Irreversible Isothermal Step
$$\Delta Q^{\mathrm{IRR}}$$

Figure 22.2 Irreversible isothermal cycle. The system moves first along an irreversible isothermal pathway from point 1 to point 2 in the entropy (S) plane and then returns along an isothermal pathway. After A. Katchalsky and P. F. Curran, in *Nonequilibrium Thermodynamics in Biophysics* (Cambridge, Mass.: Harvard University Press). Copyright © 1965 by the President and Fellows of Harvard College.

The last inequality may be written as

$$S_2 - S_1 \geq \frac{\Delta Q^{\mathrm{irr}}}{T} \tag{22-6}$$

Thus, in place of the earlier definition of dS for reversible systems, it has been shown in Eqs. (22-5) and (22-6) that

$$dS \geq \frac{\delta Q}{T} \tag{22-7}$$

for either an adiabatic or an isothermal system.

In general, the approach taken to find the entropy changes in an irreversible process is to compute the changes that would accompany reversible processes whose net effect would be to return the system to the starting point. For any irreversible process, the inequality expressed in Eq. (22-7) will be valid. It is therefore convenient to divide dS into two parts, one, dS_e, representing heat exchange with the surroundings and the other, dS_i, representing internally generated entropy. In equation form these lead to

$$dS = dS_e + dS_i$$

$$dS_e = \frac{\delta Q}{T} \tag{22-8}$$

$$dS_i \geq 0$$

In an adiabatic process, by definition,

$$dS_e = 0$$

and accordingly

$$dS = dS_i \geq 0$$

An example illustrating this is shown in Fig. 22.3. Two compartments are each in good thermal contact with large reservoirs at temperatures T_1 and a lower temperature T_2, respectively. For the entire system

$$dS = dS_{1e} + dS_{1i} + dS_{2e} + dS_{2i}$$

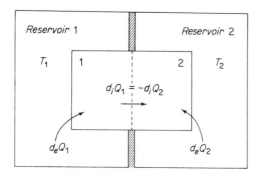

Figure 22.3   Heat transfer between two large reservoirs maintained at a higher temperature, T_1, and a lower temperature, T_2. Heat is transferred irreversibly from reservoir 1 to reservoir 2. The various symbols are defined in the discussion in the text. After A. Katchalsky and P. F. Curran, in *Nonequilibrium Thermodynamics in Biophysics* (Cambridge, Mass.: Harvard University Press). Copyright © 1965 by the President and Fellows of Harvard College.

The irreversible process occurs due to the thermal contact between the compartments. Therefore,

$$dS_i = dS_{1i} + dS_{2i} \geq 0$$

must result from heat δQ flowing from compartment 1 to 2. Hence,

$$dS_i = \frac{\delta Q}{T_2} - \frac{\delta Q}{T_1} \geq 0 \tag{22-9}$$

This confirms that δQ is positive; thus, the restatement of the second law is consistent with the statement that energy flows from a hotter body to a cooler one when the two are in contact. This is implicit in the definition of temperature and is sometimes called the *zeroth law of thermodynamics*.

22.2   Dissipation Functions

The internal production of entropy is the outstanding characteristic of a real, irreversible thermodynamic system. Such entropy production corresponds to a dissipation of the energy of the system in the form of heat energy.

From the point of view of statistical mechanics it implies the dissipation of nonequilibrium ordered forms of energy as the random motions characteristic of heat. The function Φ, which describes the rate at which internal heat energy is being produced, is called the *dissipation function*. Symbolically it is defined by

$$\Phi = T\frac{dS_i}{dt}$$

Note that Φ has units of power and must always be positive or zero. It is also possible to use, instead of Φ, the related rate of entropy production. However, the dissipation function Φ is more convenient in practical problems.

One of the simplest examples of an irreversible process occurs when a direct current flows through a resistor. In this irreversible system ordered electrical energy is dissipated as heat, thereby increasing the internal entropy. Reversing the direction of the current flow and potential difference does not reconvert the heat to electrical energy. Heating the resistor likewise fails to add electrical energy to the system to any appreciable degree. (See, however, the discussions in Sec. 22.4.) Following the notation of Chapter 21, using $\mathscr{E}$ for the potential difference across the resistor and I for the current flowing through it, one may write

$$\Phi = \mathscr{E}I$$

which, although it has different symbols, is the familiar elementary physics formula for the power dissipated in the resistor. One could write equally well

$$\frac{dS_i}{dt} = \frac{\mathscr{E}I}{T}$$

provided that T remains constant.

If there are several resistors in a circuit being considered, one may rewrite the preceding equation using the subscript j to identify the particular resistor as

$$\Phi = \sum_j \mathscr{E}_j I_j \tag{22-10}$$

This is the general approach of irreversible thermodynamics, to identify or describe a dissipation which is a sum of terms each of which is a simple product. The attempt is then made to partition these products into generalized fluxes J_j and generalized driving forces X_j. For Eq. (22-10), this process is particularly simple, since one can merely choose

$$X_j = \mathscr{E}_j \quad \text{and} \quad J_j = I_j \tag{22-11}$$

The next example of a dissipation function can be described by reference to Fig. 22.4. An adiabatic rigid box contains two compartments separated by an extremely flexible porous membrane through which gas molecules can

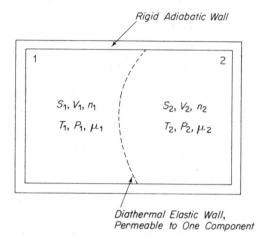

Rigid Adiabatic Wall

Diathermal Elastic Wall,
Permeable to One Component

Figure 22.4 Thermodynamically isolated system separated into two compartments by a permeable, freely extensible membrane. This is used to illustrate several examples of irreversible transfer of heat energy, potential energy, and molecules between the two compartments. After A. Katchalsky and P. F. Curran, in *Nonequilibrium Thermodynamics in Biophysics* (Cambridge, Mass.: Harvard University Press). Copyright © 1965 by the President and Fellows of Harvard College.

pass, but at a restricted rate. Subscript 1 refers to the left compartment in the figure and 2 to the right. The internal energy, E, the total volume, V, and the total number of moles, n, must all be constant. Hence, one may write for any change

$$dE_1 = -dE_2$$
$$dV_1 = -dV_2 \qquad (22\text{-}12)$$
$$dn_1 = -dn_2$$

The total entropy, however, will not in general remain constant. Hence,

$$dS = dS_1 + dS_2 \neq 0 \qquad (22\text{-}13)$$

To arrive at a formula for dS one may write out the first law of thermodynamics for each compartment as

$$dE_1 = T_1 \, dS_1 - P_1 \, dV_1 + \mu_1 \, dn_1$$
$$dE_2 = T_2 \, dS_2 - P_2 \, dV_2 + \mu_2 \, dn_2$$

Solving for dS_1 and dS_2, adding and using Eqs. (22-12) and (22-13), one finds that

$$dS = dE_1 \left(\frac{1}{T_1} - \frac{1}{T_2} \right) + dV_1 \left(\frac{P_1}{T_1} - \frac{P_2}{T_2} \right) - dn_1 \left(\frac{\mu_1}{T_1} - \frac{\mu_2}{T_2} \right) \quad (22\text{-}14)$$

Dividing both sides by dt, and using Δ for the difference between the two sides, leads to

$$\frac{dS}{dt} = \frac{dE_1}{dt}\, \Delta\!\left(\frac{1}{T}\right) + \frac{dV_1}{dt}\, \Delta\!\left(\frac{P}{T}\right) - \frac{dn_1}{dt}\, \Delta\!\left(\frac{\mu}{T}\right) \qquad (22\text{-}15)$$

Equation (22-14) is in the form desired if one identifies the derivatives as fluxes and the differences as the corresponding driving forces. Thus, one may define J_j and X_j by

$$J_1 = \frac{dE_1}{dt} \qquad X_1 = \Delta\!\left(\frac{1}{T}\right)$$

$$J_2 = \frac{dV_1}{dt} \qquad X_2 = \Delta\!\left(\frac{P}{T}\right)$$

$$J_3 = \frac{dn_1}{dt} \qquad X_3 = -\Delta\!\left(\frac{\mu}{T}\right)$$

and then reexpress the ideas in Eqs. (22-14) and (22-15) in the form of Eq. (22-10) as

$$\Phi = \left(\sum_1^3 J_j X_j\right) T \qquad (22\text{-}16)$$

Equations (22-14), (22-15), or (22-16) can be understood by considering each flux separately. Thus, if P/T and μ/T are the same on both sides but T differs, since dS must be positive, energy will flow from the hotter side to the cooler through the thin membrane. Similarly, if P and T are the same on both sides but μ differs, molecules will flow from the high-chemical-potential side to the low one. The dissipation function thus provides for this example a convenient form to describe the rate of entropy production. However, when several of the generalized driving forces are nonzero, it is not obvious which signs the fluxes will have except that the sum Φ must always be positive or zero. This problem is considered further in the next section.

Before pursuing that problem further, two other examples should be mentioned. One involves dissipation functions for chemical reactions. Thus, one may begin with a 1 mM solution of hydrogen peroxide in a nonequilibrium state. Upon adding even a minute trace of catalase, the peroxide is rapidly converted to oxygen and water. In this case there will be a flux J which is the same as the reaction velocity in Chapter 18. The corresponding generalized driving force will be

$$X = \Delta G^0 \qquad (22\text{-}17)$$

where ΔG^0 has the same meaning as in Chapter 21. All the available energy has been converted to heat and the process is irreversible. In glucose metabolism, the dissipation function is harder to specify, since most of the available energy is stored in the pyrophosphate bonds of ATP. However, the exact fraction varies considerably, depending on a variety of factors, such as ADP availability, glycolysis or pentose shunt, and temperature. Nonetheless, one can define a flux and a thermodynamic driving force which leads to internal entropy production.

Other examples involve processes that occur distributed over a membrane or throughout a three-dimensional volume. If this is the case, one has to consider the dissipation function per unit area or per unit volume, respectively. Under these circumstances the fluxes may also be measured per unit area or per unit volume. Although this is usually the convenient approach, alternative formulations partition the products in the dissipation function in other fashions.

22.3 Phenomenological Equations

The preceding two sections have summarized the general approach of irreversible thermodynamics: to compute entropy changes by considering conceptual reversible systems operating between the two end points of the system and to use this information to generate a dissipation function. This dissipation function is then separated into sums of terms, each of which is partitioned into a product of a generalized flux and a generalized force. It should be clearly noted, however, that there is absolutely nothing that guarantees that one can divide up the dissipation function in such a fashion. Success comes by an intuitive grasp of the situation, not by following well-defined algorithms. Even if one is successful, the entire process would be of limited intellectual reward if there were no way to relate the generalized fluxes and forces. However, there exist a number of examples in which approximate rules known as *phenomenological equations* can be used expeditiously. The remainder of this chapter and parts of the next one are concerned with those examples and their limitations.

To approach this problem consider an object moving (without rotation) having a mass m, an acceleration $\vec{a}$, and a velocity $\vec{v}$. In the presence of friction one can approximate external force on the object, $\vec{f}_e$, by the relationship

$$\vec{f}_e = k\vec{v} + m\vec{a} \tag{22-18}$$

where k is called the *coefficient of friction*. Much of classical physics is based on frictionless systems, where the relationship

$$k\vec{v} = 0$$

is valid. These studies proved important in the development of the concepts of physics.

However, for a real object, or any other system in a quasi-steady state, one may ignore the $m\vec{a}$ term and write

$$\vec{f}_e \simeq k\vec{v} \tag{22-19}$$

Equation (22-19) is a phenomenological equation. It is approximate not only due to ignoring ma but also because the dependence of friction on velocity is more complicated than is indicated in Eq. (22-19). Nonetheless, for a train or an airplane or a person going along on the level at somewhat constant speeds, Eq. (22-19) is a good, heuristic approximation.

A similar approximate law for diffusion is known as *Fick's law* and is discussed in Chapter 23. A widely known approximation with a similar basis is *Ohm's law* in electricity. Both Ohm's law and Fick's law assume a linear relationship between a flux and a driving force. In both of these cases, as well as in the example summarized by Eq. (22-19), multiplying the generalized flux by the corresponding driving force leads to a dissipation function.

In some cases the various fluxes and driving forces are coupled in a more complicated form. As an extremely simple example, consider the three resistors arranged as a Y shown in Fig. 22.5. It is apparent that the dissipation function can be written

$$\Phi = \mathscr{E}_1 I_1 + \mathscr{E}_2 I_2 \tag{22-20}$$

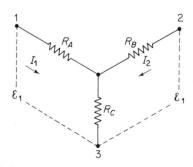

Figure 22.5 Electrical network consisting of three resistors arranged as a Y. *I* indicates electrical current; *R*, resistance; and $\mathscr{E}$, electrical potential. This figure is used to relate Ohm's law to Onsager's phenomenological equations.

so that $\mathscr{E}_1$ is the generalized force corresponding to the generalized flux I_1 and $\mathscr{E}_2$ to I_2. Now Ohm's law may be written for this circuit as two equations,

$$\mathscr{E}_1 = (R_A + R_C)I_1 + R_C I_2$$

$$\mathscr{E}_2 = R_C I_1 + (R_B + R_C)I_2 \tag{22-21}$$

These equations can be solved for the fluxes to give

$$I_1 = L_{11}\mathscr{E}_1 + L_{12}\mathscr{E}_2$$
$$I_2 = L_{21}\mathscr{E}_1 + L_{22}\mathscr{E}_2 \tag{22-22}$$

where the coefficients are given by

$$L_{11} = \frac{R_B + R_C}{D}$$

$$L_{22} = \frac{R_A + R_C}{D}$$

$$L_{12} = L_{21} = \frac{-R_C}{D} \tag{22-23}$$

$$D = R_A R_B + R_C(R_A + R_B)$$

Thus, there is cross coupling between each flux and the other driving force.

In 1931, Onsager proposed extending this type of linear relationship to coupled fluxes and generalized forces such as the ones in the example illustrated in the previous discussion. Thus, if there are three fluxes and three corresponding forces, one might write

$$J_1 = L_{11}X_1 + L_{12}X_2 + L_{13}X_3$$
$$J_2 = L_{21}X_1 + L_{22}X_2 + L_{23}X_3 \tag{22-24}$$
$$J_3 = L_{31}X_1 + L_{32}X_2 + L_{33}X_3$$

These are called phenomenological equations or *Onsager's equations*. The proportionality constants L_{jk} are called *straight coefficients* if the two indices are equal (i.e., L_{jj}) and *coupling coefficients* if they are unequal.

One may generalize to n fluxes, n driving forces and hence n equations. These may be represented by the notation

$$J_j = \sum_{k=1}^{n} L_{jk}X_k \qquad (j = 1, 2, \ldots, n) \tag{22-25}$$

The coefficients L_{jk} are fluxes per unit force, but since each type of flux may have different physical dimensions, the various L_{jk}'s will not be comparable. This is the case for the example illustrated by Fig. 22.4 in the previous section, where the fluxes were energy, volume, and mass per unit time, respectively. However, the dimensions of any force are either

$$[X_k] = \frac{[\text{power}]}{[J_k]} \quad \text{or} \quad [X_k] = \frac{[\text{power}/T]}{[J_k]}$$

Hence, the dimensions of L_{jk} and L_{kj} will be the same.

There is no a priori reason why there should be reciprocity between corresponding cross couplings, that is, that

$$L_{jk} = L_{kj} \tag{22-26}$$

However, that was the case for the Ohm's law example summarized in Eq.

(22-23). For many physical and biophysical systems, this symmetry does exist. Equation (22-26) is sometimes called *Onsager's law*, or *Onsager's reciprocal relationship.*

In general, it is possible to solve the equation set (22-25) for the forces. The result may be written

$$X_k = \sum_{j=1}^{n} R_{kj}J_j \qquad (k = 1, 2, \ldots, n) \qquad (22\text{-}27)$$

This is similar to the electrical example and the R_{kj} terms may be considered generalized resistors. It can be shown that an alternative form of Eq. (22-26), Onsager's law, is

$$R_{kj} = R_{jk} \qquad (22\text{-}28)$$

To summarize this discussion, irreversible thermodynamics leads to dissipation functions, which can in many cases be decomposed to sums of terms each of which is the product of a generalized flux times a corresponding generalized force. When this is possible, if the system is near equilibrium, or in a steady state, or changing only very slowly, one may assume a multivariate, linear relationship between each flux and all the forces. The set of relationships are called phenomenological equations. In general, whenever only small perturbations are involved, it is possible to linearize the equations describing the system. This is a standard technique in mathematics, statistics, and physics, and has many historic precedents.

An additional relationship that assumes equality of the cross-coupling reactions is also often true. Such reciprocity relationships are a characteristic of mechanical and of electrical systems. They need not be true in irreversible thermodynamics, even if the phenomenological equations are valid. However, in most of the successful biophysical applications, Onsager's law is used.

Brief attention should be directed to the case of distributed systems. Here some of the fluxes may be vector quantities. The corresponding generalized forces will also be vectors. Each component of each vector force is linearly related to all the components of all the fluxes. Thus, the problem is far more complicated. The *Curie–Prigogine principle* states that if there are scalar fluxes and vector ones, no cross-coupling exists between the scalar and the vectorial quantities. No use is made in this text of the Curie–Prigogine principle, but it may be important for more advanced applications of irreversible thermodynamics to biophysical systems.

22.4   Applications of Irreversible Thermodynamics

The preceding formulas can be subject to considerable additional analysis and manipulation. Rather than pursuing such an approach further, attention will be focused on some of the applications of irreversible thermodynamics to physics, to chemistry, and to biologically related fields. As the first example, the formulas of the preceding section will be used to describe thermoelectric phenomena in the terminology of irreversible thermodynamics. For this purpose, it will prove convenient to make repeated use of the circuit shown in Fig. 22.6. Two dissimilar metals, A and B, are joined at junctions 1 and 2

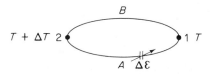

Figure 22.6   Electrical circuit consisting of two wires, A and B, made of dissimilar metals. An emf, $\Delta\mathscr{E}$, is inserted into wire A. The two junctions between metals 1 and 2 are maintained at temperatures T and $T + \Delta T$, respectively.

maintained at temperatures T and $T + \Delta T$, respectively. A variable emf, $\Delta\mathscr{E}$, is inserted into wire A. All quantities will be considered as positive if flowing (or causing flows) from junction 2 to junction 1. Fluxes both of electricity (e) and of entropy (S) must be included. Thus, the dissipation function will have the form

$$\Phi = J_e^A X_e^A + J_e^B X_e^B + J_S^A X_S^A + J_S^B X_S^B \qquad (22\text{-}29)$$

where the superscripts refer to the particular wire.

Referring to Fig. 22-6, and demanding electrical neutrality throughout, it can be shown that

$$I = J_e^A = -J_e^B \qquad \frac{dQ}{dt} = T(J_S^A + J_S^B) \qquad (22\text{-}30)$$

$$\Delta\mathscr{E} = X_e^A - X_e^B \qquad \Delta T = X_S^A = X_S^B$$

where I is the current and dQ/dt is the heat flowing from 2 to 1. (It should be noted that dQ is an exact differential in this example.) This permits rewriting Eq. (22-29) as

$$\Phi = I\,\Delta\mathscr{E} + \frac{1}{T}\frac{dQ}{dt}\,\Delta T$$

It will prove more convenient, however, to use Eq. (22-29) to develop the phenomenological equations, treating wires A and B initially as separate systems and then using Eq. (22-30).

With this approach the phenomenological equations become

$$J_S^A = L_{11}^A X_S^A + L_{12}^A X_e^A \qquad J_S^B = L_{11}^B X_S^B + L_{12}^B X_e^B$$

$$J_e^A = L_{21}^A X_S^A + L_{22}^A X_e^A \qquad J_e^B = L_{21}^B X_S^B + L_{22}^B X_e^B \qquad (22\text{-}31)$$

Now several special cases can be considered. These will be treated in an order chosen to limit the analytical manipulations rather than a historical one or one that would make the physical processes clearest.

First consider the isothermal case where ΔT, and hence, X_S^A and X_S^B all vanish. Then the equations for the current become, using Eq. (22-30),

$$I = L_{22}^A X_e^A = -L_{22}^B X_e^B$$

Therefore,

$$\Delta \mathscr{E} = \left(\frac{1}{L_{22}^A} + \frac{1}{L_{22}^B} \right) I$$

Thus, L_{22} is the reciprocal, ohmic resistance. Note that RI^2 heat will be produced. The equations for the entropy fluxes show that in the absence of a temperature gradient, J_S still need not be zero. Rather, heat may be transferred from one junction to the other. This is known as the *Peltier effect*; it is quantified by the Peltier coefficient, Π, defined as

$$\Pi = \frac{dQ/dt}{I} \bigg|_{\Delta T = 0}$$

Under these conditions Eqs. (22-31) and (22-30) can be combined to yield

$$\Pi = -T \left(\frac{L_{12}^A}{L_{22}^A} - \frac{L_{12}^B}{L_{22}^B} \right) \tag{22-32}$$

Next consider the case where $\Delta \mathscr{E}$ is adjusted to make I (and hence J_e^A and J_e^B) vanish. Under these conditions

$$X_e^A = -\left(\frac{L_{21}^A}{L_{22}^A} \right) X_S^A \quad \text{and} \quad X_e^B = -\left(\frac{L_{21}^B}{L_{22}^B} \right) X_S^A \tag{22-33}$$

Thus, it follows that for A,

$$J_S^A = \left(L_{11}^A - \frac{L_{12}^A L_{21}^A}{L_{22}^A} \right) X_S^A$$

The thermal conduction is defined for this case as

$$\lambda = \frac{1}{\Delta T} \left(\frac{dQ}{dt} \right)$$

Hence,

$$\lambda = T \left(L_{11}^A - \frac{L_{12}^A L_{21}^A}{L_{22}^A} \right)$$

Just as the RI^2 losses, this thermal conduction is an irreversible production of entropy.

The case of no electrical current flow is also the basis of the use of thermocouples to measure temperature. To relate this to the thermodynamic terms, Eqs. (22-30) and (22-33) may be combined to yield, using a bar over $\mathscr{E}$ to indicate no current flow,

$$\Delta \bar{\mathscr{e}} = X_e^A - X_e^B = -\left[\left(\frac{L_{21}^A}{L_{22}^A}\right) - \left(\frac{L_{21}^B}{L_{22}^B}\right)\right] X_S^A$$

or

$$\frac{d\bar{\mathscr{e}}}{dT} = -\left(\frac{L_{21}^A}{L_{22}^A} - \frac{L_{21}^B}{L_{22}^B}\right) \tag{22-34}$$

Equations (22-32) and (22-34) would be quite similar if Onsager's law were valid. Assuming this leads directly to the relationship

$$\Pi = T\frac{d\bar{\mathscr{e}}}{dT} \tag{22-35}$$

This is known as *Thompson's first equation* and agrees well with experimental data.

In addition, Π may be expected to vary if the temperature changes. Thus, an additional amount of heat will be found necessary to maintain the temperature gradients in Fig. 22.6. One may differentiate Eq. (22-35) with respect to temperature, obtaining

$$\frac{d\Pi}{dT} = \frac{d\bar{\mathscr{e}}}{dT} + T\frac{d^2\bar{\mathscr{e}}}{dT^2} \tag{22-36}$$

The last term may be recognized, using Eq. (22-34), as the difference between two terms σ_A and σ_B defined by

$$\sigma_A = -T\frac{d}{dT}\left(\frac{L_{21}^A}{L_{22}^A}\right) \quad \text{and} \quad \sigma_B = -T\frac{d}{dT}\left(\frac{L_{21}^B}{L_{22}^B}\right)$$

These terms represent the added energy per unit current necessary to maintain a temperature gradient in a single wire. The coefficients σ_A and σ_B are named after Thompson, who first discovered this effect. Equation (22–36) is also known as *Thompson's second equation* and can be developed without recourse to irreversible thermodynamics. The development using the latter has the advantage of relating empirical observations to general thermodynamic principles.

Chemical and biochemical applications of irreversible thermodynamics involve further approximations, as discussed in the following. The thermoelectric applications may be regarded as being especially successful because one deals with absolute temperatures and the approximations made by considering a change of 10°K as small compared to 300°K are quite appropriate. For chemical applications, one must assume that the system is close to equilibrium, which is a prohibitive assumption for many enzyme kinetic studies. However, for metabolic systems in a quasi-steady state close to equilibrium, the approximations become more tenable. Thus, the use of a cyclic process such as the citric acid cycle in Chapter 18 would be quite reasonable. However, the mathematics become quite prohibitive, even from the level of the earlier parts of this chapter. Accordingly, a simpler system illustrated in Fig. 22.7 is used.

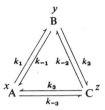

Figure 22.7 Chemical fluxes among three species, A, B, and C. The letters x, y, and z represent concentrations and the k's are rate constants.

Consider starting near equilibrium and describing the approach of the system toward equilibrium. The lowercase letters above each reactant may be used to represent its concentration, and bars to represent the equilibrium values. Then it also proves helpful to use

$$\alpha = \frac{x - \bar{x}}{\bar{x}} \qquad \beta = \frac{y - \bar{y}}{\bar{y}} \qquad \gamma = \frac{z - \bar{z}}{\bar{z}}$$

For fluxes one may use

$$J_1 = k_1 x - k_{-1} y \qquad J_2 = k_2 y - k_{-2} z \qquad J_3 = k_3 z - k_{-3} x$$

and for generalized forces the corresponding differences in chemical potentials

$$X_1 = \mu_A - \mu_B \qquad X_2 = \mu_B - \mu_C \qquad X_3 = \mu_C - \mu_A$$

For this case, the dissipation function is

$$\Phi = \sum_{j=1}^{3} J_j X_j$$

Actually only two driving forces are mutually independent, since

$$X_3 = -(X_1 + X_2)$$

Using

$$J_1^1 = J_1 - J_3$$
$$J_2^1 = J_2 - J_3$$

one can write

$$\Phi = J_1^1 X_1 + J_2^1 X_2$$

At equilibrium X_1 and X_2 vanish and J_1^1 and J_2^1 must also. An added principle of physical chemistry, called *detailed balance*, indicates that at equilibrium J_1, J_2, and J_3 must each vanish.

Far from equilibrium the linearity implied by the phenomenological equations is at variance with the law of mass action, which formed the basis for the treatment in Chapter 18. Near equilibrium, one can approximate μ_A as

$$\tilde{\mu}_A = \bar{\mu} + RT \ln \left(\frac{x}{\bar{x}} \right) \simeq \bar{\mu} + RT\alpha$$

where no subscript is needed for $\bar{\mu}$ since at equilibrium it must be the same for all three reactants. Proceeding similarly for B and C, one may write

$$X_1 \simeq RT(\alpha - \beta) \qquad X_2 \simeq RT(\beta - \gamma)$$

From this, by laborious but straightforward algebra, one may develop expressions for L_{11}, L_{12}, and L_{22} defined by

$$J_1^1 = L_{11}X_1 + L_{12}X_2$$
$$J_2^1 = L_{21}X_2 + L_{22}X_2$$

It can be shown in this fashion that Onsager's law is valid for this case also.

If one now attempts to apply this reasoning to the citric acid cycle (Chapter 18), it is found that one of the reactions is for all practical purposes irreversible. Under these conditions a different approach is required. It is assumed that some of the generalized forces are constrained and others are free to reach a steady state subject to the constraints. Under such conditions, the fluxes conjugate to the unconstrained forces may be expected to vanish, at least if the system is sufficiently close to equilibrium to justify writing phenomenological equations. In addition, if one assumes that Onsager's law is valid, it can be shown that the rate of entropy production (or dissipation of energy) is minimal.

Unfortunately, there are a large number of if's in the preceding paragraph. It is difficult to determine the extent to which any of the if clauses can be considered valid for a process such as the citric acid cycle. Thus, although all metabolic systems are irreversible in the sense discussed earlier in this chapter, the success in applying the techniques of irreversible thermodynamics has been quite limited as compared with applications such as thermoelectricity.

Probably the most successful biophysical applications of irreversible thermodynamics have been in the areas of diffusion and of membrane permeability. The assumptions underlying the phenomenological equations and Onsager's law are better realized in applications to diffusion and to permeability than they are in enzyme-related phenomena. Descriptions of those applications of irreversible thermodynamics are included in Chapter 23.

22.5   Summary of Thermodynamics

Chapters 21 and 22 have presented the basic concepts of thermodynamics, emphasizing certain features applied within biophysics. No attempt has been made to include rigorous or detailed proofs. However, the ideas and the quantitation involved are most easily described with the aid of symbolic reasoning and mathematical formulas. Thus, these chapters have a greater number of mathematical formulas than the preceding parts of the text. The mathematical formalism of thermodynamics is, however, essential for its application to real problems.

The word "thermodynamics" from its definition might be expected to apply to the motion of heat or its changes with time. While including changes of heat and its transfer from one substance to another, the word "thermodynamics" has a much more general meaning, namely, a study of the interactions and transductions between all types of energy. Many physicists have considered energy as the central concept in terms of which all systems may be described. From this point of view, the biological applications of thermodynamics might be regarded as the most fundamental and unique part of biophysics. In the two chapters summarized here, it has been implied that thermodynamics can be applied to many biological phenomena; a limited subset of these has been discussed in greater detail.

Thermodynamics as described in these chapters rests on a number of laws. Sometimes a zero law that temperature exists and can be measured is included. The first law of thermodynamics is a restatement of the law of conservation of energy applied to all types of energy, including internal energy, heat, mechanical work, electrical energy, and chemical energy. This law forms a necessary foundation for thermodynamics. In order to treat heat energy in a form amenable to mathematical analysis, a construct called entropy is introduced. It is defined by its differential, which, for reversible changes between equilibrium states, is equal to the differential heat change divided by the absolute temperature.

Entropy can be used to state the second and third laws of thermodynamics, as well as to define the added thermodynamic functions, enthalpy, Helmholtz free energy, and Gibbs free energy. The last-named function is particularly useful in biological systems. This results because most biologically related phenomena occur with external constraints that maintain temperature and pressure relatively constant. Under isothermal, isobaric conditions, Gibbs free energy tends toward a minimum, the difference from this minimum indicating the maximum amount of work other than compressional ($P \, \Delta V$) which can be obtained from the system.

Changes in Gibbs free energy of a chemical system can be related to the chemical potentials of the substances reacting with one another. In Chapter 21 such analyses are carried out to relate Gibbs-free-energy differences to chemical equilibrium processes. This is illustrated for the enzyme catalase. It is also possible to describe reaction kinetics in thermodynamic terminology, making use of Gibbs-free-energy data. Catalase kinetics is used as an example of enzyme reactions where such logic has been applied.

Entropy production in real systems in general exceeds the lower limits for reversible processes. This leads to a description of the rate of entropy production essentially as a dissipation function. The latter is described as the sum of terms each of which is itself a product of a generalized force and a conjugate generalized flux. Irreversible thermodynamics allows for forces to interact not only with the conjugate fluxes but with all possible fluxes. As a further step, linear relationships are postulated between the forces and fluxes, an assumption that is strictly valid only for small changes sufficiently close to equilibrium. Onsager's law states that the linear cross-reactions are symmetrical. This law makes it possible to apply irreversible thermodynamics to problems in physics, chemistry, engineering, and biology.

The most successful applications of irreversible thermodynamics in biology are discussed in Chapter 23. Many other biological systems although clearly irreversible are so far from equilibrium or are otherwise so constrained as to make the linear relationships and symmetry assumed by the theory untenable for the application. Other principles and different analyses must be used to study the thermodynamics of such irreversible systems.

The ideas developed in Chapter 23 are applied in Chapter 5 to analyze impulse conduction by neurons. In Chapter 25, entropy reduction is related to information. The latter is applied in Chapter 25 to the molecular mechanisms involved in protein synthesis. Thus, thermodynamic concepts form the basis for many biological studies. Others not treated include more details on bioenergetics in general, including metabolism, muscular contraction, and cardiovascular mechanics.

REFERENCES

1. PRIGOGINE, I., *Introduction to Thermodynamics of Irreversible Processes*, 2nd rev. ed. (New York: Interscience Publishers, 1961), 119 pages.
2. KATCHALSKY, A., AND P. F. CURRAN, *Nonequilibrium Thermodynamics in Biophysics* (Cambridge, Mass.: Harvard University Press, 1965), 248 pages.
3. OSTER, G. F., A. S. PERELSON, AND A. KATCHALSKY, "Network Thermodynamics: Dynamic Modelling of Biophysical Systems," *Quart. Rev. Biophys.* **6**:1–134 (1973).

CHAPTER 23

Diffusion, Permeability, and Active Transport

23.1   Introduction

In the two preceding chapters applications of thermodynamics to biological systems were discussed with frequent allusions to the importance of these concepts in relationship to diffusion and permeability phenomena. In the current chapter, these are considered in more detail, following first what might be called a classical or empirical model and then from the point of view supported by irreversible thermodynamics. The ideas of energy, Gibbs free energy, and thermodynamic equilibrium play major roles in this chapter. It is shown further that the phenomenological equations and Onsager's reciprocal relationship introduced in Chapter 22 are well suited to describe diffusion and permeability in living systems.

Diffusion is a very rapid process when it occurs within a single biological cell. However, on a macroscopic scale it may be very slow if unaided by stirring and convection. For example, if one puts several spoonfuls of sugar into a cup of coffee, the sugar will sink to the bottom. Soon there will be a thin layer of coffee which is saturated with sugar. In the absence of stirring, the sugar molecules will slowly spread, that is, diffuse, throughout the coffee.

On the gross scale of the coffee cup, it may take days to approach equilibrium. (Usually, we stir the sugar into the coffee rather than waiting for diffusion.) On the microscopic scale of biological cells, and on the still-smaller scale of reacting molecules, diffusion becomes very rapid. Molecules may diffuse throughout a cell in only milliseconds.

From the point of view of molecules, the diffusing solute may be thought of as jumping from one quasi-equilibrium site to the next, perhaps 0.1 nm away. There will be a greater probability of a molecule jumping from a region of higher concentration than vice versa. Thus, diffusion will lead toward an equalizing of concentration. When the concentration is equal throughout the container, the partial molal free energy $\tilde{G}$ will also be equal throughout. Diffusion at constant temperature and pressure leads toward the equilibrium condition:

$$d\tilde{G} = 0$$

The last paragraph emphasizes that diffusion occurs in nonequilibrium systems (although the system may be in a steady state). Accordingly, one might expect that diffusion would lend itself to description by the phenomeno-logical equations of irreversible thermodynamics. Although not necessary for simpler experiments, the phenomenological equations are the only theoretically sound basis for the interpretation of diffusion experiments.

At boundaries separated by membranes (e.g., cell membranes), the rate of diffusion may be markedly slowed. Although not really a different type of phenomenon, diffusion through a limiting membrane is called *permeability*. Many membranes are permeable only to certain substances. If the concentrations of substances to which the membrane is impermeable are different on the two sides of the membrane, the solvent will tend to move toward the greater concentration. This may be described by assigning an *osmotic pressure* to the solution.

Classical theories of permeability have difficulty incorporating observed phenomena such as solute being carried through a membrane by solvent drag. All membrane permeability experiments involve nonequilibrium systems changing along irreversible pathways. The interactions inherent in such processes are most conveniently described in the terminology of irreversible thermodynamics. This topic is treated in Sec. 23.5.

When some of the molecules that cannot pass the boundary are charged, an electrical potential may be developed across the membrane. This is called a *Donnan potential* (discussed in Chapter 24). In this case, the equilibrium concentrations of the ionic species may be different on the two sides of the membrane. Approaches of a nonequilibrium system toward equilibrium are useful examples of an application of irreversible thermodynamics to a biological system.

In Chapter 5, on the conduction of nerve impulses, it was pointed out that the ion concentrations inside nerve axons are different from those outside. These differences cannot be explained by passive diffusion through a membrane subject only to osmotic and electrical forces, although such effects are important. Rather, certain ions are actively transported at the ultimate expense of metabolic energy. Active transport is not restricted to membranes of axons. Forced diffusion in a direction different than indicated by electrical and concentration gradients probably is a common occurrence in all cells.

Many experiments have been carried out to measure diffusion rates and permeabilities, as well as to demonstrate the role of active transport against electrochemical gradients. Behind each or these experiments, indeed as an essential part of each, are the mathematical theories of diffusion and permeability. Without them the experiments would be meaningless. In this chapter, the basic mathematical development is presented. It is hoped that the reader will not be misled into feeling that the experiments are less important than the theory, for this is surely not the case. The mathematical theory can never be completely divorced from experiment or vice versa.

The development both of the individual sections and of the overall chapter illustrates the approach of the mathematical biologist as well as of the biophysicist. One feature of this approach is to start with simple idealized situations that can be described exactly by mathematical formulas. These are then gradually expanded, improved, and made to correspond more exactly to nature. In the process, the theories may become mathematically cumbersome, but throughout the entire development, the intuitive picture is colored and influenced by the simple, idealized approximation that can be exactly solved.

23.2 Diffusion Equations

The example of diffusion that is easiest to describe in mathematical terms is that in which diffusion occurs in one dimension (direction) only. This mathematical development is presented here, with the more general case of three dimensions being left to the reader. The unidimensional situation can be visualized as some substance, for example O_2, diffusing through a liquid which fills a long pipe of constant cross-sectional area A. This is illustrated in Fig. 23.1. With suitable precautions, the concentration c of the O_2 will be constant throughout any given cross section. In contrast, c will vary from one cross section to the next. At any given cross section, c will also vary in time. These variations can be described by a single partial differential equation. This equation is developed in the following paragraphs.

Consider two planes at x_1 and x_2, spaced as shown in Fig. 23.1. In a homogeneous liquid, the probability of a molecule jumping either in the $+x$ or $-x$ direction is equal. The mass per unit time of molecules jumping from plane x_1 to plane x_2 will be proportional to the concentration c_1 at x_1. Like-

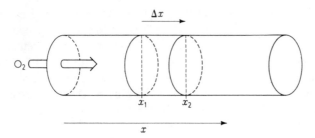

Figure 23.1   One-dimensional diffusion. In this figure, oxygen is considered to be diffusing down a long pipe such that the concentration is constant over any given plane. The concentration value at a second plane, x_2, may be different from that at the plane x_1, however.

wise, the mass per unit time going from x_2 to x_1 will be proportional to the cross-sectional area A. Analytically, one may express this as

$$\frac{\Delta m}{\Delta t} = \beta A(c_1 - c_2) = -\beta A \,\Delta c \qquad (23\text{-}1)$$

where Δm is the net mass transfer in the $+x$ direction across the surface of x_2, and β is a probability parameter.

If the two planes x_1 and x_2 are far enough apart, the probability constant β must be very low, whereas if they are close, β should be large. The exact dependence of β on the separation can be approximated by

$$\beta = \frac{D}{x_2 - x_1} = \frac{D}{\Delta x} \qquad (23\text{-}2)$$

where D is a constant. Then Eq. (23-1) may be rewritten

$$\frac{\Delta m}{\Delta t} = -DA\frac{\Delta c}{\Delta x}$$

Taking the limits as Δt and Δx go to zero reduces this to

$$\frac{\partial m}{\partial t} = -DA\frac{\partial c}{\partial x} \qquad (23\text{-}3)$$

The partial derivatives in Eq. (23-3) are necessary, since m and c both may vary with t and x.

Equation (23-3) can be derived starting from thermodynamics as well as from other points of view. Although several of these add refinements, they all contain the basic assumption made in writing Eq. (23-2). This assumption can be justified empirically because experiments with a wide variety of gases and solutes in different solvents confirm Eq. (23-3). For most purposes, it is correct; however, there is no meaning to Eq. (23-3) at distances of the order of 1 nm or less or to times comparable to the period that a molecule remains in a quasi-equilibrium state (10^{-14} sec). The cases discussed in this chapter have been restricted to those to which Eq. (23-3) can be applied.

The constant D is called the *diffusion constant* or sometimes the *Fick diffusion constant*. Equation (23-3) also can be expressed in an alternative form, which is easier to handle, although conceptually identical. To derive this alternative form, consider again the pipe of constant cross section A, redrawn in Fig. 23.2. The mass change per unit time in the volume ΔV

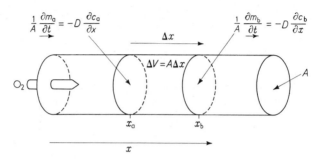

Figure 23.2 One-dimensional diffusion and continuity. This figure is used to illustrate the relationship of the change of concentration within ΔV due to the O_2 diffusing in at x_a and out at x_b.

between planes x_a and x_b is the difference between that entering at x_a and that leaving at x_b. That is,

$$\frac{1}{A} \Delta \left(\frac{\partial m}{\partial t}\right) = D \left[\frac{\partial c_b}{\partial x} - \frac{\partial c_a}{\partial x}\right]$$

or dividing both sides by Δx and taking the limit as Δx goes to zero,

$$\frac{\partial c}{\partial t} = D \frac{\partial^2 c}{\partial x^2} \qquad (23\text{-}4)$$

This is an alternative form of Eq. (23-3), valid for the case in which no oxygen is generated or destroyed.

Equation (23-4) can be readily generalized to the three-dimensional diffusion equation as

$$\frac{\partial c}{\partial t} = D \left(\frac{\partial^2 c}{\partial x^2} + \frac{\partial^2 c}{\partial y^2} + \frac{\partial^2 c}{\partial z^2}\right) \qquad (23\text{-}5)$$

Both Eqs. (23-3) and (23-5) may be written in the vector notation. By defining the mass flux $\vec{J}$ as

$$\vec{J} = \left(\frac{1}{A} \frac{\partial m}{\partial t}\right) \vec{k} \qquad (23\text{-}6)$$

where $\vec{k}$ is a unit vector normal to A, Eq. (23-3) may be rewritten as

$$\vec{J} = -D\vec{\nabla}c \tag{23-7}$$

In vector notation, Eq. (23-5) becomes

$$\frac{\partial c}{\partial t} = D\nabla^2 c \tag{23-8}$$

where ∇^2 is the Laplacian operator defined implicitly by Eq. (23-5). Equations (23-8) and (23-5) as well as Eq. (23-3) are often called *Fick's diffusion equation* or *Fick's law*. Equation (23-8) is also referred to by physicists and chemists as the *heat equation*, because its form is identical to that for the variation of temperature as a function of space and time in the absence of any sources or sinks[1] of heat energy. Although the substance considered in Eqs. (23-3) and (23-5) was called oxygen for convenience, there is nothing that restricts the use of these equations to oxygen. They are restricted, however, to the case in which the substance is neither generated nor used up chemically.

Actually, oxygen is almost always either being used or being liberated within a living cell. In most cells it is being used, and in others, which are photosynthesizing, it is being produced. If the rate of production per unit volume is called q, Eq. (23-8) must be replaced by

$$\frac{\partial c}{\partial t} = D\nabla^2 c + q \tag{23-9}$$

A negative value of q implies use of oxygen (or, at any rate, of the substance to which the equation applies). In general, the production rate q may vary with x, y, z, and t in any arbitrary manner. Equation (23-9) is known as the *inhomogeneous diffusion equation*.

For many diffusion problems, Eq. (23-9) represents an excellent starting point for the analysis. However, for examples such as diffusion limitations of enzyme reaction rates discussed in Chapters 18 and 21, a different analysis is needed. In such cases, the details of diffusion in liquids over distances comparable to molecular dimensions become important. Then it is necessary, as in Chapter 21, to introduce the idea of discrete jumps from one quasi-equilibrium site to the next. Equations (23-8) and (23-9) represent the overall macroscopic behavior; a constraint on the analysis described in Chapter 21 is that, in the absence of a reaction, the overall diffusion must reduce to Eq. (23-8).

Equations (23-8) and (23-9) also fail, even from the macroscopic point of view, if there are several species of molecules diffusing. For diffusion in gases this is rarely very important as the different molecular species tend to

---

[1] A *heat sink* is a place where heat energy is removed (and, of course, converted into some other type of energy).

diffuse independently of one another. Diffusion in liquids is quite different. Then there is at least a solvent and a solute molecule, and these do interact. Thus, for many biologically interesting situations, another approach is needed. This is provided by irreversible thermodynamics.

If one considers an isothermal, isobaric system which is not at equilibrium, diffusion will occur, converting chemical energy into heat. Thus, one may write the dissipation function in vector notation as

$$\Phi = \sum_{i=1}^{n} \vec{J_i} \cdot \vec{\nabla}(-\mu_i) \tag{23-10}$$

where the subscript i refers to the particular one of the n molecular species being considered. The dissipation function Φ is defined in Chapter 22, as is the chemical potential μ. The flux J is defined in this chapter and in Chapter 22. Note that Φ is the dissipation per unit volume. For the one-dimensional case illustrated in Figs. 23.1 and 23.2, Eq. (23-10) can be simplified to

$$\Phi = -\sum_{i=1}^{n} J_i \frac{\partial \mu_i}{\partial x} \tag{23-11}$$

Not all the driving forces in Eq. (23-10) are independent. There exists a relationship not developed in this text, known as the *Gibbs–Duhem equation*, which for isothermal, isobaric diffusion becomes

$$\sum_{i=1}^{n} c_i \vec{\nabla}(-\mu_i) = 0 \tag{23-12}$$

It is possible to solve this for the diffusional force on the solvent which will be indicated by a subscript, w, since in a biophysics application the solvent is usually water. For simplicity, the solutes can be numbered so that the water is the nth component. Then solving Eq. (23-12) leads one to

$$\vec{\nabla}(-\mu_w) = -\frac{1}{c_w} \sum_{i=1}^{n-1} c_i \vec{\nabla}(-\mu_i)$$

Substituting into Eq. (23-10) and rearranging terms gives

$$\Phi = \sum_{i=1}^{n-1} \left(\vec{J_i} - \frac{c_i}{c_w} \vec{J_w} \right) \cdot \vec{\nabla}(-\mu_i) \tag{23-13}$$

To write the phenomenological equations it is convenient to define a relative diffusional flux $\vec{J_i^d}$,

$$\vec{J_i^d} = \vec{J_i} - \frac{c_i}{c_w} \vec{J_w} \tag{23-14}$$

For the one-dimensional case,

$$J_i = c_i v_i$$

where v_i is the linear velocity of the ith molecular species. Thus, by rearranging terms, it appears that

$$J_i^d = c_i(v_i - v_w) \tag{23-15}$$

or, in other words, the basic fluxes depend on the motion of solute relative to the solvent rather than to the walls of the containing vessels.

For one solute, s, the phenomenological equation for one dimension is

$$J_s^d = -L_s \frac{\partial \mu_s}{\partial x} \tag{23-16}$$

But it has been pointed out in Chapters 21 and 22 that

$$\mu_s = \mu_s^0 + RT \ln \left(\frac{c_s}{c_0}\right)$$

whence it follows that

$$\frac{\partial \mu_s}{\partial x} = \frac{RT}{c_s} \frac{\partial c_s}{\partial x}$$

This allows rewriting Eq. (23-16) as

$$J_s^d = -\frac{L_s RT}{c_s} \frac{\partial c_s}{\partial x} \tag{23-17}$$

Comparison with Eq. (23-7) shows that if one defines

$$D' = \frac{-L_s RT}{c_s}$$

then Eqs. (23-7) and (23-17) are almost identical.

The differences are noteworthy, however. Equation (23-7) refers to motion relative to the walls of the vessel, (23-17) to motion of the solute relative to the solvent. For dilute solutes in water, this is not too important per se. However, the logic leading to Eq. (23-17) can be readily expanded to cases of net diffusion of solvent as well as solute, and to multiple solutes.

23.3   Permeability and the Red Blood Cell

Neither Eq. (23-9) nor (23-17) is in a form that is helpful to analyze the passage of molecules through biological membranes. In describing such membrane problems it is often necessary to include diffusion on the two sides of the membrane. For this the previous equations can be used. However, at the membrane itself, a different approach is needed. In this section and the following one, the classical analyses are reviewed. The added contributions of irreversible thermodynamics to studies of membrane permeability are included in Sec. 23.5.

Across the membrane there may be a very large change in concentration.

In theory, the membrane and surrounding fluids could be treated as three regions, as indicated in Fig. 23.3. In each region there will be a different

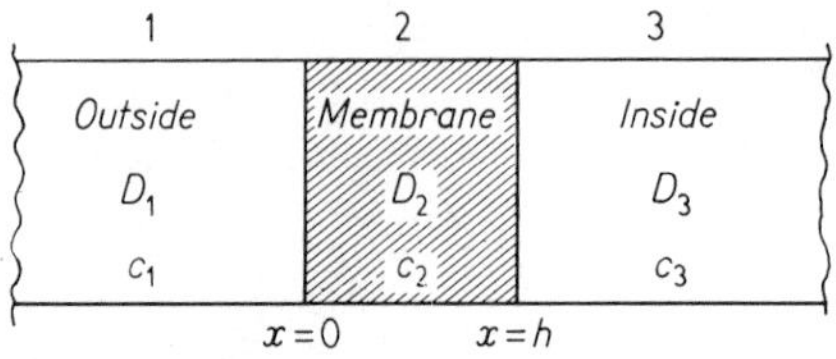

Figure 23.3   Idealized membrane. This is used in permeability discussions. See Chapter 24 for the details of membrane structure.

diffusion constant. By and large, the membranes are so thin that no empirical meaning can be assigned to the concentration c_2 within the membrane. Instead, the membrane is usually characterized by a permeability k such that the mass per unit time passing through the membrane is given by

$$\frac{1}{A}\frac{\partial m}{\partial t} = k(c_1 - c_3)|_{\text{membrane}} \qquad (23\text{-}18)$$

This, in turn, must equal the mass flux entering and leaving the membrane. The relevant equations are

$$D_1 \frac{\partial c_1}{\partial x}\bigg|_0 = -k\,(c_1|_0 - c_3|_h) = D_3 \frac{\partial c_3}{\partial x}\bigg|_h \qquad (23\text{-}19)$$

This ignores any solvent–solute or solute–solute interactions.

Equations (23-9) and (23-19), with the proper values for the diffusion coefficient D, the permeability k, and the rate of generation of the substance q, completely define the classical approach to all biological diffusion problems, from a mathematical point of view.[2] Although such problems can always be solved at least in theory, in practice they often require computer-aided numerical analysis techniques. Moreover, as noted previously, this approach ignores many features which are described by irreversible thermodynamics. Absolute rate theory, introduced in Chapter 21, has also been used to study membrane permeability. For this purpose the membrane itself in Fig. 23.3 is considered to be made up of three compartments. That analysis is not pursued further in this text.

There are two restrictions to Eqs. (23-9) and (23-19) that should be noted. First, it has been assumed that stirring did not occur. If random or turbulent stirring is present, one may include it by using appropriately larger values for

[2] Throughout this chapter it would be better to use activities than concentrations. However, there appears to be little advantage in this distinction in the examples discussed here.

D. It also has been implicitly assumed in the foregoing derivation that no electrical potential gradients are present. Most protoplasm conducts electrical charge so well that potential gradients can exist only across the membranes. If these are present, Eq. (23-19) must be modified, replacing the middle expression by

$$-k\left[c_1|_0 \exp\left(\frac{-zF\mathscr{E}}{RT}\right) - c_3|_h \right]$$

where z is the charge on the ion, F is the Faraday, R is the gas constant per mole, T is the absolute temperature, and $\mathscr{E}$ is the electrical potential difference across the membrane. This extra factor occurs because the partial molal free energies inside and outside differ by $zF\mathscr{E}$ when the concentrations are equal.

In the era before the advent of digital computers, it was customary to seek approximation methods that allow simplified solutions of the equations describing diffusion and permeability. For example oxygen diffusion into single cells may be considered a boundary-value problem. Such problems can be solved with great precision using the techniques of numerical analysis. However, the cost may exceed the resources available to the biophysicist. If such financial limits are imposed, the approximation methodology may be quite appealing. For example, an average-value approach is illustrated in Fig. 23.4. Here the specific form of the cell is ignored and the average dimensions, concentrations, and usage of oxygen inside the cell are considered. The reader is referred to the reference by Rashevsky for further details.

Figure 23.4 The average-value approach. An arbitrary cell is shown in part (a) with average radii r_1 and r_2. It is assumed that the concentration in the central part of the cell is constant, reaching the value of the average concentration $\bar{c}$ at $r_1/2$ and $r_2/2$ from the center. It is further assumed that the cell can be replaced by a rod or pillbox as shown in parts (b) and (c). The general patterns of diffusion, which do not depend on exact cell geometry, are then investigated. After N. Rashevsky, *Mathematical Biophysics* (Chicago: The University of Chicago Press, 1948).

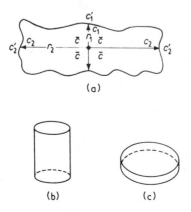

   Alternatively, a spherical cell model may be used. Here it is possible to show the existence under suitable conditions of an anoxic region within the cell using oxygen. A more detailed model, which includes glycolysis as a single process and lactic acid oxidation as a second one, is illustrated in Fig. 23.5.

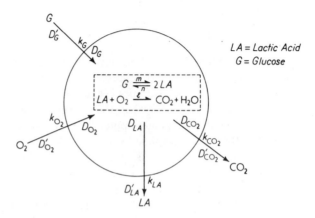

Figure 23.5 Spherical cell model. This model, with its numerous constants, can account for O_2 diffusion into living cells. The model is also supported by metabolic studies of other types.

While giving only a gross overview, the model can successfully deal with many types of experimental data. Unfortunately, it has large numbers of rate constants, all of which must be estimated from the data. Many data are needed to determine these constants, and the general treatment of Chapter 18 proves more fruitful to interpret such experimental studies. Those methodologies do permit the inclusion of diffusion and permeability in quantitative forms by treating, for example, oxygen inside and outside the cell as separate components.

   The mathematical problems of analyzing diffusion into biological cells can be greatly simplified if the rate at which a substance penetrates the cell membrane is slow compared to its rate of diffusion on either side of this membrane. Mammalian red blood cells have proved very useful for studies of this nature. They are especially convenient because the cells act as "osmometers," swelling or shrinking accordingly as their internal osmotic pressure varies. In spite of very large variations in the volume, the surface area of the erythrocytes remains almost constant. In this fashion, the volume can be used to indicate the internal concentrations, and one may compute the permeability constant k. In actual practice, the area is assumed to be constant and not measured; rather, the permeability P is used. It is defined by

$$P = kA_r \tag{23-20}$$

where A_r is the surface area of the erythrocyte.

Equation (23-18) is particularly suited to this case. If m is the mass of a penetrating substance s inside the cell, then Eq. (23-18) becomes

$$\frac{\partial m}{\partial t} = kA_r(c_s - c_i) \tag{23-21}$$

where c_i and c_s are the concentrations of s inside and outside the cell, respectively. Because diffusion occurs rapidly compared to penetration of the cell membrane, c_i may be replaced using

$$c_i = \frac{m}{V}$$

where V is the cell volume. Equation (23-18) can then be rewritten as

$$\frac{\partial m}{\partial t} = P\left(c_s - \frac{m}{V}\right) \tag{23-22}$$

As the concentration of substance s within the cell rises, there will be a flow of water into the cell. This will result, in turn, in a change of the cell volume V. The flow of water must also obey Eq. (23-18) for the mass flow through the cell wall. It takes a few algebraic manipulations to rewrite this equation for water flow as

$$\frac{dV}{dt} = P_w \left(\frac{c_0 V_0 + m}{V} - c_s - c_M\right) \tag{23-23}$$

where c_0 is the initial concentration of nonpenetrating solutes within the cell, V_0 is the initial cell volume, and c_M is the concentration of nonpenetrating solutes in the external medium. If more than one penetrating solute is present, m and c_s must be regarded as the sum of all the various values.

If the volume changes are observed, Eqs. (23-22) and (23-23) may be used to compute values for P and P_w. In the most general case, only numerical solutions are possible. There exist, however, a number of simplified conditions under which P and P_w may be estimated directly.

More exact treatments involve numerical solutions of Eq. (23-23), even replacing P_w with kA_r if independent estimates for A_r are available. However, if distilled water is used outside the cell or if measurements are made sufficiently rapidly after mixing a suspension of red blood cells with a medium of reduced concentrations, only the penetration of water need be considered. Similarly, if measurements are made sufficiently slowly so that water equilibrium is established, one can estimate P or k_r for a number of different solutes.

The values for P and D vary in very different fashions from one solute to the next. For large molecules, for example, the diffusion constant D varies roughly as the square root of the reciprocal of the molecular weight. On the other hand, the permeability for urea is

$$k = \frac{P}{A} = 7,000 \text{ cm/hr}$$

whereas for glycerol it is 54 cm/hr and for sucrose 0 cm/hr. Moreover, the times for 90 percent saturation for these three solutes would all be less than 10^{-3} sec, if only diffusion within the cell were considered. However, the permeation of the membrane is the rate-limiting step; thus, the times range from 0.5 sec to ∞, including the limitations at the membranes.

Certain general rules can be found for the relative values of P measured for erythrocytes. First, the more soluble the solute is in lipids, the greater is the value for P. For example, glycerol has a much lower value for P than does its larger, lipid-soluble ester, monacetin. Therefore, part, at least, of the red blood cell membranes appear to be of a lipid nature. This is also supported by other lines of evidence, as discussed in Chapter 24.

The second general rule for the variations of P is that, given the same lipid solubility, the smaller molecule goes through faster. For instance, the rates for ethylene glycol, diethylene glycol, and methylene glycol decrease in the order of increasing molecular weights. This supports a molecular-sieve picture of the cell membrane, in which bigger molecules, even though lipid-soluble, have a hard time going through the pores. A further interpretation of passive permeability, described in the terminology of irreversible thermodynamics, is included in Sec. 23.5.

In spite of these general rules, there exist other molecules, such as water, urea, and sodium ions, which appear to go through the cell membrane at inordinately high rates. No simple picture of the cell membrane can explain these magnitudes. One must think of the cell membrane as actively transporting certain molecular species.

23.4 Active Transport

The extremely high permeability constants of erythrocyte membranes for certain molecules suggest that in some fashion the cell membrane actively moves these molecules rather than merely permitting them to passively diffuse through the membrane, as described in Sec. 23:2. Similar evidence for active transport comes from a variety of other sources. It appears probable that all biological membranes actively transport certain substances. In Chapters 5 and 9, it is stated that the concentration of potassium ions within nerve axons and muscle fibers is higher than in the surrounding solution, whereas the concentration of sodium ions is lower. Because these membranes are charged, it is not sufficient to merely measure concentration differences; the concentrations must be compared with those predicted for the electrical potential differences measured across the membrane. This reasoning shows that the Na^+ ions, although demonstrated by tracer techniques to pass from the outside medium into the cell, must be continually pumped out against an electrochemical gradient to maintain equilibrium.

Whenever metabolic work is expended to pump ions or molecules across a membrane, the phenomenon is called *active transport*. This may result in moving the ions or molecules against an electrochemical gradient, which has been demonstrated to occur in the kidney tubules, in the epithelium of the stomach mucosa (H^+ transport), in the epithelium of the intestines (transport of ions, water, simple sugars, fatty acids, amino acids, and so on), and in frog skin. Active transport may also involve pumping to increase the net flow in the direction of an electrochemical gradient, as perhaps the entrance of urea into the red blood cell. The detailed molecular mechanisms are not known in any case studied so far, although, as discussed in more detail in Chapter 24, ATP (adenosine triphosphate) appears to be an energy source for some of them.

A mathematical theory was developed by Ussing to determine whether active transport occurred. This theory seems particularly important because the detailed mechanisms are not known on a molecular basis. The mathematical theory involves the relative rates of transport of tracer-labeled molecules across the membrane in the two directions. These are compared with actually measured values.

At equilibrium the rate of transport of molecules across the membrane in the two directions must be equal, or else the concentrations would not be at their equilibrium values. If the membrane is uncharged, the concentrations on both sides must be equal at equilibrium. If the flux from the right to the left is called J_{RL} and in the opposite direction J_{LR}, then probability considerations dictate that, in the absence of active transport or of membrane potentials,

$$\frac{J_{RL}}{J_{LR}} = \frac{c_R}{c_L} \tag{23-24}$$

where c_R is the concentration on the right side of the membrane and c_L is the concentration on the left side of the membrane. The same conclusion can be reached from considerations of Gibbs free energy, or of chemical potentials.

Likewise, if the membrane is charged, one may start from any of these bases and arrive at the formula

$$\frac{J_{RL}}{J_{LR}} = \frac{c_R}{c_L} \exp\left(\frac{zF\mathscr{E}}{RT}\right) \tag{23-25}$$

As mentioned earlier, the exponential factor appears because the partial molal free energies on the two sides of the membrane differ by $zF\mathscr{E}$ when the concentrations are equal. When the two fluxes are equal, equilibrium is established, although the concentrations need not be equal. By using two different labeled isotopes on the two sides of the membrane, one may measure the ratio of the fluxes.

Equations (23-24) and (23-25) are based on the assumptions of no coupling between molecules permeating the membrane and the existence of equilibrium on both sides of the membrane as well as within it. Subject to those restrictions the equations could be made more exact by using activities to determine the generalized forces driving the fluxes in the two directions. The latter is a far less important restriction than the absence of coupling between fluxes.

Equations (23-24) and (23-25) have been used to design experiments to test for active transport across many membranes. If the flux ratio is different than predicted, it implies that the assumption of passive transport necessary to derive Eq. (23-25) must be wrong. Somehow, the membrane must be pumping or forcing the molecules in a preferred direction. One of the simplest examples to discuss is the isolated frog skin. The experimental arrangement is diagrammed in Fig. 23.6. Perhaps this is a rather poor

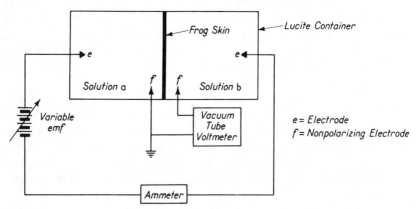

Figure 23.6 Apparatus for determining currents through and potential across frog skin. Air is bubbled into both solutions. The variable emf is adjustable over both positive and negative values. For an open-circuit potential, the variable emf is adjusted to cause the ammeter to read zero, whereas for short-circuit current determination, the variable emf is set to zero.

example, for the membrane (or membranes) responsible for the pumping action are not known. However, the frog skin can be used to demonstrate active transport because it can separate two media whose concentrations can be controlled.

If an isolated live frog skin is used to separate two containers of Ringers' solution, it develops a 60-mV potential difference between the two solutions, the outside of the skin being negative relative to the inside. Using radio-labeled tracers, the two unidirectional fluxes for sodium ions may be measured. Equation (23-25) becomes quite simple, since c_R and c_L are equal. It can be shown that there is a 100-fold difference between the experimentally determined flux ratios and that computed from Eq. (23-25), assuming no active transport. Accordingly, it is concluded that Na^+ is actively transported inwardly. It is known that frogs can take up sodium ions from their surroundings even if the external concentration is as low as 10^{-5} M. Similar tests show that Cl^- and HCO_3^- are not actively transported by frog skins, whereas Li^+ is.

Because the frog skin actively transports Na^+ and develops an electrical potential, it may be used as a battery to drive a current through an external circuit. This demands energy, which in turn must be related to some metabolic process. Inhibitors that block the active transport of Na^+ do not uniformly block O_2 consumption. In fact, one of the most effective, dinitrophenol, stimulates O_2 respiration while decreasing the production of ATP. This fact suggests that ATP may be the ultimate energy source for active transport in frog skin. As noted earlier, the detailed molecular mechanism is not known for this or any other of the numerous cases of active transport.

23.5   Irreversible Thermodynamic Interpretation of Permeability

A mathematical description of membrane permeability can be written in the terminology of irreversible thermodynamics. This allows the inclusion of added features in a natural fashion. However, the use of irreversible thermo-dynamics adds to the necessary mathematical analysis. For membrane permeability procedures involving neutral molecules at constant temperature and pressure, the dissipation function will look formally very similar to that in Eq. (23-10). However, the driving forces will depend on differences across the membrane rather than on gradients or partial derivatives.

Specifically, letting $\Delta(\mu_i)$ be the difference between μ_i outside the cell (or on the left of the membrane) minus the corresponding value inside the cell (or on the right of the membrane), the expression for Φ becomes

$$\Phi = \sum_{i=1}^{n} J_i \, \Delta(\mu_i) \tag{23-26}$$

where J_i is the net inward (or left to right) mass flow of component i. The dissipation function in Eq. (23-26) suggests describing permeation by n phenomenological equations.

Only the case of one solute, shown by subscript s, and one solvent, shown by subscript w, will be considered. It will further prove useful to allow pressure differences to exist across the membrane as well as concentration differences. Equation (23-26) can then be rearranged into an alternative form which reveals its physical implications more clearly. For this, a variety of new symbols are needed. Thus, P will represent pressure, π, osmotic pressure, and $\tilde{V}_i$, the partial molal volume of component i.

By definition, the partial molal volume is the added volume per mole due to the presence of the substance. From this it follows that the total volume V is given by

$$V = \tilde{V}_s n_s + \tilde{V}_w n_w$$

where the subscripted lowercase n's represent the numbers of moles of the substance indicated by the subscript. For dilute solutions in water, the partial molal volume of the solute, s, is much less than that of the solvent, w, and hence one may approximate V as

$$V \simeq \tilde{V}_w n_w$$

The osmotic pressure π is defined by an experiment in which a semipermeable membrane separates two chambers, one of which contains a pure solvent and the other a solute. The solvent will flow into the chamber with the solution unless an excess pressure is applied as shown in Fig. 23.7. At this time the

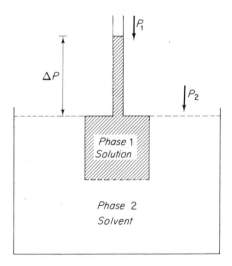

Figure 23.7 Osmotic-pressure experiment. The membrane separating the solution from the pure solvent is permeable to the solvent only. ΔP represents the equilibrium pressure difference required to prevent flow of solvent from phase 2 to phase 1.

chemical potential of the solvent on both sides of the membrane must be equal. The excess pressure necessary for equilibrium is defined as the osmotic pressure.

In general, it can be shown that the chemical potential of the solvent can be represented by

$$\mu_w = \mu_w^0 + \tilde{V}_w P - \tilde{V}_w \pi$$

where μ_w^0 depends only on temperature and π only on the chemical composition of the solutes. Hence, one may write

$$\Delta(\mu_w) = \tilde{V}_w \Delta(P) - \tilde{V}_w \Delta(\pi) \tag{23-27}$$

The driving forces $\Delta(P)$ and $\Delta(\pi)$ are more directly measurable than $\Delta(\mu_w)$ and $\Delta(\mu_s)$. Accordingly, a relationship similar to Eq. (23-27) is desired for $\Delta(\mu_s)$.

For an ideal solution, the osmotic pressure obeys a law much like the ideal gas law, namely,

$$\pi = RTc_s \tag{23-28}$$

and the chemical potential μ_s may be written as

$$\mu_s = \mu_s^0 + \tilde{V}_s P + RT \ln (c_s)$$

Since the partial molal volume is essentially constant, the last equation leads to

$$\Delta(\mu_s) = \tilde{V}_s \Delta(P) + RT \, \Delta[\ln (c_s)] \tag{23-29}$$

Provided that the concentrations on the two sides of the membrane are similar, the log can be expanded, keeping only the first term, as

$$\Delta(\ln c_s) = \ln \left(\frac{c_{s,o}}{c_{s,i}}\right) \simeq 2 \cdot \frac{c_{s,o} - c_{s,i}}{c_{s,o} + c_{s,i}}$$

where the subscripts o and i refer to the outside and inside of the membrane. From Eq. (23-28), it can be shown that

$$\Delta(\pi) = RT(c_{s,o} - c_{s,i})$$

while the average solvent concentration $\bar{c}_s$ is given by

$$\bar{c}_s = \frac{c_{s,o} + c_{s,i}}{2}$$

Combining the three preceding equations and substituting into Eq. (23-29) leads to the relationship

$$\Delta(\mu_s) = \tilde{V}_s \Delta(P) + \frac{1}{\bar{c}_s} \Delta(\pi) \tag{23-30}$$

Equations (23-27) and (23-30) can be substituted into Eq. (23-26) and rearranged as

$$\Phi = (J_w \tilde{V}_w + J_s \tilde{V}_s) \, \Delta(P) + \left(\frac{J_s}{\bar{c}_s} - \tilde{V}_w J_w\right) \Delta(\pi) \tag{23-31}$$

Now one may choose new fluxes. The first is the net volume flow,

$$J_V = J_w \tilde{V}_w + J_s \tilde{V}_s \tag{23-32}$$

The other is J_D, the flow of solute relative to solvent; this can be seen by noting that

$$\tilde{V}_w \simeq \frac{1}{\bar{c}_w}$$

and hence, using lowercase v for speed,

$$J_D = \frac{J_s}{\bar{c}_s} - \tilde{V}_w J_w \simeq \frac{J_s}{\bar{c}_s} - \frac{J_w}{\bar{c}_w} = v_s - v_w \qquad (23\text{-}33)$$

Using these definitions, the phenomenological equations and Onsager's law may be written

$$J_D = L_{DD}\Delta(\pi) + L_{DV}\Delta(P)$$
$$J_V = L_{VD}\Delta(\pi) + L_{VV}\Delta(P) \qquad (23\text{-}34)$$
$$L_{DV} = L_{VD}$$

The parameter L_{DD} is similar to the permeability coefficient of Sec. 23.3, whereas L_{VV} is a filtration coefficient giving the volume flow per unit pressure difference in the absence of a concentration gradient. The advantage of equations (23-34) is that they show, in the absence of a concentration difference, that a diffusional flow will, in general, exist and in the absence of a pressure difference, a net volume flow will be observed. The latter effect is called a *solvent drag*.

The actual solute flux J_s can be found if J_V and J_D are known. In fact, from Eqs. (23-32) and (23-33),

$$J_D + J_V = \frac{J_s}{\bar{c}_s} + \tilde{V}_s J_s \simeq \frac{J_s}{\bar{c}_s}$$

as $\tilde{V}_s$ is usually small compared to $(\bar{c}_s)^{-1}$. Thus, using Eq. (23-32),

$$J_s \simeq \bar{c}_s[(L_{DD} + L_{DV})\Delta(\pi) + (L_{VV} + L_{DV})\Delta(P)]$$

For an ideal semipermeable membrane J_s must always vanish; hence,

$$L_{DD} = L_{VV} = -L_{DV}$$

For a completely permeable membrane L_{DV} must vanish. Thus, the reflection coefficient, ν, defined by

$$\nu = -\frac{L_{DV}}{L_{VV}}$$

is an important membrane descriptor.

Another descriptor can be obtained by solving equations (23-34) for the case of no net flow. Under these conditions $\Delta(P)$ can be eliminated, giving

$$J_s = w\,\Delta(\pi)$$

where the coefficient of solute permeability, w, is given by

$$w = \bar{c}_s \frac{L_{VV}L_{DD} - L_{DV}^2}{L_{DD}}$$

The last coefficient is actually the closest to the permeability constant for the solutes discussed earlier in this chapter, whereas L_{VV} is similar to what was called the permeability constant for water. However, L_{VV}, ν, and w have the advantage of more precise definitions, which include solvent drag and selective filtration.

Equation (23-26) can be readily extended to include active transport. Conceptually, to do this, one would require a chemical reaction that occurs on one side of the membrane only. For this, one must postulate a chemical flux, J_{ch}, and a change in the chemical potential per mole of reactant X_{ch}. (The term $\Delta(\mu_{ch})$ would be ambiguous here, since the reaction may occur on only one side of the membrane.) Then the dissipation function might be written for one permeating solute as

$$\Phi = J_V \, \Delta(P) + J_D \, \Delta(\pi) + J_{ch} X_{ch}$$

Using this, there are now three phenomenological equations,

$$J_D = L_{DD} \, \Delta(\pi) + L_{DV} \, \Delta(P) + L_{Dch} X_{ch}$$
$$J_V = L_{VD} \, \Delta(\pi) + L_{VV} \, \Delta(P) + L_{Vch} X_{ch}$$
$$J_{ch} = L_{chD} \, \Delta(\pi) + L_{chV} \, \Delta(P) + L_{chch} X_{ch}$$

There is no reason per se why the chemical reaction needs to involve directly either the solute or solvent. One could imagine models in which either the solvent or solute molecules in passing through the membrane could alter the rate of chemical reaction, which in turn had an effect on their rate of passage. If the solute is to be actively transported, the only requirement is for a reaction where L_{Dch} is nonzero.

It is virtually impossible to proceed further, since so little is known about what chemical reactions one is discussing. They might involve binding the solute, steric effects or carriers in the membranes, or any of a host of other possibilities. The significant contribution of irreversible thermodynamics to discussions of active transport is a philosophic attitude: that is, cross-coupling terms such as L_{DV} are important in membrane permeability and an added cross-coupling term L_{Dch} is a natural extension. In other words, there is nothing really actively transported by some vital mechanism; rather, there is an added cross coupling. From this point of view the term "metabolically coupled transport" would be preferable to active transport.

23.6   Summary

Mathematical theories describing the diffusion of molecules and their penetration through membranes have been presented in this chapter. These phenomena are irreversible processes starting at nonequilibrium states.

Accordingly, the methodology of irreversible thermodynamics has been used to interpret certain diffusional phenomena and some of the properties of biological membranes. The approach in the current chapter has been to start with simple descriptions and build up to more precise relationships which are mathematically more complicated.

Diffusion in biological systems is important in the lungs and in other organs. However, at the cellular level, the membrane limitations prove quantitatively to be the more important effect. At the molecular level, some enzyme reactions are controlled by the rates of diffusion of the reactants, but by and large, this is not the case. When considering the utilization of oxygen by metabolizing biological cells, a combination of diffusion permeability and a reaction scheme is required.

Red blood cells are used in this chapter to illustrate permeability. They are convenient because water passes rapidly through the membrane, leading to volume changes that can be observed rapidly by spectrophotometric means. Such studies confirm the wide ranges of permeabilities for different solutes. Some of the larger values may reflect active transport in which metabolic reactions are coupled to membrane permeation. This and other types of cross coupling, such as solvent drag and selective filtration, are inherent components of the phenomenological equations of irreversible thermodynamics.

REFERENCES

1. DAVIDOVITZ, P., *Physics in Biology and Medicine* (Englewood Cliffs, N.J.: Prentice-Hall, Inc., 1975), 298 pages.
2. BREUER, H., *Physics for Life Science Students* (Englewood Cliffs, N.J.: Prentice-Hall, Inc., 1975), 453 pages.
3. RASHEVSKY, N., *Mathematical Biophysics*, rev. ed. (Chicago: The University of Chicago Press, 1948), 669 pages.
4. KATCHALSKY, A., AND P. F. CURRAN, *Nonequilibrium Thermodynamics in Biophysics* (Cambridge, Mass.: Harvard University Press, 1965), 248 pages.

CHAPTER 24

Biological Membranes

24.1 Introduction

The living cell is a very complicated structure (see Appendix E). As an extreme simplification, it may be described as a bag of protoplasm, where the word "protoplasm" is suitably defined. The bag is the *cellular membrane*. This is one of several membrane structures that separate the cell into various compartments and provide the framework that allows all living processes to occur. These most important functions of membranes may predate life itself. While biophysicists usually select proteins or nucleic acids as the initial "living" entities, most agree that these primitive cellular constituents were separated from their environment by some type of membrane before more specialization could occur.

Some aspects of membranes are discussed in Chapter 23; others are distributed through a variety of chapters, including 5, 8, 9, 10, 18, 19, and 20. Membrane biophysics has been and continues to be an active area of research. This chapter briefly summarizes much work on membranes. Topics are considered only briefly, since a detailed treatment requires more biochemical knowledge than is presupposed in this text. In addition to reviews of models of membrane structure (Sec. 24.3) and theories of membrane function (Sec. 24.4), the physical theory of Donnan membrane potentials is also included (Sec. 24.2).

24.2 Donnan Membrane Potentials

As mentioned in the introduction, membranes function as both support and barriers. The easiest explanation of their barrier role is to assume they are semipermeable, that is, that they will permit diffusion of certain molecules and restrict diffusion of others. As is mentioned in Chapter 23, electrical potentials called *Donnan membrane potentials* can develop across such a semipermeable membrane. The theory behind these is developed here and used in Sec. 24.4.

The Donnan membrane potential arises when a semipermeable membrane separates two solutions, one of which contains three ions, two of which can permeate the membrane and one of which cannot. This is pictured in Fig. 24.1. Initially, one may conceive of filling a bag made from such a semi-

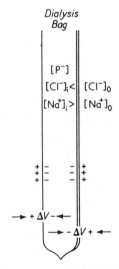

Figure 24.1 Donnan membrane potential developed as described in the text. Na^+ and Cl^- can freely permeate the membrane; the proteinate ion (P^-) cannot.

permeable membrane (a *dialysis bag*) with a solution of sodium chloride and sodium proteinate. This bag is placed in distilled water, resulting in the configuration shown in Fig. 24.1, where some of the Na^+ and Cl^- ions have left the bag to enter into the surrounding fluid. It seems intuitively clear that because there are more Na^+ ions than Cl^- ions, a few more of the Na^+ ions might permeate the membrane charging the outside positive relative to the inside. As soon as the potential difference became appreciable, it would discriminate against Na^+ ions coming out, so that the net external concentration of Na^+ and Cl^- would be almost exactly equal, and no appreciable error would be made in neglecting the difference in these two concentrations.

Thermodynamics can be used to find the magnitude of the potential developed across the membrane. According to the formulas developed in Chapter 21, equilibrium will represent a minimum in the Gibbs free energy for the system; that is,

$$dG = 0$$

In order that this be true, there must be no change in G when a few Na^+ ions are moved from one side of the membrane to the other. This implies that $\tilde{G}_{Na}$, the partial molal free energy of sodium ions, must be the same on both sides of the membrane, and $\tilde{G}_{Cl}$ must be also. Using the subscript i for inside and o for outside, one may write this as

$$\tilde{G}_{Na,i} = \tilde{G}_{Na,o}$$

$$\tilde{G}_{Cl,i} = \tilde{G}_{Cl,o}$$

(24-1)

Referring again to Chapter 21, one can show that in the presence of an electrical potential $\mathscr{E}$, the partial molal Gibbs free energy of an ionic species with concentration c is given by

$$\tilde{G} = \tilde{G}^0 + RT \ln c + zF\mathscr{E}$$

(24-2)

where z is the valence of the ion and F is the Faraday. In this expression, $\tilde{G}^0$ is the value of $\tilde{G}$ in the standard state with $\mathscr{E}$ equal to zero. For Na^+, z is $+1$, and for Cl^-, z is -1. Substituting Eq. (24-2) into Eq. (24-1) and rearranging, one obtains

$$RT \ln \left(\frac{[Na^+]_i}{[Na^+]_o} \right) = F \, \Delta\mathscr{E}$$

$$RT \ln \left(\frac{[Cl^-]_i}{[Cl^-]_o} \right) = -F \, \Delta\mathscr{E}$$

(24-3)

where $\Delta\mathscr{E}$ is the potential difference across the membrane, the outside being positive for $\Delta\mathscr{E}$ greater than zero.

Adding together the equations in (24-3) and rearranging shows that

$$[Na^+]_i[Cl^-]_i = [Na^+]_o[Cl^-]_o$$

(24-4)

Because both sides of the membrane have approximately no net charge, it is necessary that

$$[Na^+]_o = [Cl^-]_o$$

$$[Na^+]_i = [Cl^-]_i + [P^-]$$

(24-5)

where $[P^-]$ is the proteinate concentration. Because

$$[Na^+]_i > [Cl^-]_i \quad \text{whereas} \quad [Na^+]_o \simeq [Cl^-]_o$$

Eq. (24-4) allows one to conclude that

$$[Na^+]_i > [Na^+]_o \quad \text{and} \quad [Cl^-]_i < [Cl^-]_o$$

Accordingly, Eq. (24-3) indicates that $\Delta\mathscr{E}$ is positive; that is, the outside of the membrane is positively charged. This potential difference is observed in some cases; in others it is masked by other effects, as will be discussed in more detail in Sec. 24.4. The combination of Donnan membrane potentials for different ions plays a prominent role in the Hodgkin–Huxley model of the nerve axon discussed in Chapter 5.

24.3 Membrane Structure

Cellular membranes are composed of both proteins and phospholipids. The phospholipids, more specifically phosphoglycerides, contain polar phosphate head regions and nonpolar lipid tails, as shown in Figs. 24.2 and 24.3(a). When placed in an aqueous environment, these molecules will spontaneously associate into closed, ordered aggregates called *micelles* (Fig. 24.3(b)) or *bilayers* (Fig. 24.3(c)), which have many of the properties of membranes. Few current workers question the central role of the lipid bilayer in membrane structure. Rather, discussion centers on the role of the membrane proteins. All proposed structures support the Donnan membrane potentials discussed in the preceding section.

The theories of cell organization discussed in Chapter 16, Appendix E, and other places in this text are based in large part on cellular disruption and fractionation experiments. This has been compared to trying to reconstruct a precision watch by smashing it with a hammer and collecting the loose pieces. The protein pieces of this smashed cell can be loosely classed as: (a) soluble, those which stay in solution in the first fractionation step, and (b) membrane bound, those which are associated with the fraction containing the cell membranes.

The class of membrane-bound proteins is very heterogeneous. For one thing, these proteins differ quite markedly in their degree of association with the membrane. They are sometimes divided into two smaller classes: *peripheral*

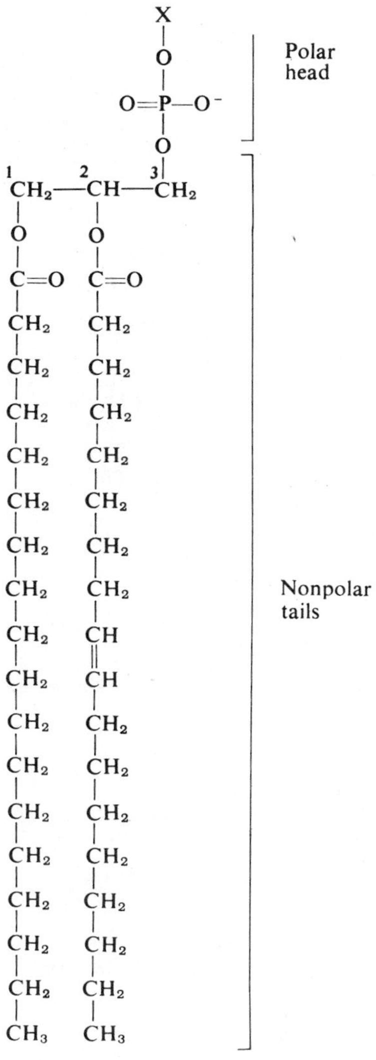

$$
\begin{array}{c}
\underset{|}{\mathrm{X}} \\
\underset{|}{\mathrm{O}} \\
\mathrm{O}{=}\mathrm{P}{-}\mathrm{O}^{-} \\
| \\
\mathrm{O}
\end{array}
\qquad
\begin{array}{l}
\text{Polar} \\
\text{head}
\end{array}
$$

$$
\overset{1}{\mathrm{CH_2}}{-}\overset{2}{\mathrm{CH}}{-}\overset{3}{\mathrm{CH_2}}
$$

O	O
C=O	C=O
CH₂	CH₂
CH₂	CH₂
CH₂	CH₂
CH₂	CH₂
CH₂	CH₂
CH₂	CH₂
CH₂	CH₂
CH₂	CH
CH₂	CH
CH₂	CH₂
CH₂	CH₂
CH₂	CH₂
CH₂	CH₂
CH₂	CH₂
CH₂	CH₂
CH₂	CH₂
CH₃	CH₃

Nonpolar tails

Figure 24.2 General structure of phosphoglycerides in a form emphasizing that they have both polar and nonpolar sections. Usually the fatty acid in the 2 position is unsaturated. The X group in the head can be any one of several alcohols, many containing positively charged amines as well. From A. L. Lehninger, *Biochemistry*, 2nd ed. (New York: Worth Publishers, Inc., 1975, p. 288).

(or *extrinsic*) *proteins*, which are easily dissociated from the membrane and are soluble in aqueous solution after dissociation, and *integral* (or *intrinsic*) *proteins*, which require the use of detergents or similar harsh agents

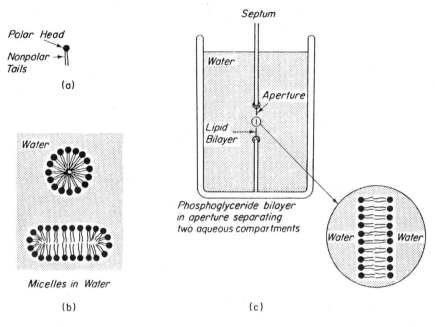

Figure 24.3 Stable phosphoglyceride–water systems. After A. L. Lehninger, *Biochemistry*, 2nd ed. (New York: Worth Publications, Inc., 1975, p. 301).

to dissociate them from the membrane and are usually insoluble in neutral aqueous buffers after dissociation.

The working hypothesis in current membrane research is that integral proteins, as the name implies, are integrally associated with the lipid bilayer. In fact, the association of integral proteins and the lipid bilayer is the membrane. Peripheral proteins, on the other hand, are seen as secondary membrane constituents. They appear to interact minimally with the bilayer, but rather seem to be attached to the membrane through some association with integral proteins.

In the late 1950's and early 1960's, the role of the integral proteins was thought resolved by the unit membrane hypothesis. This model, first proposed by Robertson based on an earlier model, assumed that the proteins were arranged in two continuous monolayers, one on each side of the lipid bilayer. Membrane size required that proteins in these monolayers be primarily in an extended or fibrillar, rather than globular, form. The protein layers were associated through their polar regions with the polar heads of the lipid bilayer, and would provide protection for the more fragile lipid bilayer to allow a stable, flexible membrane structure. The unit membrane was thus a protein–lipid–protein sandwich, which differed only in composition (proteins and lipids) from one membrane to the next. This theory unified many diverse observations, provided a framework for subsequent research, and was very well accepted for a while. By the middle 1960's, however, it became apparent that the results of many experiments on the biophysics and biochemistry of membranes were not explained by the unit membrane hypothesis.

Advances in several techniques helped overthrow the concept of the unit membrane. Spectroscopic techniques involving absorption in the infrared and optical rotatory dispersion and circular dichroism in the visible and near-ultraviolet spectrum (see Chapter 26) established that membrane proteins were primarily in the globular, not fibrillar, form. Electron microscopy was perhaps most important, however. Ordinary electron micrographs were and are of great importance in ultrastructure research (see Chapter 19), but they are two-dimensional slices through quite complex three-dimensional entities and, as such, are open to various interpretations. In the early 1960's it was found that if cellular material were frozen in the presence of glycerol, it was possible to fracture rather than slice this material along natural membrane boundaries. These surfaces could be sharpened by etching (subliming) surface ice, then replicated and examined by electron microscopy. This technique was termed *freeze-fracture* or *freeze-etching microscopy* and opened up vistas for the membrane investigator.

The fracture planes occur in most cases along the lipid bilayer, as shown in Fig. 24.4. Notice the large number of particles, of various sizes, seen on the

Figure 24.4 Electron micrograph of the concave fracture faces of the outer membrane (single arrow) and inner membrane (double arrow) of a rat liver mitochondrion, following freeze etching. × 137,500. After C. F. Hackenbrock, *Ann. N.Y. Acad. Sci.* **195**:499 (1972).

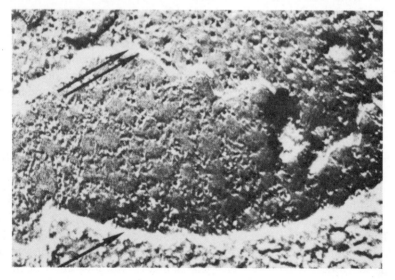

membrane surface. These were at first thought to be artifacts, but are now believed to be membrane protein complexes. It can be seen that they do not continuously cover the membrane surface.

A more satisfactory model of detailed membrane structure was first proposed in 1972 by Singer and Nicolson. This model considers the lipid bilayer as a matrix upon which the globular integral proteins float. They may be restricted to one or the other side of the bilayer, or they may extend completely through it, depending on the location of nonpolar, lipid-soluble amino acid residues, as shown in Fig. 24.5. Both the individual lipid molecules and the proteins or protein complexes are capable of movement in the plane of the membrane; thus, this model is termed the *fluid-mosaic model* of membrane structure. Most current theories for membrane function use this model for membrane structure in their hypotheses.

24.4   Membrane Function

As previously noted, membranes serve two functions; they provide a framework for living processes and separate the cytoplasm into compartments. While at times in this section each of these functions is considered separately, both are closely related not only to each other but also to membrane structure, discussed in Sec. 24.3.

The membrane is not a rigid framework. The fluid-mosaic model of membrane structure considers it more as a support medium for biochemical reactions. According to the model, groups of membrane-bound enzymes may migrate from site to site on the membrane. These groups are, however, restricted to two rather than three degrees of spatial freedom, and thus may be poorly oriented for certain reactions. Conversely, they may more closely approach other membrane-bound enzymes and be favorably oriented for other reactions. The functions of many specialized organelles, such as mitochondria, chloroplasts, and the light-sensitive segments of rods and cones depend on the precise geometry imposed on their membrane-bound reactants.

Membranes separate protoplasm into compartments, yet as was mentioned in Chapter 23, this is a selective compartmentalization. That is, membranes can be both passive barriers to the diffusion of some solutes and active transporters of others. The barrier role in most cases can be adequately explained by the chemical nature of membranes (most solutes are not lipid soluble); however, the mechanisms of active transport are a subject of much current investigation.

The potential differences observed in nerve and muscle fibers agree in sign with the Donnan potentials predicted in Sec. 24.2. That is, the outside of the membrane is positively charged. Similarly, K^+ ions are at a higher

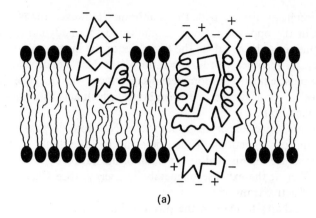

(a)

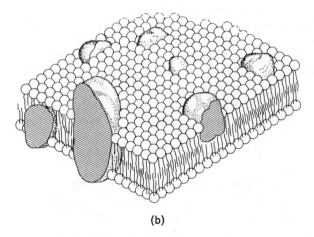

(b)

Figure 24.5 (a) Schematic representation of the cross section of the lipid–globular protein mosaic model of membrane structure. The globular proteins (with dark lines denoting the polypeptide chain) are molecules containing both polar and nonpolar groups. Their ionic and highly polar groups are exposed at the exterior surfaces of the membranes; the degree to which these molecules are embedded in the membrane is under thermodynamic control. The bulk of the phospholipids (with the filled circles representing their polar-head groups and thin wavy lines their fatty acid chains) is organized as a discontinuous bilayer. See the text for further details. (b) Fluid-mosaic model of membrane structure. The membrane consists of a fluid phospholipid bilayer with globular protein molecules penetrating into either side or extending entirely through the membrane. There is no long-range regularity in the spacing of the protein molecules, but some may be organized into complexes. Presumably the membrane is asymmetric. Adapted from S. J. Singer and G. L. Nicolson, *Science* **175**:720–731 (1972).

concentration inside than outside the membrane. However, the Na^+ ions are distributed in the opposite fashion. Equilibrium thermodynamics demands that the values of $\bar{G}$ be equal on the two sides of the membrane for all ionic species present. In other words, the ratio of the Na^+ concentration inside and outside muscle fibers is in complete disagreement with the belief that Na^+ ions are free to permeate the fiber membrane, which acts as a passive semi-permeable membrane.

Nonetheless, tracer experiments show that Na^+ ions do pass through the membrane in both directions. Thus, there can be little doubt that the Na^+ concentrations are maintained at nonequilibrium ratios by active transport out of the fiber at the expense of metabolic energy. (See Chapter 23 for a discussion of active transport.)

Most hypotheses to explain the physical mechanism of active transport can be loosely grouped as either involving carriers or pores. In one carrier hypothesis a membrane protein, floating in a lipid bilayer, could physically bind a solute molecule on one side of the membrane, move to the other side, and there release it. This hypothesis was weakened when studies which attempted to measure such rotations of membrane protein found the rates to be negligibly slow. This was to be expected if the membrane proteins are indeed bimodal, as shown in Fig. 24.5(a), with polar groups on the surface. Quite a large amount of energy would be required to transport the entire protein, including polar groups, through the bilayer. An alternative carrier hypothesis eliminates protein carriers and uses instead smaller cagelike molecules called *ionophores*, which are soluble in lipids but have a polar interior.

The pore hypotheses require physical holes in the lipid bilayer. To be stable such holes may be lined with integral proteins which extend through the bilayer. Solutes could be transported through such holes by a conformational change in the lining proteins such that a solute binding site facing one side of the membrane is shifted and reoriented to the other side.

Physical mechanism aside, a second question is where the energy for this transport comes from and how it is coupled to active transport. At a minimum, energy of active transport can be generated by (1) oxidation via the electron transport chains in mitochondria (see Chapter 18), (2) the hydrolysis of ATP by a membrane-bound ATPase, and (3) ionic gradients produced by the action of other membrane-bound enzymes. Some solutes may be transported using energy from more than one source.

There have been numerous theories as to how the energy-generating processes are linked to active transport. One, called the *Mitchell hypothesis*, or *chemiosmotic theory*, was first formulated to account for electron transport in chloroplasts and mitochondria (see Chapter 18). Mitchell postulated that electron transport resulted in proton flow from one side of the membrane to the other. This results in a *proton-motive force*, the sum of a Donnan membrane potential and a pH gradient. ATP synthesis would then occur with the help of a proton-translocating ATPase, which would attempt to restore proton equilibrium.

More recently, Blodin and Green have proposed a paired moving charge model of energy coupling in all types of active transport across membranes. This model expands Mitchell's hypothesis to paired movements of many positive and negative ions, simple or complexed with ionophores. It further specifies that the nature of the attraction between the paired charges is electrical and places a limit of 2–3 nm on the distance that can separate them.

The chemiosmotic theory, for example, hypothesizes that the movement of a potassium ion (K^+) from side A to side B of a membrane is counterbalanced by the movement of a proton (H^+) from side B to side A. The paired moving charge model regards both processes as paired ion movements; at side A, H separates into H^+ (remains at side A) and an electron (e^-), while K^+ is encapsulated into an ionophore. Both e^- and K^+ migrate across the membrane as paired ions. At side B, K^+ is released and e^- reacts with some acceptor, eliminating a positive charge. (This is equivalent to the loss of an H^+.) The net result in both cases is a gain of H^+ and loss of K^+ at side A with loss of H^+ and gain of K^+ at side B.

Thus, a membrane's functions can be related to its structural form and physical characteristics. The fluid-mosaic model of membrane structure can account for cellular compartmentalization and the generation of Donnan membrane potentials. Active transport also contributes to the transmembrane potential. Transport of molecules and electrons against an electrochemical gradient can be described by a chemiosmotic theory or by charged ion-pair movement (electron transport is further discussed in Chapter 18).

24.5 Summary

Biological membranes are important in many of the processes discussed in this text. The molecular bases of vision, hearing, photosynthesis, nerve conduction, and muscle contraction, to name a few, all depend on the structure and function of associated membranes. Membrane structure is best explained by a fluid-mosaic model, which postulates membrane proteins floating on a lipid bilayer matrix. These proteins can migrate in the plane of the bilayer and may also extend through to the other side.

Membranes function as a support medium for membrane-bound enzymes, and also separate the cytoplasm into compartments. Simple diffusion theory can account for some effects of this compartmentalization, but active transport of certain solutes is more complicated to explain. Protein or ionophore carriers may cause active transport, or there may be fixed pores or holes in the membrane. In either case, energy is required and must be coupled to the transport by means of such processes as chemiosmosis or paired ion movement.

REFERENCES

As with previous subjects, there are far too many possible references to do more here than introduce the literature.

General References

1. MANSON, L. A., ed., *Biomembranes*, Vol. 1 (New York: Plenum Press, 1971). (This multivolume series offers articles on many aspects of membrane biophysics.)
2. KORN, E. D., ed., *Methods in Membrane Biology*, Vol. 1 (New York; Plenum Press, 1974). (This is another multivolume series; Vol. 4, *Biophysical Approaches*, may be of special interest to readers of this book.)
3. GREEN, D. E., ed., "Membrane Structure and Its Biological Applications," Volume 195 (1972) of the *Annals of the New York Academy of Sciences*, reporting on a 1971 conference. (Entire volume.)

Review Articles

The *Annual Review of Biochemistry* usually has one or more articles relating to membranes, such as the following three:

4. CUATRECASAS, P., "Membrane Receptors," **43**:169–214 (1974).
5. SINGER, S. J., "The Molecular Organization of Membranes," **43**:805–833 (1974).
6. SIMONI, R. D., AND P. W. POSTMA, "The Energetics of Bacterial Active Transport," **44**:523–554 (1975).

Finally, two papers on the mechanism of energy coupling in active transport:

7. MITCHELL, P., "Chemiosmotic Coupling in Oxidative and Photosynthetic Phosphorylation," *Biol. Rev.* **41**:445–502 (1966).
8. BLONDIN, G. A., AND D. E. GREEN, "A Unifying Model of Bioenergetics," *Chem. Eng. News* **53** (45):26–42 (November 10, 1975).

CHAPTER 25

Information Theory and Biology

25.1 Languages

It is often said that mathematics is the language of physics. Perhaps, more strictly speaking, one should assign this role to calculus. Isaac Newton was a leader in the development of this branch of mathematics; calculus enabled Newton and those who followed him to more conveniently describe the physical world. Ever since the seventeenth century, calculus and its ramifications have developed side by side with physics. Some purely mathematical theorems have been applied later in physics; and some physicists have developed mathematical tools that could not be justified or made rigorous for years to follow.

In this chapter, information theory is presented as another branch of applied mathematics which has become a convenient language in certain fields. It is included in this part of the text because information theory is so closely related to the ideas of thermodynamics, particularly the concept of entropy. In biophysics, research workers describe experiments in terms of information theory when discussing sensory biophysics, molecular biology, genetics, and the operation of special equipment.

It is extremely easy to mistake the role played by a language. In literature, it is well understood that a book can be translated into many different languages and still express approximately the same ideas. Absolutely everything in physics which is discussed in mathematical terms—every proof and every theorem—could be presented without any mathematical symbols or terms whatsoever. However, the length of time and the number of words involved would be so great that most of the concepts of physics could not be understood within a lifetime. The mathematical language by itself tells us nothing new about the physical universe. A knowledge of advanced calculus is in no way synonymous with an understanding of intermediate physics. Nonetheless, it is almost inconceivable that anyone would discuss the details of quantum mechanics without considerable training in advanced mathematics.

Information theory consists of mathematical methods for assigning quantitative values to information. It is a language and as such cannot reveal anything new or unsuspected above the universe. It can help to express ideas and theorems and to realize similarities and analogies between diverse fields. Information theory is a successful language in that it can achieve an economy in thought processes and words. However, it is a far less successful language than calculus; its applications are not as forceful and its economies are not as great.

Perhaps the greatest success of information theory has been in the areas of digital computation and information storage. Information theory is the natural language to use in discussing those areas. This has occurred to such an extent that information theory is often applied without conscious recognition of its use in the analysis of computing and in information storage and retrieval systems.

25.2 Information Theory and Computing

Information theory consists of mathematical methods for quantifying information. In order to do this, one must understand what is meant by information. In the technical, restricted sense, *information* may be regarded as the removal of uncertainty or doubt. From this point of view, the ordering of randomly arranged objects is removing doubt concerning their location, and hence increasing their informational content.

Suppose that an employer wishes to know if a job candidate had a "B or better" grade average in school. If the employer has little previous knowledge of the candidate's work, either a "yes" or a "no" answer may be equally probable. After receiving an answer from the candidate, the employer's uncertainty would be less, but it would not be gone completely. In the language of information theory, such an answer is masked by the "noise" caused by a candidate's truthfulness and memory. If, instead of asking the candidate, the employer had "asked" the grade transcript, all uncertainty about grade average would have been removed.

Again, suppose a teacher asks if a student likes ice cream. It is extremely likely that the student does, so the teacher can receive only very little information if the student answers "yes." This time the answer will remove almost all uncertainties, unless, for instance, the student has such poor diction that the teacher cannot understand the answer.

From these examples, one can note that the more uncertain the prior choice, the more information there is that can be supplied by the answer. Further, the more certainty that exists after the answer, the greater the amount of information received. The quantitative form of information must reduce to zero if the initial (or *prior*) probability of the answer p_i is 1 and must be maximum if the output (or *posterior*) probability p_o is 1. The ratio p_o/p_i varies as the information but does not go to zero when p_i (and therefore p_o) are unity. A function that does behave as the information, and which is always positive, is $\log (p_o/p_i)$. For historical reasons, information, I, is defined in information theory as

$$I = \log_2 \left(\frac{p_o}{p_i} \right) \qquad (25\text{-}1)$$

The unit of I is called a *bit*, an abbreviation for binary integer. Because p_o is greater than, or at worst equal to, p_i, information I as defined by Eq. (25-1) will always be positive.

As an example of Eq. (25-1), one may ask if a given neuron is conducting a spike potential at a given time. If the two possible answers, yes and no, are equally likely, the prior probability p_i is

$$p_i = \tfrac{1}{2}$$

If the question is asked with suitable measuring equipment, the answer may be definitely yes; that is, the output probability p_o is 1. The information gained is

$$I = \log_2 \frac{p_o}{p_i} = \log_2 \frac{1}{\frac{1}{2}} = \log_2 2 = 1 \text{ bit}$$

On the other hand, if it has been observed that the neuron is conducting a spike potential 5 percent of the time and is silent for the remainder, then

$$p_i = 0.05 \quad \text{and} \quad I = \log_2 20 = 4.32 \text{ bits}$$

if conduction is observed at the instant the neuronal state is tested.

Electrical engineers were the first to use information theory. Many electronic circuits exist in one of two stable positions. In digital electronic computers, all decimal numbers are reduced to binary numbers that involve making several yes–no choices. The base 2 logarithm appears to be the natural one, not only for the electronics engineer, but also for the physiologists and biophysicists who work with nerve axons which follow a yes–no pattern. (Historically, the physiologists have preferred the words "all or none.")

Within digital computers all information is stored in an essentially binary form. The computer systems themselves have become sufficiently versatile so that most human interactions make use of symbols, number systems, and codes. The outstanding success of storing information in this format for a wide variety of disciplines is inherent verification of the basic postulate embedded in Eq. (25-1). This states, in effect, that all information can be coded as a series of binary choices.

In the presence of noise, the posterior probability may be less than 1. Noise in the sense used in information theory must be interpreted in a very general fashion. Thus, it may refer to poorly functioning electronic equipment or to electronic pickup of signals from the environment. The student's poor diction in the ice cream problem is another example of noise. In discussing diagnostic classification, noise may mean that a group reviewing all the evidence will not reach agreement on the diagnosis. Within humans, hazy memories and poor judgment contribute noise in a fashion analogous to an improperly functioning digital computer.

To recapitulate, information theory treats information as the removal of uncertainty. For this purpose, all information is encoded in a binary form. A generalized picture of an information system may be represented schematically as shown in Fig. 25.1. For a telegraph, the meaning of the boxes is

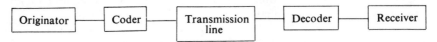

Figure 25.1   Schematic diagram of an information system.

obvious. For the process of hearing, the originator might be a piano player. The coder would be the piano. The transmission line represents the air. The decoder is the ear of the listener, and the receiver is that person's central nervous system. Similar analogies can be made for the synthesis of proteins, the genetic processes, vision, and so on.

In most examples, different amounts of information are received from one second to the next. Information theory calls the signals received in a given period of time the *message*. In a noiseless system (p_o equals 1), the information in a given message is

$$I = -\log_2 p_i \tag{25-2}$$

The average information H for a message with M possible alternatives is then

$$H = \sum_{j=1}^{M} p_j I_j \tag{25-3}$$

where p_j is the prior probability of the jth alternative and I_j is the corresponding information. Using Eq. (25-2), this may be rewritten

$$H = - \sum_{j=1}^{M} p_j \log_2 p_j \qquad (25\text{-}4)$$

Equation (25-4) is very similar to the statistical mechanics definition of entropy, except for a minus sign. Accordingly, the average information H in a message is often called *negative entropy*. In other words, H is a measure of the uncertainty (entropy), which, on the average, is removed (negative entropy) by the receipt of the message.

There are a number of other terms used in the information theory literature, some of which are included here for completeness.

A. STOCHASTIC PROCESS

This is a process that "generates" symbols, for example, words or amino acids, in a random fashion, but in which the frequency of occurrence (i.e., probability) approaches a limiting value as the number of symbols is increased. For example, a stochastic process is tossing a penny that generates the symbols "heads" and "tails." The frequency of occurrence of heads for a small number of tosses is random but approaches $\frac{1}{2}$ as the number of tosses is increased.

B. MARKOV PROCESS

This is a stochastic process in which intersymbol influences exist, so that the probability of i following j, p_{ij}, can be defined. In general, the p_{ij}'s are all different. For example, the probability of one English letter following another is measurable. A process generating letters in English words, writing for example, is a Markov process.

C. ERGODIC SEQUENCE

This is a Markov sequence of symbols in which the intersymbol influence falls off exponentially or disappears after a finite number of symbols. In English sentences, the probability of a given letter following the first one is not random. Nor is the probability of a second or third following letter determined at random. Definite intersymbol influences can be found out to eight letters. Thereafter, the probability is essentially random. Thus, letters in English sentences form ergodic sequences.

D. REDUNDANCY

If intersymbol influences exist, not all the symbols are necessary. A different coding could reduce the number of symbols. Redundancy is desirable in that it tends to increase the signal-to-noise ratio.

One example of redundancy is the written English language. The average information per letter H_1 has been computed, including various intersymbol influences. These are given in Table 25-1, which shows that a redundancy of about 1 bit (i.e., twofold) exists.

TABLE 25-1

AVERAGE INFORMATION OF ENGLISH LETTERS

Letters	H_1
Random	4.7 bits
English frequency	4.15 bits
Intersymbol influences for 2 letters	3.57 bits
Intersymbol influences for 8 letters	3.25 bits

The twofold redundancy in English greatly reduces the error rate due to noise, such as blurred printing, bad spelling, poor lighting, and so on. The error rate is reduced because there are a large number of possible messages excluded. The number of 1,000 symbol messages, N_r, if all random arrangements of letters were possible, would be

$$N_r \simeq 10^{1,400} \text{ messages}$$

If one includes all influences up to eight symbols, the number of messages N_8 is reduced to

$$N_8 \simeq 10^{975}$$

A 1,000-symbol message is about one typed page. The number N_8 is very large, but it is microscopic compared to N_r.

Redundancies also exist in many biological processes. For example, it seems likely that the endocrine systems of mammals have more interacting glands than are necessary. This particular redundancy can protect the organism against partial system failure if interacting feedback loops are present. It also appears to be a reasonable guess that the pigment myoglobin is not necessary in muscles (see Chapter 9). Its presence is a redundancy tending to further smooth out oxygen variations.

The most striking example of redundancy occurs in the higher plants. Some are *polyploids*, in which, instead of having pairs of chromosomes, all the body cells have sets of four or even eight homologous chromosomes. Each member of each set controls the same characteristics. This redundancy could markedly reduce the error rate during cell multiplication. At any rate, this possibility exists whether or not the plants use it!

25.3 Information Theory and Sensory Perception

The senses provide the link between the central nervous system and the outside world. All information reaching the central nervous system comes through the senses. It seems appropriate, therefore, that the language of information theory can be used to describe what humans perceive. This is illustrated for hearing in Sec. 25.3.A and for vision in Sec. 25.3.B.

A. HEARING

All the examples in the previous section deal with discrete phenomena. Sound is continuous both in amplitude and in time. As mentioned in Chapter 1 and in Appendixes A and C, the representation in time may be replaced by a corresponding one in frequency. Any discussion of the information content of sound must include some method of handling continuous variables. Such theories have been developed. They show that for amplitudes with no noise level, or messages of infinite length in either frequency or time, the average information per message is infinite. However, all sounds have a noise limit provided by molecular thermal motions, a finite time of duration, and a finite frequency span, the latter two being limited by the ear.

If speech is analyzed in terms of the physical noise level and the frequency response of the human ear, one may easily arrive at very large values for the information per second. Although these values represent the information that can be detected by a microphone or other physical detector, they have no meaning for human hearing. The information received is limited by the ear itself.

Humans hear about 10 octaves, from 20 Hz to 20 kHz. Few people can distinguish more than 12 tones per octave, that is, a total of 120 pure tones. If each is equally probable, all the p_j's are the same. One may write

$$p_j = \frac{1}{120}$$

and the average information H_f associated with a tone is

$$H_f = -\sum_{j=1}^{120} p_j \log_2 p_j = \frac{120}{120} \log_2 120 \simeq 7 \text{ bits/tone}$$

Similarly, from the threshold of hearing to the threshold of pain, one can distinguish about 250 steps. Again, one may assume all steps to be equally probable and find the average information associated with the sound pressure level

$$H_L = \log_2 250 \simeq 8 \text{ bits/level}$$

For a pure tone, then, the human auditory apparatus can receive about 15 bits of information. For a complex tone, this must be increased to about 20. Human auditory systems can distinguish about 10 tones per second. Therefore, the rate of receipt of useful information by the ear is approximately 200 bits/sec. This represents the ability of the ear to code information as neural impulses. The limiting factor is the ear itself. Information can be coded electronically at much higher rates.

Although 200 bits/sec may reach the brain, it in no way follows that these are recorded or used by the brain. When a person is asleep or daydreaming, most auditory information is lost. When one concentrates on reading, most auditory information is deliberately discarded. If one listens to a friend talking in a noisy room, most of the information reaching the ears is consciously or unconsciously blocked from the mind; that is, one listens only to the speech and not to the background noise.

Even with the utmost concentration under ideal conditions, it is rarely possible to use all of the 200 bits/sec reaching the brain. Numerous reasons exist for regarding 200 bits/sec as an excessively high estimate. For example, for short pure tones, most people can detect not more than one correct choice of six possibilities. Hence, one should write

$$H_f = 2.5 \text{ bits/tone}$$

of useful information.

Some investigators theorize that about 2.5 bits of information are available from the sound pressure level and perhaps another 2.5 bits from the quality of a complex tone. Adding these three, one can estimate that the human brain can receive a maximum of about 7.5 bits of auditory information each 0.1 sec, or 75 bits/sec. Estimates of information transfer by spoken words lead to values around 50 bits/sec, which agrees within the limits of the errors of the estimate. People with perfect pitch can detect up to 8 bits per piano tone. They are believed to use information of more than one frequency and of the relative intensities of the different harmonics. In addition, they must hear the tone longer than 0.1 sec.

For transmission of digital information via telephone lines, it is customary to encode the information acoustically. The rate of transmission is described in a unit called the *baud*, which may be regarded as an acronym for *b*its *audi*o encoded per second. Thus, it is possible to transmit information on a usual (voice-grade) telephone line at rates up to 4,800 baud (4.8 kilobaud). This latter number may be regarded as the *channel capacity* of the telephone circuit.

The fact that 75 bits/sec is much lower than the channel capacity of a voice-grade telephone circuit interests telephone engineers. They conclude that by suitable encoding it should be possible to increase the number of conversations on a given line. Various attempts use methods that analyze for the significant components of the spoken sounds and encode these suitably. The converse, storing in a limited amount of computer memory the encoded information necessary to reconstruct human words, is in practical use in biomedical computing.

B. VISION

A similar analysis can be carried out for the visual system. Analogous to sound information, most light information is lost in the eye. However, vision differs from hearing in that information may be received and transmitted to the brain at a much greater rate.

In the eye, there are about 10^7 receptor units, each feeding into a separate ganglion cell (see Chapter 2). Except in the most sensitive region of the eye, the *fovea centralis*, several rods and cones combine to form one receptor unit. (There are about 7×10^6 cones and 10^8 rods in the human eye.)

Each receptor unit is believed able to respond in a characteristic fashion to about 100 just noticeable differences in intensity between the visual threshold and the pain threshold. The average information H_1 per receptor unit then is

$$H_1 \simeq \log_2 10^2 \simeq 7 \text{ bits}$$

The eye sees a new picture about 10 times per second. Accordingly, the rate of useful information H_1' is

$$H_1' = 70 \text{ bits/sec/receptor unit}$$

or, for the entire eye,

$$H' = 7 \times 10^8 \text{ bits/sec}$$

This can be compared with a television channel that carries about 10^7 bits/sec.

Some alternative estimates of the channel capacity of the optic nerve are about one half of the value estimated in the preceding for H'. However, the uncertainties are so great that these two estimates are not significantly different. Still other estimates of the useful information in color and acuity vision yield similar rates. Thus, the coding of the visible information makes optimal use of the optic nerve. This contrasts sharply with the auditory system, where the channel capacity of the auditory nerve is orders of magnitude greater than the rate of transmission of useful information to the brain.

The coding of the optic nerve may be compared to a television channel. One of the limitations of television broadcasting is poor encoding of information. Many engineers have realized that a system which indicated changes of intensity only would be far more efficient in transmission of information. (In other words, narrower channels could be used.) The high efficiency of the optic nerve, as indicated by the foregoing estimates, suggests that such a system is used. Evidence from electrophysiology and histology supports this view (see Chapter 8). The approach of information theory helps to understand the histology and the electrophysiological data.

The useful information within the brain should not be confused with information stored in memory. Here one must distinguish the immediate perception of processable information described in the preceding as H' from both short-term and long-term memory. The very complicated processes involved are not understood at a neuronal or biochemical level. Rather, the transfer of such visual information to short-term and then long-term memory are processes which are being studied by scientists from various disciplines, including biophysics.

25.4 Information Theorems and Biomedical Computing

The person who first introduced information theory, Shannon, was particularly interested in the problems of sending information over telephone wires. He developed two theorems, which have implications not only for his original studies but also for biomedical computing, for protein structure, and for genetics. The theorems, along with an indication of their meaning in biomedical computing, are included in this section.

Shannon's first theorem states in effect that, provided one can transmit for a long-enough period of time, it is possible to encode information in a noise-free system so that the rate of transfer is equal to the channel capacity. Shannon proved that this is true under the most general conditions. Alternative expressions of Shannon's first theorem are that with a sufficiently complicated encoder, one can reduce the redundancy to zero, so that every bit of code carries one bit of information. For a sufficiently good coding scheme, then, one should be able to store information in a computer's memory or on magnetic tape and disk systems in a minimal number of bits, provided that the system is free of noise. The problem of suitably encoding information is not in itself solved by Shannon's first theorem or by its proof.

In the previous section, it was noted that the retina of the human eye is successful in encoding information in an optimal form with minimal or no redundancy. In the earlier days of the use of perforated Hollerith cards, a great deal of effort was spent in developing encoding rules such that a maximum amount could be stored on a single card. This could be described as recording the information in the optimal form described by Shannon's first theorem. While such encoding may still be desirable to store information within the computer, the speed and accuracy of communicating with humans is increased if the input and output of information occurs in a more natural, redundant form. Also, as perforated cards are replaced as the medium for data storage, for example by magnetic tapes, the length of a message (or record) can become variable. This can contribute to a reduction in redundancy and an increase in the efficiency of information storage and transmittal.

In the preceding paragraphs the problem of noisy information has been ignored. However, all real information is subject to errors, and, accordingly, some redundancy is desirable. *Shannon's second theorem* deals with the question of how much redundancy (or intersymbol influence) should be included in the coded information. This theorem states that, provided one is willing to transmit information at a rate less than the channel capacity, it is always possible to limit the probability of error (or the effect of the noise) to below a preassigned small number. In other words, Shannon's second theorem states that no matter how noisy the environment, it is always possible to achieve essentially error-free transmission of information provided that one is willing to transmit sufficiently slowly.

Both the first and second theorems are limited in their utility by failure to describe how to accomplish the encoding. Nonetheless, they indicate what can be achieved by sufficiently skillful coding. For biomedical computing the second theorem states that provided that one is willing to introduce sufficient redundancy by the use of added memory to store information, one can limit to any extent desired the probability of errors. Thus, the coding scheme used should represent a compromise between the least possible redundancy and the least possible chance of errors in the information stored.

Redundancy in itself does not necessarily represent a reduction in the probability of erroneous information. Thus, one might restate a major problem in designing information storage systems for medical and biological research as follows. The redundancy that is not used to reduce the error rate should be limited, to permit coming as close as possible to the optimal coding described by Shannon's first theorem. On the other hand, problem-specific and/or machine-specific redundancies should be incorporated, in order to limit the probable error rate per bit (or per message) to a preassigned small number. The solution of this problem is by no means trivial and tends to be characteristic of the particular information and the uses for which it is intended.

25.5   Information Theory and Protein Structure

Information theory, including Shannon's two theorems, can be applied to any type of information. The only restriction is that the information be encodable by some set of rules into answers to binary questions. Among the areas of its application is one most closely related to chemistry and to thermodynamics, namely, the informational content of protein structure.

To assemble a protein, it is necessary to choose the proper amino acids and arrange them in a given order with a suitable spatial configuration. There are many ways in which this ordering can be done with the same amino acids. It is too complex to illustrate this for a protein with 100 amino acid residues, but some intuitive feeling can be gained by considering a polypeptide with five residues. Suppose that these are all different: for example, one glycine, one phenylalanine, one tryptophan, one valine, one methionine. These can be arranged in 5! (i.e., 120) permutations, because the nature of the peptide bond is asymmetrical; that is, glycylphenylalanine differs from phenylalanylglycine. Thus, the information necessary to select any one permutation is 6.9 bits.

If two of the amino acids had been the same, the number of possibilities would have been reduced by a factor of 2. And if three were the same, the number of distinct possibilities would have been reduced by a factor of 3!, or 6. If only two different amino acids are present, for example, three glycines and two phenylalanines, there would only be 5!/(3! 2!) or 10 possibilities. Recognizing one of these possibilities gives $\log_2 10$, or 3.3, bits of information. By using fewer distinct amino acids, the information transmitted per message has been reduced from 6.9 bits to 3.3 bits.

In general, there are N amino acid residues of m types in a protein, such that there are n_1 of the first, n_2 of the second, and so on. The number of types m is less than or equal to 20. The number of ways of arranging these in a straight chain is

$$P = \frac{N!}{(n_1!)(n_2!)\cdots(n_m!)} \tag{25-5}$$

If all are equally likely, the information necessary to build a particular protein is

$$I = +\log_2 P = \log_2 N! - \sum_1^m \log_2 (n_i!)$$

The average information per amino acid residue is

$$H_R = \frac{I}{N} = \frac{I}{N} [\log_2 N! - \sum_1^m \log_2 (n_i!)] \tag{25-6}$$

Because N is large compared to 1, Stirling's formula,

$$\log_2 N! = (\log_2 e) \log_e N! \simeq 1.45 N \log_e N$$

can be used. Therefore, the average information per amino acid, or negative entropy per amino acid residue, is

$$H_R = 1.45 \log_e N - \frac{1}{N} \sum_1^m \log_2 (n_i!) \tag{25-7}$$

If, in addition, all the n_i are large, this expression becomes

$$H_R = 1.45 \log_e N - 1.45 \sum_1^m \frac{n_i}{N} \log_e n_i = -1.45 \sum_1^m \frac{n_i}{N} \log_e \frac{n_i}{N} \tag{25-8}$$

In a long molecule, the ratio n_i/N is the relative probability of finding an amino acid of the ith variety, so that (except for a numerical constant) the foregoing formula is identical to the previous form for H. Unfortunately, the values of n_i are so small that Eq. (25-7) must be used.

The values for I and for H_R can vary widely, even though both the total number of residues N and the number of types of residues m are fixed. For example, if there are four glycines and one phenylalanine in the pentapeptide discussed earlier, there are only five possible arrangements. The information per sequence has been reduced from 3.3 bits for three glycines and two phenylalanines to 2.3 bits for four glycines and one phenylalanine.

For larger values of N and m, the variation is much greater. In general, one can compute an I_{max} and an I_{min} for fixed N and m. Because there are usually about 20 types of amino acids within the cell, one can also compute an $I_{max}^{(20)}$ for fixed N and 20 types of residues. It is instructive to consider the ratios I/I_{max}, I/I_{min}, and $I_{max}/I_{max}^{(20)}$. It has been found for all proteins tested that I/I_{max} is greater than 0.5. For all proteins within living cells, in fact, this ratio is greater than 0.7, and for most it is greater than 0.85. The information per residue is about

$$H_R = 3.6 \text{ bits/amino acid}$$

for a typical protein. No values are less than half this or greater than 5 bits. For instance, for a protein such as albumin with over 500 residues, this is a total information of 2,000 bits needed to build the molecule. Fibrinogen has 3,400 amino acid residues; a total information of 10,000 bits is necessary to distinguish it from all other proteins with 3,400 amino acid residues.

It is not just because there are no other possibilities that these values are so high. For albumin, the ratio of I/I_{min} is about 15. Nor does replacing I_{max} with $I_{max}^{(20)}$ alter the situation very much. The ratio of the last two is not very different from 1. Thus, proteins are formed from amino acids with a coding having close to a minimal redundancy. Reduction of information content by intersymbol influences undoubtedly occurs, allowing use to be made of the concepts in Shannon's second theorem. This line of reasoning is not applied in this text to protein structure but is used in the following discussion of genetic information.

25.6   The Coding of Genetic Information

The genetic information of most cells is encoded within DNA sequences primarily within the chromosomes of the nucleus. Exceptions to the above are the bacteria; they do not have well-formed nuclei or chromosomes. Experimental evidence indicates that even in the bacteria, genetic information is coded in the form of DNA. Additional genetically active DNA is found within mitochondria and chloroplasts. As discussed in more detail in Chapter 16, the information in the DNA is transcribed to RNA, which then acts as a template for protein synthesis. The synthesized proteins include enzymes and structural proteins which combine to give the intact cell its own characteristics, including the ability to replicate the DNA molecules.

There are various approaches to the informational content of DNA. One is to note that for each base pair there are four possibilities: A-G, G-A, C-T, and T-C. (The reader may wish to refer to Chapter 14 or Appendix D for a more complete discussion of DNA structure.) Thus, each base pair could contribute 2 bits of information provided that these pairs occur in a sequence with little or no inter-base-pair influences. Using this figure and an approximate 10^9 base pairs in a human cell nucleus, one may conclude that no more than 2×10^9 bits of genetic information are stored.

A more detailed examination of the chromosomes using fluorescence staining techniques shows alternate dark and light bands. Perhaps 90 percent of the DNA is in the dark bands arranged in highly repetitive sequences having little if any informational content. The light bands then must carry the genetic information, which cannot exceed 2×10^8 bits.

Even that information is somewhat redundant, since it has been shown that it takes a series of three bases to code for one amino acid. If this represented a choice of one amino acid from a list of about 24 possibilities (including a "stop" code), the information required by an optimal coding would be about 4.6 bits. Owing to the unequal occurrences of various amino acids in natural proteins, only 3.4 bits should be needed on the average. However, within the DNA, three serial bases, or 6 bits of information, are used. Many triplet codes are redundant (see Chapter 16), accounting for the reduction to 4.6 bits, and protein characteristics reduce the useful information in 3 base pairs to 3.4 bits. Thus, the genetic information to form a human may be estimated as not exceeding about 10^8 bits.

It is possible by alternative means to arrive at far higher estimates. For example, if each atom in a human is to be specified and located, about 5×10^{26} bits would be needed. Even if water and inorganic ions are eliminated, the necessary information is only reduced by a factor of 10, to 5×10^{25} bits. This number is very large compared to the informational content of the DNA. Even attempting to specify all the synapses in the brain of a human far exceeds the information stored in the DNA of the nucleus.

A very low estimate can be obtained by a somewhat different route. It is possible to estimate the total number of genes in the human nucleus as about 30,000. If each of these has on the average 32 possible alleles, about 5 bits of information would be needed per allele, or about 10^5 bits per human. This number may be orders of magnitude too small, because it does not take into account the multitude of lethal alleles possible for each gene.

The action of the DNA may be considered as selecting suitable amino acids from a pool to synthesize desired proteins. Assuming 30,000 distinct protein types and an average of 600 amino acids per protein leads to a necessity for the information to make 1.8×10^7 choices. Assigning 3.4 bits per choice of amino acid gives 0.6×10^8 bits of information needed. This is well within the errors of estimation of the 10^8 bits available from DNA. Alternatively, one might have multiplied by the three bases used per amino acid to find 0.5×10^8 base pairs needed as compared to about 10^8 base pairs available. Such good agreement supports both the concept of DNA as the encoded genetic information and the utility of information theory in describing genetics. It is interesting also to note that a 10,000-page series of books is probably necessary to hold 10^8 bits of information.

It is possible to think of the reduction from 6 bits per triplet code to 3.4 bits as being due to redundancies in coding. From one point of view this is still a highly efficient code coming close to the optimal one defined by Shannon's first theorem, the difference of 2.6 bits per triplet representing the redundancy. If one were also to include inter-amino acid regularities in the protein sequences, or interbase regularities in the DNA sequences, a further reduction would be observed. These latter types may be thought of as the redundancy used in Shannon's second theorem to reduce noise. Gatlin has described a view of evolution in which the initial steps consist of limiting the redundancy which is unnecessary according to the first theorem, and therefore reducing the total redundancy. Thereafter, evolution proceeds by increasing both types of redundancies at a comparable rate producing, so she hypothesizes, an optimum mixture of undesired redundancy and redundancy used to limit errors of transcription and replication. The original data on which this theory was based were obtained before the technique of fluorescence staining of chromosomes and should be reevaluated with these new data. Nonetheless, the general idea is an appealing one.

25.7 Summary

It has been shown that the language of information theory can be used in various fields of biology. Information theory emphasizes the quantitative, mathematical approach appealing to the physicist. Information theory is a successful language in that it increases the rate of transmission of information from one person to another and helps focus research thoughts in new directions. Information theory has real but limited applications in biology; it is important neither to overstate its contributions nor to ignore them.

Information theory applies directly to digital computation. The discipline of computer science has incorporated information theory into all discussions of memory, information systems, and data handling. These, in turn, have become part of biomedical computation and hence of biophysics. A few other biophysical applications are discussed in this chapter. These include sensory perception, protein structure, and genetic transfer of information.

REFERENCES

1. QUASTLER, H., ed., *Essays on the Use of Information Theory in Biology* (Urbana, Ill.: University of Illinois Press, 1953), 273 pages.
2. SHANNON, C. E., AND W. WEAVER, *Mathematical Theory of Communications, and Recent Contributions to the Mathematical Theory of Communication*

(Urbana, Ill.: University of Illinois Press, 1949), 117 pages. (This is the classic book in the field.)

3. YOCKEY, H. P., R. L. PLATZMAN, AND H. QUASTLER, *Symposium on Information Theory in Biology* (New York; Pergamon Press, 1958), 418 pages.

4. JACOBSON, H., "Information and the Human Ear," *J. Acoust. Soc. Am.* **23**:463–471 (1951).

5. POLLACK, I., "The Information of Elementary Auditory Displays," *J. Acoust. Soc. Am.* **24**:745–749 (1952).

6. HALSEY, R. M., AND A. CHAPANIS, "On the Number of Absolutely Identifiable Spectral Hues," *J. Opt. Soc. Am.* **41**:1057–1058 (1951).

7. INGELS, F. M., *Information and Coding Theory* (Scranton, Pa. International Textbook Co., 1971), 229 pages.

8. ATLAN, H., *L'Organisation biologique et la théorie de l'information* (Paris: Hermann, 1972), 299 pages.

9. GATLIN, L., *Information Theory and the Living System* (New York: Columbia University Press, 1972), 210 pages.

DISCUSSION QUESTIONS—PART E

1. Describe A. V. Hill's studies of the heat changes associated with muscular contraction and relaxation. Relate these insofar as possible to the molecular process of contractions.

2. This text simply presents absolute rate theory as an empirical theory without really justifying it. Outline the types of evidence for absolute rate theory and the fashion in which it can be justified theoretically.

3. Describe transport of water and sugar across the intestinal lining in the terms of irreversible thermodynamics. Include quantitative values for the parametric descriptors to the extent possible.

4. The enzyme systems called luciferase are responsible for the bioluminescence of fireflies and various other organisms. (See Chapter 3.) These systems have been studied in detail as a function of temperature and pressure. Describe the results in terms of absolute rate theory.

5. Diffusion theory has been applied to the reaction catalyzed by the enzyme fumerase to show that it is a diffusion-limited reaction. Discuss this proof.

6. Develop in detail the application of the phenomenological equations of irreversible thermodynamics to the citric acid cycle.

7. The average-value approach to diffusion is referred to in Chapter 23. Develop this in greater detail.

8. The spherical cell approximation is referenced in the text and in Fig. 23.5. Carry out the quantitative analysis implied and graph oxygen level as a function of radius for selected parametric combinations.

9. Review the evidence for the concept that the secretion of HCl by the gastric mucosa is an example of active transport.

10. Describe the experiments which indicate that active transport occurs across the mucosa of the small intestine, resulting in the selective absorption of certain metabolites and ions.

11. Which molecules and ions have been demonstrated to be actively transported across membranes surrounding various portions of the kidney tubules? Which ions are believed to be actively transported, although the evidence is weak? How could you test to see whether active transport occurs?

12. Discuss the evidence that shows that Donnan equilibrium cannot, in itself, predict the ionic distribution within the resting neuron.

13. Describe the evidence supporting the chemiosmotic theory of membrane transport.

14. Develop in detail the application of information theory to a study of enzyme specificity.

15. Apply information theory to a discussion of echolocation by bats. Try to consider noise in such a problem. Two particular types of airborne noise are insect jamming (Chapter 3) and fog droplet background (assume the resonant water-drop frequency is approximately 20 kHz). Bats do not fly in fog conditions. Can you explain why?

PART **F**

SPECIALIZED INSTRUMENTATION AND TECHNIQUES

Introduction to Part F

It is important that a biophysicist include mathematical biophysics among the tools available without becoming lost in mathematics. It is likewise important that the biophysical scientist be familiar with special physical equipment used in biological research without becoming a gadgeteer. To present this point of view, the last five chapters deal with selected examples of biophysical instruments and techniques. The theory of their function is emphasized rather than the engineering details of their construction.

Extra emphasis is given to absorption and emission spectrophotometry; their discussion occupies the first two chapters of Part F. The important role of spectrophotometry in current biological research, coupled with the store of information potentially available from molecular spectra, make these two chapters very important. The remaining three chapters each present a different field of instrumentation: medical ultrasonography, tracer techniques, and biomedical computation. The only justification for these choices rather than any of numerous other, equally important ones is that they are ones which the authors feel are instructive to students in a variety of disciplines.

A few other specialized techniques are discussed in earlier chapters; for example, X-ray crystallography in Chapter 14, ultracentrifugation in Sec. 16.2, and chromatography in Sec. 20.6. It is hoped that this concluding part of the text added to the rest will give a balanced view of biophysical science, ranging from purely mathematical analyses to applied instrumentation, and from the characteristics of complete organisms to the form of the molecules that compose them.

CHAPTER 26

Absorption Spectroscopy

26.1   Introduction

This chapter and the one following are concerned with the specialized instrumentation and techniques associated with the measurement of absorption and emission of electromagnetic radiation. In this chapter, absorption analyses are described. These include measurement of the differential transmission of visible, infrared, and ultraviolet light through solutions of biological interest. Absorption lines and bands are a standard way of determining the chemical and physical composition of an unknown sample.

A second class of absorption analyses is known as magnetic resonance spectroscopy. Since the development of microwave sources and detectors in the 1940's, such measurements have contributed to an understanding of molecular structures. In these techniques, a magnetic and an electric field are simultaneously imposed on a specimen that contains unpaired electrons or nuclei having net magnetic moments. Characteristic absorption of microwave radiation occurs at a given magnetic intensity or electric field frequency.

Before examining either type of measurement in detail, a general discussion of quantum theory as related to molecular energy levels is helpful. Although the treatment of this subject in the next two sections is brief, it can serve as a framework for understanding this and the following chapters. Those familiar with quantum theory may wish to proceed directly to Sec. 26.4. Similarly, readers less interested in the theory of spectroscopy may also elect to skip the next two sections.

26.2 Quantum Theory

A. HISTORY

Two major theories were developed by physicists in the first quarter of the twentieth century, both of which dramatically altered the scientist's conception of the physical world. These two theories were relativity, which deals with high speeds, and *quantum theory*, which is applied to molecular and sub-molecular particles. Both have played a dominant role in many subsequent technological advances, thereby introducing major changes into human societies. Both are highly formalistic, mathematical, and have been tested and verified by a variety of experiments, but neither has played a central, direct role in the biophysical sciences.

Quantum theory must ultimately be used in describing such phenomena as molecular structures of proteins and nucleic acids, energy relationships in light reactions including photosynthesis, and membrane structure and function. Most important of all, quantum theory is the basis for a qualitative understanding of the absorption and emission of electromagnetic radiation. However, while its terminology is frequently used, the theory is too complicated to permit quantitative calculations in most of the applications that follow in this and the next chapter.

Quantum theory represents a generalization of an earlier formal, highly mathematical approach to classical mechanics developed by the Irish physicist Hamilton. He sought relationships having symmetry and simplicity and based his mechanics on pairs of variables that were said to be canonically conjugate because they yielded the desired, esthetically pleasing mathematical relationships. The function representing the energy of any system, when expressed in terms of canonically conjugate pairs of variables, is called the *Hamiltonian* and is customarily represented by H.

Hamiltonian mechanics discusses the state of any classical mechanical system. This state is specified by values for a set of canonically conjugate pairs of variables and a value for the total energy. It is determined (i.e., can be derived) if the initial conditions, including the mathematical expression for the Hamiltonian, are given. Quantum theory makes sufficient use of these mechanical ideas that it is often called quantum mechanics, even though many of its applications are in fields other than mechanics.

B. STATE FUNCTIONS AND OPERATORS

Quantum theory, however, is essentially probabilistic in nature. It defines the state of any system by mathematical operations on a function called a *state function* and customarily indicated by the symbol ψ. The latter is a complex function of space and time coordinates. The probability density of the system in space–time is given by the square of the absolute value of ψ, that is, by $|\psi|^2$. In the following it will prove more convenient to write this as $\psi^* \psi$, where the superscript indicates the complex conjugate; that is, if

$$\psi = a + jb$$

where a and b are real functions and j is $\sqrt{-1}$, then, by definition,

$$\psi^* = a - jb$$

Hamiltonian mechanics developed a deterministic function to represent each observable quantity. As mentioned above, quantum theory differs in that it is probabilistic; it assigns an *operator* to each observable quantity. This operator can be used together with ψ to find the expected value (or expectation) for the observable in a fashion to be described. In this section, the operator will be shown by a capital letter (or indicated by V for variable), and its value by the corresponding lowercase letter. A few examples are shown in Table 26-1.

TABLE 26-1[a]

QUANTUM OPERATORS

Observable quantity	Operator
x	$X = x$
x component of linear momentum, p_x	$P_x = -jh\,\dfrac{\partial}{\partial x}$
z component of angular momentum, m_z	$M_z = -jh\left(x\dfrac{\partial}{\partial y} - y\dfrac{\partial}{\partial x}\right)$
$f(x, p_x)$	$F = f\left(x, -jh\dfrac{\partial}{\partial x}\right)$

[a] In this table the symbol h is defined as $h = h/2\pi$, where h is Planck's constant.

Quantum theory expectations are calculated in a different fashion than those of statistics. In general, for the variable observable v, the expected value[1] $\langle v \rangle$ is given by

$$\langle v \rangle = \int_{\text{all coordinates}} \psi^* V \psi \, d\tau \tag{26-1}$$

where $d\tau$ is the element of "volume" in the space considered. For X the order of writing terms in Eq. (26-1) is immaterial, but for P_x, which involves partial differentiation, the order is critical.

Many operators V have a discrete set of solutions to the equation

$$V\psi_\lambda = v_\lambda \psi_\lambda \tag{26-2}$$

where the subscript λ indicates a particular value for v and the corresponding solution for ψ. Such behavior is well known mathematically; the set of solutions, ψ_λ, are called characteristic functions or *eigenfunctions* of the operator V, while the set v_λ are the corresponding characteristic values, or *eigenvalues*. ("Eigen" is a German word meaning *characteristic* or *own*.)

If ψ is an eigenfunction of V, then Eq. (26-1) is particularly simple,

$$\langle v \rangle = \int \psi_\lambda^* V \psi_\lambda \, d\tau = v_\lambda \int \psi_\lambda^* \psi_\lambda \, d\tau = v_\lambda \tag{26-3}$$

In this case the expected value of v^2 is v_λ^2, since repetitive applications of the operator V will introduce added factors of v_λ. Therefore, the value of v is exactly v_λ. In other words, if ψ is an eigenfunction of V, the corresponding experimentally observed variable can be known exactly.

Schrödinger in his original formulation placed great emphasis on the Hamiltonian operator, H, which corresponds to the total energy. For spectroscopic applications it is important to find eigenvalues for the corresponding observable, the total energy E. This is an eigenvalue of the *Schrödinger wave equation*,

$$H\psi_\lambda = E_\lambda \psi_\lambda \tag{26-4}$$

These eigenvalues are a discrete set and represent the only values that total energy may have. Any observable, such as the total energy, which can assume only discrete values is said to be *quantized*; the integer or simple fraction used to designate these values is called a *quantum number*.

C. UNCERTAINTY PRINCIPLE

A probabilistic theory deals with distributions having standard deviations rather than the exact values discussed in the preceding. In quantum theory a standard deviation for an observable is called its uncertainty and is designated by a Δ. This theory requires that the product of the uncertainties of all

[1] Statisticians customarily use the symbol $E(v)$ for the expected value.

pairs of canonically conjugate variables, such as x and p_x, must be not less than $\hbar/2$, that is, for example,

$$\langle \Delta x \rangle \langle \Delta p_x \rangle \geq \frac{\hbar}{2} \qquad (26\text{-}5)$$

Thus, if the momentum is known exactly, there can be no knowledge of where the particle is located. Conversely, if the location is known exactly, there can be no knowledge of its motion.

A second important canonically conjugate pair of variables is energy and time. If the energy state is known exactly, one cannot know when the system was in this state. Or expressed otherwise, over sufficiently short intervals of time, energy need be neither quantized nor conserved. However, for longer time intervals, many systems can be well described by eigenfunctions of the Hamiltonian operator and hence will have discrete, quantized energy levels.

Photon emission sometimes occurs when an atom or molecule falls from a higher energy level to a lower one. The time interval involved must be sufficiently short that the uncertainty in the energy may exceed the difference between the energy levels involved in the transition. Similarly, photon absorption may occur if the photon has exactly the proper energy to raise the system from a lower level to a higher one. A state function may be used to describe the transmission of the photon; in this case, that state function is mathematically identical to the classical electromagnetic wave. For this and other reasons, state functions are often called *wave functions* even when the system does not involve electromagnetic phenomena. In the older literature, quantum theory was frequently referred to as wave mechanics, for similar reasons.

Thus, the uncertainty principle, embodied in inequality (26-5), has direct bearing on spectroscopy. It requires that no state function, ψ, can be an eigenfunction for more than one of a pair of canonically conjugate variables, since both cannot be known exactly. In general, however, a given function may be an eigenfunction for more than one operator. In this case, the operators are said to *commute*, since the order of their application has no effect on the outcome. Thus if U and V commute and ψ_λ is an eigenfunction for both,

$$U \cdot V \cdot \psi_\lambda = v_\lambda \cdot U \cdot \psi_\lambda = u_\lambda v_\lambda \psi_\lambda$$
$$V \cdot U \cdot \psi_\lambda = u_\lambda \cdot V \cdot \psi_\lambda = u_\lambda v_\lambda \psi_\lambda$$

Energy states of atoms and molecules useful in spectroscopy are usually described by quantum numbers designating the eigenvalues for a group of operators, all of which commute with one another.

D. CORRESPONDENCE PRINCIPLE

A topic avoided in the preceding discussion is the set of rules by which the operator corresponding to a given observable entity is selected. A number of criteria are used. One is that operators for canonically conjugate variables do not commute, but rather (using, as examples, X and P_x) obey the equation

$$X \cdot P_x - P_x \cdot X = j\hbar$$

Another, perhaps the most critical criterion, is that the expectations computed must agree with experimental observations.

One very important test of this latter criterion is that, for larger systems (e.g., viruses) the expectations of quantum theory become the functions of classical physics. This is known as the *correspondence principle*; it is used in selecting basic operators and in choosing terms for the Hamiltonian operator, which is represented as a function of these basic operators. The correspondence principle has been used, for example, to find the nature of the added terms in the Hamiltonian for a molecule in a magnetic field (see Sec. 26.4).

26.3   Quantum Theory and Spectroscopy

Quantum theory can be applied directly to compute the expected energy levels in a variety of simple systems. In systems of even minimal complexity, such as a helium atom or a hydrogen molecule, it is necessary to use numerical methods to compute these expected values. Nonetheless, by a combination of spectroscopic observations and quantum calculations, the energy levels of single atoms can be described. These are important in atomic spectroscopy, a valuable technique for the clinical chemist.

However, the applications discussed in this and the next chapter emphasize more heavily the behavior of biological molecules, including proteins and nucleic acids, in absorbing and emitting radiation. Sophisticated theories have been constructed to allow calculation of characteristic molecular spectra. These studies require large computer programs in order to calculate numeric expected values. Such approaches have involved biophysical scientists, but it is not possible to present their theories in this text. Rather, in the following paragraphs, the applications of quantum theory to molecular spectra are reviewed in most part without derivation or proof.

A. DECOMPOSITION OF THE HAMILTONIAN

The Hamiltonian for a given system includes terms for all the types of motion and energy possible. Examples for a molecular system are given in Table 26-2. In some cases, the interactions between the various subsystems are so great that the Hamiltonian cannot be decomposed into the sum of simpler operators. Strictly speaking, this is true for all molecules.

However, for spectroscopic discussions, one may divide and subdivide the Hamiltonian of a molecule. For each subsystem, there is a partial state

TABLE 26-2

TRANSITION ENERGIES

Energy type	*Transitions*	*Approximate energy level differences*
Molecular translation	Diffusion	~ 0 neV
Nuclear magnetic moment	Microwave absorption in magnetic field	0.1 neV
Electronic magnetic moment	Microwave absorption in magnetic field	0.1 μeV
Molecular rotation	Far-infrared spectra	5 μeV
Bond vibrations	Near-infrared spectra	0.25 eV
Electronic	Visible and ultraviolet spectra	2.5 eV
Electronic	X-ray spectra	25 eV

function and unique eigenvalues specified by appropriate quantum numbers. The molecular state function is the product of these partial state functions. The total energy is the sum of these unique eigenvalues for each type of energy.

For such a decomposition, the types of energy in a molecule may be divided into several major categories. The types of motion and hence energy thus separated include molecular translation and rotation, and vibration along a bond or group of bonds. In addition, electronic energy levels and nuclear energy levels can also be separated. Symbolically, this decomposition can be represented by

$$H = H_{\text{trans}} + H_{\text{rot}} + H_{\text{vib}} + H_{\text{el}} + H_{\text{nuc}}$$

Similarly, one may write

$$\psi = \psi_{\text{trans}} \cdot \psi_{\text{rot}} \cdot \psi_{\text{vib}} \cdot \psi_{\text{el}} \cdot \psi_{\text{nuc}}$$

This is known as the *Born–Oppenheimer principle*; it is used implicitly throughout this chapter and the next.

Table 26-2 gives the order of magnitude of the energy gaps between quantized levels for these five types of energy. Molecular translation has at most minimal effects on molecular spectra, while nuclear levels are primarily important for a type of spectroscopy called *nuclear magnetic resonance*. (This is discussed in Sec. 26.6.) The remaining three types of energy are reviewed briefly in the remainder of this section.

In Sec. 26.2, it is mentioned that photons would be absorbed (or emitted) if their energy corresponded exactly to the difference between two energy eigenvalues of the molecule. However, not all transitions are equally probable. In order for a photon to be absorbed or emitted, there must be a change in the expected values of the electromagnetic characteristics of the molecule. Neither are all transitions in which electromagnetic changes occur equally probable. Instead, quantum mechanics provides mathematical formalisms for computing the relative probabilities of various transitions. These formalisms will not be detailed here, but may be found in the reference by Schiff. The transitions between electronic states most likely for photon absorption or emission result in changes due to the electrical dipole moment. Such changes are said to be permitted in the language of quantum mechanics; various criteria, termed *selection rules*, have been derived to determine which transitions are permitted.

Photons may also be absorbed or emitted in transitions due to other electromagnetic moments. These include the expected value of the electrical quadrupole moment and of the magnetic dipole moment. Such transitions are less likely and are called forbidden by quantum mechanics. There exist numerous situations in which forbidden transitions are important. Some of these are discussed in Sec. 26.6 and in several sections of Chapter 27.

Atomic emission and absorption spectra are very sharp, since they usually represent electronic changes only. (In the presence of a magnetic field, atomic transitions may also include nuclear changes.) By contrast, molecular absorption, particularly for the biologically important case of molecules in solution, occurs in broad, often overlapping, bands. These arise because vibrational, rotational, and translational energy changes all "smear out" the sharp lines caused by electronic and nuclear changes.

B. MOLECULAR ROTATION

Any molecule may rotate about an axis passing through its center of mass. Associated with the motion will be a kinetic energy of rotation which is quantized. The energy levels (eigenvalues) can be described by

$$E_{\text{rot}} = \frac{J(J + 1)\hbar^2}{2\mathscr{I}} \tag{26-6}$$

where J is the rotational quantum number and $\mathscr{I}$ is the moment of inertia of the molecule. The values of J are required to be positive integers or zero. For this pure rotation, it can be shown that the expected value of the angular momentum is $J\hbar$ and of its square, $J(J + 1)\hbar^2$.

Not only is angular momentum quantized, but its projection on any single spatial axis is also restricted to discrete values. These have the form $M_J\hbar$, where M_J is a member of the set

$$M_J = -J, -(J - 1), \ldots, -1, 0, 1, \ldots, (J - 1), J$$

This is illustrated for $J = 2$ in Fig. 26.1. Thus, there are $2J + 1$ values

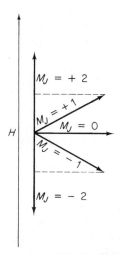

Figure 26.1   Projections of a quantum mechanical angular momentum vector along a specific axis, such as that of a hypothetical magnetic field ($\vec{H}$). The case of total angular momentum quantum number equal to 2 is illustrated. Five projections, 2, 1, 0, -1, and -2, are then permitted for the component of the total angular momentum along the specified direction. If a magnetic field were physically present, the various components would correspond to different energies.

possible for M_J. The value of M_J is especially important to describe a molecule in a magnetic field.

Selection rules for molecular rotational energy transitions include

$$\Delta J = \pm 1$$
$$\Delta M_J = \pm 1, 0$$

where Δ is used for the difference in the values for the initial and final states. These rules apply to changes in rotational energy and orientation even if vibrational and electronic energy changes also occur. In addition, pure rotational transitions are permitted only for asymmetric molecules. Such transitions produce far-infrared absorption bands in small molecules such as H_2O and HCl, but they are usually unobservable in larger molecules.

C. VIBRATION IN MOLECULES

Two parts of a molecule, joined together by a bond, may vibrate relative to one another. This usually involves much greater energy differences between the quantized levels than those mentioned in Sec. 26.3.B. Accordingly, most molecules are in the lowest vibrational level possible when at equilibrium at room temperature. This contrasts with rotational levels. There the various higher states are populated and energy can be exchanged (without radiation) between translational and rotational modes.

The energy levels of a simple vibrator are quantized so that

$$E_{\text{vib}} = h\left(k + \frac{1}{2}\right)\nu_0$$

where k is a member of the set $k = 0, 1, 2, 3, \ldots$ and ν_0 indicates the strength of the bond. Even in the lowest energy state (k equals zero), there is a vibrational energy, $\frac{1}{2}h\nu_0$, associated with the bond. The selection rule for vibrational energy transitions is

$$\Delta k = \pm 1$$

The foregoing description is qualitatively valid for all molecular bonds. As mentioned previously, however, the absorption spectrum of a complicated molecule is not sharp. Rather the lines are smeared to bands, in part by numerous rotational transitions, and in part by interactions with other parts of the molecule. These spectra can be "sharpened" by measuring them at very low temperatures, but do not reduce to a collection of lines. Much work has been done empirically associating absorption bands in the infrared "fingerprint" region of 3–20 μm with specific covalent bonds.

D. ELECTRONIC LEVELS IN ATOMS

The energy associated with the electron within an atom or molecule can be further decomposed to three or four types. The first is characterized by a principal quantum number, n, which can be thought in a very intuitive sense to indicate the average distance of the electron from the atomic center. This quantum number must be a positive integer. In addition, the electron will possess orbital angular momentum; these levels are characterized by a quantum number, l. It may assume integer values from 0 to $n - 1$. The value of l indicates the spherical asymmetry of the wave function. Finally, the electron possesses an inherent angular momentum or spin, to which a quantum number s equal to $\frac{1}{2}$ is assigned. There is no classical or intuitive analog or explanation for the spin.

The total angular momentum quantum number, j, is the vector sum of the orbital and spin angular momenta.[2] Their relative orientation is quantized so that j runs from $|l - s|$ to $l + s$. In the presence of a magnetic field, the angular momenta specified by l, s, and j behave in a fashion similar to that discussed for a molecular rotator and illustrated in Fig. 26.1. The quantum numbers representing the projections are designated m_l, m_s, and m_j, respectively. Actually, this overdetermines the electron, so that either m_l and m_s

[2] This should not be confused with the use of j to represent $\sqrt{-1}$.

may be specified or j and m_j; all four of these quantum numbers may not be included at once.

Selection rules for an electronic transition are

$$\Delta l = \pm 1$$
$$\Delta j = \pm 1, 0$$

In magnetic fields, there are added selection rules. In the limiting case of close to zero magnetic field, these become

$$\Delta m_j = \pm 1, 0$$

These rules apply to "allowed" transitions, which always involve an electrical dipole change. In Sec. 26.5, transitions are discussed which are in this sense "forbidden," since they result in a change in m_s.

For an entire atom, orbital, angular, and spin momenta are also quantized. These are represented by the same letters as those used for a single electron, except that capitals rather than lowercase letters are used; for example, S represents the total spin angular momentum and M_S its projection on an axis. The values of L, S, and J are formed from those of the constituent electrons by quantized rules for vector addition which are not developed here, except to note the permissible values. Those for L are always positive integers or zero, whereas S and J are positive integers or zero for an even number of electrons, and half-integers for an odd number. The atomic selection rules state that for a permitted electrical dipole transition

$\Delta L = \pm 1, 0$

$\Delta S = 0$

$\Delta J = \pm 1, 0$ except that a transition from $J = 0$ to $J = 0$ is forbidden

$\Delta M_J = \pm 1, 0$ except that a transition from $M_J = 0$ to $M_J = 0$ is forbidden if $\Delta J = 0$

Many other letters and selection rules are used by atomic spectroscopists. These are not detailed here. However, it should be noted that the various values of l (or L) are, for historical reasons, often indicated by the letters $s, p, d, f, g, \ldots$ (or S, P, D, F, G). Another nomenclature makes use of the number $(2S + 1)$ of J values, calling the states singlet, doublet, triplet, etc., for spin quantum numbers of 0, 1/2, 1, etc. This designation is used in Chapter 27.

As an aside, the *Pauli exclusion principle* states that no two electrons within an atom can have the same set of quantum numbers n, l, m_s, and m_l. The existence of energy shells and the form of the periodic table follows almost directly from this principle. Thus, for the first shell,

$$n = 1 \qquad l = 0 \qquad m_l = 0 \qquad m_s = \pm \tfrac{1}{2}$$

The two electrons must have their spin antiparallel. Thus, the total spin, S, due to a complete inner shell is zero. For the next shell, the set of values are

$$n = 2: \quad l = 0 \qquad l = 1$$
$$m_s = \pm\tfrac{1}{2} \qquad m_s = \pm\tfrac{1}{2}$$
$$m_l = 0 \qquad m_l = -1, 0, 1$$

Thus, eight members are present. Each time a subshell indicated by a fixed l is filled, the values for L and S in the lowest energy state of the atom are both zero. At higher values of n, the higher values of l (subshells) are not filled until after the lower values of l for the next shell.

E. ELECTRONIC LEVELS IN MOLECULES

For molecules, similar considerations apply. However, the total quantum number n for a single electron has no counterpart in the molecule. Often the lowercase Greek letters σ and λ are used in place of s and l, respectively. For an entire molecule, Σ and Λ replace S and L. The multiplicity of the state (number of J values) is given by $2\Sigma + 1$ and these are called singlet, doublet, triplet, etc., as in the atomic electronic energy levels.

By and large, states of lower multiplicity have lower-energy levels. According to the selection rules, molecules in the lowest triplet state are "forbidden" to radiate a photon to reach a still lower singlet state. Thus, energy may be "trapped" in the molecule. This effect may be important in photosynthesis (see Chapter 20). In molecular oxygen O_2, however, a triplet state, $\Sigma = 1$, $\Lambda = 0$, has the lowest energy.

The values of λ, the electronic orbital angular momentum indicate the spherical asymmetry of the electronic distribution. Electrons involved in a system of alternate single and double bonds (called a *conjugated system* by organic chemists) are believed to be shared by all atoms. Such a distribution is not spherically symmetric, hence λ must be at least 1. Electrons in a molecule are sometimes assigned the Greek letters σ, π, δ, ϕ, γ, corresponding to the values 0, 1, 2, 3, 4 for λ. In fact, conjugated systems, such as a benzene ring, have π electrons. These are responsible for the strong absorption bands of the chromophores tabulated in Sec. 26.4. Computation of the energy levels of such π electrons would make it possible to predict the absorption spectra of heme proteins, chlorophylls, and other molecules of biophysical interest (see Sec. 26.4 and all the chapters of Part D).

26.4 Light Absorption

A. THEORY

The Beer–Lambert law, mentioned briefly in Chapter 12, describes the attenuation of electromagnetic radiation by a sample. To use this law, the experimenter must establish within the experimental conditions that the system shows a variation in intensity, I, in the form of the law:

$$I = I_0 \times 10^{-\sigma c x} \qquad (26\text{-}7)$$

or

$$\log \left(\frac{I_0}{I} \right) = \sigma c x$$

where I_0 is the incident intensity, σ is the molar extinction coefficient, c is the concentration of the sample, and x is the distance through the sample at which I is measured. The coefficient σ is the only parameter that depends on wavelength. It is a function of the molecule's energy levels, which are discussed in the previous sections. The absorption spectrum of a solution or solid can be used to identify single components and quantitate their amounts.

Traditionally, analytical chemists have referred to color-inducing moieties as *chromophores*. Table 26-3 contains a list of some well-known chromophores and their characteristic absorption maxima. These maxima may be shifted if more than one such group is found in a small molecule. Several intervening carbon atoms are usually thought to guarantee that no interaction occurs between chromophores.

TABLE 26-3[a]

CHROMOPHORE ABSORPTION

Group	λ Characteristic (nm)	Solvent
(benzene ring) O	198, 255	Hexane
$\diagdown\!\!C\!\!=\!\!C\diagup$	193	Vapor
$-C\!\equiv\!C-$	173	Vapor
$-C\!\!\diagup\!\!\diagdown\!\!{}^{O}_{OH}$	204	Water
$-NO_2$	271	Methanol

[a] These data and others are found in the reference by Thompson.

Other environmental factors, for example solvents, may also shift the absorption maxima. The effects of different solvents are shown in Fig. 26.2, where the near-ultraviolet absorption of benzene vapor and benzene in cyclohexane are compared. The absorption maximum of a given molecule is also sensitive to pH and temperature. In addition, the interactions of various chromophores with each other and with other moieties in the molecule can shift the wavelength of maximum absorption.

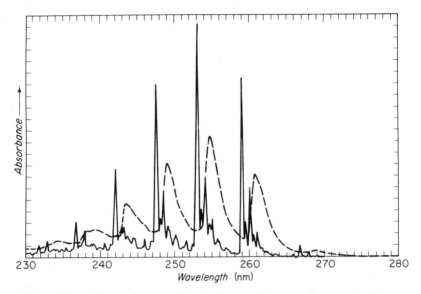

Figure 26.2   Solvent effects on the absorbance of benzene at near-ultraviolet wavelengths. Instrumental conditions are held constant for each sample, but concentrations are only roughly comparable. Benzene vapor data are shown as the solid line; benzene in cyclohexane results in the dashed curve. After R. P. Bauman, *Absorption Spectroscopy* (New York: John Wiley & Sons, Inc., 1962).

However, while no two compounds yield identical absorption data, the spectral features that distinguish between one compound and another may be difficult to determine in practice. There are various experimental techniques available to the spectroscopist; these are generally designed to maximize whatever differences are available.

B. EQUIPMENT

A block diagram of a typical spectrometer is given in Fig. 26.3. Sources include gaseous discharge tubes capable of discrete or continuous emissions.

Figure 26.3   Block diagram of a spectrophotometer. Several recording devices may be used in parallel. These can include strip charts, cathode ray tubes, and computer memories.

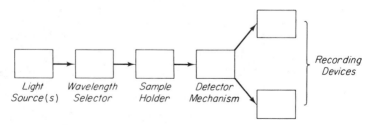

Heated filaments raised to temperatures above 3,000 °K may also be used. The spectrum of the output of the latter devices is continuous. For example, tungsten lamps are effective over the range 350–2,500 nm. Most absorption spectra are measured between 100 and 1,000 nm. The hydrogen or mercury discharge lamps can be used as sources at wavelengths below 450 nm. In the far-infrared regions of the spectrum, greater than 2,500 nm, still other light sources are needed. Generally, some form of hot glowing object is used and the visible rays are filtered from the infrared.

In any wavelength region, the electrical power operating the light source must be carefully stabilized. Otherwise, fluctuations in light intensity due to the changes in the power may be greater than the differences due to the absorption being measured. This is illustrated forcefully in the case of the incandescent filament. The power delivered to the filament is roughly proportional to the square of the applied electrical voltage. The temperature of the filament will vary proportionally to the power consumed. The light emitted, however, varies as the fourth power of the absolute temperature, and hence as the eighth power of the voltage. Thus, if the voltage is represented by V and the light intensity emitted by I,

$$I \propto (V)^8$$

Hence,

$$\frac{\Delta I}{I} = 8 \frac{\Delta V}{V}$$

and a 0.1 percent shift in voltage can cause a 0.8 percent change in the input light intensity. The long-term stability of the heater current power supply is clearly significant in practical measurements.

Wavelength selectors can be of several types. The simplest form is the optical filter used in ordinary darkroom work. It passes a relatively broad band of frequencies (up to 50 nm wide) and is only useful in relatively crude instruments termed *colorimeters*. Better instruments employ either a diffraction grating or a prism to select a very narrow bandwidth. In a grating, constructive interference occurs at different spatial positions due to the different wavelengths of light in the source. This is indicated in Fig. 26.4 for a transmission grating. Alternatively, instead of being transmitted through a grating, the light may be reflected from it. This last type is called a reflection grating.

The prism produces a spatial distribution of wavelengths due to the variation of speed of an electromagnetic disturbance in a material. This phenomenon, termed *dispersion*, depends on the nonvanishing derivative of the index of refraction with respect to the wavelength. Figure 26.5 shows a simplified system of this sort.

For either prism or grating-type monochromators, it is necessary to use

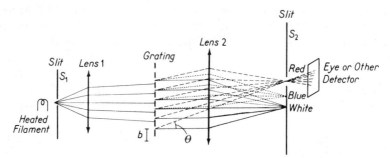

Figure 26.4 Simplified drawing of a grating monochromator. In Fig. 26.3, this device is the physical representation of the light source and wavelength selector modules.

components that will permit operation at the desired wavelengths. In the ultraviolet, all the lenses, prisms, and plates through which the light passes are usually made of quartz. Special surfaces are necessary for reflectors; thus, they are usually coated with either aluminum or silver. In the visible region of the spectrum, glass of various types is used. For infrared spectrophotometers, prisms are made from rock salt. Lenses transparent to infrared radiation are difficult or impossible to construct, so focusing must be accomplished with suitably curved mirrors.

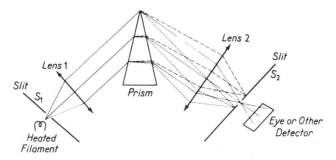

Figure 26.5 Simplified drawing of a prism monochromator. This is an alternative source of a relatively limited number of photon energies.

The sample holder or cuvette is generally a hollow rectangular solid made of quartz or glass. The nonreactive nature of these materials is essential if the sample is to be maintained under constant physical conditions. Cuvette thickness (sample path) is typically 1 cm, although values between 1 mm and 10 cm are not uncommon. The cuvette must transmit a major fraction of the incident radiation with as little absorption or reflection as possible.

Detection of the transmitted radiation may be done with photocells, photomultipliers, or photodiodes. The use of a solid-state device has the advantage that the excitation of conduction electrons requires relatively little energy compared to vacuum tubes. The diode is, therefore, the equipment of choice for infrared studies. The detector is generally inserted as part of a bridge circuit so that the current (or voltage) output may be measured with respect to some standard.

A biological specimen is often dissolved in a solvent whose absorption overlaps that of the unknown. Thus, a reference standard is necessary to help remove extraneous absorption from the unknown spectrum. Two types of spectrophotometers have been developed to compare the sample with its reference. The simpler system has a single light path. The user places the unknown and the standard sequentially into this path. Various source and detector changes during this operation can cause certain random or even systematic errors. To circumvent this problem, a beam splitter can be employed to generate two simultaneous light paths, one passing through the specimen, the other through the standard solution. A dual-beam spectrometer has the advantage that the signals from both cuvettes are detected with very little of the time delay inherent in the single-beam instrument. However, such a system is necessarily more complicated and costly.

C. APPLICATIONS

Spectroscopy is used to determine the amounts of chemical compounds in a given unknown. Perhaps the most common application, the one of importance to almost every hospital or clinic patient, is the spectroscopic determination of biochemical compounds in physiological fluids. An aliquot of the fluid is treated with reagents such that one or several related compounds in the fluid will react quantitatively in solution to produce spectral changes. Equation (26-7) can thus be used to determine the concentration, c, since σ can be measured independently and x is constant.

Another use of spectrophotometry is the determination of the types of chemical compounds in a given unknown. In Eq. (26-7), the molar extinction coefficient, σ, is the only parameter that depends on wavelength. This dependency is the key to such determinations.

In the case of n unknowns at a given wavelength, λ:

$$I = I_0(10^{-\sigma_1 c_1 x} \cdot 10^{-\sigma_2 c_2 x} \cdot \ldots \cdot 10^{-\sigma_n c_n x})$$

or

$$A = \frac{1}{x} \log_{10}\left(\frac{I_0}{I}\right) = \sigma_1 c_1 + \sigma_2 c_2 + \ldots + \sigma_n c_n \qquad (26-8)$$

where A is called the *specific absorbance*. This is one equation in n unknowns (the c's), since the σ's and x can be determined independently. The chemist may judiciously vary λ and obtain an additional $n - 1$ independent equalities of the form of Eq. (26-8). The system of equations is then

$$A_1 = \sigma_{11}c_1 + \sigma_{12}c_2 + \cdots + \sigma_{1n}c_n$$
$$\vdots$$
$$A_n = \sigma_{n1}c_1 + \sigma_{n2}c_2 + \cdots + \sigma_{nn}c_n$$

where the subscript on the A's and the first subscript on the σ's refers to the wavelength. In matrix notation, this reduces to an inversion problem, a relatively simple computer calculation for an appropriate set of wavelengths.

A third use of spectroscopy is to measure the rates of chemical reactions. Reactions with half-times on the order of minutes or hours can be observed by manually or automatically withdrawing aliquots for spectroscopic assay. Slightly faster rates may require that the reaction be performed in a cuvette; enzyme activity assays are usually carried out in this manner. The cuvette is filled with a solution containing all necessary substrates and cofactors, and the reaction is started by the rapid addition and mixing of a small amount of enzyme solution (see Chapter 18 for more information on the uses of these reaction rates).

More rapid reactions, those with half-times on the order of seconds or tenths of a second, must be measured using specially designed mixing chambers and cuvettes. In a stopped-flow apparatus, liquids are rapidly injected into a rounded mixing chamber, which then drains into a flow-through cuvette. Such stopped-flow equipment is available as an accessory to most commercial research spectrophotometers.

Reactions with half-times in the range of milliseconds require even more sophisticated equipment. The rapid-flow apparatus, shown in Fig. 26.6, is derived from the simpler stopped-flow setup. Reactants are mixed at point M and then pass through a flow-through cuvette to the spectrophotometer beam a distance d away. If the linear flow rate is V, the time t after mixing is given by

$$t = \frac{d}{V}$$

If both V and d were constant, the spectrophotometric absorbance would also be constant, since all molecules would be observed at one fixed time after mixing. To observe spectral changes as the reaction proceeds over time, d may be varied by physically moving the spectrophotometer. Since this is a rather cumbersome process, most rapid-flow instruments allow instead for varying V. Such instruments can be coupled to computers to allow rapid tabulation or graphical display of absorbance (concentration) as a function of time after mixing.

Ultraviolet, visible, and infrared absorption can be a quantitative, sensitive test. It is not, however, the only type of absorption technique available to the

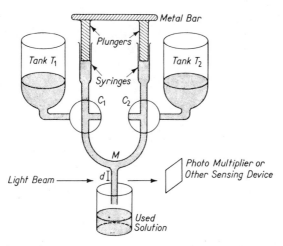

Figure 26.6   Rapid-flow apparatus for observing reactants as a function of time. The three-way stopcocks at C_1 and C_2 allow filling syringes from the storage tanks and then discharging the syringes into a mixing chamber at M. Optical density changes are measured at a distance d beyond the mixing point.

bioscientist. Both the polarization of the light (see Sec. 26.5) and lower frequency absorptions (Sec. 26.6) are also important biophysical measurements.

26.5   Optical Activity

As has been indicated in Chapter 12, the existence of electromagnetic radiation can be predicted on the basis of theroetical arguments. In essence, the radiation consists of magnetic and electric fields which are functions of time and space. At a given point in space, the motion of the end of the electric field vector is used to define polarization. If, for example, the locus of this vector describes a circle, one defines the radiation as being circularly polarized. Other loci are possible (e.g., elliptically polarized radiation). Right (R) and left (L) polarization refer to the sense of rotation of the vector.

Most biological macromolecules, such as proteins and nucleic acids, are optically active. That is, they interact differently with left and right circularly polarized light. The reason for this different interaction is complex, and the interested reader is referred to the references by Imahori and Sears. Here we briefly discuss how this interaction may be measured and used.

Optically active molecules have different molar extinction coefficients (σ_L and σ_R) and different indices of refraction (n_L and n_R) for L and R circularly polarized light. The former difference ($\Delta\sigma = \sigma_L - \sigma_R$) is termed the *molar circular dichroism*. Since this is small compared to the average extinction coefficient, $(\sigma_L + \sigma_R)/2$, it is not measurable except at wavelengths where the molecule absorbs light strongly. While circular dichroism is sometimes measured directly, many instruments report instead the related parameter, *ellipticity*. Since the incident plane-polarized light has two circularly polarized components which are differentially absorbed, the resultant light is elliptically polarized, as shown in Fig. 26.7. The ellipticity θ is

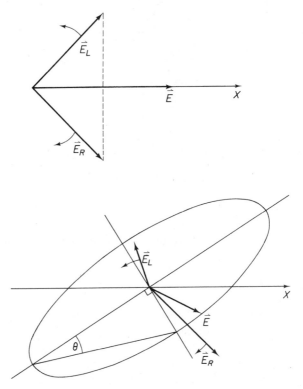

Figure 26.7 (a) A mathematical representation of linearly polarized light as a sum of oppositely polarized circular beams ($\overrightarrow{E_R}$ and $\overrightarrow{E_L}$). The total electric field is the vector sum of $\overrightarrow{E_R}$ and $\overrightarrow{E_L}$ and lies along the x-axis. (b) As above except the $\overrightarrow{E_L}$ vector has been preferentially attenuated and phase delayed with respect to $\overrightarrow{E_R}$. Summation of the two vectors now results in an elliptical locus for the total electric field $(\overrightarrow{E})$. Ellipticity (θ) is given by the arctangent of the ratio of minor to major axes.

defined as the angle whose tangent is the ratio of the minor to the major axis of this ellipse. As is shown in the reference by Beychok, it can be related to $\Delta\sigma$ by

$$\theta \simeq 3,300\Delta\sigma$$

In this form, its units are degree-cm²/decimole.

The different indices of refraction are also not themselves measured. They are most easily demonstrated as a change in the plane of polarization of the light. The angle α of this rotation is a function of both the distance that the light must travel through the solution and the concentration of the optically active compounds in the solution. It is usually expressed as the *molar optical rotation*, ϕ, per centimeter:

$$\phi = \frac{100\alpha}{c}$$

where c is the molar concentration. A simple polarimeter that can be used to measure α is shown in Fig. 26.8.

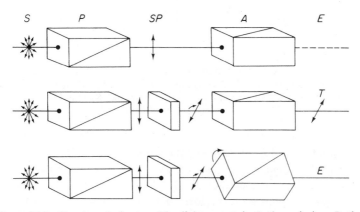

Figure 26.8 Simple polarimeter. The light source is S, the polarizer P, the specimen SP, and the analyzer A. In the top drawing, the polarizer and analyzer are aligned so that no light is transmitted (symbolized by E). With the middle configuration, the specimen rotates the plane of polarization and transmission (T) occurs. Upon rotating the analyzer by the appropriate angle α, as shown in the lower drawing, the analyzer can again extinguish the light and thereby measure the molar optical rotation. From Forrest, J. W.: Polimetry, in Glasser, O. (ed.): *Medical Physics*, Vol. 1. Copyright © 1944 by Year Book Medical Publishers, Inc., Chicago. Used by permission.

Plots of optical rotation and ellipticity versus wavelength are known as *optical rotatory dispersion* (ORD) and *circular dichroism* (CD) spectra, respectively. Like absorption spectra, these are the sum of spectral bands that relate to each individual transition. A sketch of the relationship between all three types of spectra is shown in Fig. 26.9.

While it can be shown that optical rotation and ellipticity are mathematically interconvertible and thus an ORD or CD spectrum could, in theory, be completely transformed into the other, each has its own advantages and disadvantages. They complement each other and both are used to gain information about the environment of biological macromolecules. For example, the α and β protein helical structures discussed in Chapter 15 each

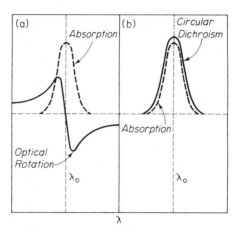

Figure 26.9   (a) The absorption (dashed) and associated optical rotary dispersion (solid) are shown for an idealized optically active transition. The central wavelength is λ_0. (b) Absorption and circular dichroism are illustrated for the same idealized transition. After S. Beychok, *Science* **154**:1288–1299 (1966).

have different optical activity. Thus, the amount of each in a protein can be approximated by simple spectroscopic measurements rather than the much more involved X-ray techniques discussed in Chapter 15. In addition, the spectroscopic techniques are performed on solutions. Thus, they measure parameters under more physiological conditions than those found in a crystal. The CD and ORD measurements are also rapid enough that it is possible to study the dynamics of conformational changes in the macromolecules.

26.6   Magnetic Resonance

A. THEORY

Chemical moieties can be detected and information on their particular electrical and geometrical forms can be obtained using the absorption of microwave radiation by a sample immersed in a magnetic field. This photon absorption is associated with energy-level transitions much as in the shorter-wavelength spectrophotometry discussed in Sec. 26.4 and 26.5. Microwave absorption is due to the transitions between various possible orientations of a magnetic dipole in an external magnetic field. The different orientations would all have the same energies in the absence of magnetic fields.

Since magnetic resonance is applied to atomic-sized and smaller particles, the methods are described in the terminology of quantum theory. From quantum theory, one may show that a particle possessing a spin (intrinsic) angular momentum[3] equal to $\vec{I}\hbar$ has an associated magnetic moment $\vec{\mu}$ proportional to the angular momentum:

$$\vec{\mu} = g\mu_0\vec{I} \tag{26-9}$$

where μ_0 is the Bohr magneton and g is a dimensionless proportionality constant that is determined from experiments. The *Bohr magneton* is defined by

$$\mu_0 = \frac{e\hbar}{2mc} \tag{26-9a}$$

where e is the charge on an electron, m is the particle mass, and c the speed of light. Equation (26-9) emphasizes the fact that the magnetic moment would vanish if the spin angular momentum were zero.

Since μ is inversely proportional to mass, relatively large moments are associated with low mass particles. It is, therefore, easier to detect an unpaired electron in a molecule than it is to detect a proton or heavier nucleus. Such electron detection is termed *electron spin resonance* (esr) or *electron paramagnetic resonance* (epr). Magnetic methods which detect nuclei are described as *nuclear magnetic resonance* (nmr); various hybrid techniques have also been developed.

The added classical energy ΔE of a magnetic dipole $\vec{\mu}$ in an imposed magnetic field $\vec{H}$ is given by

$$\Delta E = -\vec{\mu}\cdot\vec{H} \tag{26-10}$$

If $\vec{H}$ is oriented along the z axis, this may be rewritten

$$\Delta E = -\mu_z H \tag{26-10a}$$

where the vector symbolism has been dropped to indicate that magnitudes only are involved in this last relationship. From the correspondence principle mentioned in Sec. 26.2, one would expect $-\mu_z H$ to be an important term in the operator representing the orientation energy of the dipole in a magnetic field. Indeed, it turns out that this term is the only one needed for a discussion of magnetic resonances. Combining Eqs. (26-9) and (26-10a), one may write

$$\Delta E = -g\mu_0 I_z H \tag{26-11}$$

The quantization of the z component of angular momentum is discussed in Sec. 26.3, where it is shown that I_z increased in integer steps from $-I$ to $+I$; thus, there are $(2I + 1)$ values possible. Accordingly, Eq. (26-11) implies $(2I + 1)$

---

[3] Recall that I, the spin quantum number, is the magnitude of $\vec{I}$. It must be an integer divided by 2. Note that different letters are used for the spin quantum numbers in Sec. 26.2.

energy levels, owing to the orientation of a dipole in a magnetic field. This is illustrated in Fig. 26.10 for a particle such as an electron with a spin quantum

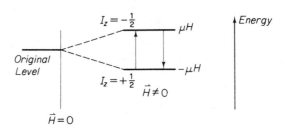

$I_z = -\frac{1}{2}$ μH Energy

Original Level

$I_z = +\frac{1}{2}$ $-\mu H$

$\vec{H} \neq 0$

$\vec{H} = 0$

Figure 26.10 Energy levels for a spin-$\frac{1}{2}$ particle with nonvanishing magnetic moment in an external magnetic field (H). The original single level, shown at the left, has been split into a doublet corresponding to the two projections ($I_z = \pm\frac{1}{2}$) of the angular momentum vector. In this figure, $g = 2$.

number, I, of $\frac{1}{2}$ in a field $\vec{H}$. Here $2I + 1$ is 2; thus, a pair of energy levels is seen. Before the field is turned on, there is only one unsplit level. The higher energy state has a negative I_z component. In a classical picture this would mean that the magnetic field and the magnetic dipole corresponding to the spin were oppositely aligned.

Absorption of a photon with the correct energy can cause a transition from the lower to the upper energy level in Fig. 26.10. The energy difference ΔE_{12} will be

$$\Delta E_{12} = g\mu_0 H \qquad (26\text{-}12)$$

where specific use has been made of the values for I_z of $\pm\frac{1}{2}$. Introducing the value for μ_0 given in Eq. (26-9a) into this gives

$$\Delta E_{12} = \frac{gehH}{4\pi mc} \qquad (26\text{-}12a)$$

The frequency of the requisite photon must be such that it has just this energy; that is,

$$\nu = \frac{\Delta E_{12}}{h}$$

or

$$\nu = \frac{geH}{4\pi mc} \qquad (26\text{-}13)$$

If H is 0.80 T in Eq. (26-12), the requisite photon frequency is 22 GHz. This is in the microwave region of the electromagnetic spectrum (see Chapter 12).

A block diagram of a typical apparatus used in spin resonance is given in Fig. 26.11. Equation (26-13) shows that the resonance may be achieved by

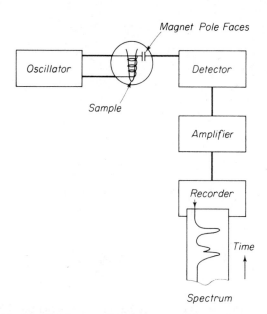

Figure 26.11   Simplified magnetic spectrometer. In nmr, the sample is literally within a coil (shown). Samples are placed within resonant cavities at the higher frequencies required in esr. In most techniques the magnetic field is varied with time, although a variation of the electric field frequency could also be used. After R. J. Myers, *Molecular Magnetism and Magnetic Resonance Spectroscopy* (Englewood Cliffs, N.J.: Prentice-Hall, Inc., reprinted by permission; 1973).

varying either H or ν. Most apparatus utilize the former technique, that is, fixed frequency. This is preferred because of the difficulties of tuning microwave circuits.

The detection of resonances in a molecule will, in general, involve the simultaneous existence of electronic *and* nuclear magnetic moments. Figure 26.9 represents an unpaired electron associated with a zero spin nucleus. It could also represent a nucleus with a spin quantum number of $\frac{1}{2}$ in a molecule with no unpaired electrons. The simplest atom is hydrogen wherein two particles each with I of $\frac{1}{2}$ coexist. This is shown in Fig. 26.12. Notice that there are now two transitions available; these obey the quantum rules of the magnetic dipole that

$$(\Delta I_z)_e = \pm 1$$

and

$$(\Delta I_z)_n = 0$$

In general, there are $2I_n + 1$ transitions for a single unpaired electron and a spin I_n nucleus. Since nuclear spins can be large and many nuclei are present

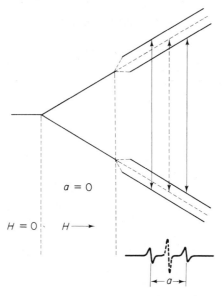

Figure 26.12 Energy levels of the hydrogen atom as a function of magnetic field at constant microwave frequency. The dotted transition would be observed if the hyperfine splitting (a) were zero. In the hydrogen atom, this splitting is, in fact, equivalent to some 506.8×10^{-4} T. The hyperfine interaction is not shown as $H \rightarrow 0$, since the energy levels become nonlinear. After H. Swartz, J. Boltan, and D. Borg, *Biological Applications of Electron Spin Resonance* (New York: Wiley–Interscience, 1972).

in a typical molecule, the nuclear magnetic structure can be quite complicated.

Magnetic moments do not exist in isolation from each other or from solvent or lattice structure. Because of these additional interactions, the excited magnetic states are inherently unstable; they can decay by giving energy to other systems. From the uncertainty principle, a finite decay time is associated with a broadening of the energy level. A particularly strong interaction implies a fast decay and, consequently, a very broad resonance.

In esr, one must consider, in addition to the electron–lattice coupling, the mutual interactions of the electronic and nuclear dipole moments. The latter is referred to as the *hyperfine interaction*. This interaction contains not only the classical dot product of the two dipoles involved, but also a term proportional to the squared magnitude of the electron wave function at the center of the nucleus.

The mathematical analysis of magnetic state decay is formulated in terms of two adjustable parameters: the relaxation times T_1 and T_2. The first is associated with spin–lattice interaction, the second with spin–spin interaction. Both relaxation times depend on temperature of the sample; as the temperature decreases, the relaxation times increase so as to freeze-in or stabilize the excited state. This should result in narrower, more easily detectable resonance bands. Thus, many magnetic resonance experiments are done in liquid nitrogen (77°K) or liquid helium ($<4°K$) baths. If T_1 remains short, bandwidths may be a few tenths of a tesla wide. These are experimentally difficult to detect.

Some transition metal ions have broad esr resonances even at low temperatures. In these cases, however, the weaker nmr resonances can be observed. The total integrated intensity of a resonance line is proportional to μ^2; this implies that esr resonances are about six orders of magnitude greater than nmr.

As is shown in Chapter 27, a net absorption of radiation is possible only if the lower level of the electromagnetic transition is more populated than the upper level. At thermal equilibrium, this is predicted to be the case. One may, however, saturate the transition by applying sufficient power, in which case the two levels could be equally populated. Selective saturation is the basis of a special technique called *electron nuclear double resonance* (endor). Feher, in 1956, proposed this method to elucidate details of very complex nmr spectra. An electronic transition is first saturated with an intense microwave field. By adding a second radiation field at a resonant frequency for the nuclear transition, the experimenter can depopulate the previously saturated electronic transition. Thus, upon striking the unknown resonant nuclear frequency, a sudden jump in the electron resonance signal is seen. Application of endor can be of great value in analyzing a complicated nmr spectrum.

B. APPLICATIONS

Both esr and nmr have been extensively applied to biomedical studies. The following few representative examples demonstrate esr as used to examine porphyrin-metal complexes and spin-labels, and nmr in ^{13}C studies of biological membranes. (Nmr is more commonly used for hydrogen nuclei.)

The porphyrin ring structure is diagrammed in Fig. 18.11 for heme and in Fig. 20.4 for chlorophyll *a*. Both molecules contain metal ions, iron for heme and manganese for chlorophyll. The unfilled *d* shells (see Sec. 26.3) in the metal atom or ion make these molecules paramagnetic and thus suitable for esr measurements.

Feher and his co-workers have shown that there is a cellular fraction of bacterial chloroplasts that exhibits a strong esr signal upon stimulation by light. The optical absorbance also shows an increase under these circumstances. The esr and optical signals rise simultaneously with similar rates. Thus, as noted in Chapter 20, these and other authors conclude that a free radical is the primary result of chloroplast illumination. This radical can be shown spectrophotometrically to be an oxidized bacteriochlorophyll.

Free-radical labeling was developed by McConnell in the 1960's. In this method a paramagnetic moiety is attached to a molecule of interest to label it. A widely used label is some variant of the nitroxide group:

$$R_1 \diagdown \overset{\displaystyle N}{\underset{\displaystyle O\cdot}{|}} \diagup R_2$$

This moiety can be attached to a molecule by either covalent or noncovalent binding.

For example, electron spin resonance techniques can then be used to detect the distinctive signals from both a labeled antigen hapten and the corresponding labeled antigen–antibody complex. Figure 26.13 depicts the case of morphine tagged with the nitroxide group. Relative magnitudes of the free and bound morphine resonance signals permit a calculation of the unknown

Figure 26.13 Summary of the spin-label immunoassay for morphine. The upper, middle, and lower tracings refer to esr signals from the indicated mixture of morphine and morphine antibody. The magnetic field is along the abscissa. After E. S. Copeland, *Ann. N.Y. Acad. Sci.* **222**:1097–1101 (1973).

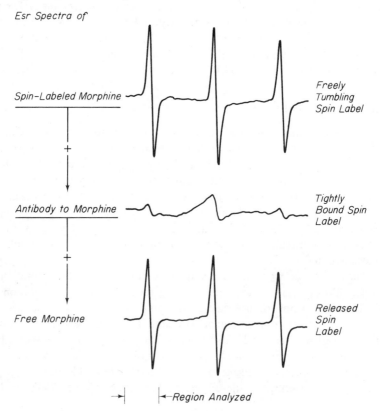

concentration of morphine in a sample. Sensitivity is on the order of 1 $\mu g/m\ell$ in clinical tests.

Theoretically, competitive binding employing esr spin labeling is identical to that using a radioactive tracer. Chapter 29 contains a description of the latter technique as well as the general mathematical analysis of competitive binding reactions. One may just as readily utilize an antibody label as an antigen label to determine antigen concentrations.

One representative application of nmr involves the study of membrane structure, as discussed in Chapter 24. Magnetic measurements can be used to study the dynamic state of lipids, lipid–lipid interactions, and lipid–protein interactions. As discussed in greater detail in the reference by James, values of both T_1 and T_2 for ^{1}H, ^{2}H, ^{31}P, and ^{13}C nuclei have been especially useful in these studies. For example, the 25.2-MHz ^{13}C nmr spectrum of mitochondrial membranes has several well-resolved peaks. These arise from lipids rather than proteins, which suggests that lipids have considerably greater mobility than the proteins in this membrane.

26.7 Summary

The quantum mechanical nature of matter implies that electromagnetic energy in certain frequency bands may be absorbed preferentially by a given material. Such specifics can be used to identify a compound, to quantitate its concentrations and, indirectly, to monitor the progress of various chemical reactions. Sources and detectors are available to cover the electromagnetic spectrum from the microwave region into the far ultraviolet. Measurement of the magnetic interactions has permitted the evaluation of the electronic and nuclear environments of particular chemical compounds. The specificity of absorption spectroscopy is one of its most important attributes.

REFERENCES

Quantum Mechanics

1. LAIDLAW, W. G., *Introduction to Quantum Concepts in Spectroscopy* (New York: McGraw-Hill Book Company, 1970), 240 pages.

2. SCHIFF, L. I., *Quantum Mechanics*, 3rd ed. (New York: McGraw-Hill Book Company, 1968), 544 pages.

Spectrophotometry

3. BAUMAN, R. P., *Absorption Spectroscopy* (New York: John Wiley & Sons, Inc., 1962), 611 pages.

A shorter text is:

4. THOMPSON, C. C., *Ultraviolet–Visible Absorption Spectroscopy* (Boston: Willard Grant Press, 1974), 90 pages.

Optical Activity

5. IMAHORI, K., AND N. A. NICOLA, "Optical Rotary Dispersion and the Main Chain Conformation of Proteins," pp. 358–444.

and

6. SEARS, D. W., AND BEYCHOK, S., "Circular Dichroism," pp. 446–593.

Both in *Physical Principles and Techniques of Protein Chemistry, Part C*, S. J. Leach, ed. (New York: Academic Press, Inc., 1973), 621 pages.

7. BEYCHOK, S., "Circular Dichroism of Poly-α-Amino Acids and Proteins," in *Poly-α-Amino Acids: Protein Models for Conformational Studies*, G. D. Fasman, ed. (New York: Marcel Dekker, Inc., 1967), 764 pages.

Magnetic Resonance

8. MYERS, R. J., *Molecular Magnetism and Magnetic Resonance Spectroscopy* (Englewood Cliffs, N.J.: Prentice-Hall, Inc., 1973), 244 pages.

9. RICHARDS, R. E., "Nuclear Magnetic Resonance Spectroscopy of Biochemical Materials," *Endeavour* **34**:118–122 (1975).

A recommended reference for endor is:

10. FEHER, G., *Electron Paramagnetic Resonance with Applications to Selected Problems in Biology* (New York: Gordon and Breach, Science Publishers, Inc., 1970), 139 pages.

11. JAMES, T. L., *Nuclear Magnetic Resonance in Biochemistry: Principles and Applications* (New York: Academic Press, Inc., 1975), 413 pages.

12. INGRAM, D. J. E., *Biological and Biochemical Applications of Electron Spin Resonance* (New York: Plenum Press, 1969), 311 pages.

13. SWARTZ, H. M., J. R. BOLTON, AND D. C. BORG, *Biological Applications of Electron Spin Resonance* (New York: Wiley–Interscience, 1972), 569 pages.

14. COPELAND, E. S., "The Effect of Ambient Temperature on Spin-Label Immunoassay for Morphine," in "Electron Spin Resonance and Nuclear Magnetic Resonance in Biology and Medicine and Magnetic Resonance in Biological Systems," S. E. LASHER AND P. MILVY, eds., *Ann. N.Y. Acad. Sci.* **222**:1097–1101 (1973).

CHAPTER 27

Emission Spectroscopy

27.1 Introduction

This chapter continues the discussions in Chapter 26 of the interaction of radiation with molecules, atoms, and subatomic systems. Here, emphasis is placed on the processes by which radiation is emitted by such systems when they are excited. The excitation may have been introduced via an electromagnetic wave, a chemical reaction, or even through radioactive decay. Ordinary fluorescence and phosphorescence are considered initially; this is followed by a description of X-ray fluorescence. The chapter concludes with a section on lasers and a section on Mössbauer spectroscopy.

Applications discussed in this chapter include visible fluorescence of aromatic amino acids, X-ray fluorescence in tissues and organs, and Mössbauer spectroscopy of iron in biological samples. Specific biological uses of lasers are summarized in Chapter 12.

27.2 Fluorescence and Phosphorescence

A. THEORY

This chapter is concerned with the emission of photons from molecules in excited states. One fashion in which this may occur is illustrated in Fig. 27.1.

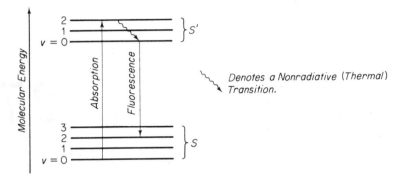

Figure 27.1   Molecular energy levels and transformations used to illustrate the underlying processes in fluorescence. In this scheme, *S* represents the group of energy levels having the lowest electronic state possible and *S'* an excited set. Both *S* and *S'* are singlet states; that is, the total electronic spin for the molecule is zero. Rotational energy levels (not shown) tend to smear out the sharp lines.

Here a molecule in the lowest electronic and vibratory energy level in *S* (*v* equals 0) absorbs a photon that raises it to a higher vibrational level in *S'*. (The groups[1] *S* and *S'* refer to lowest and first excited electronic states, respectively.) The added vibratory energy is then lost by conversion to heat through interaction with other molecules; thus, the molecule of interest falls to the lowest energy level in *S'*. After remaining here for a period of time, two different transitions may occur. The molecule may fall in energy back down to the lowest level in *S*, losing the added energy by internal rearrangements or by intermolecular interactions. Alternatively, it may emit a photon, thereby falling to a higher energy level in *S* (e.g., *v* equals 3).

Such secondary emission of a photon following absorption is called *fluorescence*. The wavelength of the fluorescent photon usually exceeds that of the absorbed one since, as is seen in Fig. 27.1, the emitted photon may have less energy. The entire cycle is completed by the molecule losing its increased vibratory energy by reacting with the surrounding molecules, thereby falling to the lowest energy level. In the case of Rayleigh emission, the secondary photon is produced before any vibratory energy is lost. The incident and emitted photons then have the same wavelength.

Not all molecules fluoresce; the spectrum of those which do depends in part on the solvent and/or other surrounding molecules. Certain molecules eliminate the fluorescence of other molecular species. This process is called *quenching*.

There is a tendency for absorption and fluorescence spectra to be mirror

---

[1] *S* is not to be confused with the spin quantum number of Chapter 26. Note also that the vibrational quantum number is called *k* in Chapter 26 and *v* here.

images of one another. A striking example of this is shown in Fig. 27.2. The energy-level diagram shown in Fig. 27.1 will account for such mirroring if two additional assumptions are made. First, the spacing of the vibrational (and rotational) levels of the lower electronic state is similar to that of the higher electronic state. Second, the relative probabilities associated with transitions ending on the different vibrational and rotational levels are the same whether one starts at level 0 or 2 in Fig. 27.1.

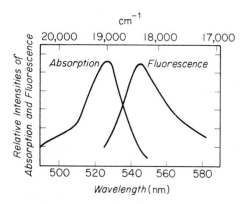

Figure 27.2 Absorption and fluorescence spectra for the dye fluorescein. Note the mirrorlike quality of the two spectra. From the energy levels in Fig. 27.1, it might be concluded that the absorption and fluorescence spectra could not cross. The observed crossing is due to the contributions of the energy associated with rotation and translation, neither of which are included in Fig. 27.1. Adapted from E. L. Nichols and E. Merritt, *Phys. Rev.* **31**:376 (1910).

In addition to the energy levels shown in Fig. 27.1 a molecule might possess other levels from which a transition to the lowest energy level was forbidden, in the sense in which this word is used in Chapter 26. An example of such levels is illustrated in Fig. 27.3. Here a typical triplet state (see Chapter 26) is included with energy somewhat lower than the upper singlet electronic level. In this case a set of transitions shown by arrows may occur with photon absorption and later photon emission. This process with an intermediate, metastable state is called *phosphorescence*.

The transition from the lowest level in T' to the third lowest level (v equals 2) in S is a forbidden one for an electronic dipole transition. Accordingly, it is very improbable, and the molecule may remain in the triplet state for periods as long as hours in extreme cases. During this time, it is possible for a transition back to the excited singlet state S' to occur, thereby resulting in a delayed fluorescence. The latter effect becomes increasingly less probable as the temperature is lowered. The true phosphorescence illustrated in the figure is relatively temperature insensitive. Since T' is lower in energy than S',

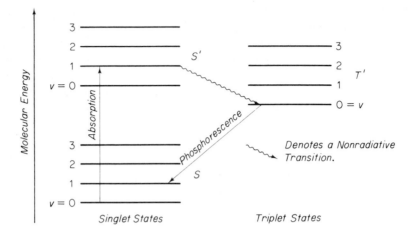

Figure 27.3 Energy-level diagram for a hypothetical molecule, showing two sets of singlet states (*S* and *S'*) for which the total electronic spin quantum number is zero and a set of triplet states (*T'*) with spin quantum number equal to 1. The arrows illustrate the transitions involved in phosphorescence.

the phosphorescent spectrum usually occurs at longer wavelengths than the fluorescent spectrum.

The decay of both fluorescent and phosphorescent excited states follows an exponential law. In differential form this can be written as

$$\frac{dn}{dt} = -An$$

where *n* is the number of molecules in the excited states and *A* is the probability of decay per unit time. The mean lifetime of a molecule in the excited level is the inverse of *A*.

Historically, phosphorescence was distinguished from fluorescence by measuring the mean lifetimes of the excited state (or states). Values around 1 ns are typical for fluorescence. Phosphorescence was defined as photon emission from states having mean lifetimes greater than 100 μsec. However, the definition based on the occupation of a metastable (triplet) energy level as illustrated in Fig. 27.3 is preferable.

B. EFFICIENCY

As noted in the preceding subsection, all absorbed photons do not result in the secondary emission of a fluorescent (or phosphorescent) photon. Accordingly, one may define the quantum efficiency, *q*, of the process by

$$q = \frac{\text{number of emitted photons}}{\text{number of absorbed photons}}$$

Since the emitted photons will, in general, have less energy, q does not represent an energy efficiency. Nonetheless, q will always be less than 1 for a substance at room temperature. The value of q is more or less independent of the wavelength of the absorbed photon provided that the latter's energy is sufficiently great to cause the requisite initial increase in the electronic energy level of the molecule.

Fluorescence and phosphorescence result in radiation in all directions. For practical applications, it is often convenient to measure such photon emissions by placing a detector along a line at 90° to the incident beam, although other angles may be used. Such a fluorescence spectrophotometer is shown in Fig. 27.4. A number of applications are discussed in the next section.

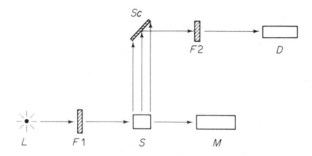

Figure 27.4   A schematic representation of a quantum yield (q) fluorescence spectrophotometer. The light source is at L with the sample labeled S. Frequency or wavelength selectors are at F1 and F2 in the incident and fluorescent beams, respectively. A fluorescent screen (Sc) is used to integrate the quantum yield and generate the input to the detector (D). The number of incident photons is monitored at M.

C. APPLICATIONS

One important application of fluorescence involves the quantitation of individual molecular species. For example, as might be anticipated from their benzene ring constituents,[2] the aromatic amino acids are fluorescent in aqueous solution and consequently can be assayed by photometric means. Quantitative analysis of amino acids such as tyrosine, tryptophan, and phenylalanine in various proteins can be accomplished following hydrolysis. Tyrosine and tryptophan are more frequently measured in this fashion, since the quantum efficiency, q, of phenylalanine is only 0.04 at a neutral pH. Both solvent and pH are known to affect q. If several aromatic amino acid residues are

---

[2] Because of the π-orbital electrons, strong $\Delta \Lambda = \pm 1$ transitions occur in molecules with conjugated double-bond systems (e.g., benzene).

associated in the same protein, spectral shifts occur. These generally are caused by nonradiative energy transfer from one amino acid to another. Many proteins in themselves have characteristic broad-band emission spectra. Because of their aromatic constituents, the majority of proteins fluoresce in the ultraviolet portion of the spectrum with q values as large as 0.20.

A second application of fluorescence involves the interaction of marker dyes with proteins. The experimenter uses a dye that will fluoresce in the visible region and whose frequency may be affected by various ligands and prosthetic groups in the protein. One particular example of such studies is the investigation of protein–ligand binding in solution. The amount of light emitted by the dye can be used to quantitate the number and type of products throughout the course of the reaction, since the fluorescent decay is much faster than any relevant rate constant. The determination of these rate constants is one of several objectives of such studies. The number and specificity of various binding sites and their position on the protein complex can also be measured using the fluorescence of the dye or the protein itself. The references at the end of this chapter may be consulted for more detailed information.

DNA phosphorescence has been reported by several experimenters. The spectrum is quite broad with onset in the UV region and a maximum extending well into the visible. A triplet state of this molecule has been associated historically with the thymine residue; in fact, thymidylic acid (TMP) has phosphorescent emissions almost identical to those of DNA itself. These emissions have not been widely used in research or diagnosis but, as in the case of the protein emissions described earlier, eventual practical application to diagnostic assays seems probable.

27.3   X-Ray Fluorescence

The measurement of characteristic X radiation from particular atoms is a method of growing importance in biomedical assays. Excitation energy is given to various stable trace elements by one of three means: gamma rays from radioactive sources, the output of an X-ray tube, or charged-particle beams. A vacancy in the innermost, or K, electronic shell (n equals 1 and $l = 0$ in the terminology of Sec. 26.3) is thereby created; subsequent electronic transitions are then possible. Figure 27.5 shows an electronic energy-level diagram. The photon emission lines labeled K_α, K_β, and K_γ originate from electronic transitions between the higher-lying shells and the vacated K shell of the element involved. The technique of X-ray fluorescence is quite specific, since elemental emission spectra are distinct and have been tabulated. One must be careful in choosing a source energy so that the Compton effect photons (see Chapter 17) do not intrude upon the expected K_α and other

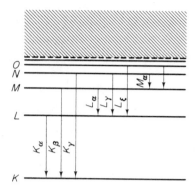

Figure 27.5   Electronic energy levels in a many-electron atom. The radiative transitions shown correspond to various X-ray energies. The crosshatched interval is the continuum of ionization levels available to the free electrons. After W. Finkelnburg, *Atomic Physics* (New York: McGraw-Hill Book Company, 1950).

X rays. Detection is accomplished with solid-state detectors (e.g., Li or Si) to provide the requisite energy resolution (~ 100 eV).

Detection of fluorescent emissions can be applied both *in vitro* and *in vivo*. A majority of the experiments have involved irradiation after a sample has been withdrawn from the subject, e.g., via a biopsy technique. A typical piece of apparatus consists of a collimated source of γ or X rays with a solid-state detector arranged in such a fashion that the excitation photons are not directly observable. For example, the 60-keV γ rays from [241]Am may be used to irradiate a tissue specimen. In that case, a solid-state detector is oriented at 90 degrees to the incident beam as shown in Fig. 27.6. The volume being observed may be several cubic centimeters in size.

For *in vivo* work, source and detector can be moved over the patient using a rectilinear or other motion (see the reference by Hoffer and Gottschalk). In the case of thyroid scanning, Hoffer has reported that X-ray fluorescence from stable [127]I in the thyroid has reduced the target-organ radiation dose by between 3 and 4 orders of magnitude as compared to a standard ^{131}I uptake study. Moreover, there was no whole-body radiation dose with the fluorescent technique. The quality of the images appeared roughly comparable in the two protocols, although the fluorescence technique actually measures a static condition, whereas the uptake study is a dynamic one. Differential tissue absorption of the incident and fluorescent photons can cause uncertain efficiency factors, however.

In vitro applications of X-ray fluorescence have included clinical laboratory evaluations of patient samples and environmental contaminant determinations. In these subject areas, the method is similar to neutron activation

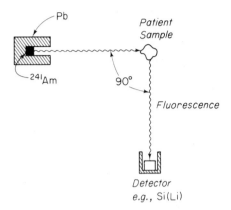

Figure 27.6   X-ray fluorescence apparatus. The excitation energy is provided by an ^{241}Am radioactive source, while fluorescent X rays are detected with a solid-state detector. Patient samples may be either *in vitro* or *in vivo*.

analysis using the (n, γ) reaction. Inexpensive γ- and X-ray sources are much more readily available and portable than neutron sources. This implies that X-ray fluorescence will continue to be the more popular biomedical assay technique.

X-ray fluorescence sensitivities are typically a few parts per million for such elements as Cr, Cs, Hg, and Pb. Because of the energy resolution of solid-state detectors, analyses can be done for almost all trace elements at the same time using the same blood, tissue, or fluid specimen. Dynamic function studies are performed with a set of samples withdrawn from the patient at prescribed intervals of time. At long times, the equilibrium tissue or blood concentrations may be used to determine volumes of various body compartments using the standard dye dilution formula (Sec. 29.5). Fluorescent determinations of volumes are quite comparable in magnitude to radiotracer values and, in addition, the patient receives no radiation dose with the *in vitro* X-ray method.

Very sensitive trace-element detection systems may have an eventual clinical impact that is hard to estimate at this time. Although some metal-associated syndromes such as Wilson's disease (a Cu metabolism error) are well known, many more such conditions may be discovered when assay procedures are standardized and normal values are determined. Different racial and geographic groups may have different trace-element concentrations in various body tissues. Measurement of these concentrations could give much information about mineral deficiencies and environmental contamination. Correlations may be established between certain trace elements and diseases such as arteriosclerosis or even certain types of cancer. Chromium deficiency, for example, has been indicated as a predisposing condition for myocardial infarction in some societies. That is, certain social groups with higher chromium levels have a lower incidence rate for myocardial infarction.

27.4 Stimulated Emission

A. THE EINSTEIN COEFFICIENTS

The previous sections of this chapter have considered the absorption of an incident photon and the subsequent spontaneous decay of the ensuing excited state. Einstein, in 1905, pointed out that a third type of process was possible under the influence of the electromagnetic field. In this additional form of decay, the incident wave causes an excited state to return to a lower energy level. This is referred to as *stimulated emission*. The frequency of the stimulator must correspond exactly to the energy difference between the two states involved. This stimulation results in the emission of a new photon without the absorption of the incident photon, thereby amplifying the input energy.

Einstein formulated the problem in terms of probabilities per unit time and per unit energy density (ρ_E). The latter assumption was required due to the wide range of frequencies present in a typical radiation source. The dimensions of ρ_E are energy per unit volume per unit frequency interval. Other density functions are sometimes used in lieu of ρ_E; these require modification of the associated Einstein coefficients.

As an example of the simplest application of the Einstein coefficients, consider a two-level scheme in which the lower energy state is labeled with the subscript i and the upper with a subscript f. Three coefficients are descriptive of the types of radiative transitions possible in this two-level system:

B_{if} = probability of stimulated absorption per unit energy density per unit time

B_{fi} = probability of stimulated emision per unit energy density per unit time

A_{fi} = probability of unstimulated (spontaneous) emission per unit time. This quantity is equivalent to the decay constant of a radioactive system (see Chapter 29)

Figure 27.7 abstractly summarizes these types of transitions.

Assuming equilibrium in the system, the rate of upward transitions must equal the rate of downward transitions:

$$n_i B_{if} \rho_E = n_f B_{fi} \rho_E + n_f A_{fi} \tag{27-1}$$

where n_i and n_f refer to the number of systems in levels i and f, respectively. Thermodynamic theory permits a determination of the ratio n_f/n_i, while the quantum mechanical form of the energy density function ρ_E is known from Planck's blackbody radiation formula. Thus, Eq. (27-1) may be solved for

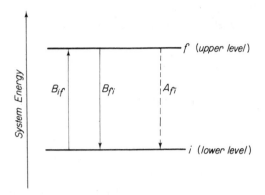

Figure 27.7 Summary of stimulated and spontaneous transitions in an hypothetical two-level (i and f) system.

the stimulated emission probability in terms of either the spontaneous decay or spontaneous absorption probabilities:

$$B_{fi} = \frac{A_{fi}}{8\pi h/\lambda^3} \qquad (27\text{-}2a)$$

$$B_{fi} = B_{if} \qquad (27\text{-}2b)$$

Equation (27-2a) reveals that stimulated emission becomes more probable relative to spontaneous decay as the wavelength increases. This result was applied to the development of low-noise amplifiers at microwave frequencies (e.g., in the detection of stellar radio sources). In that case, A_{fi} represents noise (decay) and the quotient B_{fi}/A_{fi} can be interpreted as a signal-to-noise ratio. Such a system is called a *maser* (which is an acronym for *m*icrowave *a*mplification by *s*timulated *e*mission of *r*adiation).

It can be shown that a two-level system in thermodynamic equilibrium cannot be used to amplify a signal. This is because the upward and downward transition rates are equal so that as many photons are absorbed as emitted. By various nonequilibrium means, however, a situation may be obtained whereby one (or more) excited levels is heavily populated relative to the ground state. Then the radiation energy density may cause induced emission, as indicated by Eq. (27-1). If n_f is made much larger than n_i and A_{fi} is relatively small (e.g., in a phosphorescent excited level), the stimulated emission term becomes dominant and amplification occurs. This sort of device, at optical frequencies, is termed a *laser*. (This acronym means *l*ight *a*mplification by *s*timulated *e*mission of *r*adiation.)

B. LASERS

To expedite the amplification process, a three (or more)-level nonequilibrium situation is established. There are two radiation fields in this case.

Initially, an external pumping frequency source raises the system by stimulated absorption to an excited energy level (E_1), as shown in Fig. 27.8. By non-

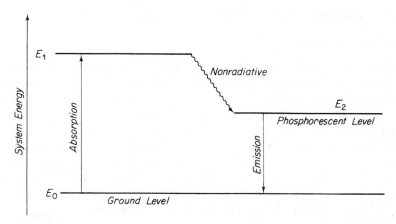

Figure 27.8 Energy-level schematic for a three-level laser system. The level with energy E_2 is phosphorescent with an appreciable half-life.

radiative transitions, a metastable or phosphorescent state E_2 is populated. If little radiation of energy ($E_2 - E_0$) is present in the pumping spectrum, a large population inversion will occur, with many systems residing in the state having energy E_2. Because this level is phosphorescent, it will have a slow spontaneous decay. Stimulated emission is then produced by irradiating the inverted system with frequency v_0 equal to $(E_2 - E_0)/h$. In most laser designs, spontaneously emitted photons of energy hv_0 are actually used to provide the stimulating energy density ρ_E of Section 27.4.A. A set of parallel mirrors (Fabry–Perot interferometer) reflect these emissions along the appropriate direction so that they repeatedly pass through the laser material. As one photon of energy hv_0 encounters an atom or molecule in a phosphorescent energy level, a second in-phase photon of that energy is emitted into the radiation field. This amplification of coherent light eventually builds up the stimulating fields. If one mirror is only partially silvered, a large burst of photons can pass through that end of the laser. The device is then said to be *lasing*.

Laser action has been achieved in a wide variety of systems, including gases and solids. Excitation is usually accomplished by optical pumping (e.g., xenon flash lamp) or current flow through the medium. Species involved have included molecules, atoms, and ionic forms. One of the most significant developments has been the ability to selectively tune the broad-band laser output frequencies by means of various dye molecules and lasing cavity geometries. These results are summarized in the reference by Schlossberg. A description of some biological applications of lasers is included in Chapter 12.

27.5 Mössbauer Spectroscopy

A. THE PRINCIPLE

Mössbauer spectroscopy uses both the emission and absorption of photons. In this method, the emitted γ-ray photon has precisely the correct energy to cause a nucleus to absorb it. This precise energy match may be achieved by moving either the emitting or absorbing nucleus. Moving the nucleus results in a Doppler shift (see Chapter 4) of the photon's frequency and hence its energy. The two nuclei are identical except that the emitter must be in an excited state which is capable of spontaneously producing a photon. A widely used Mössbauer nucleus is ^{57}Fe, produced by the decay of ^{57}Co. The energy-level diagram for these two species is given in Fig. 27.9 where the 14.4-keV excited level of ^{57}Fe is the one of specific interest.

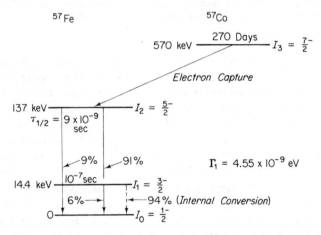

Figure 27.9 Nuclear energy-level diagram for ^{57}Co and its daughter, ^{57}Fe. Decay of the parent results in the 137-keV excited state of iron 57. This level, in turn, can decay via emission of 137 or 123-keV γ rays. The radiative transition from the first excited state of ^{57}Fe (14.4 keV) provides the Mössbauer emission. Nuclear spin and intrinsic spatial symmetry (parity) are indicated. After J. E. Maling and M. Weissbluth, *Solid State Biophysics*, S. J. Wyard, ed. (New York: McGraw-Hill Book Company, 1969), pp. 327–368.

Photon emission by such an excited state requires, by the conservation of linear momentum in an isolated system, that:

$$p_\gamma = p_r \tag{27-3}$$

where p_y is the momentum of the emitted photon and p_r is that of the recoiling emitter nucleus. For a photon of energy E_y, the momentum is E_y/c, where c is the speed of light. The conservation of energy imposes the additional constraint that

$$E_y + E_r = E_0 \qquad (27\text{-}4)$$

where E_r is the energy of the recoil and E_0 is the excitation energy of the emitting level (e.g., 14.4 keV for ^{57}Fe). From classical mechanics,

$$E_r = \frac{p_r^2}{2M} \qquad (27\text{-}5)$$

where M is the mass of the recoiling object. Thus, E_r will tend toward zero as $M \to \infty$. Under these conditions Eq. (27-4) shows that the emitted photon's energy, E_y, approaches E_0; otherwise, E_y is less than E_0.

In 1957, Mössbauer discovered an effect that depends on the relatively unusual solid-state properties of an emitting nucleus in certain crystals.[3] In these, the entire crystal rather than a single nucleus may recoil against the emitted gamma ray. The mass M is therefore a factor of approximately 10^{23} greater than for a free nucleus, and hence $E_y \simeq E_0$. Similar considerations apply for absorption. If this Mössbauer condition were not satisfied, the photon energy would be too low to raise the detector nucleus to its excited state.

The 14.4-keV energy level in ^{57}Fe has a mean lifetime τ of about 0.1 ns. Thus, the uncertainty ΔE_0 in the energy of the level can be estimated from the uncertainty principle discussed in Chapter 26 to be

$$\Delta E_0 \geq \frac{\hbar}{2\tau} = 4.5 \text{ neV}$$

Hence the Mössbauer emission will be very sharp. Indeed, the limiting value of 3×10^{-13} for the ratio $\Delta E_0/E_0$ represents one of the highest precisions obtainable in any physical measurement.

Not all excited nuclei will decay by Mössbauer emission. The relative probability, P, of this type of decay is given by the *Debye–Waller factor*,

$$P = \exp\left[\frac{-E_0^2\langle \Delta r\rangle^2}{\hbar^2 c^2}\right]$$

where Δr is the uncertainty (standard deviation) of the location of the nucleus relative to its average location during its lifetime. The uncertainty Δr results from diffusion in the liquid state or from thermal motion about the lattice

---

[3] The material need not be crystalline; the effect can be seen in viscous fluids.

site for crystalline solids. To maximize P, measurements are usually made on a sample in the solid state at reduced temperatures.

B. THE APPARATUS

Mössbauer spectroscopy quantitatively determines energy shifts in the unknown absorbing specimen. Sources are usually foils impregnated with the appropriate emitter such as the ^{57}Co parent of ^{57}Fe. Some appreciable number of ^{57}Fe nuclei in their ground state must be present in the specimen. Sample transmission is then measured as either the source or target is slowly moved. The Doppler frequency shift required to restrike the resonance is used to quantitate the small electromagnetic energy shifts away from the free atom levels (see Fig. 27.9).

A block diagram of a Mössbauer spectrometer system is shown in Fig. 27.10. Motion is often provided by an acoustical speaker driven by a pre-

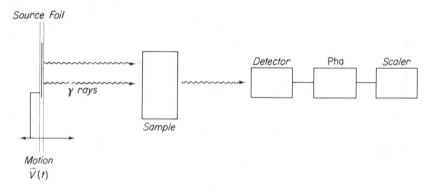

Figure 27.10   Mössbauer spectrometer. This source foil provides a γ-ray beam that is attenuated by the sample. Maximal attenuation occurs if the Doppler-induced frequency shift in the incident γ rays exactly compensates for the nuclear-level shifts in the sample. A pulse-height analyzer (pha) determines the transmitted photon's energy. The number of detected photons is counted by the scalar.

scribed voltage form. Generally, the velocity is made a linear function of time. Detectors usually are NaI(Tl) crystals. Depending on the atomic environment of the sample, one or more resonances are seen as the velocity is swept through its range of values.

C. APPLICATIONS

Known Mössbauer nuclei include the biologically interesting species ^{40}K and ^{127}I. However, ^{57}Fe is probably the most widely used emitter, particularly in heme and cytochrome investigations. At 300°K, the Debye–Waller factor, P, for ^{57}Fe is approximately 0.60, so cryogenic apparatus is not necessary. The Doppler formula for energy shift as a function of source velocity $\vec{v}$ is

$$\Delta E_\gamma \simeq E_\gamma \frac{\vec{v} \cdot \vec{p}_\gamma}{cp_\gamma} \tag{27-6}$$

where $\vec{p}_\gamma$ is the vector momentum of the photon. Equation (27-6) implies that a linear speed of 1 mm/sec is equivalent to 4.9×10^{-8} eV, or about 10 times the inherent line width of the 14.4-keV level in ^{57}Fe. The 2 percent natural abundance of this isotope has led to enrichment techniques of two types. Either the naturally occurring iron in the excised sample may be exchanged for ^{57}Fe chemically, or the latter isotope may be fed selectively to the animal or plant prior to the investigation.

Mössbauer absorption spectra have been studied for various iron-containing biomolecules. These data have augmented the esr and nmr studies (see Chapter 26) of the electronic distributions at and near the iron atom. In particular, Mössbauer spectroscopy gives information on three distinct electromagnetic interactions between the ^{57}Fe nucleus and the surrounding electronic clouds. These interactions are between the nuclear charge and the average electronic charge, the nuclear quadrupole moment and the gradient of the electric field due to the electrons, and the nuclear magnetic moment and the magnetic field due to the electrons. In addition, external magnetic fields can be used to obtain further types of interactions not detailed here.

A typical Mössbauer absorption spectrum for equine cytochrome c is shown in Fig. 27.11. The double resonance is characteristic of a ^{57}Fe nuclear quadrupole moment. Pronounced spectral differences for the more highly oxidized ferric form are illustrated in the figure. Quantitative values of the resonance positions and depths can be used to determine the ligand and valence electron distributions around the iron nucleus. Such molecular-structure information has led to a deeper understanding of biochemical reactions and their physiological consequences.

27.6   Summary

Fluorescence and phosphorescence refer to the emission of a secondary photon following the absorption of a primary one. Such techniques have been used to identify and to quantify biomolecules. In some cases, fluorescence is more effective than optical spectrophotometry for compound identification. Some molecules, particularly those with chemical structures called aromatic rings, have characteristic fluorescence spectra. Others can be labeled with

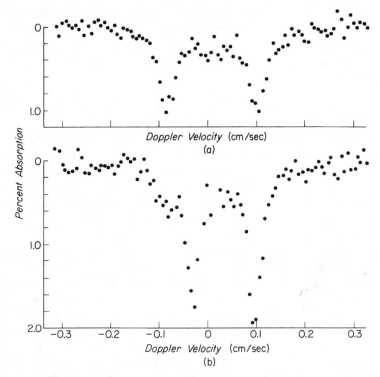

Figure 27.11 Percent absorption of ^{57}Fe 14.4-keV photons in (a) ferric and (b) ferrous cytochrome c. The double resonance follows from the nonvanishing of the nuclear quadrupole moment of ^{57}Fe. Cytochrome was in the form of a freeze-dried powder at room conditions for these experiments. After J. E. Maling and M. Weissbluth, *Solid State Biophysics*, S. J. Wyard, ed. (New York: McGraw-Hill Book Company, 1969), pp. 327–368.

appropriate dyes. Fluorescence following X-ray excitation has been used to provide organ imaging and chemical assays. The radiation doses involved are small compared to standard radiological examinations.

As an alternative to spontaneous emission, an excited molecule may be stimulated to emit a photon. In such a process the light energy is amplified, leading to a device called a laser. Applications of lasers are presented in Chapter 12.

Mössbauer techniques, a third type of emission spectrophotometry discussed in this chapter, have provided details of the bonding and electronic structures of iron-bearing compounds. In this method, the emitted photon is not frequency shifted by the recoil of the emitting atom. This produces physical measurements whose energy precision exceeds most others.

REFERENCES

Fluorescent Spectroscopy

1. PESCE, A. J., C.-G. ROSÉN, AND T. L. PASBY, *Fluorescence Spectroscopy: An Introduction for Biology Students* (New York: Marcel Dekker, Inc., 1971) 247 pages.
2. BECKER, R. S., *Theory and Interpretation of Fluorescence and Phosphorescence* (New York: Interscience Publishers, Inc., 1969), 283 pages.

Lasers

3. GOLDMAN, L., AND R. J. ROCKWELL, JR., *Lasers in Medicine* (New York: Gordon and Breach, Science Publishers, Inc., 1971), 385 pages.
4. SCHLOSSBERG, H. R., AND P. L. KELLY, "Using Tunable Lasers," *Phys. Today* 25(7):36–44 (July 1972).

X-Ray Fluorescence

5. HOFFER, P. B., AND A. GOTTSCHALK, "Fluorescent Thyroid Scanning: Scanning without Radioisotopes," *Radiology* 99:117–123 (1971).
6. BAGLAN, R. J., A. B. BRILL, AND J. A. PATTON, "Applications of Non-Dispersive X-Ray Fluorescence Techniques for *In Vitro* Studies," *IEEE Trans. Nucl. Sci.* NS-20:379–388 (1973).
7. PRICE, D. C., L. KAUFMAN, AND R. N. PIERSON, Jr., "Determination of the Bromide Space in Man by Fluorescent Excitation Analysis of Oral Bromine," *J. Nucl. Med.* 16:814–818 (1975).

Mössbauer Spectroscopy

8. DEBRUNNER, P. G., "Mössbauer Spectroscopy of Biomolecules," in *Spectroscopic Approaches to Biomolecular Conformation*, D. W. Urry, ed. (Chicago: American Medical Association, 1970), pp. 209–262.
9. MALING, J. E., AND M. WEISSBLUTH, "The Application of Mössbauer Spectroscopy to the Study of Iron in Heme Protein," in *Solid State Biophysics*, S. J. Wyard, ed. (New York: McGraw-Hill Book Company, 1969), pp. 327–368.

CHAPTER 28

Medical Ultrasonography

28.1   Introduction

Ultrasonography is one of several specialized techniques which originated in the biophysical sciences for diagnostic medical purposes. Most such ultrasonic information is derived from echo techniques roughly similar to those used by bats (see Chapter 4). Interest in medical echolocation (echosonography) followed the naval development of methods called *sonar* (an acronym for *so*und *na*vigation and *r*anging). These depend on reflection of acoustic waves at interfaces where there is a change in the specific acoustic impedance (discussed in Chapter 13).

Values for specific acoustic impedances of physiological tissues are given in Table 13-1. One conclusion from these numbers is that the presence of air or other gas is easily detected *in vivo*. Historically, in fact, reflections from the air bladders of schools of fish produced echoes that were often misinterpreted in sonar displays. This table also indicates that interfaces between bone and soft tissue should be outstanding when viewed with ultrasound. However, transitions between one type of soft tissue and another are less marked. These latter differences are difficult to detect; sophisticated equipment is required for their observation.

As mentioned in Chapter 13, the hazards of low-level sonic irradiation are usually far less than those of exposure to X rays. Accordingly, medical applications of ultrasonography are important in areas where standard X-ray techniques are dangerous to the patient, such as delineation of the fetus. Other applications have been to regions of the body where X-ray methods have been relatively ineffective, such as the imaging of the patient's eye.

28.2 Methods

A. TRANSMISSION

Ultrasonic radiation may be used to produce images by means of transmission or reflection. Because of the difficulty of obtaining adequate radiographic pictures of the brain, early ultrasonic imaging was concerned with transmission scanning of cerebral tissue. This technique relies on the synchronized movement of a pair of acoustic transducers to measure the transmissivity of the brain.

The extensive skull scatter of the radiation is the dominant problem, however. Cranial bones are of two rather dissimilar types—very uniform ivory bone and cancellous or "spongy" bone. Soft tissue masses found in the spongy bone matrix act as a rather randomized set of ultrasonic wave scattering centers which greatly influence the spatial dependence of the acoustic attenuation coefficient (see Chapter 13). In addition, interference occurs between the direct and scattered waves so as to further complicate the wave pattern detected by the receiver. The resultant detected intensity appears to depend strongly on local variations in skull bone type and consequently has little sensitivity to brain structure.

Transmission scanning of other organs, while possible in principle, has not been of great clinical interest due to the availability of X-ray methods to accomplish the same objective. In addition, computerized axial tomography using X rays (see Chapter 30) is a preferable method for evaluation of brain morphology.

B. REFLECTION

The scatter of ultrasonic radiation, while a problem in brain transmission investigations, has been turned into an advantage in various reflection techniques. These include wave scattering (echo) and Doppler reflection methods. In all of these, the spatial variation of acoustic impedance causes reflections as discussed in Chapter 13. Reflections from nonplanar surfaces will tend to diverge or scatter from the reflector. In such cases, it is sometimes best to describe the object as having a certain effective acoustic cross section σ for scattering which may differ from its geometric cross section.

Figure 28.1 shows the relative acoustic cross section as a function of relative frequency. This is plotted for a bubble of air and a rigid sphere, both immersed in an aqueous medium. The low-frequency region, where the relative cross section increases rapidly with increasing frequency, is referred to as the *Rayleigh scattering interval*. The analogous scattering of light by particulate matter in the earth's atmosphere has been used to account for the blue color of the sky. At higher frequencies the relative cross section for acoustic scattering approaches unity.

Figure 28.1 Acoustic cross sections (σ) of spherical objects in an aqueous medium. The physical radius of the sphere is given by r and the acoustic wavelength by λ. Wavelength decreases to the right. After R. S. Mackay, *Animal Sonar Systems*, Vol. II, R. G. Busnel, ed. (Jouy-en-Jonas, France: Laboratoire de Physiologie Acoustique, 1967).

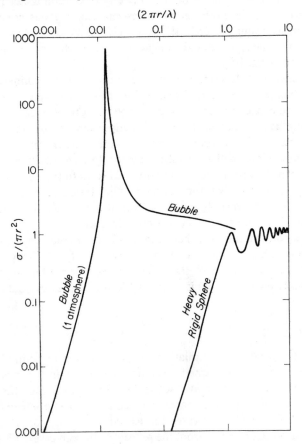

If the scattering center is immobile, the reflected wave will have the same frequency as the incident radiation. Applications of this nature are discussed in Sec. 28.3. Otherwise, Doppler frequency shifting will occur. This is outlined in Chapter 4 and is discussed in Sec. 28.4.

C. TRANSDUCERS

The principal artificial means of acoustic wave generation and detection is based on the *piezoelectric effect*. In this, a variation of an electric field imposed across a material having microscopic electric dipole moments causes the material to oscillate mechanically. Thus, pressure variations can be generated in an ambient fluid medium. Conversely, pressure variations imposed on a piezoelectric material result in a varying electrical potential across the material. This effect can be used to detect ultrasound. The reciprocity between the variation of the electric field and the physical motion along one or more axes of the piezoelectric material was described by Pierre and Jacques Curie in the 1880's.

The fundamental mechanical frequency will be equal to that of the driving oscillator; the maximum amplitude (resonance) occurs when the electrical frequency equals one of the inherent structural frequencies of the sample. Usually, one excites longitudinal waves along the length of a cylinder made of quartz, or a ceramic, such as barium titanate or lead zirconium titanate. These latter materials are ferroelectric; this means that they may be produced

Figure 28.2 Sectional view of a typical ultrasonic transducer. The electric field may be applied directly to the crystal element as shown or, at higher frequencies, this element may be located in a cavity resonant at radio frequencies (see Chapter 12). The backing block acts as a mechanical isolator which decouples the case from the high-frequency vibration of the transducer material so as to permit secure positioning of the probe on a patient. After P. N. T. Wells, ed., *Ultrasonics in Clinical Diagnosis* (Longman Group Ltd., London, England, 1972).

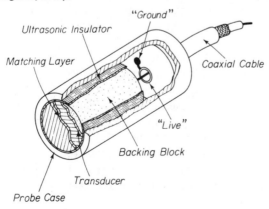

with a permanent macroscopic electric dipole moment. Titanates are extremely efficient ultrasonic radiators since the magnitude of the permanent moment can be made relatively large by addition of appropriate impurities.

The ultrasonic *transceiver* is both the source and detector of ultrasonic radiation, since it transduces energy interchangeably from electrical to mechanical forms. The transceivers used in medical diagnosis are often referred to as *ultrasonic probes*. One of these is illustrated in Fig. 28.2. It is shown in Chapter 13 that an appropriate resonant frequency for a probe of this nature is of the order of 1.5 MHz.

Frequencies usually are on the order of 1–3 MHz. If the organ to be scanned is relatively thin, a higher-frequency source may be appropriate to examine finer detail in the specimen. Eye imaging is conventionally done at 20 MHz, for example, since the relatively short half-value depth of this radiation is sufficient to examine the entire ocular volume.

D. RADIATION PATTERNS

The acoustic field distribution around the probe can be calculated using the Huygens wave superposition principle. Figure 28.3 illustrates the procedure for a hypothetical transducer which has a flat frontal surface exactly 7 wavelengths wide. Near the probe, there is a pronounced maximum in the forward direction and two secondary maxima called side lobes. As distance from the probe increases, only the central maximum remains. This transition occurs at a distance (Δ) approximately equal to r^2/λ, where r is the radius of the source. An acoustic field distribution is usually referred to as *Fresnel* or *near zone* at axial distances less than Δ, and as *Fraunhofer* or *far zone* beyond Δ.

In a typical probe, where r and λ were 6 mm and 1 mm, respectively, Δ would be approximately 36 mm. Most ultrasonic probes are used in the Fraunhofer zone, where the pressure field appears as a linear extension of the emitter itself (i.e. a beam of radiation perpendicular to the probe's surface). This arrangement simplifies the resultant echoes by eliminating side lobe scattering.

Probes may be used either singly or in groups. The former method is the more generally applied with a single transceiver being moved over the subject to produce one static image. In the second case, the image can be more rapidly formed without moving the probes. This allows a series of dynamic "moving pictures" of a living subject. In addition, by appropriately controlling phase differences between transducer elements, it is possible to selectively direct transmission and detection. This extends the range of the probes and makes the method more flexible. Such devices, termed *phased arrays*, can allow electronic scanning of an appreciable part of the human body without motion of the apparatus.

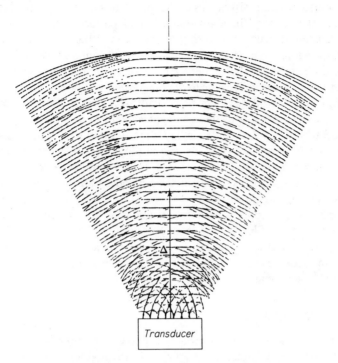

Figure 28.3 Mathematical representation of wave fronts generated by a cylindrical probe exactly seven wavelengths in diameter. Reinforcing wave fronts correspond to maxima in the acoustic pressure. After D. N. White, *Ultra Sound in Medical Diagnosis* (Ultramedicine, Box 763, Kingston, Ontario). Copyright 1976, Figure 3-5 (p. 115).

28.3   Wave Reflection Techniques

This section describes the techniques used most frequently in echosonography. Here, one measures the time delay of the reflected wave with respect to the incident pulse signal. Frequency shifts, if any, are ignored. As indicated in Chapter 4, this time delay is equivalent to a distance measurement between the acoustic source and the impedance change which produces the reflection.

A. DISPLAY MODES

Much of the notation used to describe ultrasonic data display is derived from radar terminology. Most such displays occur on cathode ray tube (crt) screens. The simplest such picture is one in which the signal amplitude is on the ordinate and time is on the abscissa. Time in such A, or amplitude, displays can be directly related to distance, given the speed of sound in the medium.

A second common display uses the B, or brightness, mode. Here, the source may be scanned in a two-dimensional pattern. The crt brightness, at a point, is made proportional to the amplitude of the echo returning from the corresponding point in the subject. Thus, the viewer sees an apparent cross section through the patient. In medical literature, such sectional images are sometimes called *ultrasonic tomograms.*

Brightness-mode displays can be difficult to interpret clinically, owing to the limited dynamic range of the crt. Typical echo signals may vary by more than 90 dB while the crt can only display some 20 dB or less. Various methods, generally called *gray-scale imaging*, are employed to alleviate the problem. One method progressively attenuates the higher-amplitude echoes resulting in a reduced signal range. A second method encodes the echo-induced electrical signals into eight gray levels.

A third type of display is used in dynamic studies primarily to picture the motion of the heart. This M, or motion, mode gives the distance of the object along the ordinate with absolute time as the abscissa. By differentiation, one may obtain velocities. Synchronization with physiologically derived signals such as the ekg may also be obtained.

B. APPLICATIONS

One example of simple static reflection studies is midline echo encephalography. As seen in Chapter 6, the brain is divided into two halves. The division between these halves, called the midline, produces a strong acoustic reflection. Midline echo encephalography is based on the premise that unilateral intracerebral masses such as tumors, metastases, and hematomas would displace the midline toward the unaffected side. For these studies a transceiver is used which produces pulsed signals a few microseconds long. After pulsing, the same probe is used to detect the echoes for periods approaching 1 msec.

Several structures occur at or near the midbrain which can act as effective acoustic scatterers. These include the left and right sides of the fluid-filled third ventricle and the lateral surface of the pineal gland. The incident radiation enters the patient perpendicular to the skull directly above an ear. At these places, the skull surface is essentially parallel to the midline so that multiple internal reflections of the input pulse will not occur.

Following a single pulse, echoes are returned to the transducer from all points at which a change of acoustical impedance occurs. The observer looks for a reflected signal occurring intermediate in time between the echoes due to the brain–skull interfaces. Figure 28.4 shows an A-mode display for reflection encephalography.

The diagnosis of displacement is based on the relative spacing between the supposed midline echo and the two reflections from the interfaces between the

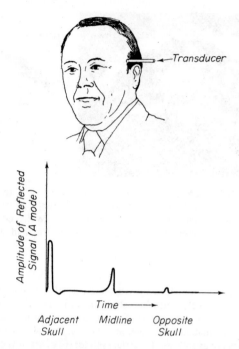

Figure 28.4 Midline echoencephalography. The lower portion of the figure shows the A mode signals as they are reflected, in turn, from adjacent skull, brain midline, and opposite skull. Transducer input is at the side of the head as shown.

soft tissue and the skull. The clinical use of A-mode midline encephalography has suffered from the difficulty of interpreting the echoes. In larger hospitals the more expensive, but more versatile, computerized axial tomography (see Chapter 30) has replaced the echo technique, since entire tomographic sections of the brain are easily obtainable in a few minutes with an X-ray source.

The second example presented here is the obstetrical evaluation of the abdomen. Obstetrical studies using ultrasonic echo detection are now standard practice in most medical centers. These examinations are quite useful in providing information to determine the presence of a gestational sac, multiple fetuses, or breech birth. Figure 28.5 illustrates a B-mode display from one of these studies. The tomograph may be a sagittal, transverse, or other section of the woman. Good acoustic coupling is obtained by placing mineral oil on the patient's skin.

The strength of any given reflected signal will be proportional to the impedance mismatch at that particular interface. As noted in Chapter 13, gas-filled spaces will be most strikingly visualized; accordingly, the patient is

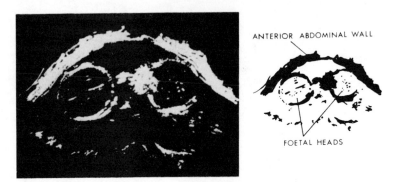

Figure 28.5 Echosonogram (B mode) of human pelvis with twins *in utero*. Gestation age of 25 weeks is established by the size of the fetal skulls. Note the midline echo of the head at the left. From P. N. T. Wells, ed., *Ultrasonics in Clinical Diagnosis* (Longman Group Ltd., London, England, 1972).

Figure 28.6 Normal M-mode echosonogram of the human mitral valve (ucg). Distance of echo-inducing structure (the mitral valve) is given as ordinate with time as abscissa. Both ecg (called ekg in text) and pcg (phonocardiogram) are included for reference points. Arrows indicate blood flow. Cardiac phases are as follows: (a) atrial systole; (b) onset of ventricular systole; (c) late in ventricular systole; (d) early ventricular diastole; (e) late ventricular diastole. After P. N. T. Wells, ed., *Ultrasonics in Clinical Diagnosis* (Longman Group Ltd., London, England, 1972).

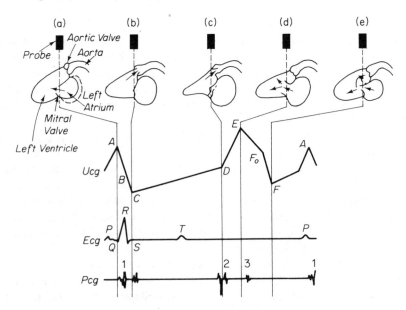

often asked to fill her bladder to push gas-filled intestines out of the field of the scanning probe. The transducer head is rocked in a small angular arc (typically $\pm\,10°$) at each contact point on the abdomen. It is also moved along a line across the abdomen to produce a complete tomograph. By switching to the A display, the clinician can determine the diameter of the fetal head, which is a good estimate of fetal age. The reduction in radiation hazards has made this technique the method of choice in following the fetus to term.

In addition to obstetrical studies, B-mode displays have been used to detect masses in various abdominal organs, including liver, pancreas, and kidneys. Soft-tissue masses and cysts in the breast and thyroid have also been examined. A water bath is often used in these latter cases to improve the acoustic coupling between transducer and the patient. A local anesthetic is usually required in eye imaging, with tears acting as the coupling agent.

The final example of reflection echosonography discussed is the determination of cardiac function (see Chapter 10). In the M-mode technique, the cardiologist might display the anterior wall, the mitral valve, the left atrium, and the pericardial surface of the heart. The mitral valve (connecting the left atrium to the left ventricle) is important in cardiac function; indeed, the determination of its motion is the most important result of the M-mode studies. Figure 28.6 shows an idealized normal mitral curve of position versus time with a corresponding ekg. The serial course of a patient pre- and post-surgery may also be followed, as shown in Fig. 28.7.

Figure 28.7   Echosonogram of patient before (a) and after (b) surgery on the mitral valve (ucg). The angle is proportional to the actual rate of the valve's motion. After E. Kelly, ed., *Ultrasonic Energy: Biological Investigations and Medical Applications* (Urbana, Ill.: University of Illinois Press, 1965).

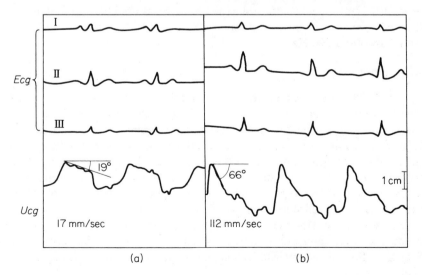

Other cardiac parameters are also observed using ultrasonic echo techniques. Many of these techniques use the M-mode display in a fashion similar to the studies of the motion of the mitral valve. Additional studies of cardiac function use methods based on Doppler reflection described in the next section.

28.4 Doppler Techniques

A. THEORY

Differentiation involves subtraction and, hence, loss of significant figures. This limits the accuracy with which velocities can be determined using the methods discussed in Sec. 28.3. As mentioned in that section, for certain applications such as cardiac function, velocity measurements are very important. Thus, there is interest in ultrasonic methods that measure velocity directly.

One such method, discussed in this section, uses the *Doppler frequency shift*, which occurs when a sound wave is reflected from a moving interface (see also Chapter 4). If v represents the frequency of the transmitted signal and v' that of the reflected signal, then the component of the velocity directed toward the sound source, V_s, is given by

$$V_s = \frac{c(v' - v)}{v' + v} \tag{28-1}$$

where c is the speed of sound in the medium.

In cardiac function measurements, the velocities to be detected are of the order of 15 cm/sec. Thus, if a value of 1.5 km/sec is used for c, the frequency ratio $(v' - v)/(v' + v)$ must be about 10^{-4}. If v is 1 MHz, then v' will differ from v by about 200 Hz.

B. APPLICATIONS

Figure 28.8 contains a diagram of a Doppler apparatus which, unlike the systems discussed previously, uses a separate transmitter and receiver. A sharp filter removes signals at the transmitted frequency v which correspond to stationary interfaces. The apparatus can directly measure the velocity of the heart valves, heart walls, and other structures. These velocities are often recorded simultaneously with the ekg to relate the motion to the heart cycle. Measurements of this type, although useful, require skill in interpretation, since the different reflections are not identified.

An alternative Doppler method uses one or more probes which can produce and interpret pulsed transmissions similar to the ones discussed in Sec. 28.3. For this Doppler technique, more elaborate circuitry is required to

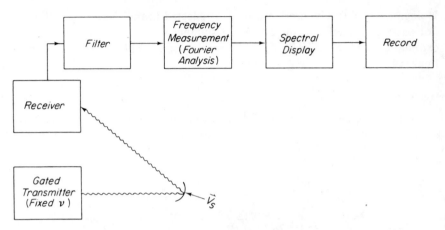

Figure 28.8 Apparatus for Doppler ultrasonic studies. The receiver signal is appropriately filtered to eliminate the transmitter's fixed frequency. If the transmitter is gated in time, the distance to the moving objects can be determined along with their velocity component along the line of sight.

eliminate all reflections except those from interfaces at a determined distance from the probe. The reflections used are then analyzed to determine $(v' - v)$ and thus V_s. Such methods are referred to as *range-gated* Doppler detection. These are convenient to measure blood flow in large vessels such as the aorta and vena cava.

In order to increase the strength of the echo signal, a catheter may be placed inside the vessel. Then small amounts of some benign material are injected into the circulatory system. Almost invariably, small air bubbles accompany such an injection; their large sonar echoes provide ideal signals for the Doppler detection system. These air bubbles are of no particular hazard to the patient.

28.5 Acoustic Holography

A. GENERAL PRINCIPLES

A *hologram* is an interference pattern used to create a three-dimensional image. For holography, two sources are needed, which must be phase- and frequency-coherent. An apparatus for acoustic holography is often simpler than that for the corresponding optical process. In the acoustic case, two transducers may be driven by the same electronic oscillator. The transducers are placed in a water bath surrounding the patient (or sample). Acoustical interference between one beam, transmitted only through the bath, and another, transmitted through the subject, is used to produce the hologram.

Holograms can also be produced from a reflected beam interfering with one not passing through the subject. With suitable illumination, holograms of extremities and of thin biological samples can yield three-dimensional images.

B. APPARATUS

A commercial system using a beam transmitted through the subject is shown in Fig. 28.9. The interference pattern is captured in the form of

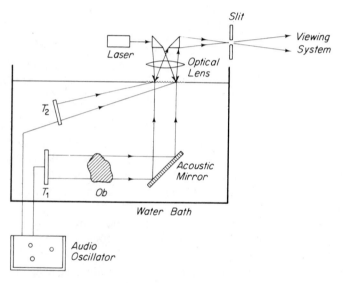

Figure 28.9 A holographic apparatus for acoustic transmission imaging. T_1 and T_2 are in-phase transducers with the object shown as Ob. The holographic interference pattern on the water surface is then transformed into a visual image by means of laser illumination. The viewing system may be a film, TV camera, or human eye.

standing surface waves on the water coupling medium. Laser light is used to illuminate the pattern and thereby recreate an image of the desired section through the subject. The hologram in this case is the surface wave pattern. It may be shown that the first-order optical interference pattern resulting from the reflection of a laser beam by this hologram is proportional to the standing wave amplitude.

A similar holographic device has been constructed in which only a single acoustic transducer is used. An acoustic grating at the surface of the water converts the parts of the beam transmitted through and around the subject into surface oscillations. The latter are illuminated with a coherent beam of light produced by a laser to result in three-dimensional images.

Instruments of both this and the preceding type appear most suitable for thin sections of the body, for two reasons. First, at this frequency, there is high attenuation of sound by tissues. Second, the imaged specimen must be completely immersed.

Both of these acoustic holographic systems can be pulsed at rates up to 100 images per second. Such a mode allows one to observe motion of a limb, for example. Acoustic holography appears to offer numerous possibilities for biophysical research and medical diagnosis. However, although the theory has been known for some years and feasibility has been demonstrated, few practical applications have resulted.

28.6 Summary

Medical diagnostic information can be obtained from analysis of echoes produced by irradiating human subjects with ultrasonic beams. The instruments that generate and receive such signals use transducers constructed from piezoelectric or ferroelectric materials. These transducers may be used in single probes or in probe arrays. The former can be moved over the subject to produce a cross-sectional body image (tomogram) while the latter can produce such tomograms while remaining stationary.

Reflection (echo) techniques use a variety of different display modes, including the one-dimensional amplitude or A mode, the two-dimensional brightness or B mode, and the motion or M mode. These displays are used as diagnostic aids in various medical specialties such as obstetrics, cardiology, and encephalography.

Other more specialized ultrasonic techniques have also proved useful. For example, Doppler reflection yields information from which a velocity can be determined. Such data are important in cardiac function studies. Ultrasound may also be used to produce acoustic holograms. While these are of interest for research and development, acoustic holography has led to few practical applications.

REFERENCES

General

1. BUSNEL, R.-G., ed., *Animal Sonar Systems, Biology and Bionics*, Symposium of the Animal Sonar System, Frascati, 1966 (Jouy-en-Josas; France: Laboratoire de Physiologie Acoustique, 1967), 1233 pages.

2. KELLY, E., ed., *Ultrasonic Energy: Biological Investigations and Medical Applications* (Urbana, Ill.: University of Illinois Press, 1965), 387 pages.

Medical

3. WELLS, P. N. T., ed., *Ultrasonics in Clinical Diagnosis* (Baltimore, Md.: The Williams & Wilkins Company, 1972), 187 pages.
4. WHITE, D. N., *Ultrasonic Encephalography* (Kingston, Ontario: Medical Ultrasonic Laboratory, Queen's University, 1970), 285 pages.
5. BAUM, G., AND OTHERS, *Fundamentals of Medical Ultrasonography* (New York: G. P. Putnam's Sons, 1975), 474 pages.
6. GOLDBERG, B. B., et al., *Diagnostic Uses of Ultrasound* (New York: Grune & Stratton, Inc., 1975), 468 pages.

Holographic

7. STROKE, G. W., W. E. KOCK, Y. KIKUCHI, AND J. TSUJIUCHI, eds., *Ultrasonic Imaging and Holography: Medical, Sonar and Optical Applications* (New York: Plenum Press, 1974), 642 pages.
8. HILDEBRAND, B. P., AND B. B. BRENDEN, *An Introduction to Acoustical Holography* (New York: Plenum Press, 1972), 224 pages.

CHAPTER 29

Tracer Methods

29.1   Introduction

Chapter 4 mentions studies of the migration patterns of birds and other animals. Information on these is often obtained by banding or otherwise labeling individuals and releasing them. The time and place of release are recorded, as are the time and place of any subsequent recapture. This labeling procedure can be applied in other areas of biophysics as well. A biological molecule or cellular organelle is labeled in such a way that it can be recognized by physical or chemical means. Its total distribution or concentration in one or more areas of special interest can then be measured at suitable time intervals. Thus, the labeled compound is traced, and techniques using these labels are often called *tracer methods*.

The label may be an atom in the compound to be traced. If the atom is unstable, its presence may be detected by its emissions upon radioactive decay. Indeed, there are several biophysical applications of this technique which can be carried out only with a radioactive label. Stable atoms can be used in other applications, e.g., they may be detected by their characteristic photon absorptions and emissions as detailed in Chapter 26 and 27. Atomic mass differences may also be used to identify such atoms. Finally, the label may be an entirely different atom or small group of atoms. The latter may be detected by physical or chemical methods.

Radioactive labels are the most common; accordingly, brief summaries of radioactivity and of those techniques which are unique to this type of tracer follow in Secs. 29.2 and 29.4. Sec. 29.3 introduces other labels. Further applications of tracer technology are discussed in Sec. 29.5.

29.2   Radioactivity

A. NUCLEAR STRUCTURE

An atom consists of a nucleus composed of neutrons and protons, which is surrounded by electrons. The nuclei of atoms are collectively known as *nuclides*. Nuclides and their corresponding atoms are symbolized as

$$_Z^A S_N$$

where S is the chemical symbol for the element, Z is the number of protons, N is the number of neutrons, and A is the nuclide mass in atomic mass units, which equals Z plus N. Since any two of the three numbers A, Z, or N will determine the third, and since S uniquely determines Z, the short forms

$$_Z^A S \quad \text{and} \quad ^A S$$

are often used. A small letter m after the A is used to distinguish metastable excited states from the ground state of the same nucleus.

Nuclides or atoms with equal values of Z are called *isotopes*. For example, isotopes of hydrogen include $_1^1 H_0$, $_1^2 H_1$, and $_1^3 H_2$. These are called protium, deuterium, and tritium, respectively. Protium is the common isotope; over 99.98 percent of all hydrogen atoms are of this form. Deuterium and tritium thus qualify as potential tracers for two reasons. First, their nuclear masses are detectably different from that of protium, and second, their natural abundance is low enough that a small artificial enrichment can be measured. Tritium is, in addition, radioactive; it decays with the emission of an electron.

B. RADIOACTIVE DECAY

For a nucleus to undergo radioactive decay, there must be lower energy levels to which it may descend. The excess energy is released in several different ways, depending on the nuclear species involved. These include the release of photons (γ rays), positive and negative electrons (β particles), and/or helium nuclei (α particles).

Given N atoms of an unstable species, the rate of radioactive decay is

$$\frac{dN}{dt} = -\lambda N \tag{29-1}$$

where λ is called the *decay constant*. Rates of decay are generally expressed in units called curies (Ci), where 1 Ci is 3.7×10^{10} disintegrations (decays) per second. The rate of decay is also called the *activity*.

Integrating Eq. (29-1) leads to

$$N = N_0 \exp(-\lambda t) \tag{29-2}$$

where N_0 is the number of unstable atoms present at time zero. At some time, $t_{1/2}$, N will equal $N_0/2$. Then, from Eq. (29-2),

$$\frac{N_0}{2} = N_0 \exp(-\lambda t_{1/2})$$

$$-\ln 2 = -\lambda t_{1/2}$$

$$t_{1/2} = \frac{\ln 2}{\lambda} = \frac{0.693}{\lambda}$$

This time, $t_{1/2}$, is called the *half-life* (or *half-time*) of the nuclide. To be useful in biophysical applications, nuclides usually must have half-lives on the order of hours or longer.

As mentioned at the beginning of this section, a nucleus may emit various particles upon decay. All pure γ-ray emitters produce these rays with fixed energies. For example, technetium in one excited state, $^{99m}_{43}\text{Tc}$, will decay to its ground state, $^{99}_{43}\text{Tc}$, with the emission of one 140-keV γ ray. Most γ emitters produce several γ rays; the energies of these rays and the proportion of them with each energy is always constant.

An electron can also be emitted by suitable radioactive nuclides. However, in this case, even when there are no associated γ emissions, a continuous distribution of energies is observed. A neutral particle called the *electronic neutrino*, $\bar{\nu}_e$, was postulated to account for this distribution of energies.

TABLE 29-1

SELECTED NUCLIDES

Isotope	Half-life	Particle(s) emitted
$^{3}_{1}\text{H}$	12.26 years	β
$^{14}_{6}\text{C}$	5730 years	β
$^{32}_{15}\text{P}$	14.3 days	β
$^{51}_{24}\text{Cr}$	27.8 days	γ
$^{99m}_{43}\text{Tc}$	6 hours	γ
$^{125}_{53}\text{I}$	60 days	γ
$^{131}_{53}\text{I}$	8.0 days	β, γ
$^{133}_{54}\text{Xe}$	5.3 days	β, γ

Although first predicted from theory, it has since been observed experimentally. For example, the most common radioactive isotope of carbon has been shown to decay as

$$^{14}_{6}C \longrightarrow \ ^{14}_{7}N + e^- + \bar{\nu}_e$$

A few representative radioactive isotopes useful in biophysical studies are presented in Table 29-1. Tritium, ^{14}C, and ^{32}P are especially important in biophysical studies at the subcellular level. Much of the work discussed in Part D, Molecular Biology, is based on these radiotracers. The γ emitters, as discussed in the next sections, are more frequently used in whole-organ or whole-organism studies.

C. ACTIVITY

The sample decay rate in disintegrations per unit time, dN/dt, which is defined in Eq. (29-1), is not usually measured. Instead, instruments report C, the counts per unit time, where

$$C = -q \cdot \frac{dN}{dt}$$

and q is a measure of the efficiency of the counting process.

While q can approach 1.0 for specially prepared samples containing low levels of strong γ emitters, it may be less than 0.5 for β emitters. There is a concentration dependence as well, owing to the slight refractory period (dead time) present in most counters. This period is the lag time between the recording of one disintegration and when the detector can next register a decay. Thus, at very high rates of decay, a significant number of disintegrations may not be recorded, and q appears lower. It may also be lowered by impurities in the sample (called *quenchers*), which absorb radiation.

At very low levels of radioactivity, background emissions such as cosmic rays, $^{40}_{19}K$ in glass, and so on, may affect sample measurements. Thus, control samples are prepared much like the experimental samples but without added radioactivity. Counts detected in these controls are subtracted from the experimental counts to yield C', corrected counts per unit time.

If the volume of the sample is V, then radioactivity concentrations can be expressed as C/V (or C'/V). This may be called the specific activity of the sample. It is also possible to divide by a number of different constants to arrive at more easily manipulated quantities. For example, the sample activity at any given time might be expected to be proportional to the total activity (dose) injected. Accordingly, the sample activity may be reexpressed as fractions of the dose per unit volume. In this form the results should be independent of dose.

Specific activity, when used in experiments to follow biochemical reactants, is defined as the activity per unit mass of the compound being traced.

Alternatively, one may replace activity by fractions of the dose. In some systems the use of fractions of the dose for activity or specific activity fails to produce dose-independent data. Such systems are called nonlinear. By and large, tracer experiments have proved to be most useful for linear systems.

29.3   Other Tracers

In addition to radioactivity, many other labels have been used in biophysics. Most, however, require special chemical or physical analyses to be performed on an isolated sample of the biological material. They are not suitable for whole-organ or whole-body work. These labels can be grouped into two classes: rare stable isotopes, which are detected by their mass differences from the more common isotopes; and different atoms or moieties, which are detected by chemical analyses or, occasionally, by physical techniques.

A. STABLE ISOTOPES

A *stable isotope*, to be useful in tracer work, must first be present in small enough quantities in the naturally occurring element that a small enrichment will greatly increase its relative percentage. Second, it must differ enough in relative mass from the common isotope to be readily separated from it. This separation must occur at least twice, once to prepare the labeled tracer compounds, and again when these compounds or their metabolites are detected in the specimen.

The second criterion implies that the most useful stable isotope labels would be isotopes of the lighter elements. For example, it is quite easy to separate deuterium, ^2_1H, from protium, ^1_1H, because the difference in their masses is 100 percent. However, separating the nitrogen isotope ^{15}N from ^{14}N is more difficult, since there is only a 7 percent mass difference between them.

A third criterion for biological tracers is that the atom in question be present in biological materials. A few representative stable isotopes used as biological tracers, with their relative abundance and percent mass difference from the most common isotope, are shown in Table 29-2.

Deuterium is mentioned in Chapter 26, since it is frequently used in proton magnetic resonance work. The carbon isotope ^{13}C was once important in investigations of carbon metabolism. Although still used in magnetic resonance spectroscopy, it has been replaced in more routine labeling by the more versatile radioactive isotope, ^{14}C.

Uses of both ^{15}N and ^{18}O are presented in other chapters of this text. The former is used to examine DNA replication as discussed in Chapter 16. In that chapter the use of density gradient centrifugation was described as the

TABLE 29-2
REPRESENTATIVE STABLE ISOTOPES

Isotope	Relative abundance (percent)	Mass difference (percent)
^2_1H	0.015	100
$^{13}_6C$	1.11	8.3
$^{15}_7N$	0.37	7.1
$^{18}_8O$	0.204	12.5

tool to separate ^{15}N-DNA from the lighter ^{14}N-DNA. Early work in photo-synthesis was based on ^{18}O research (see Chapter 20). To separate ^{18}O from ^{16}O, those workers used the technique of mass spectrometry.

The mass spectrometer, shown in Fig. 29.1, requires that any sample to

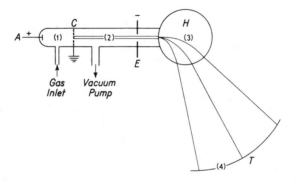

Figure 29.1 Essentials of a mass spectrometer: (1) ionizing chamber, (2) accelerating region, (3) velocity spectrometer, (4) detector. *A*, anode; *C*, cathode with holes to permit passage of "canal rays"; *E*, electrode at high negative potential; *H*, magnetic field; *T*, target.

be analyzed be a positive ion in a gaseous form. These positive ions are accelerated by a negative voltage toward a current detector. Because of conservation of energy and the fact that a moving, charged particle curves in a magnetic field, tracers with different charge/mass ratios will appear at different points on the detector target (*T*).

B. CHEMICAL GROUPS

Dyes and other molecules that can be detected by their absorption or emission of electromagnetic radiation are discussed in Chapters 26 and 27.

They are used as labels, even though their larger size weakens the assumption that the labeled and unlabeled biological compounds are treated identically by the system under study. To compensate for this, they are often detectable in tissue samples or solutions under more physiological conditions than those found in a mass spectrometer. Spin labels, discussed in Chapter 26, are an example of this class of tracers.

Chemical moieties are occasionally used as labels when it is known that they do definitely hinder normal metabolism. In this case, the amount and types of accumulated labeled compounds can help determine the course, regulation, and control of metabolism. Moieties used in this fashion include the halides F^-, Br^-, and I^-, and various aromatic groups, such as phenyl.

29.4 Radioactive Imaging

Radioactivity can be detected at a distance. If one assumes that the labeled molecule is handled exactly as its unlabeled counterparts, it can signal their concentration and position with a high degree of accuracy while it remains in the solution, tissue, organ, or organism of interest. Several techniques are used to capture this information as images.

A. AUTORADIOGRAPHY

If a specimen is thin and flat, an X-ray film can be placed in close contact with it. When this film is developed after a suitable period of time, it will produce a picture of the radioactivity. This technique is called *autoradiography*.

For example, ^{14}C-labeled biomolecules can be separated by paper chromatography (see Chapter 20) and the paper then used to produce an autoradiogram. This can nondestructively localize the compounds of interest. Tissue slices can similarly be autoradiographed. Some early work on the mechanism of DNA synthesis, discussed in Chapter 16, was based on autoradiograms of thin slices of dividing onion root tip cells supplied with radioactive thymidine.

B. GAMMA CAMERA

In the middle 1950's, a stationary gamma camera was developed by Anger of the Donner Laboratory. In this device, a hexagonal array of photomultiplier tubes records the individual scintillations produced by gamma rays striking a 25- to 35-cm-diameter NaI(Tl) crystal. Those phototubes closest to the actual point at which the scintillation occurs have a greater response than those farther away. Positional information is obtained by connecting each tube's output to a distinct electronic element. Figure 29.2

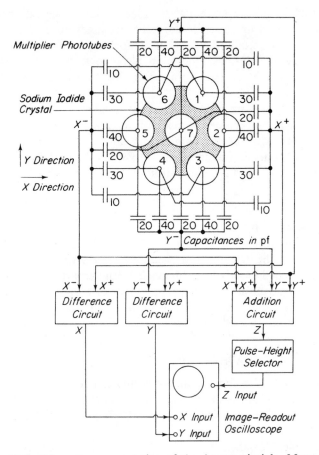

Figure 29.2 Schematic representation of the Anger principle. More recent cameras have used as many as 91 photomultipliers arranged in the hexagonal pattern shown. After H. O. Anger, *Instrumentation in Nuclear Medicine*, Vol. 1, G. J. Hine, ed. (New York: Academic Press, Inc., 1967).

shows typical capacitor networks used by Anger. Two such networks are needed in each of two dimensions.

These spatial data may then be used to cause a flash of light at the corresponding point on the face of a cathode ray tube. An example of a bone image photographed from such a crt screen is shown in Fig. 29.3. Positional resolution is usually 5 mm at 140-keV γ-ray energy.

A gamma camera remains fixed during data collection. Thus, images can be obtained serially in time. It is this capability that has made the camera the instrument of choice for most nuclear tracer dynamic imaging procedures. The only drawback of the Anger camera is its necessarily thin crystal,

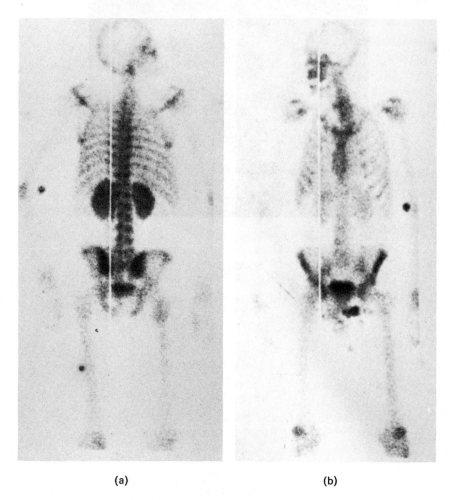

(a)   (b)

Figure 29.3   A ^{99m}Tc-pyrophosphate anterior projection bone study done with
an Anger camera. Some 15 mCi of Tc were injected intravenously for this
normal study. A large object such as the skeleton may be imaged using a
40-cm camera by moving the patient within its field of view and recording the
decay events on a long string of film. (a) Posterior view. (b) Anterior view.
Courtesy of Picker Corp.

generally 12 mm or less in thickness. The resultant lower sensitivity of these
cameras at γ-ray energies above 200 keV causes an appreciable experimental
difficulty when using high-energy emitters.

Until the middle 1970's, the Anger camera had relatively poor uniformity
of response to a planar source of activity. Digital processing of the photomultiplier outputs has greatly reduced this problem. For this purpose a

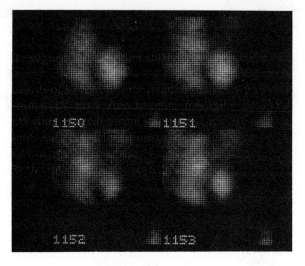

(a)

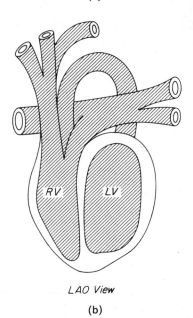

LAO View

(b)

Figure 29.4 (a) Digital images (frames) obtained from a human cardiac study using ^{99m}Tc-labeled human serum albumin as the radiotracer. Data acquisition is accomplished with computer control of the gamma camera. Frames are generated every 1/28th of the R-R wave ekg interval. The four representative images shown are:

1150	diastole (2nd frame)
1151	6th frame
1152	systole (12th frame)
1153	18th frame

By replaying the 28 frames in the appropriate sequence, the cardiologist can view a beating heart as a digital motion picture. Photos courtesy of Medical Data Systems Corporation. (b) A drawing of the human heart seen in the same projection as used above. RV and LV refer to the right and left ventricles.

microcomputer may be located inside the camera detector assembly. An additional computer is often used to digitize the camera output for more complicated processing and routine data storage. This computer allows playback of the images in real or abbreviated time and can be employed to display selected regions. A dynamic heart study using ekg gating of the computer is given in Fig. 29.4.

C. THREE-DIMENSIONAL IMAGING

The gamma camera generates two-dimensional projections of three-dimensional distributions of activity. Analog and digital methods have been devised to permit more selective detection of γ rays. A technique similar to holography is sometimes used in nuclear medicine for three-dimensional imaging. Here, one places a special collimator between the γ-ray source and the gamma camera and records a shadowgram of the aperture. The original device of this type was the Fresnel zone plate, a drawing of which is shown in Fig. 29.5. The radius of the central opening is determined by the detector resolution and the object–collimator, and collimator–detector distances. The

Figure 29.5 Fresnel zone plate collimator. This device permits sectional (tomographic) γ-ray images of relatively small emitters, such as thyroid glands. Other encoding collimators have been designed as well. After R. Collier, C. Burckhardt, and L. Lin, *Optical Holography* (New York: Academic Press, Inc., 1971).

plate's bright areas represent interference maxima, dark areas, interference minima.

In the γ-ray imaging experiments, the bright areas are made of a relatively γ-transparent metal such as aluminum, while lead is used to form the dark bands. The strategy is to allow a small-volume γ emitter such as a labeled thyroid gland to pass radiation onto the spatially sensitive detector, which then records the resultant intensity shadowgram. Upon subsequent illumination by laser light, a three-dimensional image of the source is produced and can be photographed. This technique is not suited to large-organ imaging, since the zone plate technique presumes a point source.

A second nuclear medical coded aperture is an annulus. The recorded shadowgram is not played back in analog fashion via laser illumination. Instead, a digital computer, coupled to the recording gamma camera, is used to generate the image. Such techniques record information from all depths in the emitting source simultaneously and can be of clinical significance in both static and dynamic tracer studies.

29.5 Applications

Tracers are used in a variety of applications in biology and medicine. This chapter can cover only a few selected examples; the reader may consult the references for further information. The tracer principles illustrated by these examples are used in many other applications.

To return to the animal example given in the introduction, assume that the animals are labeled and then released into a stable, closed population, e.g., one confined to an island. Subsequently, some animals are captured. The fraction that are labeled provides information on the size of the population and the rate at which individuals circulate through it. Similarly, the following examples show how molecular tracers can help determine amount and/or flow. Although radioactivity is used in these applications, they can be carried out with other labels as well.

A. RADIOIMMUNOASSAY

The concentration of any antigen, such as the hormone insulin, can be determined through a procedure involving the labeling strategy described above and called, in this context, *competitive binding*. A known amount of labeled antigen is added to the sample. Then an excess of antibody is reacted with the mixture of labeled and unknown material. The reactions are

$$Ag + Ab \;\rightleftharpoons\; Ag \cdot Ab$$
$$Ag^* + Ab \;\rightleftharpoons\; Ag^* \cdot Ab$$

where Ag is the antigen, Ab is the antibody, and an asterisk represents radioactivity. The quotient $[Ag^* \cdot Ab]/[Ag^*]$ is the ratio of the bound tracer to the free tracer (B/F). This ratio depends on the amount of stable antigen present, B/F decreasing as $[Ag]$ increases.

In experimental use a curve is determined using antigen standards; these are illustrated for four hormones in Fig. 29.6. Simultaneously with the experimental measurements of the standard curve, values of B/F are determined for several unknown samples. The corresponding concentrations of the antigen in the unknown samples is determined from the standard curve.

An antibody must be available for each hormone or other antigen that is to be assayed. The sensitivity of the method is directly attributable to the very specific nature of binding. Other strongly specific binding systems, such as enzyme–substrate pairs (see Chapter 18), can also be studied using these competitive binding techniques.

More than 100 hormones and other antigens can be quantitatively evaluated by competitive binding. This technique is widely used in clinical chemistry and endocrinology. The radioassay technique is more sensitive by several orders of magnitude than the traditional biological response method of determining hormone levels. Chemical assays might be superior but do not exist for the hormones assayed by radioimmune techniques.

Either the antibody or antigen could be labeled in this technique. The use of labeled antibody is sometimes preferred because it is easier to produce. Radiolabels used in radioimmunoassays include the γ emitters ^{125}I and ^{131}I as well as the electron emitters ^{3}H, ^{14}C, and ^{32}P. In the latter case, the samples are combined with an organic solvent and organic phosphor so as to produce scintillations that can be measured in scintillation counters.

In order to count the sample, the analyst must initially separate the bound from the free fractions. This can be done by taking advantage of the different charges, masses, or charge-to-mass ratios of Ag^* and $Ag^* \cdot Ab$. An additional technique involves the development of a second antibody, Ab', which reacts to $Ag^* \cdot Ab$ but not to Ag^*. Precipitation of the $Ag^* \cdot Ab \cdot Ab'$ complex then effectively separates the bound from the free fractions.

B. BLOOD VOLUME AND FLOW

A second application of the labeling concept is the determination of blood volume in and flow through a given organ or patient. The ideal flow evaluation agent would be injected by catheter into the primary artery supplying the system of interest and be limited to the vascular space. The second criterion has led to the use of labeled albumin molecules in some of these experiments. Both ^{131}I and ^{99m}Tc are used as radiolabels for human albumin, although technetium is preferred since it emits γ rays with energies more suitable for gamma camera work.

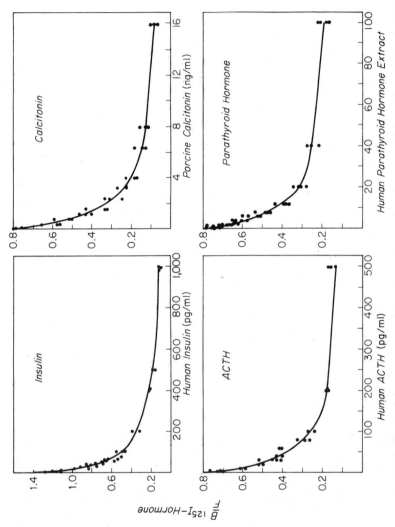

Figure 29.6 Standard curves for radioimmunoassay of four hormones. The ordinate gives B/F ratios and the abscissa shows hormonal concentrations in appropriate units (e.g., 10^{-12} g/mℓ). Hyperbolic forms for these data are anticipated from theoretical arguments. After G. Palmieri, R. Yalow, and S. Berson, *Hormones, Metab. Res.* **3**:303 (1971).

The simplest example consists of injecting a dose D of radiolabeled albumin into the bloodstream. After mixing is complete, the tracer specific activity A, in fractions of D per unit volume of plasma, is determined. The reciprocal of A is good measure of the total body plasma volume. If, in addition, the *hematocrit H* (fraction of blood volume consisting of formed cells) is known, the total blood volume V can be computed by

$$V = \frac{1}{A(1 - H)}$$

Unfortunately, this value is not too precise, since the hematocrit varies significantly from one part of the circulation to another. Thus, additional measurements are needed to determine the total cell volume. For example, some red blood cells may be withdrawn from the subject, labeled with radioactive chromium ^{51}Cr, and reinjected. The total red cell volumes in the body can be estimated in this fashion.

Blood flow studies are done in many major organs, such as the renal, cerebral, and myocardial systems. The tracer ^{133}Xe is of particular importance in these studies, because it is a gas. It thus passes out of the vascular system, via the lungs, after a single transit of the circulation. This greatly simplifies the mathematical analysis.

Cerebral blood flow studies provide a good example of this technique. A mathematical model is needed for the blood flow into and out of the brain in order to interpret the tracer data. One mathematical model sometimes used is a set of n parallel compartments, with the fractional blood flow w_i into the ith compartment. Each compartment has a characteristic probability of loss of material per unit time, k_i. Under these conditions it can be shown that the activity of the tracer in the brain at time t, $A_B(t)$, should be given by

$$A_B(t) = \sum_{i=1}^{n} \left\{ w_i k_i \int_0^t \exp\left[-k_i(t - \tau)\right] A_A(\tau)\, d\tau \right\} \qquad (29\text{-}3)$$

where τ is the time at which the activity in the arteries serving the brain, $A_A(\tau)$, is measured.

The subject can inhale the xenon or it may be injected intravenously or into the carotoid artery. Activity A_B is measured over time by a fixed probe or gamma camera. Activity A_A is monitored by an additional probe focused on the carotid artery or sampling the air flowing in and out of the lungs. Finally, various values for w_i and k_i are tried until the best fit of the model to the data is obtained. Because of the complexity of Eq. (29-3), computer programs are needed to estimate optimum values of these parameters.

A very short experiment may be modeled by only two different compartments (an n value of 2), while longer experiments may need n values of 3 or 4. The final values for w_i and k_i are used as measures of the blood flow to selected regions of the brain. The diagnostic utility of the values for w, k, and n is under study.

29.6 Summary

Tracer methods are used at the level of organelles, cells, tissues, organs, and organisms. They are used to determine amounts, rates, and distributions. The tracer compound may be labeled with radioactive isotopes, stable isotopes, and chemical moieties. Each is detected differently.

Radioisotopes can be detected at a distance by the particles they emit upon decay. This allows them to provide information on organelle, cell, or organ size and shape. Stable nuclei require more complicated processing to be detected. This usually involves a mass spectrometer. Finally, chemical moieties may be detected by their absorption or emission of electromagnetic radiation, or through more elaborate chemical analyses.

Specific applications of tracers include radioactive imaging, immunoassays, and blood volume and flow determinations.

REFERENCES

Good introductory textbooks are:

1. WANG, C. H., D. L. WILLIS, AND W. D. LOVELAND, *Radiotracer Methodology in the Biological, Environmental and Physical Sciences* (Englewood Cliffs, N.J.: Prentice-Hall, Inc., 1975), 480 pages.
2. WELCH, T. J. C., E. J. POTCHEN, AND M. J. WELCH, *Fundamentals of the Tracer Method* (Philadelphia: W. B. Saunders Company, 1972), 187 pages.

Clinical reference:

3. BLAHD, W. H., ed., *Nuclear Medicine*, 2nd ed. (New York: McGraw-Hill Book Company, 1971), 858 pages.

Instrumentation references:

4. ANGER, H. O., "Scintillation Camera," *Rev. Sci. Inst.* **29**:27–33 (1958).
5. (a) BARRETT, H. H., "Fresnel Zone Plate Imaging in Nuclear Medicine," *J. Nucl. Med.* **13**:382–385 (1972).
   (b) ROGERS, W. L., K. S. HAN, L. W. JONES, AND W. H. BEIERWALTES, "Application of a Fresnel Zone Plate to Gamma-Ray Imaging," *J. Nucl. Med.* **13**:612–615 (1972).

CHAPTER 30

Biomedical Computation

30.1 Background

For several centuries one of the aesthetically pleasing aspects of physics was its dependence on mathematics, which led to closed solutions having little need for more advanced numerical calculations. However, the practical aspects of physics and of engineering in general gave rise to numerous problems that could be solved only by complicated numerical calculations. These, in turn, proved so tedious, so prone to errors, and as a consequence so expensive that without automated computational techniques there was little possibility of realistic solutions.

Some classes of problems in physics and engineering lend themselves to solution in terms of frequently occurring functions such as square roots, sines, and cosines (or less frequently encountered ones, such as Hankel functions). One of the earliest uses of automated computation was the generation of tables of function values with greater precision than had been previously available. However, it soon became apparent that it was more convenient to generate the desired function value as part of a program than to print detailed tables of values.

In other areas of study, such as financial and social sciences, statistical tables became important as these studies advanced. The generation, analysis, storage, and later retrieval of such tables gave impetus to the design of automated machinery for data processing. The early punched card equipment, while permitting more complicated analyses than had been possible previously, also indicated the need for still more rapid and more automated computational power.

The availability of automated aids to computation and data processing has increased extremely rapidly since 1945. This growth has been marked by a continual decrease in the cost per computational step and per record processed. As this occurs, many problems that were not feasible previously suddenly become possible. Accordingly, many areas of modern physics and chemistry as well as engineering and the social sciences have grown at a pace often determined by current computer methodology.

The introduction of computers into biological and medical research tended to lag behind the earlier developments in the physical sciences and statistics. Within medical practice, except for financial functions such as billing, computers entered still more slowly. Among the first biomedical areas to make extensive use of automated computation were ones associated with biophysical science.

From one point of view, the actual contributions of automated computation are hard to point out and identify. Granted sufficient time and sufficient funds, most of the advances accomplished with the aid of electronic computers could have been realized without them. Nonetheless, computers have played a major role in many areas of biophysical science. These include the interpretation of the X-ray diffraction patterns of complex molecules discussed in Chapter 14, which was necessary to understand the structure of proteins and nucleic acids. These structures, in turn, were prerequisites for understanding the molecular bases of the transfer of genetic information and of the synthesis of proteins as discussed in Chapter 16. As a second example, knowledge of conduction of impulses by nerves has depended upon the ability to solve complicated equations such as the ones described in Chapter 5. Several other applications are reviewed in succeeding sections of this chapter.

There are, in addition, a number of computer applications in biophysics that would be impossible without electronic computational aids. Such applications are frequently referred to as on-line or real-time since the experimental instruments and sensors are directly connected to the computer system. A separate section of this chapter is devoted to on-line computation in biophysics.

The success of electronic computers within biophysics has depended to a large measure upon design of systems for their ease of use by humans with minimal additional training. In the preceding paragraphs it was noted that a major effect of electronic computation was to reduce the overall cost. The latter must include the costs associated with training as well as with personnel hours. Both the use of computer languages that are more natural for the investigator and of remote interactive terminals (described in Sec. 30.4) decrease the efficiency of the utilization of the computer time. Nonetheless, these two steps minimize the overall costs, since human effort tends to be quite expensive when compared with computer time. The convenience and flexibility of electronic computers has contributed to their widespread use in biophysics as well as in other subdisciplines within biology and medicine. As costs per computation have dropped, it has repeatedly become necessary to adapt computer systems to people, to achieve maximum benefits.

In progressing to more specific aspects of biomedical computing, it is helpful to define a few added terms. Electronic computer systems may be classified as digital, analog, and hybrid. Computers all operate in one sense or another by analogies. For historic reasons, those which depend on discrete analogies are called *digital computers*, while those dependent on continuous analogies are termed *analog computers*. *Hybrid computers* combine subsystems using discrete analogies with others using continuous analogies.

Digital computer systems have proved to be better adapted to humans and less expensive per computation than either analog or hybrid systems. Thus, they are more widely used in biophysics and are emphasized within this chapter. In digital computer systems, electronic configurations are used to represent numbers, special characters, strings of alphabetic and numerical symbols, or optical images. The concept that such computer systems only work with numbers or only carry out arithmetic is quite misleading. Actually, digital computers can equally well manipulate strings of arbitrary symbols, for example to parse sentences.

Digital computer systems may, as illustrated in Fig. 30.1, consist of a number of subsystems. Each of the boxes shown may function as an independent computer and may use different analogies for the numbers and symbols it manipulates. Within one subsystem the analogy may depend on a string of pulses, in another the direction of magnetization, and still another the modulation of a carrier frequency. The designer and the maintenance engineer must be aware of the actual physical analogies employed, but the user may think of the operation of the digital computer system in terms of the discrete symbols being represented. All that the user must accept is the basic idea presented in Chapter 22 that information can be quantified and encoded.

For many purposes, large centralized computer systems have proved to be most effective. This has been particularly true for problems involving large computational times, such as those which arise in applications of quantum mechanics and for projects which create and use large data bases, such as population-based and epidemiological studies. There exists another large class of applications prominent in biophysics in which smaller computers, called minicomputers and microcomputers have proved preferable. *On-line*

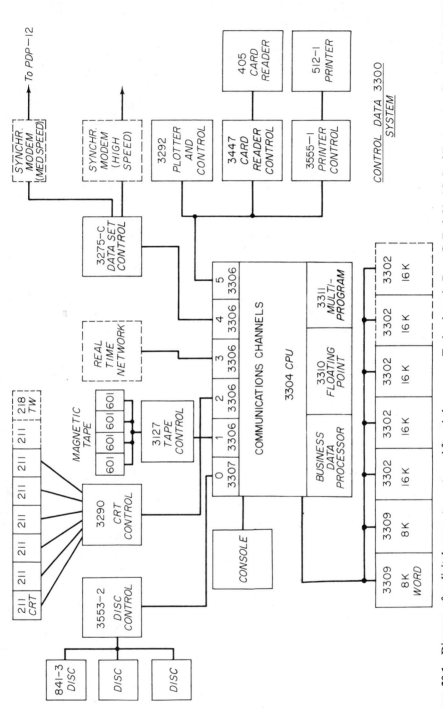

Figure 30.1 Diagram of a digital computer system. After Ackerman, E., and Gatewood, L.: Biomedical Engineering Aspects of Computer Technology, in Ray, C. D. (ed.): *Medical Engineering.* Copyright © 1974 by Year Book Medical Publishers, Inc., Chicago. Used by permission.

uses, where the timing and availability to the individual laboratory are most important, fall into this latter category. An extreme of this type of mini-computer usage is illustrated by systems custom built for a particular set of applications, such as the gamma camera in Chapter 29. Such computers are described in the reference by Williams.

The computer actually operates by carrying out, often in parallel, operations (or micro-instructions) on individual circuits that logically represent 1 bit. Such micro-instructions are grouped together into complete instructions which the user may request. The computer actually operates so fast that the user must first prepare a list of these instructions called a *program*, which is then read in as a group into the computer's memory. This program could be a sequence of bits. In actual practice, the cost of programming at such a level is excessive. Accordingly, there exist programming systems which save the user's time at the expense of the computer's time. Such systems, called *programming languages*, can exist at a number of different levels of abstraction. The ones closest to the machine level are called *assembly languages*. Others, more natural for the user but requiring greater amounts of computational time to translate to a machine-usable form, are called *higher-level languages*.

A major limitation of minicomputer systems is the cost of writing programs, which may exceed manyfold the equipment costs. Even on mini-computer systems which can execute higher-level languages, the programs that make such languages useful, called *compilers* and *interpreters*, are usually less efficient than those on a larger system. Thus, there is more of a tendency to do assembly language programming for minicomputers.

One of the most widely used higher-level languages within biophysical science is called BASIC. An example of a program in this language is shown in Fig. 30.2. Other popular higher-level languages for scientific applications include FORTRAN and MUMPS. Since it is not the intention to teach programming, the focus of the remainder of this chapter is directed to biomedical applications rather than computer hardware or programs.

30.2 Data Processing

The term *data processing* is used here in its most general sense, namely the manipulation of data from a wide variety of sources. For a very simple example, one might measure temperature in °F and then convert (or transgenerate) this to °C. In this case the computer would be given a set of values and repeat the same calculation on each member of the set. The temperatures might be coupled with corresponding times of measurement; the computer could also be programmed to select coordinate axes and scales appropriate for the data and then plot this information on a graphics terminal. In such a case the computer acts just as a human being would except that the far greater

```
LIST
  1  REM
  2  REM FIND SUM, AVERAGE, SUM SQUARED, STANDARD
  4  REM DEVIATION, MAX, AND MIN OF N NUMBERS
  5  REM BETWEEN -1000 AND +1000
  8  REM
 10  READ N,X1,X2,S,S2
 15  FOR I = 1 TO N
 20  INPUT X
 25  LET S = S+ X
 30  LET S2 = S2 + X↑2
 35  REM X1 WILL BE MIN, X2 MAX
 40  IF X<=X2 GOTO 50
 45  LET X2 = X
 50  IF X>=X1 GOTO 60
 55  LET X1 = X
 60  NEXT I
 65  PRINT "SUM = ",S," SUM SQUARED =",S2
 67  PRINT "AVERAGE = ", S/N
 70  PRINT "SIGMA =",SQR((N*S2-S↑2)/(N*(N-1))), "N=",N
 75  PRINT "MAX =",X2," MIN=",X1
 76  REM
 77  REM FIRST DATA IS N
 80  DATA 25
 81  DATA 1000, -1000,0,0
100  END
```

Figure 30.2 Short program written in a higher-level language, BASIC. After Ackerman, E., and Gatewood, L.: Biomedical Engineering Aspects of Computer Technology, in Ray, C. D. (ed.): *Medical Engineering.* Copyright © 1974 by Year Book Medical Publishers, Inc., Chicago. Used by permission.

speed of the computer allows processing of larger sets of data. The computer also makes graphic output easier to modify and redisplay than would be the case with a manual graph.

In other problems it is desired to transgenerate the data from a linear to a nonlinear scale. An example is the conversion from drug dosage to the logarithm of the dosage. Were this to be done without automated assistance, one could look up values in a table of logarithms. For a computer it is faster, easier, and less demanding of its memory to compute each logarithm numerically. Most computer compilers recognize functions like logarithms and automatically select the appropriate programs from a common store called the *system library.* Other more specialized functions may be written as special programs and kept in a library accessible only by the individual user.

A related type of data processing involves biostatistical analyses. Most empirical biophysical research produces data that are subjected to some type of biostatistical processing. This can usually be expressed in an algebraic form, readily translatable into a number of higher-level computer languages, such as BASIC, FORTRAN, or APL. Many of the biostatistical operations employed within biophysics are simple, for example finding means and standard deviations. These can be readily computed using a desk or hand calculator. Such calculations can also be carried out using an electronic computer system. This allows storage of the data and computational results in a machine-readable form, which can be very helpful for data editing and for further calculations.

At some institutions a limited set of biostatistical analyses have been used repetitively. Generalized programs for such calculations have been written, carefully tested, and then placed into computer libraries. Such a collection of programs for biostatistics is known as a set of *packaged programs.* The use of packaged biostatistical routines can limit redundant programming efforts, allowing the research investigator to concentrate on the scientific import of the results rather than on the tedious details essential to the analysis of the data.

While packaged biostatistical programs have numerous advantages, there often are disadvantages such as the failure of the available routines to perform exactly the biostatistical maneuvers desired by the user. Accordingly, a major effort in many biomedical computing centers is involved with writing specialized biostatistical routines to process the available data in the fashions selected by the investigators. Such programs may be most efficient if they are disposable, that is, used for a limited period of time with minimal documentation and then discarded.

Biostatistical procedures form a large component of biomedical data processing; however, a larger amount involves other mathematical, numerical analyses. The time saving and the economic benefits possible in some biophysical experiments is quite impressive. For example, a whole day spent once a week by two persons in analyzing sets of tracer data concerned with membrane permeability can be replaced by a few seconds of computer time (worth a few dollars) using programs that may take one or two days to develop.

One type of mathematical data processing frequently employed in biophysical research consists of determining frequency and power spectra of continuous signals. The general process is discussed in Chapter 1. It is pointed out there that signals which are truly periodic or ones that are defined for only a finite interval can be represented by a Fourier series, whereas ones that are aperiodic and not restricted to a finite interval can be represented in the frequency domain by a continuous Fourier transform, which gives an amplitude (or power) per frequency interval.

When processing continuous data with a digital computer, it is necessary first to represent it electronically as a discrete set of values measured at specified times. Such data streams are called *time series.* In any real application, the time series must be limited to a finite interval. The frequency spectrum resulting from applying Fourier techniques to a real time series consists of a finite number of values at discrete frequencies. Since at each frequency there is an in-phase and an out-of-phase component, there are essentially half as many points in frequency space as in the time domain. The resulting spectrum is called a *discrete Fourier transform.* With suitable precautions it approximates the continuous Fourier transform and hence the modifier "discrete" is often dropped. A technique for calculating such discrete Fourier transforms rapidly is summarized in Appendix C.

Such Fourier transforms are important not only in processing data from biophysical experiments but also in assessing the performance of electronic equipment. An example of this is discussed further in Sec. 30.4. Fourier transforms have also played important roles in analyzing holographic representations and in the interpretation of X-ray diffraction patterns from complex molecules, including proteins and nucleic acids. Thus, the availability of high-speed, electronic, digital computers has made economically feasible the processing of biophysical data while helping to assure that such data processing was as free of errors as possible. Some of the studies intermix data processing and model simulation in such a fashion that it is hard to draw a distinctive line between these two types of biomedical computation. Simulation of deterministic mathematical models is discussed in the following.

30.3 Simulation

Computer-aided *simulation* of mathematical models occupies a major place in biophysical computation. The general ideas of simulation far predate the appearance of electronic computers. Physicists have sought mathematical relationships between observables which they built into specific laws and then more general theories. In applying the latter to a new situation, that is, in testing the theory, it is necessary to determine the quantitative form predicted from the theory. This process might be termed *model simulation*.

Within biophysics, simulation of models has played a key role in developing new ideas and testing new hypotheses. Many of these models can be best developed and studied using computer simulation techniques. The models of the X-ray diffraction patterns of complex molecules referred to previously are characteristic of biophysical models in this regard. The models associated with the conduction of impulses by nerves discussed in Chapter 5 are other examples. Still other mathematical models have been simulated to predict the surface potentials associated with the electrical activity of the heart. These models are referenced in Chapter 10. None of the simulations above would have been practical to complete without the assistance of automated electronic equipment.

The mathematical models simulated may achieve one or more of several possible goals, such as data reduction, testing alternative hypotheses, diagnostic classification, and the design of new experiments. Simulations that confirm the investigator's beliefs are most satisfying, whereas ones that indicate the need for revised concepts produce the most information. An added use of simulation may be to select the parameters of the model in such a fashion as to give optimal agreement between simulated and empirical values. Such parameter estimation usually requires computer aid to select the values for the parameters, that is, to conform the model to the data.

At one time it was felt that analog computers were better suited for simulation studies than digital computers. Indeed, analog computers do have a number of advantages, such as continual display of simulation results and easy parameter and model changes. However, these advantages can be obtained with special higher-level digital computer languages for simulation. These have several advantages, including ease of use and of rerunning a simulation, the ability to deal with a wider variety of logical relationships, and the availability of alphanumeric labels for graphs and output data.

Simulations can be carried out using general-purpose higher-level languages, such as FORTRAN and BASIC. However, the simulation languages such as CSMP and MIMIC put fewer restrictions on the user. This means that such programs are more natural for the modeler, permitting a greater concentration of effort on the model. It also implies a greater amount of computer time to analyze the process and decide on the order in which to execute the various statements. The choice of a special simulation language rather than a more general language is somewhat arbitrary, depending on the size and nature of the model and on the time and training of the investigator.

Other more specialized simulation languages have been written to approach specific problems or classes of problems. An example of this type is the biochemical simulation language developed by Garfinkel and discussed briefly in Chapter 18. That language allows the user to enter data in the form of stoichiometric equations most natural for someone with chemical training. Other special systems, such as Rand's BIOMOD, can use as input a cathode ray tube screen on which the user may draw, with a light pen, boxes and arrows representing the model configuration. The user can enter rates, initial values, and so forth through a keyboard terminal. Thus, higher-level simulation languages coupled with advanced technology represent a still further step in adapting the computer system for the convenience of the user.

One further example of biological simulation, used by bioengineers as well as biophysicists, occurs in control theory. In its nonbiological, engineering form, control theory is concerned with the problem of designing feedback networks so that, given a quantitatively characterized system and an input to the system, the output can be specified to obey selected rules. In biological control, this problem is generally inverted. That is, inputs to the system, such as the light to the eye as a function of time, are known and the system response that is, the time course in the variation of the pupil diameter, is observed. Now, instead of wanting to design a feedback network (or a system), the problem is to find the characteristics of the biological system being studied.

Other models attempt to combine knowledge of the behavior of component parts of society to allow simulations of future changes. For this special computer simulation languages such as DYNAMO can be used.

Various parameters, rates, and relationships within the world model can be altered to seek solutions that would lead to a more stable world. The existence of convenient computer systems and of special, higher-level programming languages makes such studies possible.

30.4   On-Line Applications

As mentioned in the introduction, another major area of biophysical computation deals with situations called on-line, in that the experimental equipment is in some sense connected directly to the computer system. Such on-line connections are particularly important when the computer is expected to carry out computations as the experiment progresses. This is sometimes called *real time*, in that the analyses and data processing are completed while the data are being acquired and at a rate comparable to the rate of data acquisition.

A real-time application may involve data signals, in an analog form, from an array of electrodes placed on the surface of the body. These signals are led into the computer system, where they are digitized. This must be accomplished sufficiently rapidly that only small changes in each signal occur between digitizations. The necessary frequency of digitization can vary from a few times per microsecond to a few times per second.

Systems capable of digitizing a few times per second are also able to operate remote input-output terminals in an investigator's laboratory. Suitable terminals may consist of a teletypewriter, a cathode ray tube and keyboard, a printer, or a touch-sensitive device. When using such terminals, response times of the order of 1 sec are usually desired. For many purposes, it is most economically feasible for several such low-speed terminals to share the time of a computer. Such a system is referred to as *time-sharing* rather than real time. Both real-time and time-sharing systems are on-line, directly connected to the computer system. While both types of on-line systems have proved to be efficacious in biophysical science, the real-time laboratory system is in some sense unique and thus is emphasized in the remainder of this section.

In processing of laboratory data a number of considerations must be borne in mind. First is that the analog-to-digital converters (digitizers) work only within a limited range of voltages. Accordingly, some type of preamplifier is necessary. Second, if too few points are taken in a given time interval, the resulting digitized series of values will not be a satisfactory representation of the analog signals. In terms of the Fourier transform, unless one digitizes at a rate twice the highest frequency at which any significant amount of signal occurs, the higher-frequency components of the signal will be "aliased" onto the lower frequencies. To avoid this, the preamplifiers often include analog filters for processing.

Another important consideration is that if the computer system is to serve as an integral part of the experimental setup, it must be waiting and ready for the investigator's use. To achieve these goals along with real-time computation, it is often convenient to use a small dedicated computer within the laboratory. The PDP-12 pictured in Fig. 30.3 is such a computer system.

Figure 30.3   A PDP-12 computer.

These laboratory computers are called minicomputers, although this may be a misleading term. In general, minicomputer systems are characterized by low cost, small physical size, and resources limited to those necessary for the particular application. They are relatively expensive to program and to maintain. There also is the very human tendency to keep augmenting the minicomputer system so that one which was originally inexpensive increases severalfold in cost. Nonetheless, minicomputer systems serve successfully the need of the investigators for real-time, dedicated computation.

The results described in several chapters of this book have depended on minicomputer systems located in the laboratory. For example, the processing of electrocardiographic data as well as vectorcardiographic records for purposes of diagnostic classification, as discussed in Chapter 10, can be accomplished economically using a laboratory computer system. In general, any scheme that would make use of such ekg data to alter the individual's behavior or stimuli requires real-time computation.

Another important example discussed in Chapter 6 involves the transient response to stimuli. These evoked responses are superimposed on the normal electroencephalogram, and part of the computer's task is to extract the transient, which is often small compared to the background signals. The earliest attempts to perform this task were done manually. This proved quite unfeasible; a number of response averages based on analog computer principles were then designed and used. However, the entire field took a major step forward with the introduction of small, special-purpose digital computers such as the computer of average transients. Even those proved excessively restrictive in their capabilities by contrast with the general-purpose minicomputer systems that now dominate the laboratory computer field.

Small laboratory computers are also used in conjunction with laboratory instruments. Most laboratory instruments of any degree of complexity contain at least minimal analog processing components. The trend since the middle 1960's has been to incorporate small general-purpose digital computers directly on-line as an essential component of the laboratory instrument. For example, such computers have increasingly been used with spectrophotometers, mass spectrometers, and gamma cameras. On-line applications have played a critical role in the progress of biophysical science and may be expected to be even more ubiquitously employed in the future.

Computers more directly aid medical diagnosis in several areas. One of these involves the construction of cross-sectional images (tomograms) from X-ray data. In this technique, called *computerized axial tomography*, a beam of X rays is passed through the subject. A detector system on the other side measures the transmitted intensities in discrete positions.

The computer analysis is based in part on an algorithm that relates the measured intensities to the absorption within the subject. This results in several thousand simultaneous equations which are solved by the computer in real time, while the X-ray source is in motion. The computer output is a digital representation of relative attenuation parameters in a matrix format. Typically, these square matrices containing tens of thousands of numerical values are graphed before being presented to the physician. Such matrices correspond to an axial tomographic patient section approximately 1 cm thick. An example is shown in Fig. 30.4.

The theory of calculated axial sections was developed in the early 1960's, and the availability of inexpensive dedicated computers made the method practical by the middle 1970's. At present, patient motion must be reduced as much as possible. However, as instrumentation improves and scanning times decrease, image artifacts due to patient motion will become less important. Eventually, dynamic studies should be possible.

In the middle 1970's, a striking reduction in computer size was made

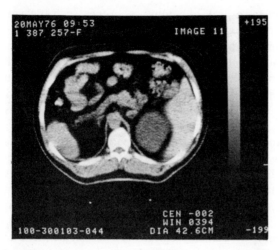

Figure 30.4 Midabdominal tomogram obtained with a whole-body X-ray transmission scanner. Anterior is at the top. Note the sections through both kidneys and spine. A renal cyst is seen on the right kidney. Courtesy of Ohio Nuclear, Inc.

possible by placing the entire central process unit on a silicon chip several centimeters long. The small size and low cost of these *microprocessors* have allowed their utilization in almost any piece of technical equipment. An example of their use in the Anger camera has been given in Chapter 29. The biophysical experimenter is now in the position where each apparatus can be given a microprocessor for data acquisition and computation. Digital memories of these various dedicated processors can be interrogated by a control microprocessor so as to efficiently generate the several experimental results.

30.5 Information Systems

The last area of biomedical computing included here—the acquisition, processing, storing, and retrieval in logically selected groups of significant information sets—has had only minimal impact on biophysical science. Nonetheless, this is a major part of biomedical computing, and accordingly is reviewed, albeit briefly, to give a better perspective on the applications of computer technology in biology and medicine. Perhaps the most important application from the point of view of the biophysicist is to bibliographic information systems. National systems are available in all large medical libraries, which make it possible to retrieve lists of publications by subjects, key words, or authors. In addition, large data bases are important for studies of clinically related aspects of biophysical science, such as electrocardiography and electroencephalography.

Large numbers of experiments, large numbers of animals, and large numbers of individuals are often employed in order to assure that results have statistical significance. Unfortunately, increasing these numbers can have exactly the opposite effect. An investigator dealing with five experiments can study each carefully to make certain that the numbers are recorded correctly, the proper technique is followed, and internal inconsistencies are eliminated. Increasing the number of experiments from 5 to 100 decreases the care that can be given to each experiment. Still another increase, to 1,000, virtually precludes a careful screening of the data by the investigator. Computer information systems can improve data quality by programs that check the data for limits, internal consistency, and so forth. Even within library information systems, similar problems arise. The existence of detailed, well-organized quality-assurance programs is essential before a computer information system can be of maximum vlaue.

Information systems involving large data bases have been used to develop and test various techniques for automated diagnostic classification. Such applications were among the primary motivating factors for the support of biomedical computation in the years shortly following 1960. The research accomplishments enabled by that support have from many points of view far exceeded the dreams of the early investigators. However, with a few notable exceptions, computers have not been used for diagnostic classification. A variety of factors have undoubtedly contributed to this, perhaps the biggest one being that given sufficient laboratory, physical, and historical data the human physician is an effective diagnostician. On the other hand, without these data, no computer system can generate the desired diagnostic classification with reasonable precision. Thus, computers have proved to be far more helpful in preprocessing selected subsets of the necessary data into a form that is easy for the diagnostician to use rather than completing the classification.

The most notable exception to the preceding negative views is the classification of abnormal electrocardiograms. Computer-based systems for such classification are in regular clinical use. Another exception is in computer-aided instruction programs. These allow students to study a subset of possible symptom signs and diseases. Certain students can learn diagnostic techniques faster by using these programs.

Numerous computer information systems are also employed to assist in patient care. One of the most successful types is used in the clinical laboratory. The functions of the chemistry, hematology, and microbiology laboratories are to take in small specimens of fluids, tissues, and so forth and to return information to the requesting physician or nurse. In order to continue to exist, clinical laboratories must also prepare timely and accurate billing information. No matter how precise the actual laboratory determinations, the utility of the laboratory can be destroyed by an unsatisfactory information system. Accordingly, automated clinical laboratory information systems, by preparing a wide variety of reports, can increase the benefits obtained from the laboratory while actually decreasing the cost per determination in the laboratory. A sample report from such a system is shown in Fig. 30.5. Other reports include quality-control, laboratory utilization, and diagnostic interpretations.

UNITS	95% INTERVAL OR NORMAL RANGE	TEST	SPEC	8:00A 1 MAY 1977	4:00A 2 MAY 1977	8:00A 2 MAY 1977	1:00A 3 MAY 1977	8:00A 3 MAY 1977	8:00A 4 MAY 1977
		BICARB/CL	B						
MEQ/L	26.0-29.0	HCO3		32. *		32. *		32. *	
MEQ/L	101.-112.	CL		101.		103.		101.	
		NA/K	B						
MEQ/L	138.-147.	SODIUM		141.		141.		140.	140.
MEQ/L	3.70-5.20	POTASS.		3.1 *		3.2 *		3.2 *	3.2 *
		CREAT/UREA N	B						
MG/DL	0.50-1.30	CREAT.		1.8 *		1.6 *		1.7 *	1.7 *
MG/DL	9.00-23.0	UREA N		38. *		29. *		28. *	32. *
		COLLECTION	U						
HR	–	DURATION					24.		
ML	–	VOLUME					3240.		
		CR CLEARANCE	U						
MG/DL	–	SERUM CR					1.6		
MG/DL	–	URINE CR					58.		
ML/MN	–	RAW CL.					82.		
ML/MN	110.-153.	STD.CL.					73.*		
		PROTEIN. QT.	U						
G/SP	0.00-0.15	PROTEIN					0.2 *		
		URINE SCREEN	U						
	–	SG			A				
					1.013				
	–	PH			6.0				
	–	PROT.SCN			TRACE				
	–	PROTEIN			TRACE				
	–	GLUC.SCN			NEG				
	–	KETONES			NEG				
	–	BILI			NEG				
	–	BLOOD			NEG				
	–	EHRLICHS			NEG				
		MICRO SCREEN	U						
	–	CONC			10:1				
	–	RBC			2+				
	–	PMN			NEG				
	–	CASTS			NEG				

A ASCORBIC ACID PRESENT

Figure 30.5 Sample report from a clinical laboratory information system. Computers in the clinical laboratory are used to generate a variety of different types of lists and reports, some for the technologist, others for the physician, the laboratory director, and the hospital. This sample report is a cumulative chemistry record. Note the asterisks, which indicate values outside predefined limits, and the natural language comment in the footnote.

At one time it was felt that computers in the clinical laboratory would be useful to reduce the computations done by the technologists. Thus, on-line instrumentation was heavily emphasized. In actual practice it was found that the benefits obtained from the information system outweighed the advantages of on-line instrumentation. Where the laboratory device contains a small, dedicated, special purpose computer as an integral part, this may advantageously be connected directly to the central laboratory information system computer. However, the primary justification for the latter remains the accurate and convenient storage of information and the inexpensive preparation of a diversity of reports.

30.6   Summary

Electronic computers are specialized instruments that have critically augmented progress in biophysical science. One classification separates computers according to the types of analogies used between the electronic states and the external world. Those which use continuous variables are called analog computers, whereas those depending on discrete states are called digital. During the 1950's it appeared that analog computers had more to contribute than digital ones to biophysical science; they did, indeed, play a prominent role in earlier biomedical computing. However, by the 1970's, digital computers dominated all scientific applications and are emphasized in this chapter.

The three major modes of computer utilization in biophysical science reviewed are data processing, model simulation, and on-line computation. All three modes are illustrated by examples selected from the remainder of the text. Well over half of the other chapters in this edition contain illustrations of the essential role of automated computation in biophysical science. This chapter also includes a very brief overview of information systems, which, although impacting biophysical science to only a limited extent, are major components of biomedical computing.

REFERENCES

1. STACY, R. W., AND B. D. WAXMAN, eds., *Computers in Biomedical Research* (New York: Academic Press, Inc.) Vol. I, 1965, 562 pages; Vol. II, 1965, 363 pages; Vol. III, 1969, 288 pages; Vol. IV, 1974, 325 pages.
2. RAY, C. D., ed., *Medical Engineering* (Chicago: Year Book Medical Publishers, Inc., 1974), 1256 pages.

3. (a) KEHL, T. H., C. B. MOSS, AND T. S. KECK, "Interactive Continuous System Simulation," *Computers Biomed. Res.* 7:71–82 (1974).

   (b) KEHL, T. H., P. MOREAU, A. HENKINS, M. TERRY, AND T. C. RUCH, "A Teaching Machine Approach to a Retrieval System for Special Libraries and Information Analysis Center," *Computers Biomed. Res.* 7:266–277 (1974).

4. COLLEN, M., ed., *Hospital Computer Systems* (New York: John Wiley & Sons, Inc., 1974), 768 pages.

5. HICKS, P., in N. Teetz, *Fundamentals of Clinical Chemistry* (Philadelphia: W. B. Saunders Company, 1976).

6. LINDBERG, D. A. B., Chairman, "Conference on the Computer as a Research Tool in the Life Sciences," *Fed. Proc.* 33:2311–2422 (1974).

7. HEINMETS, F., ed., *Concepts and Models of Biomathematics* (New York: Marcel Dekker, Inc., 1969).

8. WILLIAMS, L., AND M. LOKEN, "Considerations on the Choices and Utilization of Computers on Radiology and Nuclear Medicine," *CRC Critical Reviews in Clinical Radiology and Nuclear Medicine* 6:1–30 (1975).

DISCUSSION QUESTIONS—PART F

1. Show for a general wavefunction ψ, that the X and P_x operators do not commute, that is,

$$\langle XP_x \rangle \neq \langle P_x X \rangle$$

where $\langle V \rangle$ is defined by Eq. (26-1). This type of result forms the mathematical basis for the Heisenberg uncertainty principle. Find at least two other quantum mechanical operator pairs that do not commute.

2. Use the energy–time form of the uncertainty principle to derive the half-time of a typical excited electronic state with excitation energy $(\Delta E) \simeq 1$ eV. Find the actual half-life of one such state. Discuss why your calculated results might differ from the experimental data.

3. One assumes that the rate of a quantum mechanical electromagnetic transition is proportional to

$$\left| \int \psi_f^* O \psi_i d\tau \right|^2$$

where O is an operator, ψ_i is the initial wavefunction, and ψ_f is the final wave function. Let O be the electric (or magnetic) dipole operator and derive the selection rules cited in Chapters 26 and 27.

4. Review the literature on transmission-computerized tomography scanning. Discuss the various computer algorithms used to generate the transaxial image. In particular, can you find applications of spatial frequency analysis and of the fast Fourier transform in these analyses?

5. Show how radionuclides that emit positrons can be used to produce tomographic images of a patient. What are the limitations of this methodology?

6. Discuss neutron (or photon) activation as a technique for trace-element analysis. What sensitivity is expected? Can you quantitate the radiation dose to living animals subjected to such irradiations? (Chapters 11 and 17 may be helpful.)

7. Using the Planck blackbody radiation formula and thermodynamics equilibrium conditions, prove the results for the Einstein probabilities shown in Eq. (27-2).

8. Describe the various radiation detectors (e.g., NaI(Tl), Si(Li), and Ge(Li)). What are the advantages and disadvantages of each for the detection of γ and X-rays?

9. Because of the tragic early history of X-ray imaging, many clinicians are concerned with possible biological damage caused by ultrasonic irradiation. Review this field and provide appropriate safety guidelines to a physician using an ultrasonic transducer in clinical practice. Consider transducer pulse rate, emitted intensity, and frequency in these recommendations.

10. Holographic imaging and data storage are becoming important aspects of information processing in medicine. Consider one such system and describe its use of three-dimensional data. How are these data recorded, stored, and replayed?

11. If the physical half-life is given by T_p and the biological half-life is defined as T_b, show that the effective half-life is given by

$$T_f = \frac{T_p\, T_b}{T_p + T_b}$$

The physical and biological clearance rates $(0.693/T)$ are considered to be mutually independent for this argument.

12. Demonstrate that the Fresnel zone (Fig. 29.2) is the interference pattern between a spherical and plane wave. Show that this result implies that the zone plate only produces a pseudo-hologram for a point source of radioactivity.

13. Radioimmunoassay often involves the specificity of antigen–antibody reactions. Describe how a particular assay was developed. Define the sensitivity of that test.

14. Many pieces of biophysical apparatus now contain one or more microprocessors. Describe one such system; include in your description the hardware (interfacing) necessary as well as the specific algorithm employed. What improvements in data accuracy and/or precision have resulted from utilization of silicon chip processors?

15. The uptake of radiotracers by a physiological system is accomplished by a number of biochemical and biophysical mechanisms. List several of these mechanisms and show how they are used in medicine.

16. Why do resonance phenomena, as described for heme groups in Chapter 18, give rise to absorption bands in the visible and near-ultraviolet regions of the spectrum?

17. How were the very fast half-times for rhodopsin intermediates, shown in Fig. 19.5, determined?

APPENDIX A

Acoustics

The purely physical part of hearing is included within the field referred to as acoustics; it is discussed in this appendix in more detail than it is presented in Sec. 1.2. *Acoustics* is defined as the study of vibration and sound. *Sound* refers to the propagation of elastic disturbances in a continuous (three-dimensional) medium, whereas vibration often is restricted to elastic disturbances in simpler systems, such as springs or loudspeakers. In either case, what happens is that certain particles are displaced from their equilibrium positions, thereby developing potential energy. Later, these particles are restored to their equilibrium positions and exhibit an equivalent kinetic energy. These motions can be easily handled mathematically only if they are sufficiently small. Theories using the approximation of very small vibrations are called *infinitesimal amplitude acoustics*. They describe most of the properties of sound which are important in a study of hearing. However, finite amplitude effects are significant in the absorption of higher-intensity ultrasound by tissues. A compilation of acoustical terms is given in Table A-1.

Any vibration or elastic disturbance may occur only once, or the phenomenon may be repeated. If it is repetitive, one can distinguish a certain frequency,

T<small>ABLE</small> A-1

T<small>ERMS</small> U<small>SED IN</small> P<small>HYSICAL</small> C<small>HARACTERISTICS OF</small> S<small>OUND</small>

Quantity	*Symbol*	*Quantity*	*Symbol*
Displacement	$\vec{\xi}$	Sound pressure level[a]	L
Time	t	Density	ρ
Frequency[a]	ν	Absolute pressure[a]	P
Particle velocity	$\vec{v}$	Equilibrium pressure[a]	P_0
Local acceleration	$\vec{a}$	Bulk modulus	B
Wave velocity	c	Specific heat ratio	γ
Wavelength[a]	λ	Distance	x
Sound pressure[a]	p	Angle	ϕ
Intensity[a]	I	Laplacian operator	∇^2
Specific acoustic impedance[b]	Z	$\sqrt{-1}$	j
Characteristic impedance[b]	ρc	Amplitudes	A_1, A_2, C

Subjective Equivalents

Pitch or tone[a] $\sim$ Frequency
Loudness[a] $\sim$ Sound intensity level
Quality or timbre[a] $\sim$ Harmonic content

[a] Discussed in Sec. 1.2.
[b] Discussed in Chapter 13.

that is, the number of times per second that the particle has the same displacement and velocity. A very simple case arises if the motion of a given particle can be described by

$$\xi = A_1 \cos 2\pi\nu t + A_2 \sin 2\pi\nu t \qquad\qquad \text{(A-1)}$$

where ξ is the displacement, t is the time, ν is the frequency, and A_1 and A_2 are constants (either of which may be zero). This is referred to as a *simple harmonic motion* or *pure tone*. The resolution of a complicated motion into simple harmonic terms is illustrated in Fig. 1.1. This type of synthesis, known as *Fourier analysis*, can be applied to any time-dependent phenomena. Because a speech pattern or any other sound can be represented as a sum (or integral) of simple harmonic terms, most of the following discussion is restricted to single frequencies.

In discussing sound waves, it is easiest to start from the vector particle displacement $\vec{\xi}$, which represents the distance a particle is displaced from equilibrium. Because $\vec{\xi}$ is a function of time, its first and second derivatives, the particle velocity $\vec{v}$, and the local acceleration $\vec{a}$ will, in general, be different from zero. For many acoustic analyses, $\vec{v}$ is slightly easier to use than is $\vec{\xi}$.

Particularly, if $\vec{\xi}$ is simple harmonic, it is convenient to use the *complex notation*. In this procedure, $\vec{\xi}$ is represented by a complex number which is easier to manipulate than the real part. Only the latter represents the experimental value. To illustrate this, one may write the following one-dimensional example:

Real part	Complex notation

$$\xi = A_1 \cos 2\pi\nu t + A_2 \sin 2\pi\nu t \qquad\qquad \xi = Ce^{j2\pi\nu t} \qquad C = A_1 - jA_2$$

$$\text{(A-2)}$$

$$v = \frac{d\xi}{dt} \qquad\qquad\qquad\qquad v = \frac{d\xi}{dt} = 2\pi\nu j C e^{j2\pi\nu t} = 2\pi\nu j\xi$$

$$= 2\pi\nu(-A_1 \sin 2\pi\nu t + A_2 \cos 2\pi\nu t)$$

$$\text{(A-3)}$$

$$a = \frac{d^2\xi}{dt^2} \qquad\qquad\qquad\qquad a = \frac{d^2\xi}{dt^2} = -(2\pi\nu)^2 C e^{j2\pi\nu t}$$

$$= -(2\pi\nu)^2(A_1 \cos 2\pi\nu t + A_2 \sin 2\pi\nu t) \qquad = -(2\pi\nu)^2\xi$$

$$= -(2\pi\nu)^2\xi$$

$$\text{(A-4)}$$

No matter what type of object is vibrating, be it a piano wire, an organ pipe, or a part of the ear, the same relationships are valid.

In addition to a local particle velocity, the wave group speed c is often used in acoustics. When a displacement is transmitted, the rate at which a wave front moves through the medium is called the wave group speed. For non-dispersive media such as air, water, and most tissues, c is independent of frequency. For any nonviscous fluid,

$$c = \sqrt{B/\rho_0} \qquad\qquad \text{(A-5)}$$

where B is the adiabatic bulk modulus and ρ_0 is the average (or equilibrium) density. For gases, B is related to the average pressure P_0 by the equation

$$B = \gamma P_0 \quad \text{and hence} \quad c = \sqrt{\gamma P_0/\rho_0}$$

In this expression, γ is the ratio of the specific heat at constant pressure to the specific heat at constant volume. For ideal gases, the ratio P_0/ρ_0 is proportional to the absolute temperature. Hence, the wave speed c will increase as the square root of the Kelvin temperature.

It may be shown that the partial differential equation

$$\frac{\partial^2 v}{\partial t^2} = c^2 \nabla^2 v \qquad\qquad \text{(A-6)}$$

describes the motion of a nonviscous medium subject to infinitesimal displacements. The operator ∇^2 has the following meaning in Cartesian coordinates:

$$\nabla^2 f = \frac{\partial f^2}{\partial x^2} + \frac{\partial^2 f}{\partial y^2} + \frac{\partial^2 f}{\partial z^2}$$

Any phenomenon described by Eq. (A-6) is known as a *wave phenomenon*, and this type of equation is called a *wave equation*. For the mathematically initiated, this linear equation, with its few symbols, expresses the wide variety of physical properties, such as interference and diffraction, associated with wave motion.

For plane acoustic waves in a fluid, the particle velocity is parallel to the direction of the propagation of the wave. Such waves are called *longitudinal* (or *compressional* or *irrotational*). These properties are not expressed by Eq. (A-6).

Most acoustic experiments do not measure the particle velocity, but rather the sound (or acoustic) pressure p defined by the relationship

$$p = P - P_0$$

where P is the instantaneous total pressure and P_0 the average (or equilibrium) pressure. For a plane wave traveling in the positive x direction, one can show that

$$p = Ae^{j2\pi v(t - x/c)}$$

and

$$p = \rho c v$$

The sensation of pitch, it has been noted, is associated with frequency and the sensation of loudness with the square of the sound pressure amplitude. The quality of a musical note is recognized by the number and relative intensity of the harmonics present. Not only are the harmonics important, but in a few cases the relative phases are important. Qualitatively, one may think of the relative phase as the indication of the displacement and velocity at the time $t = 0$. Symbolically, p at a given place is represented by

$$p = Ae^{j\phi}e^{j2\pi vt} \quad [\text{or } p = A\cos(2\pi vt + \phi)]$$

where A is a real number and $2\pi vt + \phi$ is called the *phase angle*. If the acoustic pressure wave is made up of two frequencies v_1 and v_2, one may represent p at a specific place by the expression

$$p = A_1e^{j\phi_1}e^{j2\pi v_1 t} + A_2e^{j\phi_2}e^{j2\pi v_2 t}$$

The difference, $\phi_2 - \phi_1$, is called the *phase difference*. Under most conditions, the ear is insensitive to phase differences, but for clicks, drum beats, piano attacks, and so on, these phase differences between the various audible components are very important. Phase differences also play a central role in acoustic holography, discussed in Chapter 28.

To those familiar with electromagnetism, the analogies between acoustics and circuit theory are striking. Several of the many analogies that are possible are presented in Table A-2. These correspondences help the person trained in physics or engineering apply to acoustics the mathematical symbolism and proofs developed for electrical circuits and transmission lines.

TABLE A-2

ELECTROACOUSTIC ANALOGS

Acoustic	*Electric*
Pressure, p	Voltage
Particle velocity, v	Current
Intensity, I	Power dissipated
Characteristic impedance, ρc	Resistance

As an example of the use of this analogy, one may define a specific acoustic impedance Z,

$$Z = \frac{p}{v}$$

which corresponds to the electrical impedance Z_{el}, where

$$Z_{el} = \frac{\text{voltage}}{\text{current}}$$

Thus, for plane waves going in the $+x$ direction:

$$Z = \rho c$$

This value of Z is called the *characteristic impedance* of the medium, and is often indicated by z.

When an electrical transmission line is joined to another element having the same impedance, a maximum power transfer will occur. If the two have an impedance ratio much different from unity, relatively less power transfer takes place. Likewise, if two acoustic media with very different characteristic impedances are in contact, a plane wave will be primarily reflected at the interface. For example, air has a characteristic impedance

$$\rho c_{air} = 420 \text{ SI units}$$

whereas water and tissue have characteristic impedances

$$\rho c_{H_2O} \simeq \rho c_{tissue} \simeq 1.5 \times 10^6 \text{ SI units}$$

Thus, sound energy is transmited readily from water to tissue, but very little is transmitted from air to tissue. The difference in characteristic impedance means that only a very small portion of the power incident on the ear can be used in hearing. However, the sound pressure amplitude remains approximately the same in the tissues of the ear as in the incident air. Impedance ratios are also used in Chapters 13 and 28.

It is generally assumed in acoustics that if a generator sends out waves of a given frequency, then only these will be transmitted by the medium and received by the ear. This is a consequence of the wave equation (A-6), which applies not only to velocity (as shown) but also to pressure and to excess density. Actually, no real medium propagates sound in the fashion predicted by Eq. (A-6), but the deviations from it are often so small as to be unimportant. At high amplitudes, and in certain special cases even at low amplitudes, the shape of a propagated wave may change in a fashion that cannot be predicted from wave equation (A-6). An extreme example is the surface waves on the ocean. In this case, the wave speed, c, depends on the frequency. As a consequence, waves rise up to a maximum at some places and then seem to disappear; others actually double back and form breakers. Similarly, in the inner ear, incident waves pile up to give a maximum amplitude at a location which depends on their frequency.

The wave form is distorted by the production of harmonics in the transmission of waves in air at sound pressure levels in excess of about 120 dB in the audible range. These effects are referred to as *nonlinear*; they contradict the assumptions of infinitesimal amplitudes made in traditional acoustics. Similar production of harmonics can be observed in the ear. However, in most cases the harmonic distortion due to nonlinearities is not important in hearing. Nevertheless, it is quite possible that nonlinearities produce some of the important effects when tissue is subjected to sonic irradiation.

There are many other terms used to describe acoustics. It is our belief, however, that a familiarity with the words and symbols of Table A-1 is sufficient to understand most journal articles and more advanced texts dealing with the physical aspects of hearing and sonics.

REFERENCES

1. MORSE, P. M., *Vibration and Sound*, 2nd ed. (New York: McGraw-Hill Book Company, 1948), 468 pages.
2. BLITZ, J., *Fundamentals of Ultrasonics* (London: Butterworth & Co. Ltd., 1963), 214 pages.
3. LAMB, H., *The Dynamical Theory of Sound*, 2nd ed. (London: Edmund Arnold and Company, 1931), 307 pages.
4. RSCHEVKIN, S. N., *A Course of Lectures on the Theory of Sound*, translated from Russian by O. M. Blunn (New York: Macmillan Publishing Co., Inc., 1964), 464 pages.
5. SKUDRZYK, E., *The Foundations of Acoustics: Basic Mathematics and Basic Acoustics* (New York: Springer-Verlag, 1971), 790 pages.

APPENDIX **B**

Electrical Terminology

There are a number of electrical terms which are used in the discussion of nerve and muscle action. The more important ones are included in Table B-1. This appendix is devoted to a discussion of some of these quantities and their units.

The concept of electrical charge, q, may be defined mechanically in terms of Coulomb's law, giving the magnitude of the force of interaction between two charges separated by a distance r; that is,

$$F = \frac{q_1 q_2}{Kr^2} \qquad \text{(B-1)}$$

If K is given the appropriate value for SI units, $4\pi\epsilon_0$,[1] and F is in newtons and r in meters, the unit for the q's will be the coulomb. The force is found to be repulsive if the two charges have the same sign and attractive if the signs are opposite.

---

[1] The symbol ϵ_0 stands for the dielectric constant of free space; it is equal to 8.85×10^{-12} F/m.

TABLE B-1

ELECTRICAL TERMS AND UNITS

Term	Symbol	SI units[a]	Defining equations
Charge	Q, q	coulomb	(B-1)
Electron charge	e	1.6×10^{-19} coulomb	—
Potential difference	$V, \mathscr{E}$	volt	$\Delta V = \dfrac{dW}{dq}$
Field strength	E	volt/meter	$\vec{E} = \dfrac{d\vec{F}}{dq} = -\vec{\nabla}V$
Current	I	ampere	$I = \dfrac{dq}{dt}$
Current density	J	ampere/meter^2	$J = \dfrac{dI}{dA}$
Resistance	R	ohm	$R = \dfrac{V}{I} \;\; \left(R = \dfrac{dV}{dI}\right)$
Capacitance	C	farad	$C = \dfrac{Q}{V}$
Inductance	L	henry	$\mathscr{E} = -L\dfrac{dI}{dt}$
Impedance	Z	ohm	$Z = \dfrac{V}{I}$ (for single frequency)
Reactance	X	ohm	$Z = R + jX$
Power	P	watt	$P = \dfrac{dW}{dt} \quad P = RI^2$
Dielectric constant ratio	ϵ		$\epsilon = \dfrac{C}{C_0}$

Other symbols used above and in text

Force	$\vec{F}$	newton	—
Work	W	joule	$W = \int \vec{F} \cdot d\vec{x}$
Distance	$\vec{x}, \vec{r}$	meter	—
Area	A	meter^2	—
Time	t	second	—
	j		$j = \sqrt{-1}$
Vector operator del	$\vec{\nabla}$	meter^{-1}	$\vec{\nabla}V = \vec{i}\dfrac{\partial V}{\partial x} + \vec{j}\dfrac{\partial V}{\partial y} + \vec{k}\dfrac{\partial V}{\partial z}$
Unit vectors	$\vec{i}, \vec{j}, \vec{k}$	None	
Free space	Subscript $_0$		
Complex amplitude	Subscript $_0$		

[a] The SI (meter–kilogram–second–ampere) units are used in most of this text for electrical quantities.

Because there exist forces between charges, it will in general require work to bring a new charge into any region of space. The work, W, necessary to move a unit charge from the point r_1 to the point r_2 is called the *potential difference*, V. Symbolically, this is expressed as

$$V_2 - V_1 = \Delta V = \frac{dW}{dq} \qquad \text{(B-2)}$$

If a potential difference exists between two points, there will be a tendency for charge to flow. In many bioelectrical experiments, the physical quantity measured is a potential difference in volts.

In intermediate courses in electricity, the potential V is defined in terms of the electric field strength $\vec{E}$ by the equation

$$\vec{E} = -\vec{\nabla}V \qquad \text{(B-3)}$$

The electric field strength is the magnitude and direction of this force per unit charge at a given place in space. Thus, symbolically, $\vec{E}$ may also be defined by

$$\vec{E} = \frac{d\vec{F}}{dq}$$

The value of E determines, among other things, the point at which an insulator will "break down" passing a current to ground or the surrounding medium. This effect is used in spark plugs to cause a discharge when the field strength at their points becomes sufficiently great. In dry air, the critical field strength is about 3×10^6 V/m. If the spark plug is sufficiently worn away, the separation between the two points becomes too large, and accordingly, the field strength never attains this value.

If a direct (unidirectional) current flows between two points, one may compute the ratio R called the *resistance*, defined by

$$R = \frac{V}{I} \qquad \text{(B-4)}$$

Ohm's law states that R is independent of I. It is approximately true for many substances over large ranges of electrical current.

In an ohmic conductor, the resistance is given by $\rho l/A$, where ρ is the resistivity, l the length, and A the cross-sectional area. Thus, the material's resistance to flow is contained in the parameter ρ. Values of resistivity vary widely in nature. Insulators, for example, exhibit values of nearly 10^{15} ohm·m, while a metal will have a resistivity around 10^{-2} in the same units. The resistivity of superconductors approaches a zero value below a certain (low) temperature. By contrast, pure water at room temperature has a resistivity of only 0.5×10^4 ohm·m; it is a surprisingly good electrical conductor. The presence of electrolytes can reduce the value still further.

If other forms of energy are reversibly converted to electrical energy when a current flows, the generated potential difference is called an *electromotive force* (emf). It is often distinguished by the symbol $\mathscr{E}$. Electrochemical cells, electromechanical transducers such as motors and microphones, and junctions between dissimilar metals all give rise to an emf. When the emf causes the current to flow in an external circuit nonelectrical energy is converted to electrical energy. If the current is caused to flow in the opposite direction, electrical energy is removed by the emf. Some direct current circuits are illustrated in Fig. 5.1.

A term used frequently in discussing biological tissues is capacity. If two conductors, charged $+Q$ and $-Q$, respectively, are separated by an insulating medium, across which there is a potential difference V, the ratio Q/V is called the capacity, C. A current cannot flow from one of the conductors to the other through the capacitor. However, while the capacitor is being charged, positive charges flow to one conductor and from the other. In an alternating-current circuit, capacitors are continually charged, discharged, and then charged in the reverse direction, thereby effectively transmitting an alternating current. Because the charge on the capacitor is proportional to the voltage across the capacitor, the current through the capacitor must precede the voltage change across it. A sinusoidal current through a capacitor leads the voltage across it by a phase angle of 90°.

An element in which a sinusoidal current lags the voltage by a phase angle of 90° is called an *inductance*, L. Thus, if a coil is formed, it is found that an emf $\mathscr{E}$ is induced across it if the current through it changes. This emf is so directed that it opposes the current change (Lenz's law). The self-inductance L is defined by

$$\mathscr{E} = -L\frac{dI}{dt} \tag{B-5}$$

where t is time. An inductance does not alter a steady current but hinders the flow of an alternating current.

Many of the interesting bioelectric phenomena involve currents and potentials that change in time. As is mentioned in Chapter 1, any complicated function of time can be represented as a sum of terms at single frequencies, or at worst, as an integral of a frequency distribution function (see also Appendix C). Thus, if one can describe the behavior of any circuit element as a function of frequency, one can compute its response to a transient. Accordingly, it is useful to be familiar with the terminology applied to sinusoidal alternating currents.

Such an alternating current, just as an alternating acoustic pressure, may be described by an expression such as

$$I = A \cos (\omega t + \Phi) \tag{B-6}$$

where A and Φ are constants and ω is 2π times the frequency. It is often more convenient to treat the measured current as the real part of a complex current

$$I = I_0 \exp (j\omega t)$$

where the complex current amplitude I_0 is related to the parameters in Eq. (B-6) by

$$I_0 = A \exp (j\Phi) \tag{B-7}$$

Using a similar complex notation for the potential V, one may form the complex ratio

$$Z = \frac{V}{I} = \frac{V_0}{I_0} \tag{B-8}$$

which is called the electrical *impedance*. Equation (B-8) has meaning only for sinusoidal currents and potentials or for Fourier transforms of currents and potentials. The impedance can be represented in the form

$$Z = R + jX \tag{B-9}$$

where both R and X are real numbers. If the impedance Z is real (i.e., $X = 0$), the voltage and current are said to be *in phase*. In this case, at all times, the instantaneous ratio of V to I is the same constant. The real part of Z is called the *resistance*, R. On the other hand, if R is zero, but the reactance X is different from zero. V and I will be 90 degrees out of phase. An alternating circuit is illustrated in Fig. 5.2.

Only the resistance, R, contributes to the power, P, dissipated by an element in an alternating circuit. In a direct-current circuit, one may write

$$P = VI \tag{B-10}$$

The same formula may be used for an alternating-current circuit, provided that one uses the real instantaneous values of V and I and averages over a period. For a sinusoidal current,

$$P = \frac{|I_0|^2 R}{2} = \frac{|V_0|^2 R}{2|Z|^2}$$

where the vertical lines indicate the absolute values of the complex quantities. In the neuron, the power dissipated is so small compared to metabolic heat losses that it is in general unimportant.

Other electrical terms and symbols are discussed in Secs. 12.2 and 12.3. For additional references, see Chapters 5 and 12.

REFERENCES

1. DUFFIN, W. J., *Electricity and Magnetism*, 2nd ed. (New York: John Wiley & Sons, Inc., 1973), 423 pages.
2. KIP, A. F., *Fundamentals of Electricity and Magnetism*, 2nd ed. (New York: McGraw-Hill Book Company, 1969), 630 pages.
3. ROJANSKY, V., *Electromagnetic Fields and Waves* (Englewood Cliffs, N.J.: Prentice-Hall, Inc., 1971), 464 pages.
4. PANOFSKY, W. K. H., AND M. N. PHILLIPS, *Classical Electricity and Magnetism*, 2nd ed. (Reading, Mass.: Addison-Wesley Publishing Company, Inc., 1962) 494 pages.

APPENDIX C

The Fast Fourier Transform

C.1 Theory

A number of mathematical techniques can be used to transform functions to representations in different coordinate systems. Each type of transform is named after its originator; examples are the Fourier and Laplace transforms. All these transforms are used to simplify analyses, the independent variable being transformed into one with reciprocal dimensions. For example, a complicated acoustic signal may be more easily studied as its Fourier transform, which is a frequency distribution.

Each type of transform is most convenient for certain characteristic types of problems. However, the Fourier transform, which is introduced in Chapter 1, lends itself most readily to intuitive descriptions. Probably for this reason, it is the transform most widely used in the biophysical sciences.

Essentially, the Fourier technique transforms a function of time into a function of frequency called a *spectrum*. The same mathematical technique can also be used for other independent variables; for example, a function of distance can be Fourier-transformed into a function of wave number. Applications of Fourier transforms are mentioned in several chapters of the text. Among these examples are discussions of acoustics, electromagnetic spectra, and X-ray diffraction. An application to tracer kinetics is included in this appendix.

The Fourier transform $F(\omega)$ is defined for the corresponding function of time $f(t)$ by the equation

$$F(\omega) = \int_{-\infty}^{+\infty} f(t) \exp (j\omega t) \, dt \tag{C-1}$$

The inverse transformation is described by a similar equation:

$$f(t) = \frac{1}{2\pi} \int_{-\infty}^{+\infty} F(\omega) \exp (-j\omega t) \, d\omega \tag{C-2}$$

In modern experiments, particularly those which employ a digital computer, it is customary to measure only at discrete points or to convert a continuous signal into a series of discrete values. If there are N such values f_i spaced evenly in an interval of length T, it is necessary to use a related form called the *discrete Fourier transform*, which is defined by the equation set[1]

$$F_k = \sum_{i=0}^{N-1} f_i \cdot \exp \left(\frac{j2\pi ik}{N} \right) \qquad k = 0, 1, 2, \ldots, N - 1 \tag{C-3}$$

If, in addition, all f_i are real, all the F_k are not independent. Specifically, equation set (C-3) can be used to show that

$$F_{N-k} = F_k^* \tag{C-4}$$

where the asterisk is used to denote the conjugate complex. The corresponding inverse discrete Fourier transform is described by the equation set

$$f_i = \frac{1}{N} \sum_{k=0}^{N-1} F_k \cdot \exp \left(\frac{-j2\pi ik}{N} \right) \qquad i = 0, 1, 2, \ldots, N - 1 \tag{C-5}$$

The systems of equations described by (C-3) and (C-5) involve numerous complex multiplications as well as calculations of exponentials. However, when N is factorable, the computational effort for either set can be reduced. The algorithm to accomplish this is called the *fast Fourier transform* (FFT). Although used for many years by persons who calculated discrete Fourier transforms either manually or with a desk calculator, the FFT was not known to most of the early numerical analysts who programmed digital computers. It was rediscovered by Cooley and Tukey and often bears their names, although numerous earlier references exist.

The FFT is particularly simple if N contains a factor of 2. The algorithm is illustrated for that case, although a comparable discussion can be written

---

[1] Note that F_k as defined differs dimensionally from $F(\omega)$.

for factors other than 2. If N is even, that is, if it contains a factor of 2, one may rewrite set (C-3) as a sum of even terms plus a sum of odd terms as

$$F_k = \sum_{i=0}^{(N/2)-1} f_{2i} \exp\left(\frac{j4\pi ik}{N}\right) + \exp\left(\frac{j2\pi k}{N}\right) \sum_{i=0}^{(N/2)-1} f_{2i+1} \exp\left(\frac{j4\pi ik}{N}\right)$$

$$\text{(C-6)}$$

Finally, this is rewritten as

$$F_k = A_k + \exp\left(\frac{j2\pi k}{N}\right) B_k \qquad\qquad \text{(C-7)}$$

where

$$A_k = \sum_{i=0}^{(N/2)-1} f_{2i} \exp\left(\frac{j4\pi ik}{N}\right) \quad \text{and} \quad B_k = \sum_{i=0}^{(N/2)-1} f_{2i+1} \exp\left(\frac{j4\pi ik}{N}\right)$$

$$\text{(C-8)}$$

From the properties of exponentials it follows for $k \geq N/2$ that

$$A_k = A_{k-N/2} \quad \text{and} \quad B_k = B_{k-N/2}$$

while

$$\exp\left(\frac{j2\pi k}{N}\right) = -\exp\left[\frac{j2\pi(k-N/2)}{N}\right]$$

Thus, only the first $(N/2) - 1$ values of A_k and B_k need be computed. The final FFT formulas corresponding to equation set (C-3) may be written

$$F_k = \begin{cases} A_k + \exp\left(\dfrac{j2\pi k}{N}\right) B_k & 0 \leq k \leq (N/2) - 1 \\[2ex] A_{k-N/2} - \exp\left[\dfrac{j2\pi\{k-N/2\}}{N}\right] B_{k-N/2} & N/2 \leq k \leq N - 1 \end{cases}$$

$$\text{(C-9)}$$

This reformulation saves computer time in two ways. First, if separate complex exponentials are computed, the same ones appear in both A_k and B_k. The second is that there is a reduction in the necessary multiplications. There are N multiplications for each of the N members of set (C-3), for a total of N^2 multiplications. On the other hand, the sums A_k and B_k each have $N/2$ multiplications, but there are only $N/2$ of each type needed in (C-9). This leads to a total of $N^2/2$ multiplications, which can represent a significant reduction in some biophysical problems.

Much greater savings of computer time can be realized if N is a power of 2, say 2^n. In this case, by repetitively applying the FFT algorithm, the number of multiplications as well as the number of complex exponentials required can be reduced by a factor of 2^n. The saving in multiplications can be shown to be approximately a factor of N/n. For an n of 10, N is 1,024, and the reduction in multiplications is over 100-fold. Thus, the use of the FFT can reduce the computational time to an amount which is quite small compared to that necessary to apply equation set (C-3) directly. It should be emphasized that although (C-3) is used in the preceding examples, completely analogous reasoning can be used to derive the FFT for equation set (C-5).

C.2 Applications

A. CONVOLUTION INTEGRALS

In the use of tracers (see Chapter 29), a mathematical form known as the *convolution integral* frequently arises. For example, let $w(t)$ be a dye concentration measured at one site in the circulatory system and $h(t)$ be the distribution of transit times to a second location. Then the concentration $f(t)$ at the second location is given by

$$f(t) = \int_{-\infty}^{+\infty} h(\tau)w(t - \tau)\,d\tau \qquad (C\text{-}10)$$

This is the convolution integral.

In actual practice, there is no meaning to negative transit times, so the lower limit on the integral may be replaced by zero. Moreover, the concentration at the first site will be zero until some time, conveniently chosen as zero. Accordingly, the upper limit of the integral may be replaced by t. Further, if there are N measurements made within the time T, necessary for $f(t)$ to become negligible, Eq. (C-10) may be replaced by a discrete analog,

$$f_i = \sum_{m=0}^{i} h_m w_{i-m} \qquad i = 0, 1, 2, \ldots, N - 1 \qquad (C\text{-}11)$$

The subscripted terms are related to the quantities in (C-10) by

$$f_i = f(t_i)$$
$$h_i = h(t_i)\Delta t = h(t_i) \cdot \left(\frac{T}{N}\right)$$
$$w_i = w(t_i)$$

The problem is often to find $h(t)$ given measurements for $f(t)$ and $w(t)$. Equivalently, one can ask for the set h_i given values for the data sets f_i and w_i. These problems can be simplified using Fourier transformations and computed efficiently using the FFT. The simplification occurs because the transform of a convolution integral is the product of the transforms of the functions convolved.

To illustrate this, capital letters are used for the Fourier transforms. Thus, in the continuous case, Eq. (C-10) becomes

$$F(\omega) = H(\omega) \cdot W(\omega) \qquad (C\text{-}12)$$

This, in turn, can be formally solved as

$$H(\omega) = \frac{F(\omega)}{W(\omega)} \tag{C-13}$$

Then $h(t)$ can be found from $H(\omega)$ computed according to Eq. (C-13), if one can carry out by table lookup or other means the inverse transform (Eq. (C-2)).

In the discrete case, Eq. (C-13) becomes

$$H_k = \frac{F_k}{W_k} \tag{C-14}$$

Both F_k and W_k can be computed using FFT. Then, using the values of H_k found in Eq. (C-14), one can compute h_i by using the FFT a third time.

B. POWER SPECTRA

In some examples, the choice of zero for time is arbitrary. This is particularly true in discussing eeg spectra (see Chapter 6). A frequency distribution independent of this arbitrary choice is desired. One may show that if

$$g(t) = f(t + \tau)$$

then

$$G(\omega) = \exp(j\omega\tau)F(\omega) \tag{C-15}$$

By taking the absolute value of both sides of Eq. (C-15), one obtains

$$P(\omega) = G(\omega)G^*(\omega) = F(\omega)F^*(\omega) \tag{C-16}$$

This is called the *power spectrum*. Its value is independent of the choice of time zero. Power spectra are used in eeg studies, in descriptions of tissue irradiation by a sonic beam, and in absorption and emission spectroscopy, discussed in Chapters 26 and 27.

REFERENCES

1. COOLEY, J. W., AND J. W. TUKEY, "An Algorithm for the Machine Calculation of Complex Fourier Series," *Math Comput.* **19**:297–301 (1965).
2. BRIGHAM, E. O., AND R. E. MORROW, "The Fast Fourier Transform," *IEEE Spectrum* (December 1967):63–70.
3. COCHRAN, W. T., J. COOLEY, D. L. FAVIN, H. D. HELMS, R. A. KAENEL, W. W. LANG, G. C. MALING, D. E. NELSON, C. M. RADER, AND P. D. WELCH, "What Is the Fast Fourier Transform?" *IEEE Trans. Audio Electroacoust.* **AU-15**: 45–55 (1967).
4. GENTLEMAN, W. M., AND G. SANDE, "Fast Fourier Transforms—for Fun and Profit," *Fall Joint Computer Conf.* **29**:563–578 (1966).
5. BRIGHAM, E. O., *The Fast Fourier Transform* (Englewood Cliffs, N.J.: Prentice-Hall, Inc., 1974), 252 pages.

APPENDIX D

Biochemistry

Since this text does not presuppose an extensive background in the life sciences, yet refers to elements of biology in almost every chapter, certain aspects of biochemistry and cellular physiology are introduced for the less biologically knowledgeable reader. More specialized topics in biochemistry are included where needed in the various chapters. This appendix summarizes the structure and definitions of the basic building blocks of life and their use in taxonomy. Their function in the cell is considered in Appendix E.

D.1 Amino Acids and Proteins

Proteins are one of the major classes of compounds found in all living matter. Enzymes are protein catalysts that control the rates of many biological reactions (see Chapter 18). Muscular contraction depends on proteins, and active transport across cell membranes appears to be a lipoprotein function (Chapters 9 and 24). Many physical means have been used to study protein structure; current work leans heavily on the results obtained from X-ray diffraction (see Chapter 14). Here the chemical composition of proteins is considered.

Proteins are natural high polymers built from small monomers called *amino acids*. There are molecules with an average molecular weight of about 120, each of which has an organic acid group

$$-C\diagup^{O}_{\diagdown OH}$$

and a basic amino group, $-NH_2$. The acid and basic groups are attached to the same carbon, leading to the form

$$R-\underset{\underset{H}{|}}{\overset{\overset{NH_2}{|}}{C}}-C\diagup^{O}_{\diagdown OH}$$

where R is either H or any of a number of different organic radicals. This form is called an α-amino acid.

If one makes a three-dimensional model of such an α-amino acid, there are two steric arrangements of the α-carbon, which cannot be rotated into one another (except when R is a proton). A carbon atom, which is sterically asymmetric in this fashion, is called *optically active* because in solution it rotates plane-polarized light (see also Chapter 26). The two stereoisomers are labeled D and L, which historically stood for dextrorotary and levulorotary, respectively. Now these letters indicate geometric or steric relationships and do not necessarily indicate the direction of rotation of plane-polarized light. Test-tube syntheses usually yield equal amounts of the D and L isomers, but most living cells produce only one variety or the other. With very few exceptions, the amino acids polymerized into proteins in the living cell are all L-α-amino acids.

Two amino acid molecules may react to eliminate a molecule of water, thereby forming a peptide bond. Schematically, this can be represented as

TABLE D-1

AMINO ACIDS

Amino acid	Structure	Molecular weight	Abbreviation for residue in proteins
1. Glycine	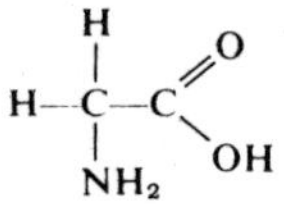 (structure)	75	—gly—
2. Alanine	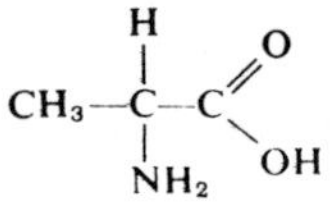 (structure)	89	—ala—
3. Valine	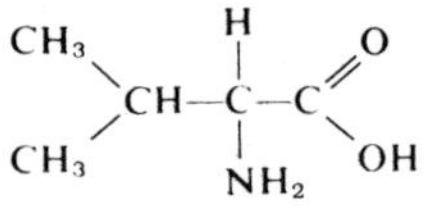 (structure)	117	—val—
4. Leucine	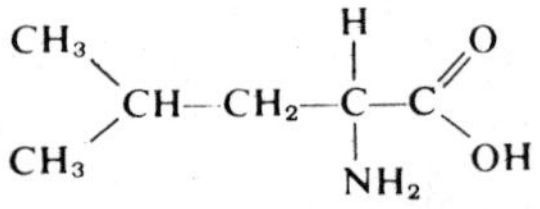 (structure)	131	—leu—
5. Isoleucine	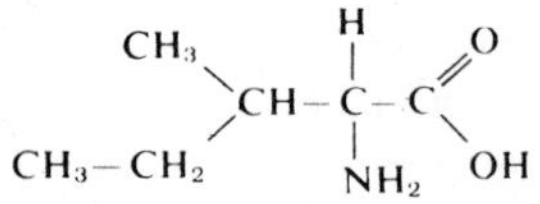 (structure)	131	— ileu—
6. Serine	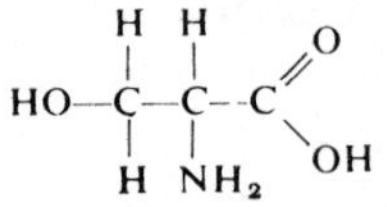 (structure)	105	—ser—
7. Threonine	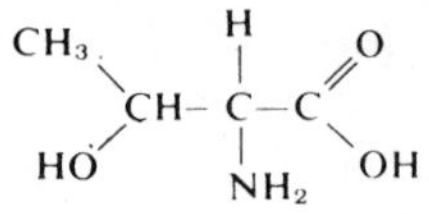 (structure)	119	—thr —
8. Phenylalanine	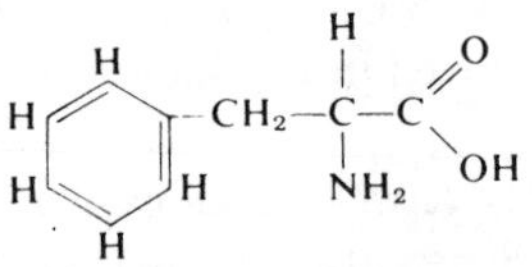 (structure)	165	—phe —
9. Tyrosine	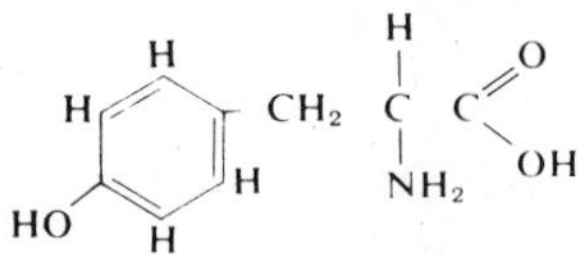 (structure)	181	tyr

TABLE D-1 (*Continued*)

Amino acid	Structure	Molecular weight	Abbreviation for residue in proteins
10. Tryptophan	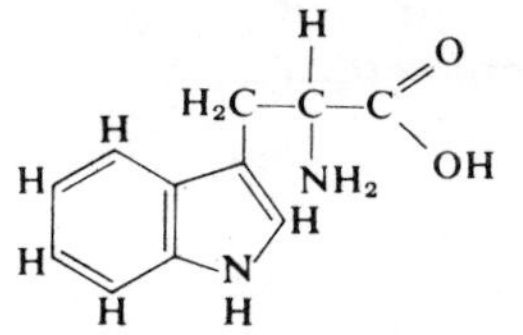	204	—try—
11. Cysteine	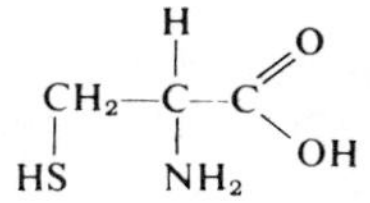	121	SH ǀ —cy—
11a. Cystine	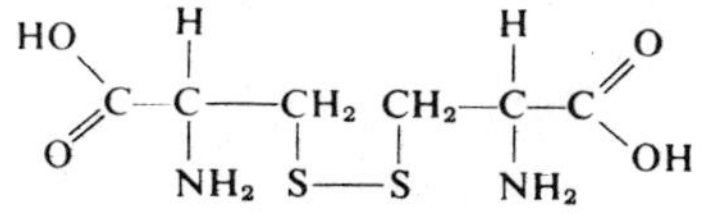	240	—cy— ǀ S ǀ S ǀ —cy—
12. Methionine	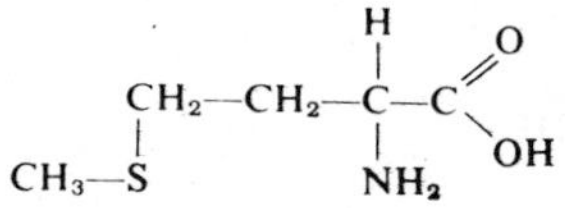	135	—met—
13. Aspartic acid	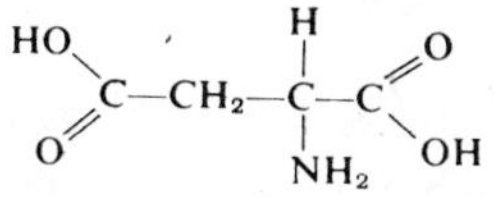	133	—asp—
14. Glutamic acid	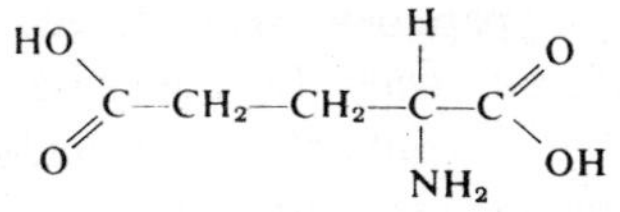	163	—glu—
15. Lysine	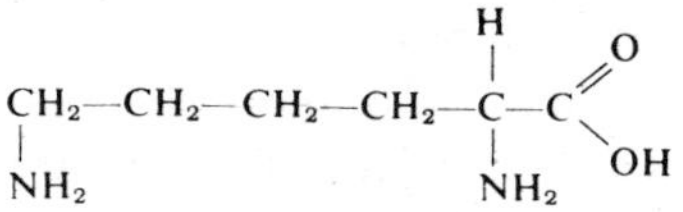	146	—lys—
16. Arginine	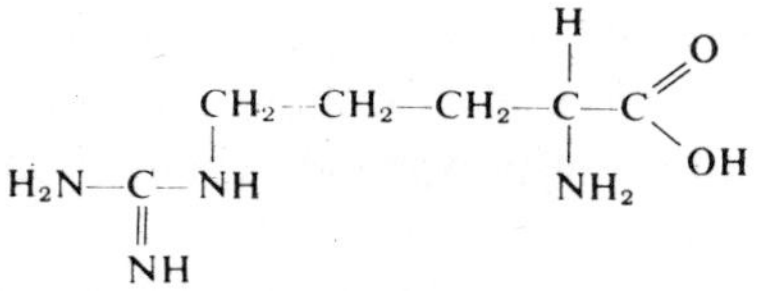	174	—arg—

(*Continued*)

TABLE D-1 (*Continued*)

Amino acid	Structure	Molecular weight	Abbreviation for residue in proteins
17. Citrulline		175	—cit—
18. Histidine		155	—his—
19. Proline		115	—pro—
20. Hydroxyproline		131	—hypro—

The peptide bond so formed is very stable and the molecule is called a *dimer* or *dipeptide*. It is possible to attach this molecule to other amino acid molecules, forming chains, or polypeptides. When these chains include 50 or more amino acids, they are called *proteins*. In some cases, the chains may be branched or cross-linked. Proteins contain amino acids and may contain other molecules as well (see Chapter 18). Molecular weights of proteins vary from several thousand into the millions.

The number of different amino acids conceivable has no known limit. A very large number have been synthesized. Of these, approximately 20 L-α-amino acids make up almost all of the proteins of living cells. Other amino acids occur in nature, especially in bacteria, but they are the exceptions rather than the rule. These 20 amino acids are shown in Table D-1. The physical form and arrangement of these polypeptide chains is discussed further in Chapter 14.

The various proteins differ from one another in the number and order of the amino acids in the polypeptide chains and in the configuration of these chains. Although the detailed order is known for only a few proteins, pieces of many others have been studied to determine this order. Figure D-1 gives the amino acid sequence for the protein insulin.

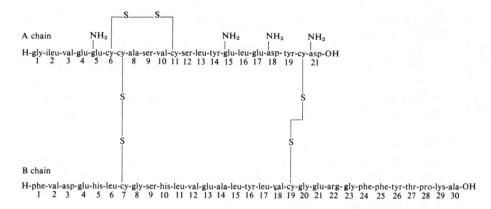

Figure D.1   Beef insulin. The molecular weight of the beef insulin molecule is 5,733. Its amino acid sequence is shown above as it was worked out by Sanger; for symbols, see the table of amino acids. In solution, it probably exists as dimers, trimers, and tetramers of molecular weight about 11,500, 17,000, and 23,000, respectively. Horse, sheep, whale, and pork insulin differ from beef insulin only in the residues 8-9-10 of the A chain. These are: horse, $-thr_8-gly_9-ileu_{10}-$; sheep, $-ala_8-gly_9-val_{10}-$; and whale and pork, $-thr_8-ser_9-ileu_{10}-$.

D.2 Purines, Pyrimidines, and Nucleic Acids

As late as 1950, scientists regarded proteins as the fundamental "stuff" of life, controlling reactions, contracting, transmitting genetic information, and reproducing themselves. This view has given way to one that assigns to proteins a more restricted role and emphasizes the importance of nucleic acids in protein synthesis and genetic transmission of information (see Chapter 16). The nucleic acids are high polymers, just as are proteins, but the monomers from which nucleic acids are built are not amino acids. Since they consist of different structural units, the nucleic acids differ significantly from proteins in their physicochemical properties as well as in their biological action.

The monomers from which nucleic acids are polymerized are called *nucleotides*. Each nucleotide consists of a *nucleoside* condensed, by the elimination of a water molecule, with a phosphate group. A nucleoside in turn is the condensation product of a five-carbon *sugar* (pentose), plus an organic *base* derived from a purine or pyrimidine ring. Symbolically, this may be represented as

$$\text{base} + \text{sugar} \xrightarrow{\;-H_2O\;} \text{nucleoside}$$

$$\text{nucleoside} + H_2PO_4^- \xrightarrow{\;-OH^-\;} \text{nucleotide}$$

$$n\ \text{nucleotide} \xrightarrow{\;-(n-1)H_2O\;} \text{nucleic acid}$$

The number n is very large, because the molecular weight of nucleic acid molecules usually runs in the millions.

There are two types of sugar molecules included in nucleic acids. Throughout any one nucleic acid molecule, all the sugar residues are the same. One of these sugars is called D-*ribose*, in which case the polymer is ribonucleic acid (RNA). The other sugar is D-2-deoxyribose, in which case the polymer is deoxyribonucleic acid (DNA). The structures of ribose and deoxyribose are shown below, in a ring form. The ring with carbon atoms at four of the corners is supposed, in this illustration, to lie in one plane, with the hydrogen and hydroxyl groups on bonds at right angles to the plane. Although this type of structure is often drawn, it is believed that the ring is not restricted to one plane but rather is pleated, up and down, about the plane, as illustrated in the chair model in Fig. D.4.1(e).

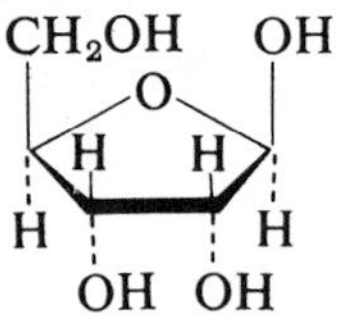

D-ribose 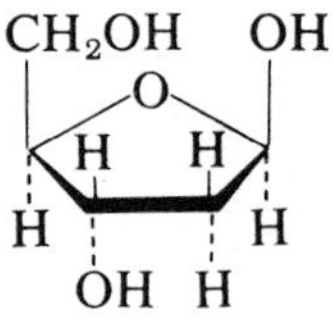   D-2-deoxyribose

The organic bases referred to above are derived from purine and pyrimidine rings. These rings have structures that can be shown as

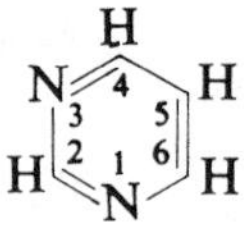

pyrimidine 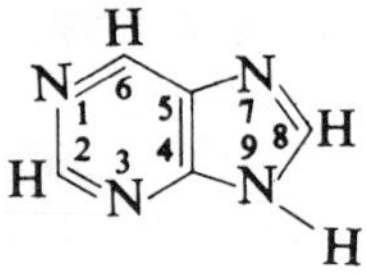   purine

(where the corners without letters in the ring are to be interpreted as carbon atoms). The two pyrimidines found in DNA are *cytosine* (C) and *thymine* (T), represented structurally by

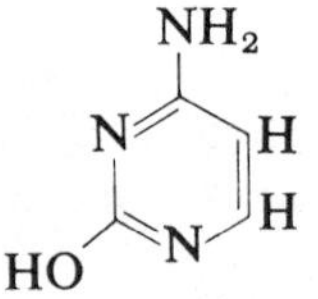

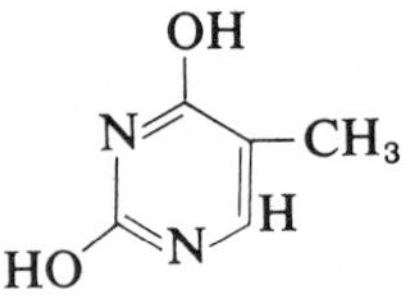

cytosine   thymine

In RNA, two pyrimidines are also found. These are cytosine and *uracil*. The latter has the structure

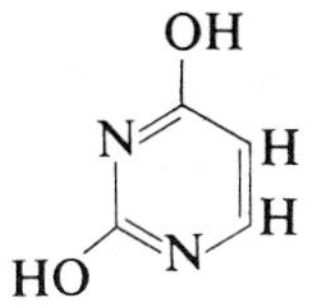

uracil

In both RNA and DNA, the same two purine-derived bases occur. These are called *adenine* (A) and *guanine* (G). The structural formulas of adenine and guanine are

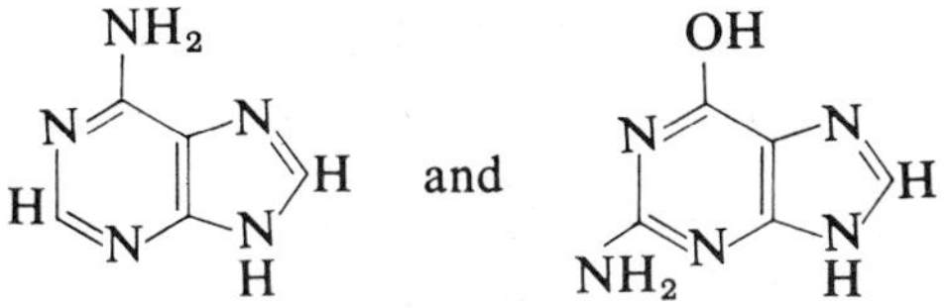

adenine   and   guanine

As mentioned above, most genetic information in plant and animal cells is believed to be coded in DNA. This has only four monomers; these are nucleotides containing, respectively, A, T, C, and G. If it seems surprising that this alphabet is sufficient, it should be remembered that any English sentence can be written in Morse code, which has three basic letters: a dot, a dash, and a pause. Indeed, information theory (see Chapter 25) states that two symbols could be sufficient.

Besides their role in nucleic acids, purine and pyrimidine bases have an additional place as components of several low-molecular-weight cofactors which are necessary for many living processes to occur. For example, the nucleotide adenine ribose phosphate (or adenosine monophosphate, abbreviated AMP) can be condensed with additional phosphate groups to form the energy transport compounds, ADP and ATP. The structure of ATP is as follows:

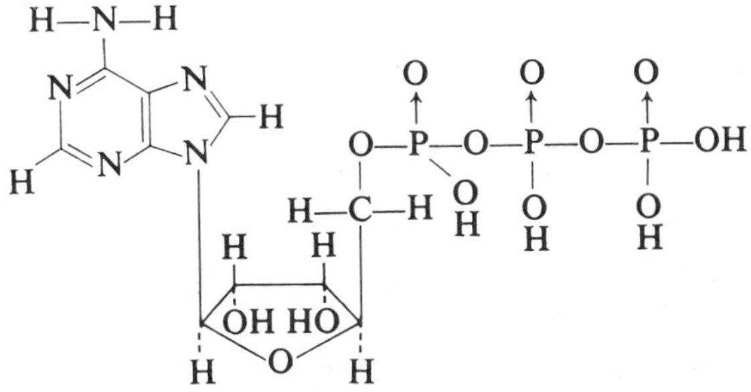

Energy is released when the condensations above are reversed, that is, when ADP is split to AMP and inorganic phosphate or ATP is split to ADP and inorganic phosphate.

Another important class of cofactors is involved with electron-transport in oxidations and reductions. The structures of the oxidized forms of two such cofactors, nicotinamide adenine dinucleotide (NAD) and flavin (6,7-dimethyl-9-*d*-ribitylisoalloxain) adenine dinucleotide (FAD), are shown in Fig. D.2.

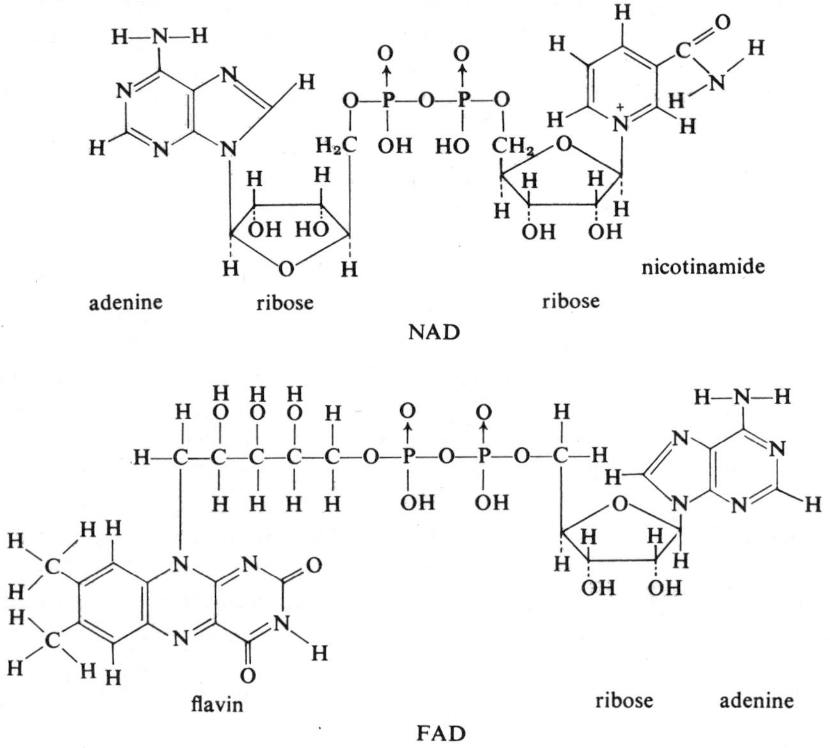

Figure D.2   Structure of NAD and FAD.

D.3   Lipids and Carbohydrates

There are many other types of molecules within the cell besides proteins and nucleic acids. In the typical living cell, there are more molecules of water than of any other compound. Water makes up as much as 80 percent of the cell weight in some cases. The nonpolar lipid molecules are mentioned in connection with membranes in Chapter 24 but are by no means restricted to the membranes. Rather, lipids are found in various roles throughout the cell. Those in fat globules are used to store energy, and a few lipids are hormones. However, the role of some lipids is unknown. Structures of a few lipids are

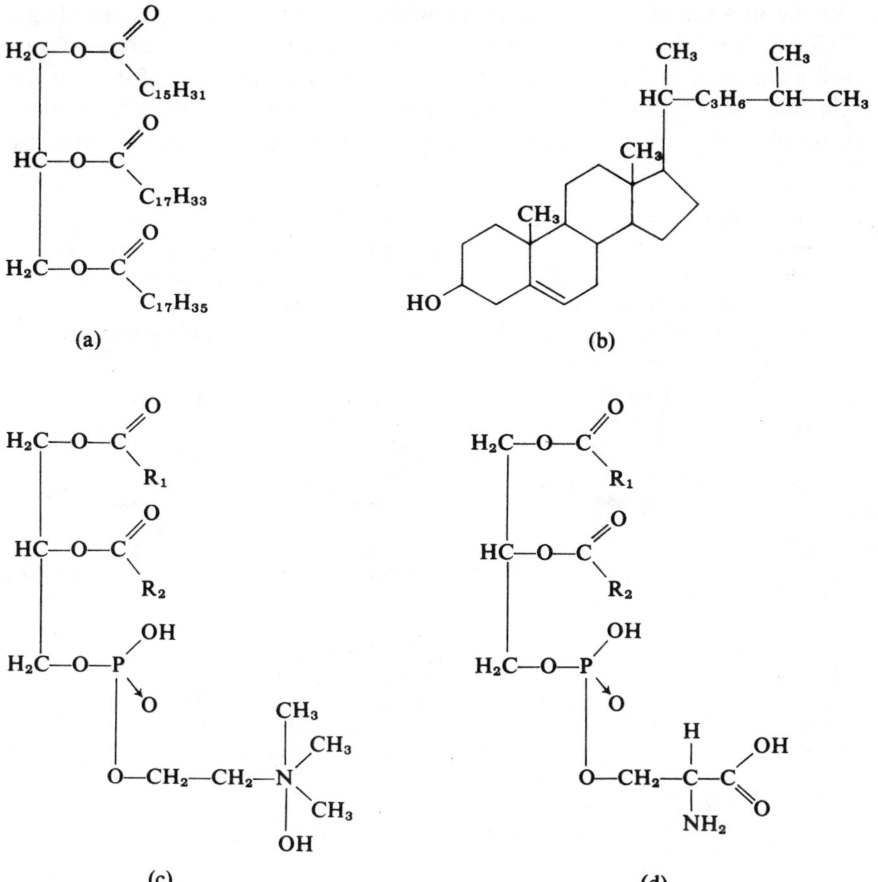

Figure D.3 A few lipids. Lipids are one general class of molecules found in all cells. The phospholipids are a part of many cell structures. (a) Typical fat molecule; (b) cholesterol (a steroid); (c) α-lecithin (a phospholipid); (d) phosphatidyl serine (another phospholipid). Many steroids act as hormones. The fats serve as storage depots for chemical energy, which is not rapidly available to the cells.

shown in Fig. D.3. Typical lipids have molecular weights between 100 and 1,000, which are small compared to those of proteins and nucleic acids.

Other molecules, found in all living cells, are called *carbohydrates*. These consist of carbon, hydrogen, and oxygen atoms, the latter two always occurring in the same ratio as in water. The carbohydrates sometimes are found as small ring or chain structures called *monoses*, or simple sugars, having 3–7 carbon atoms. They are given class names consisting of the Greek prefixes for the number of carbon atoms, to which is attached the suffix -ose (e.g., "triose" for a three-carbon sugar). Two pentoses, ribose and 2-deoxyribose, are mentioned in Sec. D.2 as parts of nucleic acid. Hexoses, for example glucose and fructose, are the most common sugars in living organisms. Carbohydrates are also found as compound sugars (or dimers), such as

Figure D.4 Typical carbohydrates found in living cells. All hexoses can exist in solution in at least five forms: a straight chain, two six-membered (pyranose) rings, and two five-membered (furnose) rings. For glucose, most of the molecules are in the pyranose forms. The chair model is closer to the actual molecular arrangement but is harder to draw. The di- and polysaccharides exist in only one form.

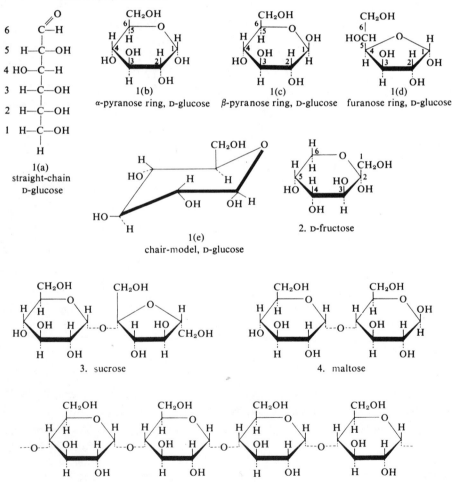

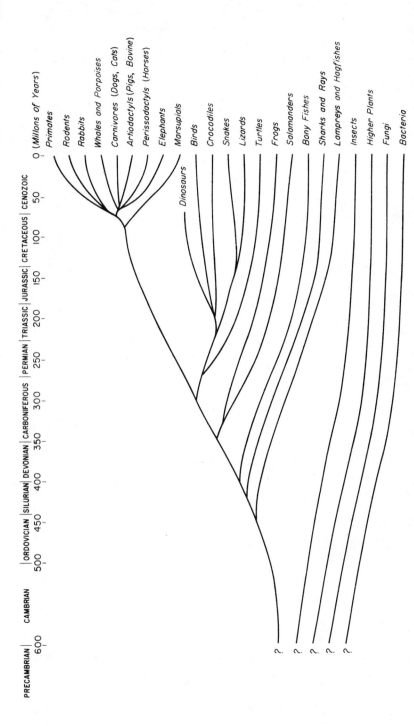

Figure D.5 Divergence of the vertebrate groups based on geological and biological evidence. The details are not known with as much confidence as the sharp lines seem to indicate. After M. O. Dayhoff, ed., *Atlas of Protein Structure and Research*, Vol. 4, Natl. Biomed. Res. Fdn., Silver Springs, Maryland, 1969, p. 40.

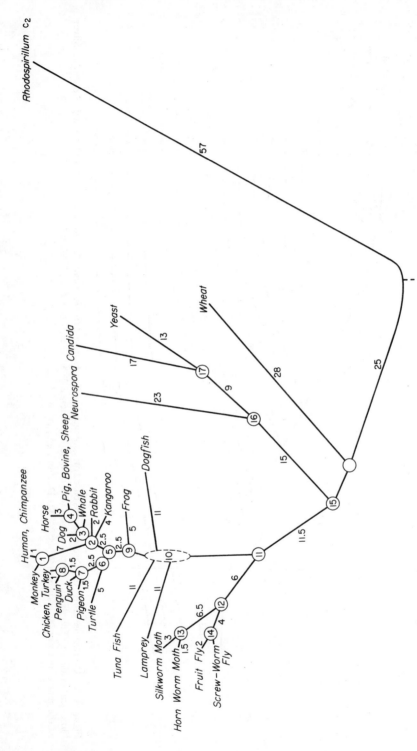

Figure D.6 Phylogenetic tree of cytochrome c. The topology has been inferred from the sequences as explained in the text. The numbers of inferred amino acid changes per 100 links are shown on the tree. The point of earliest time cannot be determined directly from the sequences; it is placed by assuming that, on the average, species change at the same rate. After M. O. Dayhoff, ed., *Atlas of Protein Structure and Research,* Vol. 4, Natl. Biomed.-Res. Fdn., Silver Springs, Maryland, 1969, p. 9.

sucrose, and as long polymerized chains of monoses, such as starches and celluloses. Several carbohydrates are indicated in Fig. D.4.

D.4   Biochemical Taxonomy

Throughout this text specific plants and animals are discussed, and although much of what is said can be generalized to all organisms, some is only relevant for the organism under discussion and its near relatives. To help define near as opposed to far relatives in the taxonomic sense, Fig. D.5 shows one scheme for relating all organisms, but especially the vertebrates, in terms of the geologic time at which they separated from a common ancestor. According to this scheme, for example, rodents are more closely related to primates (including man) than are carnivores.

Taxonomy might be considered the branch of biology least relevant to the biophysicist, yet the field has changed quite a bit since the days of expeditions to far off lands to bring back specimens and fossils to be cataloged by museum curators. One of the most interesting new techniques (at least to a biophysicist) is the use of *primary-sequence analysis* of proteins and nucleic acids to determine phylogenetic information.

Briefly, this technique assumes that all forms of a common protein (e.g., cytochrome c) or nucleic acid (e.g., val-*t*RNA) had a common ancestral sequence. It further assumes that the less proteins or nucleic acids of two species differ in sequence, the more closely related the species are. The method allows certain residues (e.g., those necessary for catalysis) to be immutable and others to be changed randomly. In addition, protein sequence changes are either easy (those involving mutation at one base of the DNA) or hard (those involving two or more base pairs). A computer is necessary to find the best fit to the sequence information, and the phylogenetic tree resulting from such a computer simulation is shown for the protein cytochrome c in Fig. D.6.

REFERENCES

References for Appendix D are found at the end of Appendix E.

APPENDIX **E**

Cellular Physiology

E.1 The Cell

Cellular physiology is defined as the science dealing with the functions common to most living cells and their parts. A composite cell is shown in Fig. E.1, illustrating the features of many different types of cells. The structures shown in this figure are based on the results obtained with a variety of techniques. There are two major subdivisions to the cell: the nucleus and its contents, sometimes called *nucleoplasm*; and the remainder of the cell, often called *cytoplasm*. Both the nucleus and the cytoplasm are surrounded by membranes, as are also the smaller organelles such as the *mitochondria* and the *Golgi bodies*. These membranes consist partly of proteins and partly of lipids. The membranes not only give physical form to the structures they surround, but also can use energy to transport molecules actively. (Active transport is discussed in Chapter 23, membranes in Chapter 24.)

Proteins are found in abundance within the organelles, the contractile elements, the nucleus, and the endoplasmic reticulum shown in Fig. E.1. Some of these proteins serve structural purposes and others act as catalysts for the numerous biological reactions which must occur at controlled rates if

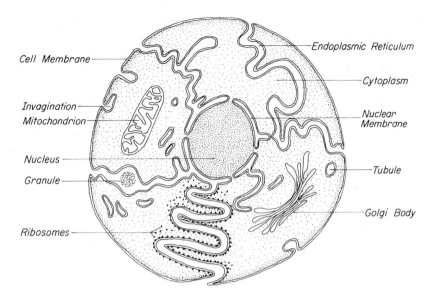

Figure E.1 Reconstructed cell, showing general cellular components as revealed by electron microscopy. After C. P. Swanson, *The Cell*, 3rd ed., © 1969, p. 20. Reprinted by permission of Prentice-Hall, Inc., Englewood Cliffs, New Jersey.

the cell is to live. Other proteins are found in the liquid parts of the cytoplasm and nucleoplasm; these are probably mostly enzymes, although some may be concerned with the osmotic balance of the cell.

Nucleic acids are also found in both the cytoplasm and in the nucleus. RNA's are in both areas, while DNA's are restricted for the most part to the nucleus. (Some organelles, such as mitochondria and chloroplasts, also have a small amount of DNA.) Chapter 16 discusses the way DNA controls the synthesis of RNA, which in turn controls the synthesis of proteins. Both DNA and RNA thus act as biological catalysts, ultimately controlling the synthesis of protein enzymes. These, in turn, control the rates of most other chemical reactions within the cell (see Chapter 18). As stated in Chapter 15, the genetic information necessary to build a new virus particle is sometimes carried by RNA rather than DNA.

All cellular organelles are specialized compartments for special cell functions, and they are best described in terms of these cellular functions:

1. Energy intake and waste disposal: The cell membrane is permeable to a large class of energy-rich compounds (food) and can engulf others through the processes of *pinocytosis* and *phagocytosis*. The engulfed particles reside in *vesicles* which fuse with *lysosomes* containing digestive enzymes to break down the food into its more useful constituents. Lysosomes are synthesized

(packaged) by the *Golgi apparatus*. Used vesicles are "unengulfed" and returned to the surrounding medium.

2. Energy transformation: The energy in certain food constituents is released for the use of the cell in the mitochondria. Chloroplasts can also generate energy, but they use light rather than food to do so.

3. Movement and irritability: The membrane plays an important part in cell movement. *Flagella* or *cilia*, specialized organelles for movement, can extend beyond the membrane and propel the cell in various directions.

4. Reproduction: During cell division the DNA in the nucleus of most plant and animal cells is organized into long threads called *chromosomes*.

E.2   Cellular Reproduction

The process of cellular reproduction or cellular division is referenced in several chapters. In somatic cells, it occurs in a series of characteristic steps called *mitosis*; in germ cells the analogous process is termed *meiosis*. The major exceptions to the universal nature of mitosis and meiosis are the bacteria, which do not possess a clearly defined nucleus and divide in a less organized fashion.

Figure E.2 illustrates the process of mitosis. The chromosomes within the cell nucleus are believed to carry most of the genetic information of the cell, controlling its form, metabolism, and function. However, as shown in Fig. E.2(a), the chromosomes do not exist as such in the nucleus during most of the cell cycle. Rather, during the period between divisions, they appear dispersed into heavily staining birefringent granules called *chromatin*. This portion of the cell life is called *interphase*.

As the cell prepares to divide, the chromatin is organized into long filaments which pull together to form chromosomes. These are double filaments at this point in mitosis, which is called *prophase*. Simultaneously, a spindle starts to form, the nuclear membrane starts to dissolve, and the nucleoli disappear. This stage is shown in Fig. E.2(b).

In the next stage, *metaphase*, the chromosomes attach to the spindle at a specific point, the centromere (also called the kinetochore), and line up at the center of the cell. As shown in Fig. E.2(c), the nuclear membrane is completely gone.

The chromosomes then each pull apart into two separate fibers and follow the spindles to the cell centers. In the absence of a spindle (which results from certain types of irradiation), the chromosomes do not divide. The forces causing the chromosomes to adhere to the spindle at the centromere, to separate, and to migrate are not clearly understood. The observed phenomena known as *anaphase* are shown in Fig. E.2(d). Note that each half of the

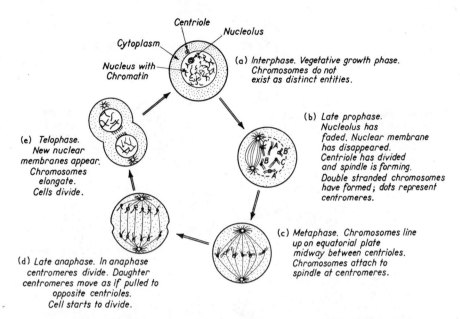

Figure E.2   Diagrammatic outline of mitotic cycle. Each cell starts with two homologous chromosomes, distinguished in the diagram by showing their centromeres as dots and squares. Adapted from G. G. Simpson, C. S. P. Hendrigh, and L. H. Tiffany, *Life: An Introduction to Biology* (New York: Harcourt Brace Jovanovich, Inc., 1957).

cell now has the same number and types of chromosomes as the original one in Fig. E.2(b).

As illustrated in Fig. E.2(e), new nuclear membranes then form and the cell pinches in two during the final stage, called *telophase*. In normal mitosis, one ends up with two duplicates of the original cell. These duplicates are sometimes referred to as *daughter cells*.

In the normal cells, the chromosomes occur as *homologous pairs*. The two members of each pair have similar shapes and control the same characteristics. This control is exercised via segments of the chromosome called genes, as described in Chapter 16. If two genes in homologous chromosomes are not identical, one may be said to *dominate* the other. The cell and the individual usually reflect this dominant character. During mitosis, each chromosome replicates and, therefore, the daughter cells have the same character as the original cell.

During *meiosis*, however, the two homologous chromosomes line up together, entwine about one another, and then separate along the spindle to opposite poles. Thus, the germ cells end up with half the number of chromosomes as the normal body cells. This division is not completely random

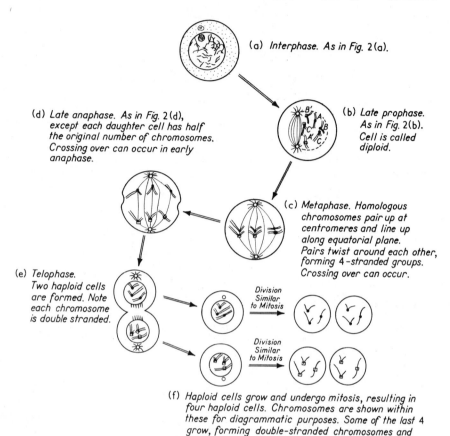

(a) *Interphase. As in Fig. 2(a).*

(d) *Late anaphase. As in Fig. 2(d), except each daughter cell has half the original number of chromosomes. Crossing over can occur in early anaphase.*

(b) *Late prophase. As in Fig. 2(b). Cell is called diploid.*

(c) *Metaphase. Homologous chromosomes pair up at centromeres and line up along equatorial plane. Pairs twist around each other, forming 4-stranded groups. Crossing over can occur.*

(e) *Telophase. Two haploid cells are formed. Note each chromosome is double stranded.*

Division Similar to Mitosis

Division Similar to Mitosis

(f) *Haploid cells grow and undergo mitosis, resulting in four haploid cells. Chromosomes are shown within these for diagrammatic purposes. Some of the last 4 grow, forming double-stranded chromosomes and becoming the active cells of sexual reproduction.*

Figure E.3 Meiosis. Note that if there are pairs of homologous chromosomes, the cell is called *diploid*, whereas if there are only half this number of chromosomes, the cell is called *haploid*. For a discussion of crossovers, see Fig. E.4. Adapted from G. G. Simpson, C. S. P. Hendrigh, and L. H. Tiffany, *Life: An Introduction to Biology* (New York: Harcourt Brace Jovanovich, Inc., 1957).

because each egg or sperm contains one member of each pair of chromosomes. When the sperm eventually fertilizes an egg cell, the normal number is reformed. Figure E.3 illustrates diagrammatically the chromosome changes in meiosis.

By and large, the different characteristics are segregated during meiosis according to the member of the homologous pair on which they are located. However, occasionally pieces of the chromosomes break off during meiosis. The broken pieces then rejoin the same homologous pair, but often a part of chromosome A will join the remainder of A′ and vice versa. This breaking and re-forming is known as *crossing over*. This is hard to observe in large animals which reproduce slowly because crossing over is a rare event. However, in fruit flies, wasps, microorganisms, and viruses, the rate of reproduction is so large that the frequencies of crossing over between two loci can be accurately measured. Figure E.4 shows a possible crossing over during meiosis.

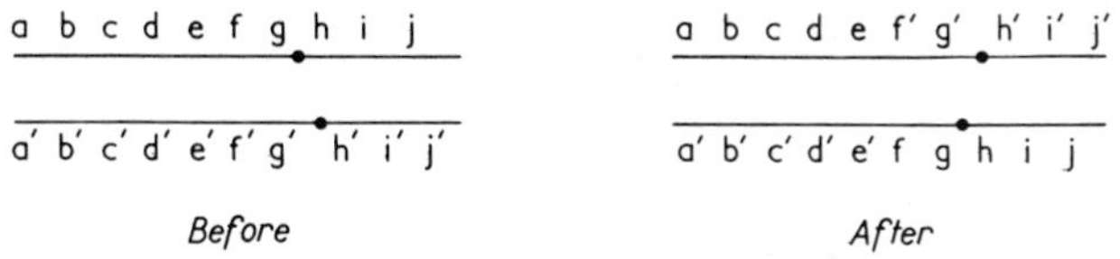

```
a  b  c  d  e  f  g  h  i  j              a  b  c  d  e  f′ g′  h′ i′ j′
              •                                          •
──────────────────────              ──────────────────────
a′ b′ c′ d′ e′ f′ g′  h′ i′ j′          a′ b′ c′ d′ e′ f  g  h  i  j
              •                                          •

        Before                                  After
```

Figure E.4   One type of crossover. Letters show locations along chromosomes. This is schematic only; most chromosomes are not straight lines and are always twisted when crossovers occur. Dot indicates centromere. For instance, *a* might represent blue eyes, *a′*, brown; and *g* might represent tall, *g′* short. Then the offspring with no crossover would always have blue eyes and be tall. With the crossovers shown, blue eyes can occur with short. After H. J. Mueller, in *Radiation Biology*, Vol. I, Part 1, A. Hollander, ed. (New York: McGraw-Hill Book Company, 1954).

REFERENCES

Biochemistry and Cellular Physiology

Most biochemistry and cellular physiology texts devote several chapters to the topics in Appendices D and E:

1. McELROY, W. D., *Cell Physiology and Biochemistry*, 3rd ed. (Englewood Cliffs, N.J.: Prentice-Hall, Inc., 1971), 152 pages.

2. LEHNINGER, A. H., *Biochemistry*, 2nd ed. (New York: Worth Publishers, Inc., 1975), 1104 pages.

3. YOST, H. T., *Cellular Physiology* (Englewood Cliffs, N.J.: Prentice-Hall, Inc., 1972), 925 pages.

Biochemical Taxonomy

4. DAYHOFF, M. O., et al., *Atlas of Protein Sequence and Structure 1965* (first of a continuing series) (Silver Springs, Md: National Biomedical Research Foundation, 1965).

Index

A

A band (muscle), 154–155
Absolute rate theory, 243, 408–412, 440
Absorption:
    ionizing radiation, 306–311
    sonic, 234–242, 251
Absorption spectroscopy, 484–514
Absorption spectrum, 344, 364, 377–380,
    492, 496–503, 516–517
Ac (*see* Alternating current)
Accommodation, lens, 34
Acetylcholine, 99, 160
Acetylcholinesterase (*see* Cholinesterase)
Acetyl CoA, 350–351
ACh (*see* Acetylcholine)
Achromatic lens, 45
Acoustic encoding, 132–135
Acoustic holography, 543–545
Acoustic pressure, 4–11, 234, 243–244, 584
    amplification, 20
    amplitude, 4–9, 243, 586–587
    level, 8–12, 471–472, 584

Acousticolateral system, 4, 68
Acoustics, 4–9, 583–588
Actin, 155, 163–167
Action current (*see* Action potential)
Action spectrum, 378–379, 381
Action (spike) potential, 79–101, 119,
    142–143, 146, 159–161, 174–183
Activated complex, 409–412
Activation energy, 406–412
Activator, enzyme, 341
Active complex, 166
Active site of an enzyme, 270
Active transport, 432–452, 462–463
    models, 462–463
Activity (*see* Radioactivity)
Acuity, visual, 44–46, 365, 473
Adenine, 271–274, 297, 323, 607
Adenosine diphosphate (*see* ADP)
Adenosine monophosphate (AMP), 323–324,
    607
Adenosine triphosphate (*see* ATP)
Adiabatic process, 414–418
ADP, 166–167, 348–349, 373–376, 607
After potential, 88, 177

Alarm reactions, 103
Algae, photosynthetic, 371–372, 376–377
Aliasing, 572
All-or-none law, 88, 104, 468
Alpha protein configuration, 263
Alpha helix, 263–269, 505–506
Alpha particle, 197, 308, 310, 548
Alpha wave (eeg), 112–114
Alternating current (ac), 82–83, 592–593
Amino acid, 298–305, 320, 475–478, 519–520,
    600–605
  table, 602–604
Amino group, 601
Amphibian heart, 173
Anaphase, 616–618
Anger camera (*see* Gamma camera)
Angular momentum (*see* Momentum,
    angular)
Anisotropic band (*see* A band)
Antagonist theory, 140–142
Antibody and antigen, 320–322, 512–513,
    558–559
Aorta, 173–175, 180
Apoenzyme, 329
Aqueous humor, 34
Areal capacity, 219
Areal conductance, 93
Arrhenius constant, 407–409
Arrhenius plot, 402, 407–408
Artery, 171–173
Astigmatism, 35–36
Atomic structure factor, 260
ATP, 348–350, 373–376, 383–387, 462–463,
    607
ATPase, 167, 462–463
Attenuation of radiation:
  acoustic, 236–241, 251
  electromagnetic, 222–223, 309–310
      496–497
Audiometer, 12–14
  Békésy, 12–14
  speech, 14
Auditory meatus, 16, 17
Auditory system pathways, 130–132, 135
Auricle:
  heart, 173–176, 183
  outer ear, 16, 17
Auricular wall, 176–177
Auriculo-ventricular (A-V):
  bundle, 176–183
  node, 176, 183
  valve, 174–175
Autonomic nervous system, 85, 104, 146
Autoradiography, 385, 553
Axon, 39, 79–80, 83–101
  membrane, 86–87, 460
  properties, 90

B

Bacteria, as viral host, 276–289
Bacteriochlorophyll, 377
Bacteriophage (*see* Phage)
Basilar membrane, 21–22, 120–129, 135
Bat, 56–57, 59–66
Baud (unit), 472
Bee, 51–53, 56–57
Beer's law, 222, 496–497
Békésy wave, 121–130, 133, 135
Bergonie and Tribondeau law, 202, 204–205
Bernoulli's equation, 179
Beta protein configuration, 263–265, 505–506
Beta emitter, 548–550
Beta wave (eeg), 112–114
Binary integer (*see* Bit)
Binocular vision, 148
Biochemical taxonomy, 611–613
Biochemistry review, 600–613
Biological clock, 52, 72–73
Bioluminescence, 50–51
Bipolar electrode, 160
Birds, 64, 71–73, 363, 377
  migration and homing, 71–73
Birefringence, 154
Bit, 51, 97, 467, 474
Blackbody radiation, 29, 523
Black–white vision, 141
Blind spot, 32–33
Blink reflex, 37, 146
Blood, 169–172, 189–190
  density, 171
  flow measurement, 559–562
  pressure, 170–172, 175, 178
  viscosity, 171
  volume, 559–561
Bohr magneton, 507
Bombykol, 54–55
Born–Oppenheimer principle, 491
Bragg peak, 308
Bragg relationship, 258–259
Brain, 472–474, 533, 538, 561
  electrical potentials, 102–118
  waves (*see* eeg)
Brain stem, 104, 106, 131–132
Bulk modulus, 585
Bundle of His, 176

C

Camera eye, 30
Cancer:
  radiation induction, 209–211
  therapy, 201, 228, 252–254

Candela (unit), 138
Canonical conjugates, 486, 489–490
Capacity, 82, 217–218, 590, 592
    areal, 219
Capillary, 171–172
Carbohydrate, 608–610, 613
Carbon cycle, 381, 387
Carbon dioxide fixation, 370, 373–376,
    383–384, 387
Carbon dioxide transport, 169, 174
Carbon-14, 198–199, 385, 549–551, 553, 559
Carbon-13, 513, 551–552
Carbonic anhydrase, 326
Cardiac function, 541–542, 556
Cardiac muscle, 154, 174–177
    model, 178–179
Cardinal points of a lens, 26–27, 34–35
Carotenoids, 359, 362–364, 376–380, 387
    (*see also* Retinal)
Catalase, 341–347, 403–405, 411–412, 420
Catalatic reaction, 342
Catalyst, 325–356, 404, 408, 600
Cataract, radiation, 225
Catecholamine, 99
Cavitation, ultrasonic, 242–254
Cell, 219, 614–619
    cycle, 201–202, 616–618
    electrical properties, 219–222
    irradiation, 201–209
    mechanical properties, 234
    membrane (*see* Membrane, biological)
    physiology, 614–619
    reproductdion, 616–619
    surface models, 250–251
Central nervous system (CNS), 84–85,
    103–107, 161
    hearing, 129–135
    radiation effects, 204–205
    vision, 146–148
Centromere, 205, 616–619
Cerebral cortex, 104–110, 137, 146–148
Cerebral hemisphere, 104, 157
Cerumen, 17
Channel capacity, 472–475
Characteristic function, 488
Characteristic value, 488, 491–492
Charge, electrical, 81, 589–592
Charge transfer, 316, 382
Charged–particle beam, 307–308
Chemical:
    energy, 395–396, 399, 430
    flux, 428–429, 451
    kinetics, 502
    potential, 395–396, 399, 419–420, 447–451
    tracer, 552–553
Chemiosmotic theory, 462–464
Chemoreceptor, 55

Chemotaxis, 228
Chirp sonar, 61
Cholesterol, 609
Chlorolabe, 363–365
Chlorophyll, 343, 372, 376–382, 387, 511
Chloroplast, 371–376, 381–383, 460, 511
Cholinesterase, 99, 160
Choroid layer, 31–32
Chromatic aberration in the eye, 45–46
Chromaticity, 138–139
Chromatin, 616–617
Chromatography, 385
Chromophore, 359, 363, 496–497
Chromosome:
    cell division, 616–619
    information theory, 470, 477–479
    radiation effects, 201–205
Chromatid, 202, 204
Cilia, 616
Ciliary muscles of the eye, 34
Circular dichroism, 459, 503–506
Circulatory system, 169–170 (*see also* Heart)
Cis isomer, 359–363, 378
Cisterna (muscle), 155, 163, 166
Cistron, 288–289 (*see also* Gene)
Citric acid cycle, 348–352, 428–429
Clamped nerve, 91–97
Click signals, 61–62, 64–66
Clone, 286
CNS (*see* Central nervous system)
CoA, 350–352
Cochlea, 21–22, 131, 135
    model, 121–128
Cochlear microphonics, 130, 135
Codon, 300–305
Coenzyme, 329, 348–350
Cofactor, 329, 608
Collision theory, 405–409
Color, 137–138, 146, 473
    blindness, 364–367
    discrimination, 138–142, 148
    vision, 138–142, 146, 357–367
Colorimeter, 499
Communication, ultrasonic, 66–67
Communication, insect, 48–58
Compartment, 460, 463, 561
Competitive binding, 512–513, 558–559
Complementary color, 138
Complementary mutation, 288
Complementary strand, 297–299
Compound eye, 31, 49–50
Commutation, operator, 489
Compiler, 567
Complex notation, 584–585, 593
Compton effect, 309
Computed tomography, 539, 574–575
Computer, 466–475, 556–558, 563–579

Computer *(cont.)*
  analog, 565, 571, 574, 578
  dedicated, 573
  digital, 565–566, 571, 574, 578
  hybrid, 565
  laboratory, 573, 576–578
  on-line, 564–567, 572–575, 578
Computer language, 567–570
  assembly, 567
  BASIC, 567–568, 571
  FORTRAN, 567–568, 571
  higher-level, 567–568, 571–572
  MUMPS, 567
Computer terminal, interactive, 565, 571–572
Conductor, 591–593
Cone, 32, 37–46, 141–142, 358, 363–365, 473
  blindness, 365–366
  opsin *(see* Photopsin)
  pigment, 363–364
Conjugated bond system, 496, 519
Conjugation, bacterial, 285
Contraction, muscular, 155–159, 174
  thermodynamics, 397
Convolution integral, 598–599
Corepressor, 304–305
Cornea, 31, 36–37, 40, 46
Correspondence principle, 489–490, 507
Cortical potential, 103, 148 *(see* Eeg)
Cosmic radiation, 198–199
Coulomb's law, 589
Coupling coefficients, Onsager's, 423, 451
Craniosacral division, 85
Critical point, 279–280
Critical volume, 282, 289, 317–324
Cross bridge, muscle, 155, 164–166
Cross-link, 314–315, 320, 604
Cross-section, 207–208, 317–322
  acoustic, 534
Crossing over, 619
Crystal structure, 256–263, 269
  factor, 260–261
Curie (unit), 549
Curie-Prigogine principle, 424
Current, electrical, 81–83, 216, 590–593
  density, 217, 590
  source, 184–187, 190
Cuvette, 498, 500–502
Cyanolabe, 363–364
Cyclic phosphorylation, 383–384
Cysteine, 320, 323, 603
Cytochrome, 329, 343, 352, 383–384, 529
Cytochrome c, 343, 352–353, 529–530
Cytochrome chain, 343, 352–355
Cytoplasm, 219–221, 614–615
Cytosine, 271–274, 297, 606

D

Data processing, 567–570, 578
Dc *(see* Direct current)
Debye effect, 221
Debye-Waller factor, 527, 529
Decibel, 8–10
Dehydrogenase, 352
Delay line, 70
Delta wave (eeg), 112–114
Denaturation, protein, 361, 411
Dendrite, 39, 83–86, 99
Density gradient method, 296–297, 551–552
Deoxyribose, 299, 606, 610
Depth of focus, 33
Detailed balance, 429
Deuterium (^{2}H), 321, 548, 551–552
Diabetic retinopathy, 227
Diathermy, 224–227, 241–242, 254
Dichromatism, 365–366
Dielectric constant, 216–217, 219–222, 589–590
Difference limen, hearing, 14–15
Differential, exact, 395–396
Differentiating electrode, 160
Diffraction, 6, 27–28, 235
  grating, 499–500
Diffusion, 295, 408, 432–452, 454
  average value model, 441
  constant, 408–409, 436–443
  equations, 422, 437–439
  flux, 438–439, 449–450
  spherical cell model, 442
Diffusion-independent reaction, 409–411, 452
Diffusion-limited reaction, 408–409, 452
Digitization, 572
Dinitrophenol, 352
Diopter (unit), 27, 35
Diploid cell, 618
Dipole moment:
  electric, 183–190, 221, 492, 517
  magnetic, 492, 506–511
Direct current (dc), 81–82, 591–592
Discrete Fourier transform, 569–570, 596
Displacement, particle, 584
Display mode, ultrasonic:
  A (amplitude), 537–539, 541
  B (brightness), 538–541
  M (motion), 538, 540–542
Dissipation function, 417–431, 438, 451
DNA, 269–275, 284–290, 297–305, 477–479, 615
  radiation effects, 201–204, 313, 323–324
  replication, 297–298

DNase, 321
Dolphin, 65–67
Dominator, neural, 143
Dispersion, 499
Dominant gene, 617
Donnan potential, 93–97, 454–456, 460,
    462–464
Doppler effect, 61–65
    Mössbauer application, 526–529
    ultrasonography, 542–543
Doppler sonar in bats, 64
Ducts of the inner ear, 19–22

E

Ear, 2–23, 471 (*see also* Hearing)
    inner, 16, 19–22, 119–135
    middle, 16–20
    outer, 16–18, 20
Eardrum (*see* Tympanum)
Ecg (*see* Ekg)
Echolocation, 61–67, 532
Echosonography, 532–546
Eclipse (virus), 284
Escherichia coli, 278–280, 283–289
Eddy currents, 125, 242
Eeg, 102–118
Eigen (*see* Characteristic)
Einstein (unit), 42, 375
Einstein coefficients, 523–524
Einthoven's triangle, 183–184, 187
Ekg, 180–190, 573, 576
Electric fish, 68–71
    ac and dc signals, 70
    pacing center, 70
Electrical:
    conductivity, 68 (*see also* Resistivity)
    current, 593
    energy, 395–396, 418, 592
    field, 216–217, 441, 503–504, 590–591
    flux, 425–427
    force, 434, 589–591
    impedance (*see* Impedance)
    potential, 81–83, 455, 590–592
    transmission, 98
Electricity, 80–83, 589–594
    "let-go" threshold, 223
Electrocardiogram (*see* Ekg)
Electrochemical gradient, 434, 444
Electrocution, 223
Electrocyte (electroplaque), 70–71
Electrodes, scalp, 111 (*see also* Eeg)
Electroencephalography (*see* Eeg)

Electromagnetic spectrum, 215–216, 227, 485
Electromagnetic waves, 25, 27–29, 515
Electromagnetism, 215–232, 589–594
Electromotive force (emf), 82–83, 592
Electromyogram (emg), 114
Electron, 81, 197, 308–311, 548–549
    carrier, 373, 381, 383–384
    density, 261
    transport, 347–355, 381–384, 462–463
    unpaired, 311–312, 507–509
Electron microscopy, 86, 152–154, 278–280
    freeze fracture, 459
Electron nuclear double resonance (endor),
    511
Electron spin resonance (esr), 507–513
    application, 266–267, 362, 511–513
Electron volt (unit), 199
Electronic energy level, 494–496, 516–518,
    521
Electronic quantum numbers, 494–496
Electrooculogram (eog), 113–114
Electrophoresis, 282
Electroretinogram, 144
Em (*see* Electromagnetic spectrum)
Emf (*see* Electromotive force)
Emg (*see* Electromyogram)
Emission spectroscopy, 515–531
Endocrine, 79–80, 470
Endolymph, 21
Endoplasmic reticulum, 155, 614–615
Endothelial cell in the cornea, 36
Energy:
    conversion, 346, 373–378, 387, 616
    internal, 395, 405–408, 430
    level, 489–497, 506–510, 516, 526
    shell, 495–496, 520–521
    transfer, 316, 320–322, 380–381
Enthalpy, 398, 430
    activation, 411–412
    partial molal, 400
Entropy, 395–400, 420, 469
    activation, 411–412
    flux, 425–427
    partial molal, 400
Environmental radiation, 198, 201
Enzyme, 325–356, 360, 386–387, 403–405,
    408–409, 431, 600
    adaptive, 362
    classification, 327–328
    kinetics, 325–356, 344–347, 400, 427–431,
    502
Enzyme–substrate complex, 270, 322,
    331–333, 337–347
Eog (*see* Electrooculogram)
Epilepsy, 103, 112–113

Epithelial radiation syndrome, 204
Equilibrium constant, 400–405
Equilibrium, chemical, 325–326, 396,
    400–405, 428–429
Equilibrium state, 395–397, 401, 424, 428,
    444–445, 523–524
Ergodic sequence, 469
Erythrocyte (*see* Red blood cell)
Erythrolabe, 363–365
Esterases, 328
Euglena, 228, 369, 371–372, 387
Eustachian tube, 19–20
Evoked response average, 115–117, 574
Excited state, 515, 526, 548
Expectation value, 487–490
Extinction coefficient, 497, 501–503
    (*see also* Attenuation)
Eye, 24–47, 357, 472–474
    anatomy and histology, 30–40

F

FAD, 352–353, 608
Fast Fourier Transform (FFT), 595–599
Fat (*see* Lipid)
Feedback, 19, 80, 107–109, 161, 571
    negative, 80, 107–109, 375–376
    positive, 107–109
Fick constant (*see* Diffusion constant)
Fick's law (*see* Diffusion equation)
Filtration coefficient, 450
Fish heart, 173
Flagella, 616
Flavin adenine dinucleotide (*see* FAD)
Flavoprotein, 350, 353
Fluorescence, 316, 379–380, 515–522
Fluorescent decay, 518, 520
Fluoroscope, X-ray, 199
Flux, generalized, 418–431, 447
Focal length, 27, 34–35
Focal point, 26–27, 35
Force, driving, 418–431, 447
Fourier analysis, 5, 261, 269, 569, 584
Fourier transform, 5, 569–570, 595–599
Fovea, 32, 45–46, 146
Fragility, cell, 248–251
Fraunhofer (far) zone, 536
Free radical, 245, 311–316, 382, 511
Frequency, 4, 583, 586
    audible, 3, 6, 14–15
    angular, 218
    characteristic (eigen), 6–7
    shift (*see* Doppler effect)
    sweep, 61–64
Frequency analysis, 110–113, 129–130
    (*see also* Fourier analysis)

Frequency response curve, 10, 63–64
Frequency spectrum, 569, 595
Fresnel (near) zone, 536
Fresnel zone plate, 557

G

Gamma camera, 553–558, 561, 574–575
Gamma emitter, 549–550, 559
Gamma ray, 216, 526–529, 548–549, 554–559
Gamma wave (eeg), 112–113
Ganglion, 83, 85, 131, 141
Gene, 212, 287–289, 478
    code, 302–303, 478–479
    doubling dose, 212–213
    information, 297–299, 477–479
    mapping, 287–289
    recombination, 286–289
Giant axon in squid, 89–96
Gibbs–Duhem equation, 438
Gibbs free energy, 398–405, 409–413, 455
    partial molal, 399–404, 433, 455, 462
Globulin, 163, 263
Glucose, 385, 610
Glucose oxidation, 325–326, 347–354, 420
Glyceraldehyde-3-phosphate, 349, 386
Glycogen, 349
Glycolysis, 348–350, 352, 442
Golgi body, 86, 614–616
Granum, 371–376
Gray matter (CNS), 104
Gray-scale imaging, 538
Guanine, 271–274, 297, 607

H

H zone (muscle), 153–155
Hair cells, 22, 84, 130–133
Half life, 549
Half-value layer, 309
Hamiltonian, 486–492
Hanes–Woolf plot, 336, 339
Hearing, 2–22, 471–472, 587–588
    bone conduction, 12, 18
    limits, 9–11, 59–60
    neural mechanisms, 119–136
    nonlinearity, 7, 19, 242, 588
    spatial theory, 120–129
    temporal theory, 120, 129–130
    tests, 9–16
    threshold, 9–17
Heart, 169–191, 540–542, 556 (*see also*
    Cardiac)
    dipole (equivalent), 183–189

Heart *(cont.)*
  output, 179
  valves, 174–176, 540–542
  vector, 183–188
Heartbeat, 169–191
Heat, 394–397, 406, 417–420
  equation, 437
  exchange, 394, 397, 416–417
Helicotrema, 21, 125–126
Helmholtz free energy, 398
Helmholtz theory, 120–121, 125, 127
Hematocrit, 171, 561
Heme group, 266–269, 329, 342–343
Hemoglobin, 245, 266–269, 329, 343
Hexose, 349, 374–376, 381–387, 610
High-energy phosphate bond, 350–352
Hill reaction, 381
Hippocampus, 113
Hodgkin–Huxley model, 94–97, 456
Holography, 543–545, 557–558, 586
Homeostasis *(see* Feedback, negative)
Hormone, 79–80, 558–560, 608–609 *(see also* Insulin)
Hue, 138, 357
Hydrodynamic wave *(see* Békésy wave)
Hydrogen acceptor, 370, 373
Hydrogen bond, 263–266, 271–275, 297
Hydrogen donor, 342–345, 370
Hydrogen peroxide, 311–313, 326, 342–347, 403–405, 411–412
Hydrolase, 328, 330, 335, 347
Hydrolytic reaction, 330–336, 347
Hyperfine interaction, 510

I

I band (muscle), 153–155
Impedance:
  acoustic, 235–237, 532–533, 584, 587
  bioelectric, 215–222
  characteristic, 235–237, 584, 587
  electrical, 82, 216–218, 590, 593
Incus, 18
Index of refraction, 26, 503–504
Inducer, protein synthesis, 304–305
Inductance, 82, 217, 590, 592
Induction period of a virus, 284
Information, 51, 431, 466–470, 565
  acoustic, 53–54, 132, 369, 471–472
  average, 468–471, 476
  visual, 50–53, 149, 472–474 *(see also* Genetic information)
Information rate, 471–474
Information system, 468, 575–578
Information theory, 465–480, 607
Information transmission, 48–58, 78–101 *(see also* Communication)

Infrared radiation, 67–68, 500 *(see also* Diathermy)
Inhibition, enzyme, 337–341
  competitive, 337, 339–340
  noncompetitive, 338, 340–341
  uncompetitive, 338
Inhibition, neural, 100, 142–146
Initiation marker, 299, 304
Insect, 48–58
  anatomy and physiology, 49–50
  communication, 48–58
  eye, 49–51
  repellent, 57
Insulin, 322, 560, 604–605
Intensity:
  acoustic, 8, 234, 584
  electromagnetic, 140, 222–223, 308
Intermediate complex *(see* Enzyme-substrate complex)
Interphase, 201–202, 616–618
Interpreter, computer, 567
Intersymbol influence, 474–476
Ion pump, 166, 445
Ionic flux, 62–67, 92–97, 445–447, 462
Ionization, 197
  dosimeters, 199–200
  sensitivity, 323–324
Ionophore, 462–463
Iris, 32–34, 109–110, 146
Irreversible system, 413–417
Irreversible thermodynamics, 413–433, 438–439, 447–452
Isobaric system, 399, 401, 438
Isoelectric point, 320
Isomerase, 328
Isometric process, 157–159
Isomorphic replacement, 262, 266–269
Isothermal process, 399, 401, 414–416, 426, 438
Isotonic process, 157–159
Isotope, 548
Isotropic band *(see* I band)

J

Junction potential, 81

K

Kidney, 169
Kinetic energy, 179, 394
King–Altman rules, 354
K_M, 334–340

L

Label (*see* Tracer methods)
Lactic acid, 348–349, 442
Lambert equation, 222, 496
Lambda wave (eeg), 113
Laplace equation, 185, 584–585
Laser, 524–526
    application, 17, 67, 227–231, 544, 558
Lateral line receptor, 4, 68
Lateral tubule, 155, 161, 166
Laue pattern, 258–259
Lead systems, 181–182, 188–189
Leakage conductance, 93–97
Lecithin, 609
Lens:
    acoustic, 252
    arthropod, 30–31
    crystalline, 34, 37–38, 46, 224–225
    thick, 26
    thin, 26
Lens opacity, 225
Lens strength, 27, 34–35
Lenz's law, 592
LET (*see* Linear energy transfer)
Leuciferin, 50
Ligase, 328
Light, 24–30
    absorption, 496–506
    detector, 498, 500–501
    source, 498–499
Limbic system, 113
Limulus, 42, 129, 142, 144–145
Linear energy transfer, 307–308, 311
Linearity hypothesis, 210–211
Lineweaver–Burk plot, 336, 339–340
Lipase, 328
Lipid, 608–609, 614
Lipid bilayer, 456–463
Liquid structure, 408–409
Longitudinal wave, 586
Loudness, 2–8, 14–15, 584, 586
Lumen (unit), 138
Luminosity, 43–44, 138, 146
Lysase, 328
Lysis, 285–289
Lysogenic virus, 284–285
Lysozyme, 267, 270, 333

M

Mach band, 127, 145–146
Macula luteau (*see* Fovea)
Magnetic:
    energy, 430
    field, 73, 216–217, 224, 506–510

Magnetic *(cont.)*
    moment, 312, 507–509
    permeability, 216–217
    resonance, 485, 506–513 (*see also* Nuclear magnetic resonance, Electron spin resonance)
    spectrometer, 508–509
Malleus, 18–20
Mammalian heart, 173–174 (*see also* Heart, Cardiac)
Map-compass theory, 72–73
Markov process, 469
Maser, 524
Masking, 128–129
Mass action law, 326, 354, 429
Mass:
    flux, 435–447
    spectrometer, 552–553, 574
    tracer, 297, 373, 547–548, 551–553
    transfer, 435
Maximum velocity of enzyme (*See* V_{max})
Maximum velocity of shortening, 179
Maxwell's equations, 29, 222
Maxwell–Wagner effect, 221
Medical diagnosis, 532–544, 570–577
Medical radiation, 198–201
Meiosis, 616–619
Melanin, 230
Membrane, biological, 82, 217–222, 433, 440–464
    equivalent circuit, 94
    fluid mosaic model, 460–461
    impedance, 91
    permeability, 94, 99–100
    polarization, 174, 460 (*see also* Ionic flux)
    potential, 87–97, 100–101
Membrane, permeable, 450
Membrane, semipermeable, 450, 454, 462
Memory, 473–475, 479
Meninges, 104
Message, 468, 470, 474–475
Metabolism, 294–305, 348–354, 383–387
    heat output, 224–225, 251 (*see also* Work, metabolic)
Metaphase, 201–204, 616–618
Methemoglobin, 245
Micelle, 456, 458
Michaelis constant (*see* K_M)
Microcomputer, 557, 565, 575
Microelectrode, 87, 152, 160–161
Microwave radiation, 224–227, 506–510
Midbrain, 146–147
Midline echo encephalogram, 538–540
Migration, 52, 71–73
Miller indices, 258–260
Minicomputer, 565, 567
Mitchell hypothesis, 462–464
Mitochondrion, 350–355, 614–616

Mitosis, 201–202, 205, 616–619
Mitral valve (*see* Heart valve)
Molecular:
  energy level, 516–518
  rotation, 491–494
  spectrum, 490
  vibration, 491–494, 516–518
Momentum:
  angular, 494–496, 507–509, 516
  linear, 526–527
Monazite, 198, 212
Monochromatism, 365–366
Monochromator, 498–501
Monose, 609–610, 613
Mössbauer:
  applications, 17, 123, 528–530
  source, 528
  spectrometer, 528
  spectroscopy, 526–530
Motor end plate (*see* Muscle end plate)
Mu wave (eeg), 113
Multiple hit model, 207
Multitarget model, 206–207, 211
Muscle, 151–168
  activation, 164–167
  chemistry, 162–167
  fatigue, 161
  heat, 161–162, 393
  mechanics, 151, 157–159
  structure, 152–157
Muscle end plate, 84, 98–99, 160, 167
Muscle fiber, 90–91, 97–98, 153–155,
  159–161, 460
Muscle types:
  cardiac (*see* Cardiac muscle)
  smooth, 152–157
  striated, 152–167
Mutation, 211–212, 283–289, 322, 613
  noncomplementary, 288
Myelin sheath, 83, 86
Myelinated axon, 84, 86, 97
Myofibril, 153–156, 161–167, 205
Myoglobin, 156, 163, 266–269, 343, 470
Myoneural junction (*see* Neuromuscular
  junction)
Myosin, 155, 163–167

N

NAD, 329, 348–350, 608
NADP, 373–376, 380–387
Negative entropy, 468–471, 476
Neoplasm, 209–211
Nerve, 78–80, 83–101
  auditory, 3, 21, 131
  optic, 31–32, 38–39, 42, 142–143, 473

Nerve cell body, 83, 85–86, 104
Nerve fiber (*see* Axon)
Nervous system, 78–80, 84–101
  central (*see* Central nervous system)
  parasympathetic, 85
  peripheral, 84
  sympathetic, 85
Neural processing (*see* Neural sharpening)
Neural sharpening, 127–129, 133, 144–145,
  148
Neural spike (*see* Action potential)
Neurochemical transmitter, 80, 98
Neuromuscular junction, 98–100, 160–161
Neuron, 39, 78–101, 104, 133–134, 148 (*see*
  also Nerve)
  afferent, 84–85, 99, 104
  auditory, 134
  efferent, 84–85, 99, 104
  frequency (*see* tuned)
  motor (*see* efferent)
  sensory (*see* afferent)
  tuned, 129, 133, 135
Neuronal network, 107–110, 146–148
Neutron, 197–199, 310, 548
Nicotinamide adenine dinucleotide (*see* NAD)
Nicotinamide adenine dinucleotide phosphate
  (*see* NADP)
Nissl substance, 84, 86
Nodal point, 26–27, 35
Node of Ranvier, 83, 86, 97
Noise, 468, 471–475, 479, 524
  physiological, 12
Nonequilibrium system, 404, 413–417, 462,
  524
Nonlinear parameter estimation, 337, 345
Nonmyelinated axon, 91
Nuclear magnetic resonance (nmr), 491,
  507–513, 551
Nuclear quadrupole moment, 529–530
Nucleation theory, 243
Nucleic acid, 281–289 (*see also* DNA, RNA)
  crystal, 269–275
  function and synthesis, 294–305
  radiation effect, 322–324
  structure, 269–275, 605–610
Nucleoplasm, 614
Nucleus of an atom, 309–310, 526–529, 548
Nucleus of a cell, 201–206, 614–619
Nucleus of the CNS, 83, 104, 131–132
Nuclide, 548–549

O

O₂ (*see* Oxygen)
Occipital lobe, 146–148
Octave, 11

Ohm's law, 81–82, 95, 422–423, 591
Ommatidium, 30, 42, 142, 144–145
Onsager's equations (*see* Phenomenological equations)
Onsager's law, 424, 427, 429
Operator, genetic, 304
Operator, quantum, 487–490
Opponent theory (*see* Antagonist theory)
Opsin, 358–363, 366
Optic nerve (*see* Nerve, optic)
Optical:
    activity, 503–506, 601
    pumping, 525
    rotary dispersion, 459, 505–506
    wave guide, 141–142
Optics, 25–30
    geometrical, 25–27, 34–36
    wave, 25, 27–29
Organ of Corti, 21–22, 132–133
Organic acid, 601
Organic base, 605
Orientation, 59–73, 148
Osmotic pressure, 433–434, 442, 447–451
Ossicles, 18–20
Oval window, 19–22, 120, 122, 124–126
Overtone, 7
Oxidation, biological, 341–343, 347–355, 373–374, 608
Oxidative phosphorylation, 350–354
Oxidoreductase, 328, 341, 361 (*see also* Catalase *and* Peroxidase)
Oxygen:
    diffusion, 207, 434–437, 441–442
    effect, 209, 297, 312, 315–316, 320, 323
    production, 370, 373–376, 380–384
Ozone, 229–230

P

P wave (ekg), 181–182, 188
Pacemaker, 70, 173 (*see also* S-a node)
Pacinian corpuscle (*see* Proprioceptor)
Pair production, 309
Pauli exclusion principle, 495–496
Peltier effect, 426
Pentose, 605, 610
Pentose phosphate shunt, 350, 421
Peptide bond, 475, 601
    synthesis, 300
Perilymph, 21
Periodicity pitch, 129
Permeability, 432–452
    constant, 440–444, 450–451
    irreversible thermodynamics, 447–451
Peroxidase, 329, 341–347
Peroxidatic reaction, 342, 344–347

Phage, 276–290 (*see also* Virus)
    genetics, 286–289
    T series, 277–280, 283–284, 286–289
Phase angle, 261–262, 586, 592
Phase difference, 586
Phenomenological equations, 421–430, 433, 450–451
Pheromone, 54–57
Phonons, 15–16
Phosphatase, 328
Phosphene effect, 224
Phosphoglyceride, 456–458
Phospholipid, 456–458, 461, 609
Phosphorescence, 515–520, 524–525
Phosphorescent spectrum, 518
Photocoagulation, 227–229
Photoelectric effect, 29, 308–309
Photon, 25
    absorption, 226, 489, 491–492, 496–513, 516, 523, 526–530, 547
    emission, 489, 491–492, 496, 515–530, 547–548
Photopic vision, 43–44
Photopigments, 357–367, 375–382
Photosynthesis, 369–388
    dark reaction, 373–375, 381, 383–387
    light reaction, 373–374, 381–383
    path of carbon, 383–387
    phosphorylation, 375–376, 381, 383
    reaction, 370, 373–375
Photosystems, 373, 380–384
Phycobilin, 376, 380
Phylogenetic information, 613
Phylogenetic tree, 611–612
Pi electron, 496, 519
Piezoelectric effect, 535
Pigeon, homing, 71–73, 224
Pitch discrimination, 3, 120, 123, 128–129, 133–135
Planck's constant, 16, 30
Plant histology, 371–373
Plaque (virus), 283–289
Plastid, 372
Plexus, 144
Poisson distribution, 41, 206, 222
Polarimeter, 505
Polarization of light, 52, 503–505, 601
Polyethylene, 313–315
Polyisobutylene, 313–316
Polyploid, 470
Porphyrin, 342–343, 376–377, 511
Posterior probability, 467–468
Potential energy, 179, 394, 419
    barrier, 405–406, 408–412
    well, 406
Power, 418, 587, 590, 593
    heart output, 180

Powerline, biological effects of exposure to, 224
Power spectrum, 569, 599
Pressure:
  absolute, 7, 170, 243, 584, 586
  atmospheric, 11, 244
  blood (*see* Blood pressure)
  negative, 243
  sound (*see* Acoustic pressure)
Primary sequence analysis, 613
Principal point of a lens, 26–27, 35
Prior probability, 467–471
Prism, 499–500
Probability density, 487
Probability of error, 474–475 (*see also* Noise)
Program, computer, 567–569
Promotor, 299, 304
Prophage, 284–285
Prophase, 616–618
Proprioceptor, 104, 161
Prosthetic group (enzyme), 329, 342–343 (*see also* Heme)
Protein and polypeptides, 614–615
  crystal, 262–269
  fluorescence, 520
  radiation effects, 230, 313, 319–324
  structure, 263–270, 475–479, 564, 600–605, 613
  synthesis, 294–305
Protein types (*see also individual proteins and enzymes*)
  fibrous, 263–265, 458–459
  globular, 263–266, 458–461
  membrane, 456–462
  muscle, 163
  soluble, 456
Proteolytic enzyme, 328
Protium (^{1}H), 548, 551
Proton beam, 197, 205, 308, 310
Proton, in solution, 347, 462–463
Proton-motive force, 462
Protoplasm, 453, 460, 614–616
Pulmonary valve (*see* Heart valve)
Purine, 271, 605–608
Purkinje fiber, 176
Pyrimidine, 271, 605–608
Pyruvate, 348–350

Q

QRS complex, 181–183, 188–189
Quality factor, 197–198, 211, 311, 317
Quantum, 29
Quantum efficiency in fluorescence, 518–519
Quantum mechanics, 25, 29–30, 486–496, 506–513
Quantum number, 488, 491–496, 508–509
Quencher, 516, 550
Q_{10}, 409

R

Rad (unit), 197–198
Rad equivalent mammal (*see* Rem)
Radiation:
  charged, 307–308
  electromagnetic, 215–216, 223, 503
  ionizing, 196–214, 278, 282, 306–324
  microwave, 215–216, 224–227
  protection agent, 323–324
  solar, 230–231
  ultraviolet, 229–230
  uncharged, 308–311
  visible light, 227–229
Radiation dose, 197–201, 317–318
  background, 198–199
  medical, 199–200
  threshold, 206–207, 211
  whole-body, 201
Radiation effects:
  direct, 311–316, 319–322
  genetic, 211–213
  indirect, 322–324
  low-frequency, 223–227
  macrobiological, 209–213
  protein film, 319–323
  somatic, 196, 209–211
  synthetic polymer, 312–317
Radioactive imaging, 553–558, 561–562
Radioactive tracer, 547–551
  applications, 383–385, 446, 553–562
Radioactivity, 198–199, 548–551
Radioimmunoassay, 558–560
Range:
  charged particle, 307–308
  sound, 53–54
Rapid eye movement (*see* REM)
Rapid flow apparatus, 502–503
Rate constants, 326–327, 442
  chemical, 405–409, 428
  enzymic, 326–347, 400–405
  in absolute rate theory, 409–412
  temperature dependence, 400–401, 407–412
Rayleigh:
  criterion, 29
  emission, 516
  scattering interval, 534
RBC (*see* Red blood cell)

Reactance, 82, 217, 590, 593
  acoustic, 235
Receptor potential, early, 366–367
Receptor site of a virus, 284
Recessive gene, 212
Recovery from radiation damage, 208, 211
Red blood cell, 205, 240, 245, 248–249, 561
  permeability, 439–444
Reduction, biological, 347, 370, 373–374
Redundancy, 469–470, 474–475, 477–479
Reflection coefficient, 450
Reflection densitometry, 364–365
Refractory period, 90, 550
Rein control, 109–110
Relaxation phenomena, 219–221, 239–240,
    251, 510
REM (sleep stage), 114
Rem (unit), 197–198
Repressor protein, 304–305
Reproduction:
  cell, 202, 204
  viral, 278
Reptile heart, 173
Resistance, 81–82, 217–218, 418, 590–593
  acoustic, 236
Resistivity, 218–221, 591
  water, 592
Resolution limit, 257
Resolving power of the eye, 28–29
Resonance, 6, 17, 221, 509, 528, 535
  acoustic cellular, 249–251
  (*see also* Magnetic resonance, Nuclear
    magnetic resonance *and* Electron spin
    resonance)
Respiratory enzyme, 328–329, 347–355
Resting potential, 87, 160–161, 177
Retina, 32, 46, 140–148, 357–363
  information, 472–474
Retinal, 359–363, 366, 377
Retinol, 359–363
Rhabdom, 31
Rhodopsin, 358–363, 366–367
Ribose, 299, 606
Ribosome, 86, 299–305, 615
Rigor complex, 166
RNA, 273–275, 281, 298–305
  aminoacyl *t*RNA, 274, 300–301
  messenger (*m*RNA), 273, 300–305
  peptidyl *t*RNA, 300, 302
  ribosomal (*r*RNA), 273, 300–305
  transfer (*t*RNA), 273–274, 300–305, 613
RNA polymerase, 299
RNA transcription, 298–299, 615
RNA translation, 299–302
Rod, 32, 37–44, 141–142, 358–367, 473
Rod opsin (*see* Opsin)
Roentgen (unit), 317
Round window, 19, 22

S

S-a node of the heart, 173, 176, 183, 190
Sarcolemma, 153, 160–161
Sarcomere, 154, 156
Sarcoplasmic reticulum, 155–156, 166
Satellite cell (*see* Schwann cell)
Saturation (color), 138
Scala (*see* Ducts of the inner ear)
Schiff's base, 362
Schrödinger wave equation, 488
Schwann cell, 83, 86
Scission, 315, 320
Sclera, 31, 36
Scotopic vision, 43–44, 365
Scotopsin (*see* Opsin)
Sedimentation, 294–297
  coefficient, 295–296
  velocity, 295–297
Selection rules, 492–496, 509
Semiconservative replication, 297–298
Semilunar valve of the heart, 175
Sensitivity coefficient, 206
Septum of the heart, 176, 183
Set point, 109, 170
Shannon's theorems, 474–475, 477, 479
Shearing stress, 246
Shock wave, 242
Sight (*see* Vision)
Simple harmonic motion, 82, 584
Simulation, 346, 352–355, 570–572
Single hit model, 206–208, 282, 320
Singlet state, 495–496, 516–518
Sino-auricular node (*see* S-a node)
Sinus venosus, 173
Skeletal muscle, 154, 159–164
Skin pigmentation, 230
Sleep, 103, 110
Sleep stages, 113–115
Smooth muscle, 152–153, 155–157
Snake, 67–68, 377
Snap lysis of a virus, 284
Solar arc theory, 72–73
Solvent drag, 433, 450
Sonar, 65, 532 (*see also* Echolocation)
Sonic destruction of tissue, 251–254
Sonic heating, 241–244, 251–254
Sonic irradiation, high intensity, 251–254, 588
Sonics, 233–255, 588
Sonograph of the heart, 175
Soret band, 342
Sound, 2–23, 583
  detection, 53, 64
  guide, 53
  information, 471–472
  pressure (*see* Acoustic pressure)
  pressure level (*see* Acoustic pressure level)
  production, 53, 64

Specific absorbance, 501–502
Specific activity, 550
Specific heat, 397
Specific ionization, 318–319
Spectral intensity, 140
Spectrophotometer:
  absorption, 498–503
  fluorescence, 519
Spectrophotometry, 485–531
Speech, 471–472
Speed (*see* Velocity)
Spike, eeg, 113 (*see also* Epilepsy)
Spike potential (*see* Action potential)
Spin labeling, 511–513
Spiral ganglion, 131
Spinal cord, 104
Stable isotope tracer, 551–552
Standard state, 400–403, 411, 455
Stapes, 18, 19, 125–126
Starch, 349, 385–386, 613
State function, 487–491, 510
Steady-state approximation, 334–341,
  422–424, 428
Stimulated transition, 523–525
Stereoisomer, 601
Stiff differential equations, 346, 354
Stochastic process, 469
Stopped-flow apparatus, 502
Stress, 170, 178
Straight coefficients, 423
Stridulation, 53
Stroke volume, 179
Substrate, 329–341, 344–346
Succinate, 351–353
Sugars, 269–273, 370, 373, 386, 605–610
Sulfhydryl group, 320, 323
Sun-compass theory, 52
Superconductor, 591–592
Surgery, sonic, 234, 252–254
Survival curve, 206–209, 230
Survival fraction, 206
Svedberg (unit), 295–296
Synapse, 84–85, 97–100, 146–147
Synaptic conduction, 97–100
Systolic pressure, 171

T

T wave (ekg), 181–182, 188
Target theory, 206–208, 282, 317–322
^{99m}Tc, 549, 555–556, 559
Tectorial membrane, 22, 133, 135
Telophase, 617–618
Terrestrial radioactivity, 198–199
Terminator (DNA), 299, 302

Tetany, 158–159, 161
Thalamus, 106, 112
Thermal conduction, 426
Thermal energy (*see* Heat energy)
Thermocouple, 81, 427
Thermodynamic isolation, 398, 419
Thermodynamic state, 395
Thermodynamics, 391–431, 523–524
  first law, 395–397, 415, 419
  second law, 397
  third law, 397–398
  zeroth law, 417
Thermoelectricity, 425–427
Thermoluminescent dosimeter (*see* TLD)
Theta wave (eeg), 112–114
Thoracolumbar division, 85
Three-dimensional imaging, 557–558 (*see
  also* Tomography)
Threshold:
  cavitation, 242–248
  "let-go," 223
  pheromone concentration, 55
  radiation dose (*see* Radiation dose)
  visual (luminosity), 43–44, 144–145
  visual (quantum), 40–43
Thylakoid disc, 372–373
Thymine, 271–274, 297, 520, 606
Timbre, 4, 584
Time series, 569
Tissue conductivity, 185
TLD (dosimeter), 199–200
Tobacco mosaic virus (TMV), 277, 280–282,
  298
Tomography, 557, 574–575
Tone, 4–5, 10, 17, 122, 471–472, 584
Trace element detection, 522
Tracer methods, 547–562, 546, 598–599
Transducer, 37, 119, 152, 179
  ultrasonic (*see* Ultrasonic transducer)
Transferases, 328
Trans isomer, 359–363, 378
Transit time distribution, 598–599
Transition:
  energy, 491–492, 506–508, 516–518
  forbidden, 492–496, 517–518
  permitted, 492–496
Transverse tubule, 155–156, 163–166
Trichromatism, 365–366
Tricolor theories, 140–142
Triplet state, 495–496, 517–518
Tritium (^3H), 548–550, 559
Trochoidal motion, 124–125
Tropomyosin, 155, 163–167
Troponin, 155, 163–167
Tumor therapy (*see* Cancer therapy)
Twitch, 29, 157–158
Tympanic membrane (*see* Tympanum)
Tympanum, 10, 16–21

U

Ultracentrifuge, 282, 294–297
Ultrasonic:
    cardiogram (ucg), 540–541
    tomogram, 538–539, 541
    transducer, 252, 535–536, 541, 543
Ultrasonography, 532–546
    radiation patterns, 536–538
    reflection (*see* Echosonography)
    transmission, 533
Ultrasound, 60, 64–65, 227, 233–238, 532–546
Ultraviolet radiation, 229–231
Uncertainty principle, 488–489, 510
Uncharged particle, 307–308, 311
Uracil, 299, 606–607
UV (*see* Ultraviolet radiation)

V

Vector cardiography, 187–190, 573
Vegetative phase of a virus, 284
Vein, 171–173
Velocity:
    action potential, 89–90, 96
    blood, 170–172
    electromagnetic wave, 26, 220, 222
    particle, 6, 234, 584–585
    reaction, 330–336, 339–341
    sound, 5–8, 234, 584–585, 587–588
Ventricle of the brain, 104
Ventricle of the heart, 173–180, 183
    fibrillation, 223
    motion, 177–180
    muscle, 176–178
Vesicle, 86, 100
Vestibular canal, 124
Virus, 276–290, 297–299, 322, 615, 619
    crystallized, 280
    reproduction, 276, 278, 285
Vision, 24–26, 472
    color (*see* Color vision)
    molecular basis, 357–368
    neural aspects, 137–150
    neural responses, 142–146
Visual:
    pathway, 146–148, 366
    pigment (*see* Photopigments)

Visual *(cont.)*
    purple (*see* Rhodopsin)
    receptor, 37–40, 140–142, 352, 453
    threshold, 30, 40–44
Vitamin A, 359–363
Vitamin D, 230
Vitreous humor of the eye, 34
V_{max} of an enzyme, 334–336, 338–340
V_{max} of a ventricular muscle, 179
Volley principle, 133–135
Voltage clamp, 92–97, 177
Volume flow rate, 171
Volume, partial molal, 447–450

W

Waggle dance of bees, 50–53
Water:
    flux, 443
    radiation effects, 311–312, 322–323
    splitting, 383
Wave, 4–7
    equation, 235, 488, 585–586
    function (*see* State function)
Wavelength, 5–7, 29, 138–140, 235, 357, 584
Whales, 65–67
White matter (CNS), 104
Work:
    mechanical, 394–396, 414
    metabolic, 445, 462
    muscular, 157–159, 179

X

Xenon-133, 549, 561
X ray, 257, 520–522, 574
    diffraction, 256–275, 281, 333
    fluorescence, 520–523
    irradiation, 199–201, 207–209, 312
    scanner, 574–575

Z

Z line (muscle), 154–156, 161, 165